Maths 30 Days Wonder

Full-fledged and complete maths package in a simple language can be understood even by a layman

[For CLAT, MBA (CAT/MAT), Banking, Hotel Management, NIFT, CDS, NDA, GRE/GMAT, IAS (GS) Campus recruitment tests and to overcome maths phobia]

KJS Khurana
R Markanday

S. CHAND PUBLISHING

S Chand And Company Limited
(ISO 9001 Certified Company)

S Chand And Company Limited

(ISO 9001 Certified Company)

Head Office: D-92, Sector–2, Noida 201301, U.P. (India), Ph. 91-120-4682700
Registered Office: A-27, 2nd Floor, Mohan Co-operative Industrial Estate, New Delhi – 110 044
Phone: 011-49731800

www.**schandpublishing.com**; e-mail: **info@schandpublishing.com**

Marketing Offices:

Chennai	:	Ph: 044-2363 2120, chennai@schandpublishing.com
Guwahati	:	Ph: 0361-4066 369, guwahati@schandpublishing.com
Hyderabad	:	Ph: 040-4018 6018, hyderabad@schandpublishing.com
Jalandhar	:	Ph: 0181-4645 630, jalandhar@schandpublishing.com
Kolkata	:	Ph: 033-2335 7458, 2335 3914, kolkata@schandpublishing.com
Lucknow	:	Ph: 0522-4003 633, lucknow@schandpublishing.com
Mumbai	:	Ph: 022-2500 0297, mumbai@schandpublishing.com
Patna	:	Ph: 0612-4011400, patna@schandpublishing.com

First Edition 2010
Revised Edition 2013
Reprints 2014, 2015
Third Revised Edition 2020
Reprint 2021 (Twice), 2022, 2024, 2025
Reprint 2026

ISBN : 978-93-528-3755-7

Product Code: H4MAD49MATH40ENAC19O

PRINTED IN INDIA

By Shri Shyamlal Printing Press Private Limited, Plot 20/4, Site-IV, Industrial Area Sahibabad, Ghaziabad – 201 010 and Published by S Chand And Company Limited, A-27, 2nd Floor, Mohan Co-operative Industrial Estate, New Delhi – 110 044

PREFACE TO THE THIRD EDITION

It gives us immense pleasure to present the third edition of **Maths 30 Days Wonder**. Quantitative Aptitude has been an important part of every entrance examination. Therefore we felt that it was necessary to explain, otherwise thought as difficult, topics in a simple and lucid manner.

Revision and salient features include:

1. Vedic Maths supplement (for fast calculations and understanding of advance concepts)
2. Each concept is explained in detail and examples and chapter-wise practice questions provided, strengthen the concepts explained.
3. "Test Yourself" segment carries questions from latest competitive exams which add to the conviction of the aspirant as they provide rigorous revision.
4. Language of the book is again so simple that even a layman can understand it.

Care has been taken to ensure that even a minor concept should not go unexplained. Short-cuts are given wherever necessary and useful. It is hoped that this course will serve the purpose for which it is made. Suggestions will be received with gratitude. For suggestions and feedback feel free to write to us at khuranamarkanday@gmail.com

AUTHORS

ACKNOWLEDGEMENTS

Any accomplishment requires efforts of a lot of people and this work is no exception. First of all we would like to thank the Almighty who has given us the strength, guidance and enlightenment to start this work. We are also thankful to our parents and family members who have given all the support and encouragement. We are also thankful to our teachers for their kind blessings. We would also like to thank all our students without whose valuable feedback in the form of suggestions this book would not have seen the light of the day. At last but not the least we are thankful to the management of S. Chand And Company Limited for the faith they have shown in us.

AUTHORS

About the Course

This course has been divided in five sections. At end of each section question bank from latest competitive exams is given for self evaluation. The five sections are given below:

SECTION–I (ARITHMETIC - I)

Day 1. Introduction and Concepts

Day 2. Decimal Fraction

Day 3. Simplification

Day 4. The Number System & Concepts of Divisibility

Day 5. HCF and LCM

Day 6. Percentage

Day 7. Profit and Loss–I

Day 8. Profit and Loss–II

- Test yourself (Section – I)

SECTION–II (ARITHMETIC - II)

Day 9. Time and Distance

Day 10. Average

Day 11. Partnership

Day 12. Ratio and Proportion and Unitary

Day 13. Time and Work

Day 14. Alligation or Mixture

Day 15. Simple Interest

Day 16. Compound Interest

- Test yourself (Section – II)

SECTION–III (MENSURATION AND GEOMETRY)

Day 17. Mensuration (2D)

Day 18. Mensuration (3D)

Day 19. Geometry–1

Day 20. Geometry–2

- Test yourself (Section – III)

SECTION–IV (ALGEBRA)

SECTION–V (ADVANCED MATHS)

CONTENTS

SECTION–IV

SECTION–V

SECTION-I

ARITHMETIC-I

Day 1

Introduction and Basic Concepts

Dear Student,

The 30 days Wonder Package is in your hand. It's your first day to this 30 day wonder course. Before starting this course let us discuss a few basic facts about **Maths**. In your life you might have met with people who are very good at Maths.

Have you ever thought why these people have very fast & accurate Maths?

These people are very good at Maths because:

1. Their **calculations** are very **fast**.

2. Their **concepts** are very **clear**.

3. They know **shortcuts**.

Shortcuts are useless if your **calculations** are **slow** and your **concepts** are not **clear**. Mostly a question comes in the **twisted form** and shortcuts are **not helpful**. First you have to bring that **question** to **shortcut form** then to apply the shortcut. We will discuss each of above points, i.e., **calculation**, **concept** and **shortcuts** one by one. First comes the calculation.

1. CALCULATION

Calculations are the most important part of the question. Mostly students say that they can do the questions but it takes time. Sometimes they are aware of concept and when they try to solve it they are never near to the answer. They get **puzzled** in the **calculation**. So, we should think why the **calculations** of some people are **fast**? Do they have better **memory**? No, it's not so. We had seen in the past that our elders had very fast calculations. The reason is that their **foundations** were **solid**. In childhood they were taught tables. Due to that their **tables** were very **fast**. You should also make **sure** to have **tables** on your **fingertips**. How fast are your calculations can be judged by the following test :

TEST-1

1. 19 × 13 = ...
2. 19 × 137 = ...
3. 12 × 11 = ...
4. 18 × 19 = ...
5. 19 × 29 = ...
6. 13 × 39 = ...
7. 17 × 59 = ...
8. 16 × 37 = ...
9. 14 × 28 = ...

(Answers to above test are given in the Solutions at the end of this chapter)

You should be able to do this exercise in 35-45 seconds. If you can do it, it's okay. **If not** then come to the point.

If we consider minutely we will see that all these calculations require tables from 12 to 19. Make sure you have these tables on your fingertips. You are to be sure that you don't leave this part because it had been left in your childhood so don't leave it this time. There will be enough of practical questions later on. But don't leave this part. Now you will be helped to learn these tables, i.e., from 12 to 19. Before that two tests are given to check how well you know these tables in sequence. For that purpose you have to just fill in the blanks given in these tests and to note down the time taken to do this exercise.

TEST-2

12 × 1 = ...
12 × 2 = ...
12 × 3 = ...
12 × 4 = ...
12 × 5 = ...
12 × 6 = ...
12 × 7 = ...
12 × 8 = ...
12 × 9 = ...
12 × 10 = ...

TEST-3

16 × 1 = ...
16 × 2 = ...
16 × 3 = ...
16 × 4 = ...
16 × 5 = ...
16 × 6 = ...
16 × 7 = ...
16 × 8 = ...
16 × 9 = ...
16 × 10 = ...

(Answers to above test are given in the Solutions at the end of this chapter)

Ideal Timings for Test-2 are less than 6 seconds

Ideal Timings for Test-3 are less than 6 seconds

Now the question arises how to **learn these tables**, whether by **simple reading** or **simple cramming**? At your stage, the best ever practical method to learn these tables is by **writing**. Now I will help you to learn the table of 12.

Given below is a table of 12 followed by ten practice tables of 12 with blanks. Your task is to fill the blanks with appropriate answer and note down the time taken to complete this exercise.

12 × 1 = 12
12 × 2 = 24
12 × 3 = 36
12 × 4 = 48
12 × 5 = 60
12 × 6 = 72
12 × 7 = 84
12 × 8 = 96
12 × 9 = 108
12 × 10 = 120

12 × 1 = ...
12 × 2 = ...
12 × 3 = ...
12 × 4 = ...
12 × 5 = ...
12 × 6 = ...
12 × 7 = ...
12 × 8 = ...
12 × 9 = ...
12 × 10 = ...

12 × 1 = ...
12 × 2 = ...
12 × 3 = ...
12 × 4 = ...
12 × 5 = ...
12 × 6 = ...
12 × 7 = ...
12 × 8 = ...
12 × 9 = ...
12 × 10 = ...

12 × 1 = ...
12 × 2 = ...
12 × 3 = ...
12 × 4 = ...
12 × 5 = ...
12 × 6 = ...
12 × 7 = ...
12 × 8 = ...
12 × 9 = ...
12 × 10 = ...

12 × 1 = ...
12 × 2 = ...
12 × 3 = ...
12 × 4 = ...
12 × 5 = ...
12 × 6 = ...
12 × 7 = ...
12 × 8 = ...
12 × 9 = ...
12 × 10 = ...

12 × 1 = ...
12 × 2 = ...
12 × 3 = ...
12 × 4 = ...
12 × 5 = ...
12 × 6 = ...
12 × 7 = ...
12 × 8 = ...
12 × 9 = ...
12 × 10 = ...

12 × 1 = ...
12 × 2 = ...
12 × 3 = ...
12 × 4 = ...
12 × 5 = ...
12 × 6 = ...
12 × 7 = ...
12 × 8 = ...
12 × 9 = ...
12 × 10 = ...

12 × 1 = ...
12 × 2 = ...
12 × 3 = ...
12 × 4 = ...
12 × 5 = ...
12 × 6 = ...
12 × 7 = ...
12 × 8 = ...
12 × 9 = ...
12 × 10 = ...

12 × 1 = ...
12 × 2 = ...
12 × 3 = ...
12 × 4 = ...
12 × 5 = ...
12 × 6 = ...
12 × 7 = ...
12 × 8 = ...
12 × 9 = ...
12 × 10 = ...

12 × 1 = ...
12 × 2 = ...
12 × 3 = ...
12 × 4 = ...
12 × 5 = ...
12 × 6 = ...
12 × 7 = ...
12 × 8 = ...
12 × 9 = ...
12 × 10 = ...

12 × 1 = ...
12 × 2 = ...
12 × 3 = ...
12 × 4 = ...
12 × 5 = ...
12 × 6 = ...
12 × 7 = ...
12 × 8 = ...
12 × 9 = ...
12 × 10 = ...

You have seen that it takes at the most 5 secs to write a table of 12 and to write it 10 times it will take less than a minute and if you do this exercise daily for a week you will have this table on your fingertips. You have learnt the method of learning tables.

Now you are to do the same exercise for the tables of 13, 14, 15, 16, 17, 18 & 19 and believe it, it will take less than 10 minutes to do this exercise daily for a week and it will completely change your attitude towards Maths.

Similarly, you must know the squares and cubes of certain numbers given below. These should also be at your fingertips.

SQUARES

1^2 = **1**	2^2 = **4**	3^2 = **9**
4^2 = **16**	5^2 = **25**	6^2 = **36**
7^2 = **49**	8^2 = **64**	9^2 = **81**
10^2 = **100**	11^2 = **121**	12^2 = **144**
13^2 = **169**	14^2 = **196**	15^2 = **225**

16^2 = **256**	17^2 = **289**	18^2 = **324**
19^2 = **361**	20^2 = **400**	21^2 = **441**
22^2 = **484**	23^2 = **529**	24^2 = **576**
25^2 = **625**	26^2 = **676**	27^2 = **729**
28^2 = **784**	29^2 = **841**	30^2 = **900**
31^2 = **961**	32^2 = **1024**	33^2 = **1089**
34^2 = **1156**	35^2 = **1225**	

CUBES

1^3 = **1**	2^3 = **8**	3^3 = **27**
4^3 = **64**	5^3 = **125**	6^3 = **216**
7^3 = **343**	8^3 = **512**	9^3 = **729**
10^3 = **1000**	11^3 = **1331**	12^3 = **1728**
13^3 = **2197**	14^3 = **2744**	15^3 = **3375**

2. CONCEPT

Each day is started by the explanation of the concept of the topic to be covered on that day with the help of daily life simple examples. Your task is to understand that concept before doing examples and exercise.

3. SHORTCUT

Shortcuts are given wherever they are applicable and useful. It has been seen that useless shortcuts take more time than normal methods. In this course shortcuts are supported by lot of practical questions to have better understanding and practice of the shortcuts.

Now as discussed you should learn the tables of 12-19 and **squares** and cubes of numbers given above, i.e., **squares** from 1-35 and cubes 1-15. Let us start this exercise just now. Now being first day this exercise may take more time. **You have started this course because you want to succeed in life. For that you have to work desperately and enthusiastically in that direction.** Your first task is to make your foundation solid as "**No big structure can be built on weak foundations**". So, now write the tables of 12 to 19 on your copy at least ten times. Each table should be written **one by one** means first write the **table of 12 ten times** then **table of 13 ten times** and so on. After tables you have to write the squares and cubes of numbers discussed above. The method of learning the squares and cubes is same as that of learning the tables.

⋮

Break for learning tables (1-12), squares (1-35) and cubes (1-15)

⋮

It is hoped that you have followed the instructions and have done the exercise explained above. Now based on above learning solve the following exercise :

EXERCISE

Q.1. (i) $1178 \times 19 = ?$
(ii) $135 \times 17 = ?$
(iii) $132 \times 11 = ?$
(iv) $155 \times 25 = ?$
(v) $1225 \times 15 = ?$
(vi) $812 \times 12 = ?$
(vii) $9195 \times 14 = ?$
(viii) $1812 \times 16 = ?$
(ix) $8112 \times 13 = ?$
(x) $81250 \times 15 = ?$

Q.2. Solve the following :
(i) $32^2 + 21^2 + 27^2 + 34^2 + 31^2 = ?$
(ii) $11^2 + 24^2 + 26^2 + 28^2 + 35^2 = ?$
(iii) $5^2 + 10^2 + 12^2 + 19^2 + 22^2 = ?$
(iv) $3^2 + 16^2 + 22^2 + 29^2 + 31^2 = ?$
(v) $8^2 + 15^2 + 17^2 + 23^2 + 30^2 = ?$

Q.3. Solve the following :
(i) $1^3 + 7^3 + 13^3 = ?$
(ii) $1^3 + 8^3 + ?^3 = 9^3$
(iii) $6^3 + 8^3 + 10^3 = ?^3$ (iv) $3^3 + 5^3 + 15^3 = ?$
(v) $1^3 + 2^3 + 14^3 = ?$

Q.4. Solve $\dfrac{15^3 - 5^3 - 10^3}{15^2} = ?$

Q.5. Solve $\dfrac{2^2 + 4^2 + 6^2 + 8^2 + 10^2}{1^2 + 2^2 + 3^2 + 4^2 + 5^2} = ?$

Q.6. A man engaged a servant on the condition that he would pay him Rs. 90 and a turban after the service of one year. He worked for nine months and got Rs. 65 and turban. Find the cost of turban.
(a) Rs. 10 (b) Rs. 15 (c) Rs. 20 (d) none

Q.7. A road is 2 km in length. We have to supply the lamp-post. One post at each end. The distance between two consecutive lamp-posts is 20 m. Find the number of lamp-posts required.
(a) 51 (b) 100 (c) 101 (d) none

Q.8. A monkey climbs a pole 34 m long. In first minute he climbs 6 m and in second minute he slips down 2 m. In third minute he again climbs 6 m and in fourth minute he slips down 2 m. If he continue to do like this then find the time taken by him to reach the top.
(a) 15 min (b) 17 min (c) 18 min (d) none

Q.9. Find the difference between the greatest number of five digits and smallest number of five digits obtained by using 0, 1, 2, 3, 4 all but once.
(a) 32976 (b) 32966 (c) 32906 (d) none

Q.10. A certain number when multiplied by 13 the product consists entirely of fives. The smallest such number is ....

(a) 42735 (b) 42375 (c) 42365 (d) none

Q.11. A number when added to itself 14 times gives 75 as result. Find the number.

(a) 10 (b) 15 (c) 5 (d) none

Q.12. A number when added to itself 17 times gives 162 as result. Find the number.

(a) 9 (b) 10 (c) 11 (d) none

SOLUTIONS

Test 1

1. 247 **2.** 2603 **3.** 132 **4.** 342 **5.** 551 **6.** 507

7. 1003 **8.** 592 **9.** 392

Test 2

12, 24, 36, 48, 60, 72, 84, 96, 108, 120

Test 3

16. 32, 48, 64, 80, 96, 112, 128, 144, 160

Exercise

1. (i) 22382 (ii) 2295 (iii) 1452 (iv) 3875

(v) 18375 (vi) 9744 (vii) 128730 (viii) 28992

(ix) 105456 (x) 1218750

2. (i) 4311 (ii) 3382 (iii) 1114 (iv) 2551

(v) 2007

3. (i)
$$1 + 343 + 2197 = 2541$$

(ii)
$$\begin{aligned} 1 + 512 + x^3 &= 729 \\ x^3 &= 216 = 6^3 \\ x &= 6 \end{aligned}$$

(iii)
$$\begin{aligned} 216 + 512 + 1000 &= x^3 \\ 1728 &= x^3 \\ 12^3 &= x^3 \\ x &= 12 \end{aligned}$$

(iv)
$$27 + 125 + 3375 = 3527$$

(v)
$$1 + 8 + 2744 = 2753$$

4. $\dfrac{3375 - 125 - 1000}{225} = \dfrac{2250}{225} = \mathbf{10}$

5. $\dfrac{220}{55} = \mathbf{4}$

6. (a)

$$\begin{aligned} \text{12 months wages} &= 90 + T\,[T \rightarrow \text{Turban}) \\ \text{9 months wages} &= 65 + T \\ - \quad & \quad - \quad - \\ \hline \text{3 months wages} &= 25 \\ \hline \text{9 months wages} &= 25 \times 3 \\ &= 75 \\ \text{Therefore } 65 + T &= 75 \text{ Rs.} \\ T &= 75 - 65 \\ T &= \textbf{Rs. 10} \end{aligned}$$

7. (c) $\frac{2000}{20} + 1 = \mathbf{101}$

8. (a) Distance covered after two minutes $= 6 - 2 = 4$ m

Time taken by him to climb 4 m $= 2$ minute

Time taken by him to climb 1 m $= \frac{2}{4}$

Time taken by him to climb $34 - 6 = 28 \text{ m} = \frac{2}{4} \times 28 = 14$ min

So in 15th min he will climb remaining 6 m and once he reaches top he will not slip.

Therefore total time taken $= 14 + 1 =$ **15 min.**

9. (a)

Greatest number = 43210

Smallest number = 10234

(01234 is a four digit number)

Therefore required diff. = 43210 – 10234 = **32976.**

10. (a) This question can be done by two ways. First method is by checking the answer. We multiply 13 by all the choices and get the required choice **(a)**

Second method, we start dividing 55 by 13 and go one putting 5 in the dividend and start solving further till we get a zero remainder. The quotient obtained is the required answer. In this case it is **42735.**

11. (c)

$$\begin{aligned} \text{Let number} &= x \\ \text{Its 14 times} &= 14x \end{aligned}$$

According to statement

$$\begin{aligned} x + 14x &= 75 \\ 15x &= 75 \\ x &= \frac{75}{15} = \mathbf{5} \end{aligned}$$

Therefore required number is **5**

12. (a)

$$\begin{aligned} \text{Let number} &= x \\ \text{Its 17 times} &= 17x \end{aligned}$$

According to statement

$$\begin{aligned} x + 17x &= 162 \\ 18x &= 162 \\ x &= \frac{162}{18} = \mathbf{9} \end{aligned}$$

Therefore required number is **9**

Day 2

Decimal Fraction

On day 2, the concept of **decimal fraction** is covered.

Q.1. What is **decimal fraction** ?

Ans. A fraction in which denominator are 10 or powers of 10. eg. $\mathbf{\frac{7}{10}, \frac{7}{100}}$ they can be written as **.7** or **.07**.

First concept to be understood clearly is "**How to convert decimal fraction into vulgar fraction** ?" This can be understood by the following example.

Example 1. Convert .525 into vulgar fraction.

Sol : $$.525 = \frac{525}{1000} = \mathbf{\frac{21}{40}}$$

Explanation : Put **1** for **decimal** in **denominator**. Then **suffix** as many **zero's** with **1** as are **digits** of the **decimal fraction**. So, in above question there are three digits after decimal so denominator becomes 1000.

Example 2. Convert **.0525** into vulgar fraction.

$$.0525 = \frac{525}{10000} = \mathbf{\frac{21}{400}}$$

[As there are **four digits** after **decimal** so **denominator** becomes **10000**]

Now we will do the **addition, subtraction, multiplication** and **division** of decimal fraction.

Before doing operations of **decimal fraction** one thing is to be noted that "**Suffixing zero's to the extreme right of decimal fraction does not change its value**." It means that **5.21** and **5.210** or **5.2100** have the same value.

Now let us understand addition & Subtraction of fraction with following example

ADDITION AND SUBTRACTION

Example 3. Solve 521 + 52.1 + 5.21 + .521 + .0521

Sol :

$$\begin{array}{r} 521.0000 \\ 52.1000 \\ 5.2100 \\ 0.5210 \\ 0.0521 \\ \hline \mathbf{578.8831} \\ \hline \end{array}$$

Example 4. Solve 7.31 + 237.1 + 317.22 + 5.217

Sol :

$$\begin{array}{r} 7.310 \\ 237.100 \\ 317.220 \\ 5.217 \\ \hline \mathbf{566.847} \\ \hline \end{array}$$

Example 5. Subtract 3.202 from 5.212

Sol :

$$\begin{array}{r} 5.212 \\ -3.202 \\ \hline \mathbf{2.010} \\ \hline \end{array}$$

Example 6. Solve 13.215 – 2.131

Sol :

$$\begin{array}{r} 13.215 \\ -2.131 \\ \hline \mathbf{11.084} \\ \hline \end{array}$$

Example 7. Solve 17.071 + 3.023 + 5.115

Sol :

$$\begin{array}{r} 17.071 \\ 3.023 \\ 5.115 \\ \hline \mathbf{25.209} \\ \hline \end{array}$$

MULTIPLICATION

Example 8. $3.21 \times 5.1 = ?$

Sol : Step :

Multiply 321 × 51. It comes out to be 16371. We have three digits after the decimal points (two in 3.21 and one in 5.1). So we will put decimal after three digits starting from right. So our answer will be 16.371.

Example 9. $15.25 \times 1.2 = ?$

Sol : $1525 \times 12 = 18300$

We have to put **decimal** after **three digits** starting from right, Therefore, 18.300 or 18.3 is the answer.

DIVISION

Example 10. $22.5 \div .15$

Sol :

$$22.5 \div .15 = 22.5 \times \frac{1}{.15}$$

$$= \frac{22.5}{.15}$$

$$= \frac{225}{10} \times \frac{100}{15} = \mathbf{150}$$

Example 11. Divide 12.25 by 3.5

Sol :

$$12.25 \div 3.5 = \frac{12.25}{3.5}$$

$$= \frac{1225}{100} \times \frac{10}{35} = \frac{35}{10} = \mathbf{3.5}$$

So, it is now hoped that you have clearly understood the **addition, division, subtraction and multiplication** of **decimal fraction.**

RECURRING DECIMAL

To understand recurring decimal let us divide 1 by 3.

```
92)10000(.333
    9
   10
    9
   10
    9
    1
```

It goes on like this.

In this 3 is repeated in the quotient again and again infinitely.

So $\frac{1}{3}$ = **.3333........** We represent it as putting a **dot** on **3**. Therefore $\frac{1}{3} = .33333 = \mathbf{.\dot{3}}$

We take another example in which we divide 135 by 99.

$$\frac{135}{99} = 1.36\ 36\ 36\ 36\ 36\ 36.....$$

If a set of digits repeats itself continuously then it is represented by putting a bar (–) on it.

Therefore $\frac{135}{99} = \mathbf{1.\overline{36}}$

So, we can say that in a decimal fraction if a figure or set of figures repeat continuously then such a number is called recurring decimal number. **Recurring decimal** numbers are of two types:

(i) **Pure**
(ii) **Mixed**

(i) **Pure Recurring Decimal :** A **decimal fraction** in which **all the figures** after **decimal point** are repeated is called **pure recurring** decimal. e.g., $\frac{2}{3}$ = **.6666666......** = $\mathbf{.\dot{6}}$

(ii) **Mixed Recurring Decimal :** A **decimal fraction** in which after **decimal point some figures** are repeated and **some are not,** is called a **mixed recurring** decimal. e.g., 1.311111111.........

TO CONVERT PURE RECURRING INTO VULGAR FRACTION

Pure recurring **decimal fractions** are converted into vulgar fractions by the following method:

Let us suppose we have to convert $\mathbf{.\dot{3}}$ in vulgar fraction.

We write $x = \mathbf{.\dot{3}}$

$x = .3333......$ **...(1)**

Multiply both sides by 10

$$10x = 3.33333..... \quad ...(2)$$

Subtract (1) from (2)

$$9x = 3$$

$$x = \frac{3}{9} = \mathbf{\frac{1}{3}}$$

Example 12. Convert $1.\overline{36}$ into vulgar fraction.

Sol :

$$x = 1.36\ 36\ 36\ 36\ 36.......... \quad ...(1)$$

$$100x = 136.36\ 36\ 36\ 36.......... \quad ...(2)$$

Subtract (1) from (2)

$$99\,x = 135$$

$$x = \mathbf{\frac{135}{99}}$$

TO CONVERT MIXED RECURRING INTO VULGAR FRACTION

Let us suppose we are to convert 1.31 into vulgar fraction.

Let

$$x = 1.3\overline{1}$$

$$10\,x = 13.\overline{1} \quad ...(2)$$

$$100x = 131.\overline{1} \quad ...(3)$$

Subtract (2) from (3)

$$90x = 118$$

$$x = \frac{118}{90} = \mathbf{\frac{59}{45}}$$

Example 13. Convert $1.5\overline{7}$ into vulgar fraction.

Sol : Let

$$x = 1.5\overline{7}$$

$$10x = 15.\overline{7} \quad ...(2)$$

$$100x = 157.\overline{7} \quad ...(3)$$

Subtract (2) from (3)

$$90x = 142$$

$$x = \frac{142}{90} = \mathbf{\frac{71}{45}}$$

COMPARISON OF FRACTION

Decimal fraction is very useful in comparing large number of fractions. In this we convert each one of the fraction in the decimal form and arrange them as per the condition.

Example 14. Which is greater $\frac{6}{7}$ or $\frac{17}{21}$.

Sol :

$$\frac{6}{7} = .857$$

$$\frac{17}{21} = .809$$

$$\therefore \quad \mathbf{\frac{6}{7} > \frac{17}{21}}$$

Example 15. Arrgange in the ascending order $\frac{7}{9}, \frac{17}{20}, \frac{21}{25}$.

Sol :

$$\frac{7}{9} = .77$$

$$\frac{17}{20} = .85$$

$$\frac{21}{25} = .84$$

Ascending means increasing order which is as follows :

$$\mathbf{\frac{7}{9}, \frac{21}{25}, \frac{17}{20}}$$

Example 16. Arrange in descending order $\frac{2}{3}, \frac{5}{7}, \frac{11}{14}$.

Sol :

$$\frac{2}{3} = .66$$

$$\frac{5}{7} = .71$$

$$\frac{11}{14} = .78$$

Descending order means decreasing order which is as follows

$$\mathbf{\frac{11}{14}, \frac{5}{7}, \frac{2}{3}}$$

Example 17. Which is smaller $\frac{5}{9}$ or $\frac{15}{29}$.

Sol.

$$\frac{5}{9} = .55$$

$$\frac{15}{29} = .51$$

$$\mathbf{\frac{15}{29}} \text{ is smaller than } \mathbf{\frac{5}{9}}.$$

EXERCISE

Q.1. 6666.6 + 666.6 + 66.66 + 6.66 + .66 = ?

(a) 7201.7 (b) 7407.18 (c) 7307.18 (d) none

Q.2. 2.310 + 2.031 + 2.130 + 2.003 + .2130 = ?

(a) 8.687 (b) 8.677 (c) 8.667 (d) none

Q.3. 4545 + 454.4 + 44.55 + 5.544 = ?

(a) 5094.49 (b) 5050.494 (c) 5049.94 (d) none

Q.4. $4 \times .4 \times .04 \times .004 \times 400 = ?$

(a) 1.024 (b) 10.24 (c) 102.4 (d) none

Q.5. $23.50 \times 1.4 = ?$

(a) 33.9 (b) 39.2 (c) 43.9 (d) none

Q.6. $2.1 \times 3.2 \times 4.5 = ?$

(a) 302.4 (b) 3.024 (c) 30.24 (d) none

Q.7. $\frac{.000044}{.22} = ?$

(a) .002 (b) .0002 (c) .02 (d) none

Q.8. $\frac{2.5}{1.25} = ?$

(a) .02 (b) .2 (c) 2 (d) none

Q.9. $\frac{?}{1.7} = 2.04$

(a) 34.68 (b) 3.468 (c) 346.8 (d) none

Q.10. $34.40 - ? = 13.998$

(a) 2.040 (b) 204.02 (c) 20.402 (d) none

Q.11. $? - 701.17 = 232.55$

(a) 1833.72 (b) 933.72 (c) 943.72 (d) none

Q.12. $202.4 + 1125.02 - 117.099 = ?$

(a) 1210.321 (b) 1110.321 (c) 1210.221 (d) none

Q.13. If $\frac{1}{2.514} = .3977$ then the value of $\frac{1}{.0002514}$

(a) 3.977 (b) 3977 (c) 39770 (d) .3970

Q.14. If $\sqrt{5} = 2.24$ then what is the value of $\frac{3\sqrt{5}}{2\sqrt{5} - .48}$

(a) 1.68 (b) .168 (c) 16.8 (d) none

Q.15. $\sqrt{3136} = 56$ then the value of $\sqrt{31.36} + \sqrt{.3136} + \sqrt{.003136} + \sqrt{.00003136}$ is ....

(a) 17.22 (b) 6.22 (c) 5.91 (d) none

Q.16. $\frac{\sqrt{.00289}}{\sqrt{.00144}} = ?$

(a) 1.7/12 (b) 17/12 (c) 17/1.2 (d) none

Q.17. $.\overline{47} = ?$

(a) $\frac{47}{99}$ (b) $\frac{47}{90}$ (c) $\frac{46}{90}$ (d) none

Q.18. $.\dot{6} + .\dot{7} + .\dot{8} = ?$

(a) $2.\dot{3}$ (b) $2.\dot{1}$ (c) $2.\dot{2}$ (d) none

Q.19. $1.2\overline{3} = ?$

(a) $\frac{110}{99}$ (b) $\frac{111}{90}$ (c) $\frac{111}{99}$ (d) none

Q.20. $.\overline{47} + .\overline{53} = ?$

(a) 1.00 (b) $1.\overline{01}$ (c) $1.\overline{10}$ (d) none

Q.21. $5.\overline{12} = ?$

(a) $\frac{508}{99}$ (b) $\frac{512}{99}$ (c) $\frac{512}{90}$ (d) none

Q.22. $.26\overline{60} + .23\overline{38} = ?$

(a) $.49\overline{98}$ (b) .5000 (c) $.50\overline{01}$ (d) none

Q.23. Which of the following four is greatest

(a) $\frac{7}{9}$ (b) $\frac{9}{11}$ (c) $\frac{11}{13}$ (d) $\frac{14}{17}$

Q.24. which of the following four is smallest

(a) $\frac{3}{7}$ (b) $\frac{7}{9}$ (c) $\frac{9}{10}$ (d) $\frac{11}{13}$

Q.25. Arrange $\frac{3}{8}, \frac{2}{3}$ and $\frac{16}{25}$ in descending order ?

(a) $\frac{3}{8} > \frac{2}{3} > \frac{16}{25}$ (b) $\frac{2}{3} > \frac{16}{25} > \frac{3}{8}$ (c) $\frac{16}{25} > \frac{2}{3} > \frac{3}{8}$ (d) none

SOLUTIONS

1. (b)

$$\begin{array}{r} 6666.60 \\ 666.60 \\ 66.66 \\ 6.66 \\ .66 \\ \hline 7407.18 \\ \hline \end{array}$$

2. (a)

$$\begin{array}{r} 2.3100 \\ 2.0310 \\ 2.1300 \\ 2.0030 \\ .2130 \\ \hline 8.6870 \\ \hline \end{array}$$

3. (d)

$$\begin{array}{r} 4545.000 \\ 454.400 \\ 44.550 \\ 5.544 \\ \hline 5049.494 \\ \hline \end{array}$$

4. (d) $4 \times \frac{4}{10} \times \frac{4}{100} \times \frac{4}{1000} \times 400 = \frac{1024}{10000} = \mathbf{.1024}$

5. (d) $23.50 \times 1.4 = \frac{2350}{100} \times \frac{14}{10} = \frac{3290}{100} = \mathbf{32.9}$

6. (c) $2.1 \times 3.2 \times 4.5 = \frac{21}{10} \times \frac{32}{10} \times \frac{45}{10} = \frac{30240}{1000} = \mathbf{30.24}$

7. (b) $\frac{.000044}{22} = \frac{44}{1000000} \times \frac{100}{22} = \frac{2}{10000} = \mathbf{.0002}$

8. (c) $\frac{25}{10} \times \frac{100}{125} = \mathbf{2}$

9. (b)
$$\frac{x}{1.7} = 2.04$$
$$x = 2.04 \times 1.7$$
$$= \frac{204}{100} \times \frac{17}{10} = \mathbf{3.468}$$

10. (c)
$$34.40 - x = 13.998$$
$$34.400 - 13.998 = x$$
$$20.402 = x$$
$$x = \mathbf{20.402}$$

11. (b)
$$x - 701.17 = 232.55$$
$$x = 232.55 + 701.17 = \mathbf{933.72}$$

12. (a)
$$202.4 + 1125.02 - 117.099 = x$$
$$1327.420 - 117.099 = \mathbf{1210.321}$$

13. (b)

14. (a) $\frac{3 \times 2.24}{2 \times 2.24 - .48} = \frac{6.72}{4.48 - .48} = \frac{6.72}{4} = \mathbf{1.68}$

15. (b) $5.6 + .56 + .056 + .0056 = \mathbf{6.22}$

16. (b)
$$\sqrt{\frac{289}{100000} \times \frac{100000}{144}} = x$$
$$x = \mathbf{\frac{17}{12}}$$

17. (a)
$$x = .47\ 47\ 47 \quad \text{...(1)}$$
$$100\,x = 47.47\ 47\ 47 \quad \text{...(2)}$$

Subtract (1) from (2)
$$99x = 47$$
$$x = \mathbf{\frac{47}{99}}$$

18. (a)
$$.\dot{6} + .\dot{7} + .\dot{8} = x$$
$$\frac{6}{9} + \frac{7}{9} + \frac{8}{9} = x$$
$$\frac{21}{9} = x$$

$$x = \mathbf{2.\dot{3}}$$

19. (b)

$$x = 1.2333 \quad ...(1)$$
$$10x = 12.3333 \quad ...(2)$$
$$100x = 123.3333 \quad ...(3)$$

Subtract (2) from (3)

$$90x = 111$$
$$x = \frac{\mathbf{111}}{\mathbf{90}}$$

20. (b)

$$.\overline{47} + .\overline{53} = \frac{47}{99} + \frac{53}{99} = \frac{100}{99} = 1.\overline{01}$$

21. (d)

$$x = 5.12\ 12 \quad ...(1)$$
$$100\,x = 512.12\ 12 \quad ...(2)$$

Subtract (1) from (2)

$$99x = 507$$
$$x = \frac{\mathbf{507}}{\mathbf{99}}$$

22. (a)

$$x = .26\overline{60}$$
$$100x = 26.\overline{60} \quad ...(1)$$
$$10000x = 2660.\overline{60} \quad ...(2)$$

Subtract **(1)** from **(2)**

$$9900x = 2634$$
$$x = \frac{2634}{9900}$$

$$\text{Similarly } .23\overline{38} = \frac{2315}{9900}$$

$$\therefore \quad .26\overline{60} + .23\overline{38} = \frac{2634}{9900} + \frac{2315}{9900}$$
$$= .49\ 98\ 98\ 98\ 98\ 98\ 98 \ ...... = \mathbf{.49\overline{98}}$$

23. (c)

24. (a)

$$\frac{3}{7} = .42, \frac{7}{9} = .77, \frac{9}{10} = .90, \frac{11}{13} = .84 \qquad \therefore \frac{3}{7} \text{ is smallest}$$

25. (b)

$$\frac{3}{8} = .375, \frac{2}{3} = .6666, \frac{16}{25} = .64$$

$\mathbf{\frac{2}{3} > \frac{16}{25} > \frac{3}{8}}$ is correct descending order.

Day 3

Simplification

On **Day 3** we will do the topic of **simplification** of fraction. **Simplification** means to **simplify a complicated expression** to get a **single answer**. These are simplified by a set of rules. To understand it more clearly let us solve the following example.

Example. Solve $6 + 4 \div 2$.

Solution. The correct solution is as follows

$$= 6 + 4 \times \frac{1}{2}$$
$$= 6 + 2 = \mathbf{8}$$

Note : Many students solve it as follows :

$6 + 4 \div 2 = 10 \div 2 = 5$, which is wrong. To solve these types of questions we must follow the rule of **BODMAS**. According to this rule of multiplication should be done after division.

RULE OF BODMAS

B = **Brackets**

Brackets are solved in order of ( ), { } and [ ].

O = **Of**

D = **Division**

M = **Multiplication**

A = **Addition**

S = **Subtraction**

It means that while solving we must follow the above sequence. So in above given example we have to first divide 4 by 2 and add it to 6. Now solve the following example.

Example 1. Solve $\frac{2+2\times 2\div 2}{2-2+2\times 2}$.

Sol :
$$\frac{2+2\times 2\times\frac{1}{2}}{2-2+4} = \frac{2+2}{6-2}$$
$$= \frac{4}{4} = \mathbf{1}$$

Example 2. Solve $2 + 2 \text{ of } 4 \div 2 - \frac{17}{4}$.

Sol :
$$2 + 8 \div 2 - \frac{17}{4}$$

$$2 + 8 \times \frac{1}{2} - \frac{17}{4}$$

Note : Of must be solved before ÷

$$= 2 + 4 - \frac{17}{4}$$

$$= \frac{6}{1} - \frac{17}{4}$$

$$= \frac{24 - 17}{4} = \mathbf{\frac{7}{4}}$$

Example 3. Solve $13 - [4 - \{3 - (5 - \overline{6 + 3})\}]$.

Sol : This is an example in which the brackets are given. Brackets are solved after Bar. The order of solving the brackets is (), {}, and [ ]. So solution of above example is as follows.

$$13 - [4 - \{3 - (5 - \overline{6 + 3})\}] = 13 - [4 - \{3 - (5 - 9)\}]$$
$$= 13 - [4 - \{3 - (-4)\}]$$
$$= 13 - [4 - \{3 + 4\}]$$
$$= 13 - [4 - 7] = 13 - [-3]$$
$$= 13 + 3 = \mathbf{16}$$

Example 4. $\dfrac{5 + 5 \times 5}{5 \times 5 + 5} \div \dfrac{\frac{1}{5} \div \frac{1}{5} \text{ of } \frac{1}{5}}{\frac{1}{5} \text{ of } \frac{1}{5} \div \frac{1}{5}}$.

Sol :
$$= \frac{5 + 25}{25 + 5} \div \frac{\frac{1}{5} \div \frac{1}{25}}{\frac{1}{25} \div \frac{1}{5}}$$

$$= \frac{30}{30} \div \frac{\frac{1}{5} \times \frac{25}{1}}{\frac{1}{25} \times \frac{5}{1}} = 1 \div \frac{\frac{5}{1}}{\frac{1}{5}} = \mathbf{\frac{1}{25}}$$

Example 5. $\dfrac{2 + 2 \times 2}{2 \times 2 + 2} + \dfrac{\frac{1}{3} \div \frac{1}{3} \text{ of } \frac{1}{3}}{\frac{1}{3} \div \frac{1}{3} \times \frac{1}{3}}$

Sol :
$$\frac{2 + 4}{4 + 2} + \frac{\frac{1}{3} \times \frac{9}{1}}{\frac{1}{3} \times \frac{3}{1} \times \frac{1}{3}} = \frac{6}{6} + \frac{\frac{3}{1}}{\frac{1}{3}}$$
$$= 1 + 9 = \mathbf{10}$$

QUESTIONS BASED ON FORMULAE

Now second type of questions which are quite often asked in the exams are the questions based on algebric formulae. Here below is a simple example given which shows how formulae are helpful while solving the question [Where these are applicable}

Solve. $(1000)^2 - (999)^2$

Sol : Normal solution is as follows

$$(1000)^2 - (999)^2 = 1000000 - 998001 = \mathbf{1999}$$

Solution with the help of formula

$$(1000)^2 - (999)^2$$

Let $a = 1000 \quad b = 999$

So $(1000)^2 - (999)^2$ becomes $(a)^2 - (b)^2$

But in algebra we know that

$$a^2 - b^2 = (a + b)(a - b)$$

Therefore
$$(1000)^2 - (999)^2 = (1000 + 999)(1000 - 999) = 1999 \times 1 = \mathbf{1999} \quad \textbf{Ans.}$$

Similarly there are other formulae of algebra which are very helpful in solving the question. Following is the list of formulae which are helpful in solving these types of questions. You have to learn these formulae by heart.

1. $a^2 - b^2 = (a + b)(a - b)$
2. $(a + b)^2 = a^2 + b^2 + 2ab$
3. $(a - b)^2 = a^2 + b^2 - 2ab$
4. $\boldsymbol{a^3 - b^3 = (a - b)(a^2 + ab + b^2)}$
5. $\boldsymbol{a^3 + b^3 = (a + b)(a^2 - ab + b^2)}$
6. $(a + b)^3 = a^3 + b^3 + 3ab(a + b)$
7. $(a - b)^3 = a^3 - b^3 - 3ab(a - b)$
8. $a^3 + b^3 + c^3 - 3abc = (a + b + c)(a^2 + b^2 + c^2 - ab - bc - ca)$

Following are some examples based on above formulae.

Example 6. Solve $\dfrac{.89 \times .89 \times .89 - .64 \times .64 \times .64}{.89 \times .89 + .89 \times .64 + .64 \times .64}$

Sol : Let $a = .89$ and $b = .64$

Therefore question becomes

$$\frac{a^3 - b^3}{a^2 + ab + b^2} = \frac{(a - b)(a^2 + ab + b^2)}{a^2 + ab + b^2} = a - b$$

Substitute the value of a and b $= .89 - .64$

$= \mathbf{.25}$

Example 7. Solve $\dfrac{(.471)^3 + (.529)^3}{(.471)^2 - (.471 \times .529) + (.529)^2}$

Sol : Let $a = .471$ and $b = .529$

Therefore question becomes

$$= \frac{a^3 + b^3}{a^2 - ab + b^2}$$

$$= \frac{(a+b)(a^2-ab+b^2)}{a^2-ab+b^2}$$

$$= a+b$$

$$= .471 + .529 = \mathbf{1.000}$$

Example 8. Solve $\dfrac{.1\times.1\times.1+.2^3+.3^3-3\times.1\times.2\times.3}{.1^2+.2^2+.3^2-.1\times.2-.2\times.3-.3\times.1}$

Sol : Let $a = .1, b = .2$ and $c = .3$

Therefore question becomes

$$\frac{a^3+b^3+c^3-3abc}{a^2+b^2+c^2-ab-bc-ca} = \frac{(a+b+c)(a^2+b^2+c^2-ab-bc-ca)}{(a^2+b^2+c^2-ab-bc-ca)}$$

$$= a+b+c$$

$$= .1+.2+.3 = \mathbf{.6}$$

Example 9. Solve $(1.75)^2 + (.75)^2 - 2 \times 1.75 \times .75$

Sol : Let $a = 1.75$ and $b = .75$

Therefore question becomes

$$a^2+b^2-2ab = (a-b)^2$$

$$= (1.75-.75)^2 = \mathbf{1}$$

CONTINUED FRACTION

This is explained by following example

Example 10. Simplify $1+\cfrac{1}{1+\cfrac{1}{1+\cfrac{1}{1+\cfrac{1}{3}}}}$

Sol : Starting from the bottom

$$1+\cfrac{1}{1+\cfrac{1}{1+\cfrac{1}{\frac{4}{3}}}} = 1+\cfrac{1}{1+\cfrac{1}{1+\frac{3}{4}}}$$

$$= 1+\cfrac{1}{1+\cfrac{1}{\frac{7}{4}}}$$

$$= 1+\cfrac{1}{1+\frac{4}{7}} = 1+\cfrac{1}{\frac{11}{7}}$$

$$= 1+\frac{7}{11} = \mathbf{\frac{18}{11}}$$

Example 11. Simplify $2+\cfrac{1}{2+\cfrac{1}{1+\cfrac{1}{1+\cfrac{1}{6}}}}$.

Sol : Starting from the bottom :

$$= 2+\cfrac{1}{2+\cfrac{1}{1+\cfrac{1}{\frac{7}{6}}}} = 2+\cfrac{1}{2+\cfrac{1}{1+\frac{6}{7}}}$$

$$= 2+\cfrac{1}{2+\cfrac{1}{\frac{13}{7}}} = 2+\cfrac{1}{2+\frac{7}{13}}$$

$$= 2+\cfrac{1}{\frac{33}{13}} = 2+\frac{13}{33} = \mathbf{\frac{79}{33}}$$

QUESTIONS BASED ON STATEMENTS

These types of questions are very important and are mostly asked in the exams. A few examples are given below.

Example 12. 2/3 of an estate is Rs. 44000. What is 3/11 of it?

Sol : Let the value of estate be x

According to the given condition

$$\frac{2}{3} \text{ of } x = 44000$$

$$2x = 44000 \times 3$$

$$x = \frac{44000 \times 3}{2} = 66000$$

Therefore $x = 66000$ Rs.

Now we are asked 3/11 of x which is equal to

$$= \frac{3}{11} \times 66000 = \mathbf{18000 \text{ Rs.}}$$

Example 13. A man had 2 sons. To the elder son he left 5/11 of his property, to the younger 5/11 of remainder and rest to widow. Find the share of son if widow gets Rs. 3600.

Sol : Let the total share $= x$

$$\text{Elder's share} = \frac{5x}{11}$$

$$\text{Remaining} = x - \frac{5x}{11}$$

$$= \frac{6x}{11}$$

$$\text{Younger's share} = \frac{5}{11} \text{ of } \frac{6x}{11}$$
$$= \frac{30x}{121}$$
$$\text{Remainder} = \frac{6x}{11} - \frac{30x}{121}$$
or
$$\text{widow's share} = \frac{66x - 30x}{121}$$
$$= \frac{36x}{121}$$

According to condition
$$\frac{36x}{121} = 3600 \text{ Rs.}$$
$$x = \frac{3600 \times 121}{36}$$
$$x = \textbf{Rs. 12100}$$
$$\text{Elder's share} = \frac{5x}{11}$$
$$= \frac{5 \times 12100}{11} = \textbf{Rs. 5500}$$
$$\text{Younger's share} = \frac{30x}{121}$$
$$= \frac{30 \times 12100}{11} = \textbf{Rs. 3000}$$

Example 14. In a family the father took 1/4th of the cake and he had 3 times as much as the other had. Find the total number of family members.

Sol : Let the total number of family members = x

Share of cake of father = 1/4

Total share of remaining $(x - 1)$ member = $1 - \frac{1}{4} = \frac{3}{4}$

$$\text{Each member gets} = \frac{3}{4\,(x-1)}$$

According to condition : Father's share = 3 times the share of each member

Therefore
$$\frac{1}{4} = 3 \times \frac{3}{4\,(x-1)}$$
$$\frac{1}{4} = \frac{9}{4\,(x-1)}$$
$$4(x - 1) = 9 \times 4$$
$$4x - 4 = 36$$
$$4x = 40$$
$$x = \mathbf{10}$$

Example 15. A man spends 1/5th of his salary to meet pocket expenses and 4/5th of the remainder to meet the other expenses. If his monthly saving amounts to Rs. 2500. Find his monthly salary.

Sol :

$$\text{Let monthly salary} = x$$

$$\text{Amount spent on pocket expenses} = \frac{x}{5}$$

$$\text{Remainder} = x - \frac{x}{5}$$

$$= \frac{4x}{5}$$

$$\text{Other expenses} = \frac{4}{5} \times \frac{4x}{5}$$

$$= \frac{16x}{25}$$

Monthly saving = Total salary – (pocket expenses + other expenses)

$$x - \left(\frac{x}{5} + \frac{16x}{25}\right) = x - \frac{21x}{25}$$

$$= \frac{25x - 21x}{25} = \frac{4x}{25}$$

According to condition

$$\frac{4x}{25} = 2500$$

$$x = \frac{2500 \times 25}{4}$$

$$x = \mathbf{15625\ Rs.}$$

Direct Method : He spends **1/5th** of the salary means he saves **4/5**. Now of **4/5** he spends **4/5** means he saves **1/5** of the remainder that is of **4/5**. Hence we can directly write as

$$\frac{4}{5} \times \frac{1}{5} \text{ of income} = 2500 \text{ Rs.}$$

$$\text{Therefore income} = \frac{2500 \times 25}{4} = \mathbf{15625\ Rs.}$$

Example 16. A student was asked to multiply a certain no. by 8/17. He divided the number by 8/17 and got the answer which was 225 more than the correct answer. The given number was :

Sol :

$$\text{Let the given number} = x$$

$$\text{Correct answer} = x \times \frac{8}{17}$$

$$= \frac{8x}{17}$$

$$\text{Wrong answer} = \frac{\frac{x}{8}}{17}$$

$$= \frac{17x}{8}$$

According to condition

$$\frac{17x}{8} - \frac{8x}{17} = 225$$

$$\frac{289x - 64x}{136} = 225$$

$$\frac{225x}{136} = 225$$

$$x = \mathbf{136}$$

Example 17. The highest score in an innings was 3/11 of the total and next highest was 3/11 of the remainder. If the scores differ by 9, find the total score.

Sol : Let the total score $= x$

$$\text{Highest score} = \frac{3x}{11}$$

$$\text{Remainder} = x - \frac{3x}{11} = \frac{8x}{11}$$

$$\text{Next highest} = \frac{3}{11} \text{ of } \frac{8x}{11}$$

$$= \frac{24x}{121}$$

According to given condition

$$\frac{3x}{11} - \frac{24x}{121} = 9$$

$$\frac{33x - 24x}{121} = 9$$

$$\frac{9x}{121} = 9$$

$$x = \mathbf{121}$$

Example 18. A man gave 1/3 of his property to his son. 1/2 of his remaining property to his wife and 3/5 of the remaining property to his daughter. He still has Rs. 3200 which he gave in charity. What was the total value of his property?

Sol : Let the total property $= x$

$$\text{Son's share} = \frac{x}{3}$$

$$\text{Remaining} = x - \frac{x}{3}$$

$$= \frac{2x}{3}$$

$$\text{Wife's share} = \frac{1}{2} \text{ of } \frac{2x}{3} = \frac{x}{3}$$

$$\text{Remaining} = \frac{2x}{3} - \frac{x}{3} = \frac{x}{3}$$

$$\text{Daughter's share} = \frac{3}{5} \text{ of } \frac{x}{3} = \frac{x}{5}$$

$$\text{Remaining} = \frac{x}{3} - \frac{x}{5} = \frac{2x}{15}$$

According to question

$$\frac{2x}{15} = 3200 \text{ Rs.}$$

$$x = \textbf{Rs. 24000}$$

EXERCISE

Q.1. Simplify $\dfrac{.727 \times .727 - .273 \times 273}{.727 - .273}$

(a) 2 (b) 1 (c) 0 (d) none

Q.2. Simplify $\dfrac{304 \times .002 \times .18}{.009 \times .038}$

(a) 320 (b) 32 (c) 40 (d) 0

Q.3. Simplify $\dfrac{\frac{3}{4} \div \frac{4}{5} \times \frac{6}{7}}{\frac{3}{4} \div \frac{8}{5} \text{ of } \frac{3}{7}} + \dfrac{1}{3.5} \times 4\dfrac{4}{7} - \dfrac{1}{1.4}\left(\dfrac{1}{5} - \dfrac{1}{7}\right)$

(a) 2 (b) 1 (c) 4 (d) none

Q.4. Simplify $\dfrac{\frac{1}{2} \div \frac{1}{4} \text{ of } \frac{1}{8}}{\frac{1}{2} \div \frac{1}{4} \times \frac{1}{8}}$

(a) 1/64 (b) 1 (c) 1/4 (d) none

Q.5. $\dfrac{(272 - 32)(124 + 176)}{17 \times 15 - 15} = ?$

(a) 1 (b) 5 (c) 300 (d) none

Q.6. $\dfrac{24 \times 8 - 2 \times 15}{12 \times 12 + 18} = ?$

(a) 2 (b) 2/5 (c) 1 (d) 4

Q.7. $\dfrac{25 \times 25 + 13}{22 \div 11 \times 5} = ?$

(a) 638 (b) 63.8 (c) 6.38 (d) none

Q. 8. $\dfrac{1000^2 + 100}{100} = ?$

(a) 100 (b) 10000 (c) 10001 (d) none

Q.9. $121 \times 5 + 5 \div 5 \times 5 - 5 = ?$

(a) 605 (b) 525 (c) 510 (d) 1515

Q.10. $\dfrac{1}{25}$ of $100 \times 2.55 = ?$

(a) 10.02 (b) 102 (c) 5.20 (d) none

Q.11. $\dfrac{(811)^3 + (189)^3}{(811)^2 + (189)^2 - 811 \times 189} = ?$

(a) 622 (b) 1000 (c) 100 (d) none

Q.12. $\dfrac{(547)^2 - (353)^2}{(446)^2 - (146)^2} = ?$

(a) $\dfrac{295}{297}$ (b) $\dfrac{297}{295}$ (c) $\dfrac{297}{291}$ (d) none

Q.13. $242 \times 242 + 121 \times 121 - 2 \times 242 \times 121 = ?$

(a) 14641 (b) 12123 (c) 13213 (d) 32131

Q.14. $\dfrac{11^3 + 13^3 + 15^3 - 3 \times 11 \times 13 \times 15}{11^2 + 13^2 + 15^2 - 11 \times 13 - 13 \times 15 - 165} = ?$

(a) 39 (b) 53 (c) 17 (d) none

Q.15. $1 + \cfrac{1}{1 + \cfrac{1}{1 + \cfrac{1}{5}}} = ?$

(a) 11/6 (b) 6/11 (c) 6/17 (d) none

Q.16. $2 + \cfrac{2}{1 + \cfrac{1}{1 + \cfrac{1}{7}}} = ?$

(a) 46/15 (b) 46/7 (c) 15/46 (d) none

Q.17. $\dfrac{243 \times 243 \times 243 + 357 \times 357 \times 357}{243 \times 243 - 357 \times 243 + 357 \times 357} = ?$

(a) 800 (b) 600 (c) 8600 (d) none

Q.18. One third of Rajesh's marks in maths exceeds half of his marks in social studies by 30. If he got 240 marks in two subjects together. How many did he get in social studies?

(a) 40 (b) 60 (c) 80 (d) none

Q.19. The expansion of $\left(1-\frac{1}{2}\right)\left(1-\frac{1}{3}\right)\left(1-\frac{1}{4}\right)...\left(1-\frac{1}{m}\right)$ gives

(a) $\frac{m-1}{m}$ (b) $\frac{2}{m}$ (c) $\frac{1}{m}$ (d) none

Q.20. The expension $\left(2-\frac{1}{3}\right)\left(2-\frac{3}{5}\right)\left(2-\frac{5}{7}\right)...\left(2-\frac{999}{1001}\right)$ gives

(a) $\frac{991}{1001}$ (b) $\frac{1001}{13}$ (c) $\frac{1003}{3}$ (d) none

Q.21. In a school 1/6 of the girls and 1/7 of the boys took part in NCC camp. Total number of students in NCC camp are :

(a) 700 (b) 800
(c) data inadequate (d) none

Q.22. On a sports day if 30 children were made to stand in columns, 16 columns could be formed. If 24 children were made to sit in the column how many columns will be formed ?

(a) 45 (b) 20 (c) 12 (d) none

Q.23. A student was asked to multiply a number by 32. By mistake he multiplied the number by 23 and get the answer which was 27 less than the correct one. What is the correct answer ?

(a) 66 (b) 69 (c) 96 (d) none

Q.24. 1/3 of a number is 330. What is 5/11 of that number ?

(a) 550 (b) 150 (c) 450 (d) none

Q.25. In a library 1/2 of the books are story books, 3/4 of the remaining are reference books and the remaining 300 are physics books. How many of them are reference books ?

(a) 480 (b) 180 (c) 380 (d) none

SOLUTIONS

1. (b)

$$a = .727,\ b = .273$$
$$= \frac{a^2-b^2}{a-b} = \frac{(a+b)(a-b)}{a-b}$$
$$= a+b = .727 + .273 = \mathbf{1.000}$$

2. (a)

$$\frac{304\times\frac{2}{1000}\times\frac{18}{100}}{\frac{9}{1000}\times\frac{38}{100}} = \frac{304\times2\times18\times1000\times1000}{1000\times100\times9\times38} = \mathbf{320}$$

3. (a)

$$\frac{\frac{3}{4}\times\frac{5}{4}\times\frac{6}{7}}{\frac{3}{4}\times\frac{5}{8}\times\frac{7}{3}}+\frac{1}{\frac{7}{2}}\times\frac{32}{7}-\frac{10}{14}\left(\frac{1}{5}-\frac{1}{7}\right) = \frac{\frac{90}{112}}{\frac{35}{32}}+\frac{2}{7}\times\frac{32}{7}-\frac{5}{7}\times\frac{2}{35}$$
$$= \frac{90}{112}\times\frac{32}{35}+\frac{64}{49}-\frac{2}{49}$$
$$= \frac{36}{49}+\frac{64}{49}-\frac{2}{49}$$
$$= \frac{36+64-2}{49} = \frac{98}{49} = \mathbf{2}$$

4. (d) $$\frac{\frac{1}{2} \div \frac{1}{4} \text{ of } \frac{1}{8}}{\frac{1}{2} \div \frac{1}{4} \times \frac{1}{8}} = \frac{\frac{1}{2} \div \frac{1}{32}}{\frac{1}{2} \times \frac{4}{1} \times \frac{1}{8}} = \frac{\frac{1}{2} \times \frac{32}{1}}{\frac{1}{4}}$$

$$= \frac{16/1}{1/4} = 16 \times 4 = \mathbf{64}$$

5. (c) $$\frac{240 \times 300}{240} = \mathbf{300}$$

6. (c) $$\frac{192 - 30}{144 + 18} = \frac{162}{162} = \mathbf{1}$$

7. (b) $$\frac{625 + 13}{22 \times \frac{1}{11} \times 5} = \frac{638}{10} = \mathbf{63.8}$$

8. (c) $$\frac{1000000 + 100}{100} = \frac{1000100}{100} = 10001$$

9. (a) $$605 + 5 \times \frac{1}{5} \times 5 - 5 = 605 + 5 - 5 = \mathbf{605}$$

10. (d) $$\frac{1}{25} \times 100 \times 2.55 = \mathbf{10.20}$$

11. (b) $$a = 811, b = 189$$

$$= \frac{a^3 + b^3}{a^2 + b^2 - ab}$$

$$= \frac{(a+b)(a^2 + b^2 - ab)}{a^2 + b^2 - ab}$$

$$= a + b = 811 + 189 = \mathbf{1000}$$

12. (d) $$\frac{(547 + 353)(547 - 353)}{(446 + 146)(446 - 146)} = \frac{900 \times 194}{592 \times 300} = \mathbf{\frac{291}{296}}$$

13. (a) $$(a - b)^2 = (242 - 121)^2 = (121)^2 = \mathbf{14641}$$

14. (a) $$\frac{a^3 + b^3 + c^3 - 3abc}{a^2 + b^2 + c^2 - ab - bc - ca} = a + b + c = 15 + 13 + 11 = \mathbf{39}$$

15. (d) $$1 + \frac{1}{1 + \frac{1}{1 + \frac{1}{5}}} = 1 + \frac{1}{1 + \frac{1}{1 + \frac{6}{5}}} = 1 + \frac{1}{1 + \frac{5}{6}} = 1 + \frac{1}{\frac{11}{6}}$$

$$= 1 + \frac{6}{11} = \mathbf{\frac{17}{11}}$$

16. (a) $$2 + \frac{2}{1 + \frac{1}{\frac{8}{7}}} = 2 + \frac{2}{1 + \frac{7}{8}} = 2 + \frac{2}{\frac{15}{8}}$$

$$= 2 + \frac{16}{15} = \mathbf{\frac{46}{15}}$$

17. (b)

$$\frac{a^3+b^3}{a^2+b^2-ab} = \frac{(a+b)(a^2+b^2-ab)}{a^2+b^2-ab}$$

$$= a+b = 243+357 = \mathbf{600}$$

18. (b)

$$\frac{M}{3}-\frac{SS}{2} = 30 \quad \text{Or} \quad \frac{2M-3SS}{6}=30$$

$$2M-3SS = 180 \quad ...(1)$$

$$M+SS = 240 \quad ...(2)$$

Solving (1) and (2) we get

$$SS = \mathbf{60}$$

19. (c) $\frac{1}{2}\times\frac{2}{3}\times\frac{3}{4}..........\frac{m-1}{m}$

(2 gets cancelled by 2, 3 by 3 and so on and so forth)

$$= \frac{\mathbf{1}}{\mathbf{m}}$$

20. (c)

$$\frac{5}{3}\times\frac{7}{3}\times\frac{9}{7}..........\frac{1003}{1001} = \frac{1003}{3}$$

21. (c) More information is needed.

22. (b) Total number of students 30 × 16 = 480. So required columns = $\frac{480}{24}$ = **20**

23. (c)

$$32x-23x = 27$$

$$9x = 27$$

$$\text{Therefore correct number} = 32\times 3 = \mathbf{96}$$

24. (c)

$$\frac{1}{3}\text{ of } x = 330$$

$$x = 990$$

$$\text{Req. number} = \frac{5}{11} = \mathbf{450}$$

25 (d) Let total number of books $= x$

$$\text{story books} = \frac{x}{2}$$

$$\text{remaining} = x-\frac{x}{2}=\frac{x}{2}$$

$$\text{reference books} = \frac{3}{4}\text{ of }\frac{x}{2}=\frac{3x}{8}$$

$$\text{remainder or physics books} = \frac{x}{2}-\frac{3x}{8}=\frac{x}{8}$$

But $\frac{x}{8} = 300$ **(Given)**

Therefore, $x = 2400$

$$\text{Reference books} = \frac{3x}{8}$$

$$= \frac{3\times 2400}{8} = \mathbf{900}$$

Day 4

The Number System & Concepts of Divisibility

THE NUMBER SYSTEM

Odd No : A number which is not divisible by 2 is called odd number. e.g., 3, 5, 7, 9, 11 etc.

Even No : A number which is divisible by 2 is called even number. e.g., 2, 4, 6, 8, 10 etc.

Prime No : Numbers, other than 1 which are divisible by only themselves and 1 are called prime numbers.

THE PRIME NUMBERS UPTO 100 ARE

Between :

1-10	:	2,3,5,7
11-20	:	11,13,17,19
21-30	:	23,29
31-40	:	31,37
41-50	:	41,43,47
51-60	:	53,59
61-70	:	61,67
71-80	:	71,73,79
81-90	:	83,89
91-100	:	97

Composite Numbers

The numbers which are not prime are called composite numbers. The fact is to be remembered that no even number can be prime except 2.

RULES OF DIVISIBILITY

(i) **Divisibility by 2 :** A number is divisible by 2 if its last digit is even, e.g., 2852, 2860, 281252 all are divisible by 2

(ii) **Divisibility by 3 :** A number is divisible by 3 if the sum of digits is divisible by 3, e.g., 3567934212. The sum of the digits is $3 + 5 + 6 + 7 + 9 + 3 + 4 + 2 + 1 + 2 = 42$. Further $4 + 2 = 6$, which is divisible by 3.

(iii) **Divisibility by 4 :** A number is divisible by 4 if its last two digits are divisible by 4, e.g., 22024 is divisible by 4.

(iv) **Divisibility by 5 :** A number is divisible by 5 if its last digit is 0 or 5. e.g., 250, 1200 etc.

(v) **Divisibility by 6 :** A number is divisible by 6 if the number is divisible by 2 and 3 both. This means the last digit is even and the sum of the digits are divisible by 3. e.g., 132. Here some of the digits is 1 + 3 + 2 = 6 which is divisible by 3 and the last digit of the number is even. Hence 132 is divisible by 6.

(vi) **Divisibility by 8 :** A number is divisible by 8 if its last three digits are divisible by 8. e.g., 125888, 225008, 425032 etc.

(vii) **Divisibility by 9 :** A number is divisible by 9 if the sum of digits in a number is divisible by 9. e.g., 31221. Some of the digits is 3 + 1 + 2 + 2 + 1 = 9. Which is divisible by 9.

(viii) **Divisibility by 11 :** A number is divisible by 11 if the difference between **sum** of its digits in the **odd places** and **sum** of its digits in the **even places** is **0** or divisible by **11**. This is explained below with an example.

Example. Find whether 203170 is divisible by 11 or not.

Solution : Digits at odd places are 2, 3, 7 and their sum is (2 + 3 + 7) = 12

Digits at even places are 0,1,0 and their sum (0 + 1 + 0) = 1

Difference of the sums 12 – 1 = 11

∴ number is divisible by 11.

Now solve the following exercise

EXERCISE 1

Q.1. How many numbers are odd in the following numbers ?

121, 212, 414, 373, 124, 2288, 2377, 2444, 2666, 2773, 2444, 2662, 3888.

Q.2. How many numbers are prime between 1 and 70 ?

Q.3. Is 87 prime ?

Q.4. Is 37338951 divisible by 3 ?

Q.5. Is 37733895217 divisible by 3 ?

Q.6. Is 5892157424 divisible by 4 ?

Q.7. Is 377238942 divisible by 6 ?

Q.8. Is 3752400 divisible by 8 ?

Q.9. Is 593779901 divisible by 11 ?

Q.10. How many composite numbers are there upto 100?

Q.11. Which is the smallest prime number ?

Q.12. A number which is divisible by 9 is divisible by 3 also. Say True or False ?

SOLUTIONS

1. There are 4 odd numbers i.e., 121, 373, 2377, 2773.
2. 19 [2, 3, 5, 7, 11, 13, 17, 19, 23, 29, 31, 37, 41, 43, 47, 53, 59, 61, 67]
3. No. [87 is product of 3 and 29]
4. Yes [because the sum of digits 3 + 7+ 3 + 3 + 8 + 5 + 9 + 1 = 39 and 3 + 9 = 12, 1 + 2 = 3]
5. No.
6. Yes. (Because last 2 digits are 24 which is divisible by 4.)
7. Yes. [Because the last digit is even and the sum of the digits is divisible by 3, so the number is divisible by both 3 and 2. Hence it is divisible by 6.]
8. Yes. [The last three digits are 400 and are exactly divisible by 8.]
9. Yes. [The sum of the digits at odd places is 5 + 3 + 7 + 9 + 1 = 25 and the sum of the digits at even places is 9 + 7 + 9 = 25. Difference 25 – 25 = 0. Therefore the number is divisible by 11.]
10. 74. [Because the prime numbers between 1 and 100 are 25 and 1 is neither prime nor composite. Therefore the composite numbers are 74 {100 – 26}.]
11. 2.
12. True.

Now we will clear some more basic concepts.

CONCEPT I

What is **dividend, divisor, quotient** and **remainder** ? Let us understand this terminology by following simple example.

$$\begin{array}{r|l} 7) & 58\ (8 \\ & \underline{56} \\ & \underline{\ 2} \end{array}$$

In above division, 58 is **divided** by 7. Here, 58 is called **dividend** {the number to be divided}, 7 the **divisor** {by which **dividend** is **divided**}, 8 is **quotient** [the number which tells how many times the **divisor** is contained in **dividend**} and 2 is **remainder.**

It is clear that **58 = 7 × 8 + 2**

or	Dividend = divisor × quotient + remainder
or	**D** = d × **Q** + **R**

Further for convenience we will write **dividend** as **D**, **divisor** as **d**, **quotient** as **Q** and **remainder** as **R.**

Now we will do a few examples.

Example 1. The quotient in a division is 403, the divisor is 100 and the remainder is 58. Find the dividend.

Sol :

$$Q = 403$$
$$d = 100$$

$$
\begin{aligned}
R &= 58 \\
D &= d \times Q + R \\
&= 403 \times 100 + 58 \\
&= 40300 + 58 \\
D &= \mathbf{40358}
\end{aligned}
$$

Example 2. Find divisor when quotient, dividend and remainder are 457, 98793 and 81.

Sol :

$$
\begin{aligned}
Q &= 457 \\
D &= 98793 \\
R &= 81 \\
D &= d \times Q + R \\
98793 &= d \times 457 + 81 \\
98793 - 81 &= 457d \\
98712 &= 457d \\
d &= 98712/457 \\
d &= \mathbf{216}
\end{aligned}
$$

CONCEPT II

The 2nd concept in this exercise is as follows.

Suppose we divide 52 by 7.

$$
\begin{array}{r|l}
7 & 52 \;(7 \\
 & \underline{49} \\
 & \underline{\;\;3}
\end{array}
$$

The Q is 7 and R is 3.

Now three types of questions can be asked.

Q. 1. Find the least number which must be subtracted from 52 in order to make it exactly divisible by 7.

Q. 2. Find the least number which must be added to 52 so that it is exactly divisible by 7.

Q. 3. Find the number nearest to 52 which is exactly divisible by 7.

Before solving these questions consider the following number series.

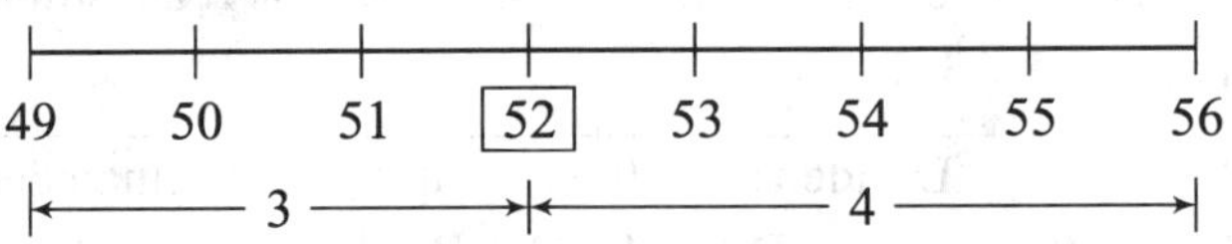

52 is the **number** and **nearer** to it there are **two numbers** which are **divisible** by **7**. One is **49** and another is **56**. Now we will solve the above three questions one by one.

Ans 1.

$$
\begin{array}{r|l}
7 & 52 \;(7 \\
 & \underline{49} \\
 & \underline{\;\;3}
\end{array}
$$

In this question we are to find what should be subtracted from 52 so that it is exactly divisible by 7. It is clear from the number series that 3 i.e. remainder should be subtracted.

***So Remember** : **R** is to be **subtracted** from **dividend** to make it exactly divisible by that divisor.*

Ans 2. Number 7 divides 52 upto 49 and in order to complete the next interval of 7, we have 3 as remainder and should add 4 more to make it 56 or we are to add (7 – 3) i.e., (divisor – remainder).

***So Remember : d – R** is to be **added** to **dividend** to make is exactly divisible by that **divisor**.*

Ans 3. The number nearest to **52** exactly divisible by **7** is **49**. Because another number could be **56**, but it is **4 steps ahead** of **52** and **49** is **three steps backwards**. So, **49** is the **nearest**.

Third type of questions are mostly asked in exams. To do this type of questions do the following steps:

(*i*) *Ist Step : **Divide** the given **number** by **divisor** and find the **remainder R**.*

(*ii*) *IInd Step : Find **(d – R)***

(*iii*) *IIIrd Step : Compare **R** and **d – R**. If **R < (d – R) then directly subtract remainder from the number.** If **R > (d – R)** then add (d – R) to the number.*

We will do a few more examples to understand this concept more clearly.

Example 1. Find the number nearest to 56586 which is exactly divisable by 552.

Sol : Ist step : Divide 56586 by 552.

```
552)56586(102
     552
     ----
      1386
      1104
      ----
       282
```

∴ $\quad R = \mathbf{282}$

IInd Step : Find $d - R$

$$d - R = 552 - 282 = \mathbf{270}$$

IIIrd Step : Compare **R** and ***d* – R**

$$R = 282$$

$$d - R = 270$$

So $\quad \boldsymbol{d - R < R}$

∴ We get the desired number by adding d – R to the dividend.

So the number nearest to 56586 which is exactly divisible by 552 is = 56586 + 270 = **56856**

Example 2. Find the number nearest to 77693 exactly divisible by 718.

Sol : Ist Step

Divide 77693 by 718

```
718)77693(108
     718
     ----
      5893
      5744
      ----
       149
```

∴ $\quad R = \mathbf{149}$

IInd Step : Find $d - R$

$$d - R = 718 - 149 = \mathbf{569}$$

IIIrd Step : Compare **R** and $\boldsymbol{d - R}$

$$R = 149 \text{ and } d - R = 569$$

$\therefore$ $\quad R < d - R$

So we get the desired number by subtracting **R** from dividend.

$\therefore$ Required number is $= 77693 - 149$

$= \mathbf{77544}$

CONCEPT III

In this we will discuss how many divisors are their in a composite number. e.g., 6 can be written as 1 × 6 and 2 × 3. So it has 4 divisors that is 1, 2, 3, 6. Now if the number is large and we have to find its number of divisors it can be found by following method.

The number of divisors of a composite number : If D is a composite number in the form $D = a^p \times b^q \times c^r$, where a, b, c are primes,

then the no. of divisors of D, represented by n is given by

$$n = (p + 1)(q + 1)(r + 1).$$

Example 1. What is the total number of divisors of 200 ?

Sol : 200 can be written as $2^3 \times 5^2$.

The value of p & q in this case are 3 & 2 respectively.

Thus the total number of divisors or factors are (3 + 1) × (2 + 1) = 12.

So 200 can be written as 1 × 200, 2 × 100, 4 × 50, 5 × 40, 8 × 25, 10 × 20.

So it has 12 divisors or factors that are **1, 2, 4, 5, 8, 10, 20, 25, 40, 50, 100 and 200.**

Next concept we will discuss is *'how to find the sum of all the divisors of a composite number'*. e.g., if we have to find the sum of divisors of 6 then we can find it manually as follows : factors of 6 are 1, 2, 3, 6 and there sum is 1 + 2 + 3 + 6 = 12.

But if a number is large the sum of factors can be found by the following method.

The sum of all the divisors of a number S_n is given by the following formula

$$S_n = \frac{(a^{p+1} - 1)(b^{q+1} - 1)(c^{r+1} - 1)}{(a - 1)(b - 1)(c - 1)}$$

where a, b, c are the prime factors of the number n and p, q and r be their powers. This can be clearly understood by following example

Example 2. What is the sum of all the factors of 200 ?

Sol : As per the above formula $S_n = \dfrac{(2^{3+1} - 1)(5^{2+1} - 1)}{(2 - 1)(5 - 1)} = \dfrac{15 \times 124}{4} = 15 \times 31 = 465.$

EXERCISE 2

Q. 1 to 5.

Fill in the blanks in the following table where

$D =$ Dividend $\qquad Q =$ Quotient

$R =$ Remainder $\qquad d =$ Divisor

Q. No.	d	Q	R	D
1.	254	104	210	—
2.	—	125	25	3900
3.	77	100	—	7737
3.	610	126	51	—
5.	—	147	36	37521

Q. 6 to 10. Based on the concept 2 solve the following table. Find the number nearest to S which is exactly divisible by P. S and P are given in the following table:

Q. No.	S	P	Number
6.	58701	567	—
7.	16200	425	—
8.	2800	130	—
9.	53250	140	—
10.	5125	150	—

Q.11. Find the greatest number of five digits which is exactly divisible by 279.

Q.12. Find the smallest number of five digits which is exactly divisible by 279.

Q.13. Find the least number by which 4851 must be multiplied or divided so as to make it a perfect square.

Q.14. Find the least number which is a perfect square and contain 252 as its factor.

Q.15. A certain number when divided by 899 gives a remainder 63. Find the remainder when same number is divided by 29.

SOLUTIONS

1.

$$\begin{aligned} D &= d \times Q + R \\ &= 254 \times 104 + 120 \\ &= 26416 + 210 = \mathbf{26626} \end{aligned}$$

2.

$$\begin{aligned} D &= d \times Q + R \\ 3900 &= d \times 125 + 25 \\ 3900 - 25 &= 125\,d \\ 3875 &= 125\,d \\ d &= 3875/125 = 31 \end{aligned}$$

3.

$$\begin{aligned} d \times Q + R &= D \\ 77 \times 100 + R &= 7737 \\ 7700 + R &= 7737 \\ R &= 7737 - 7700 = \mathbf{37} \end{aligned}$$

4.

$$D = d \times Q + R$$

$$D = 610 \times 126 + 51$$
$$= 76860 + 51 = \mathbf{76911}$$

5.
$$d \times Q + R = D$$
$$d \times 147 + 36 = 37521$$
$$147d = 37521 - 36 = 37485$$
$$d = 37485/147 = \mathbf{255}$$

6. **Ist Step:** Find R

$$567 \overline{)58701}(103$$
```
567)58701(103
    567
    ----
     2001
     1701
     ----
      300
```

$$R = \mathbf{300}$$

IInd Step : Find $d - R$

$$d - R = 567 - 300 = \mathbf{267}$$

IIIrd Step
$$R = 300$$
$$d - R = 267$$

$\therefore$ $$\boldsymbol{d - R < R}$$

$$\text{Required No.} = 58701 + 267 = \mathbf{58968}$$

7. **Ist Step :** Find R

```
425)16200(38
    1275
    ----
     3550
     3400
     ----
       50
```

$$R = \mathbf{50}$$

IInd Step : Find $d - R$

$$d - R = 425 - 50 = \mathbf{375}$$

IIIrd Step
$$R = 50$$
$$d - R = 375$$
$$\boldsymbol{R < d - R}$$

$\therefore$ $$\text{Required No.} = 16200 - 50 = \mathbf{16150}$$

8. **Ist Step :** Find R

```
130)2800(21
    260
    ---
     200
     130
     ---
      70
```

$$R = \mathbf{70}$$

IInd Step $$d - R = 130 - 70 = \mathbf{60}$$

IIIrd Step $$d - R = 60$$
$$R = 70$$
$$\boldsymbol{d - R < R}$$

Required No. = 2800 + 60 = **2860**

9. **Ist Step**

$$\begin{array}{r} 140\overline{)53250}(380 \\ \underline{420} \\ 1125 \\ \underline{1120} \\ \underline{50} \end{array}$$

R = **50**

IInd Step : Find $d - \text{R}$

$d - \text{R}$ = 140 – 50 = **90**

IIIrd Step

R = 50

$d - \text{R}$ = 90

R < *d* – R

∴ Required no. = 53250 – 50 = **53200**

10.

$$\begin{array}{r} 150\overline{)5125}(34 \\ \underline{450} \\ 625 \\ \underline{600} \\ \underline{25} \end{array}$$

R = **25**

$d - \text{R}$ = 150 – 25 = **125**

R < *d* – R

∴ Required No = 5125 – 25 = **5100**

11. Greatest No. of Five digits = 99999

Now we will proceed as in above question

Ist Step : Find R

$$\begin{array}{r} 279\overline{)99999}(358 \\ \underline{837} \\ 1629 \\ \underline{1395} \\ 2349 \\ \underline{2232} \\ \underline{117} \end{array}$$

R = **117**

Now we don't need to check ***d* – R** in this question because ***d* – R** is always to be added, but here if we add even **1** then it will become a **six digit number**. So we can't **go ahead**

So we have to move backward in such a case means R is to be subtracted.

No. = 99999 – 117 = **99882**

12. Smallest No. of Five digits = 10000

Find R

$$\begin{array}{r} 279\overline{)10000}(35 \\ \underline{837} \\ 1630 \\ \underline{1395} \\ \underline{235} \end{array}$$

Now we can only move forward because if we move even one step backward the number will become a four digit number.

Now $d - R = 279 - 235 = 44$

$\therefore$ Required No. $= 10000 + 44 = \mathbf{10044}$

13. A perfect square is a number whose square root is a complete number 4 etc. In this question Ist we have to make the factors of 4851.

3	4851
3	1617
7	539
7	77
	11

$$4851 = 3 \times 3 \times 7 \times 7 \times 11$$
$$= 3^2 \times 7^2 \times 11$$

Here 11 is not in pair. So if we divide 4851 by 11 or multiply it by 11 we will get a number which is perfect square.

14.

2	252
2	126
3	63
3	21
	7

$$252 = 2 \times 2 \times 3 \times 3 \times 7$$
$$= 2^2 \times 3^2 \times 7$$

Except 7 rest factors are in pairs so if we multiply 252 by 7 we will get a number which is a perfect square and contain 252 as its factor.

$\therefore$ Required Number $= 252 \times 7 = \mathbf{1764}$

15. Here we have been given $d = 899$

$R = 63$

Let $Q = Q$

Number $= d \times Q + R$

$= 899 \times Q + 63$

Now we have this information available and we have to divide this number by 29.

$$\frac{\text{Number}}{29} = \frac{899 \times Q + 63}{29} = \frac{899 \times Q}{29} + \frac{63}{29} = 31Q + \frac{63}{29}$$

Now the remainder will be obtained by dividing 63 by 29 as first part is completely divisible by 29.

$$29\overline{)\,63\,}(2$$
$$\underline{58}$$
$$5$$

so 5 is the remainder.

Day 5

HCF and LCM

Now on day 5 we will learn **HCF** and **LCM** and questions asked in the exams on the basis of this topic. Before discussing the types of question let us first understand what is **HCF**

H → Highest

C → Common

F → Factor

As above shown **HCF** means **Highest Common Factor**. It is between two or more numbers. To understand it more clearly consider two numbers **12** and **16**.

In order to find **HCF** between **12** and **16** we will have to first factorize **12** and **16**.

$$12 = 2 \times 2 \times 3$$

So, factors of 12 are **[2, 2, 3]**

$$16 = 2 \times 2 \times 2 \times 2$$

So factors of 16 are **[2, 2, 2, 2]**

Therefore, **Highest Common** between them is **2 x 2** as shown below.

$$12 = \boxed{2 \times 2} \times 3$$
$$16 = \boxed{2 \times 2} \times 2 \times 2$$

So, **HCF** of **12** and **16** is **4**.

But this method is not convenient when the two numbers are large. Then we find the **HCF** by following method.

[DIVISION METHOD]

Step 1: Divide the greater number by smaller number and find the remainder. In this case16 is divided by 12 and 4 is the remainder.

```
12) 16 (1
    12
    ---
     4) 12 (3
        12
        ---
         X
```

Step 2: Now repeat this process with 1st remainder as divisor and 1st divisor as dividend and continue this process until we get zero remainder. Last divisor is the **HCF**. In this 12 is exactly divisible by 1st remainder i.e., 4, so 4 is the **HCF** of **12** and **16**.

But in exams you will rarely be asked directly.

"What is the **HCF** of **12** and **16** ?"

or

"Find the **HCF** of **12** and **16**."

The question will be asked in the following two forms :

1st Form : Find the greatest number that divides 12 and 16 exactly.

Solution: In above question we are to fulfil two conditions.

1. Required number should divide both 12 and 16 exactly.
2. It should be greatest.

These two conditions are met if we find HCF of 12 and 16.

2nd Form : Find the greatest number that divides 12 and 16 without leaving any number.

Solution: Obviously we are to find **HCF**.

Now, it is quite sure that you will be able to understand the statements in which **HCF** is to be found.

HCF OF THREE NUMBERS: To find the **HCF** of three numbers we first have to find the **HCF** of the two numbers (we can take any two numbers of the three) and then we will find the **HCF** of the **HCF** so found and the third number. This can become more clear by the following example.

Example 1. Find the greatest number that divides 221,1156 and 1462 without leaving any remainder.

Sol : Now, it is quite clear that we have to find the **HCF** of above three numbers i.e., 221, 1156 and 1462. For this we will first of all find the **HCF** of 1156 and 1462. It comes out to be 34. [Illustration shown below] Then we will find the **HCF** of 34 and 221 which is 17. So, 17 is the **HCF** of 221, 1156 and 1462.

```
1156 ) 1462 ( 1
       1156
       306 ) 1156 ( 3
              918
             238 ) 306 ( 1
                   238
                    68 ) 238 ( 3
                         204
                          34 ) 68 ( 2
                               68
                                X
```

HCF of 1156 and 1462 is 34.

Now **HCF** of 34 and 221 is 17 (shown below)

```
34 ) 221 ( 6
     204
      17 ) 34 ( 2
           34
            X
```

So, 17 is the required number.

Now we will solve some more typical problems of **HCF** that have been asked in the exams.

Example 2. The **HCF** of two numbers is 67 and their sum is 670. How many pairs of such numbers are possible?

Sol : As two numbers are always a factor of their HCF. Here 67 is the HCF. So two numbers can be written as 67 x and 67 y.

According to given condition :

$$\begin{aligned} 67x + 67y &= 670 \\ 67(x+y) &= 670 \\ x+y &= 670 \div 67 \\ x+y &= 10 \end{aligned}$$

So, possible combinations of ***x* and *y*** are

1, 9 $\quad \because$ (1 + 9 = 10 and so on)
2, 8
3, 7
4, 6
5, 5

[After this the values start repeating, e.g., 6, 4, 7, 3 etc.] x and y have five possible combinations

So, numbers can be

$67 \times 1, 67 \times 9$ **... (1)**
$67 \times 2, 67 \times 8$ **... (2)**
$67 \times 3, 67 \times 7$ **... (3)**
$67 \times 4, 67 \times 6$ **... (4)**
$67 \times 5, 67 \times 5$ **... (5)**

Above shown are the five possible pairs. But in these pairs only two pairs 1st and 3rd are valid. Rest pairs ($67 \times 2, 67 \times 8$), ($67 \times 4, 67 \times 6$), ($67 \times 5, 67 \times 5$) are invalid.

Reason: Take the second pair i.e.,

67×2 & 67×8

It can be written as $\underline{67 \times 2}$ & $\underline{67 \times 2} \times 4$

In this case the **HCF** becomes 67×2. (**HCF** is changed because 2 and 8 have common factor 2). Hence we take only those pairs which are **coprime** [**Coprime pairs** are the **pairs** of numbers which **don't** have any **common factor** in between them] and on these conditions only 1st and 3rd pairs are possible. **So only two pairs are possible.**

Example 3. The product of two numbers is 4732 and their GCD is 13. How many pairs of such numbers are possible?

Sol : Let the numbers be $13x$ and $13y$ According to the given condition

$$13x \times 13y = 4732$$

or
$$xy = 4732 \div 169$$
$$xy = 28$$

This is possible as 1×28 **...(1)**
2×14 **...(2)**
4×7 **...(3)**

and from these three pairs only 1st and 3rd are coprime. Hence two pairs are possible.

Example: Find the greatest number that divides 187, 233 and 325 leaving the same remainder in each case.

Explanation: This is a very important question and is based on the following property :

"The greatest number that divides three or more numbers leaving the same remainder in each case is the **HCF** of the possible positive difference of these numbers. This is clearly understood by the following example.

```
17) 55 (3        17) 72 (4        17) 89 (5
    51               68               85
    --               --               --
     4                4                4
```

We have divided 55, 72 and 89 by 17 and the remainder is 4 in each case. According to above property 17 should be HCF of their positive possible differences as shown

```
 72        89        89
 55        72        55
 --        --        --
 17        17        34
```

HCF of 17, 17 and 34 is 17.

So, the solution of above given question is as follows :

1st step: Find the possible positive difference.

```
233       325       325
187       233       187
---       ---       ---
 46        92       138
```

2nd step: Now find the HCF of 46, 92 and 138

```
92) 138 (1
     92
    ---
    46) 92 (2
        92
        --
         x
```

HCF of 92 and 138 is 46. Now **HCF** of 46 and 46 is also 46. Therefore, **46** is the required number.

CONCEPT – II

LCM

LCM means **Least Common Multiple**. To understand it more clearly consider the two numbers, say again, **12** and **16**.

Multiple of	12 are 12 and	16 are 16
	24	32
	36	48
	48	64
	60	80

So, least common in between them is 48 as shown above. But in actual practice we are to find the **LCM** of many numbers this method is quite lengthy. There we adopt the following method [factor method]

2	12 – 16
2	6 – 8
	3 – 4

$$\text{LCM} = 2 \times 2 \times 3 \times 4 = \mathbf{48}$$

Example. Find the **LCM** of 9, 12, 16, 18, 20.

1st step: Write the numbers as shown below.

2nd step: We first divide the numbers by 1st prime number i.e., 2 and continue with next prime number until we have at least two numbers which are divisible by that prime number. Then multiply all the divisors.

2	9-12-16-18-20
2	9-6-8-9-10
3	9-3-4-9-5
3	3-1-4-3-5
	1-1-4-1-5

$$\textbf{LCM} = \mathbf{2 \times 2 \times 3 \times 3 \times 4 \times 5 = 720}$$

As in **HCF**, in **LCM** also question is not directly asked. The question is asked in the following forms.

"Find the least number which is exactly divisible by 12 and 16."

OR

"Find the least number which when divided by 12 and 16 leaves no remainder."

[It is quite clear in above two forms that **LCM** is to be found]

So, it is now hoped that you will now not misunderstand the statements. Some more questions related to **LCM** :

Example. "Find the least number which when divided by 12 and 16 leaves 3 as **remainder**."

Sol : It is very much clear that it has been asked "What is the least number which when divided by 12 and 16 leaves no remainder." Then we are to find the **LCM** of these two numbers which is 48. But, in this case we **need 3 as remainder.** So, we must add 3 to **LCM** in order to meet this condition.

So, 48 + 3 = **51** will be our answer. So, in this type of questions we add the required remainder to **LCM**.

Example. "Find the least number which when increased by 3 is exactly divisible by 12 and 16.

Sol : So in above question we know that least number which is exactly divisible by 12 and 16 is 48 i.e., their LCM. So, in order to have the required number we subtract 3 from 48. (48 – 3) = **45** will be the required number. In which if we add 3, it becomes 48 i.e., LCM which is exactly divisible by 12 and 16.

CONCEPT – III

LCM of 12 and 16 is 48

HCF of 12 and 16 is 4

$$\textbf{LCM} \times \textbf{HCF} = 48 \times 4$$

$$\text{1st no.} \times \text{2nd no.} = 12 \times 16 = 192$$

1st number × 2nd number = LCM × HCF

CONCEPT – IV

HCF AND LCM OF FRACTIONS

Find the **HCF** of $\frac{a}{b}, \frac{c}{d}, \frac{e}{f}$.

For this find the **HCF** of numerator (In $\frac{a}{b}$, **'*a*'** is **numerator** and **'*b*'** is **denominator**) and **LCM** of denominator i.e., **HCF** of *a*, *c* and *e* and **LCM** of *b*, *d* and *f*

In order to have **LCM** we have to find **LCM** of numerator and **HCF** of denominator

Example. Find the **LCM** of 1/2, 2/3 and 4/5

Sol : Find the **LCM** of 1,2,4 i.e.,

2	1-2-4
	1-1-2

$$\textbf{LCM} = 2 \times 1 \times 1 \times 2 = \mathbf{4}$$

and **HCF** of 2,3,5 is **1** Therefore, the required LCM is $\mathbf{\frac{4}{1}}$.

Example. Find **HCF** of $\mathbf{\frac{1}{2}, \frac{2}{3}, \frac{4}{4}}$

Sol : In this case we have to find **HCF** of 1,2,4 which is **1** and LCM of 2, 3, 5 which is **30**.Therefore, the required **HCF** is $\mathbf{\frac{1}{30}}$.

EXERCISE

Q.1. The HCF of 3003 and 7007 is

(a) 101 (b) 202 (c) 1001 (d) none

Q.2. The HCF of 1804 and 2132 is

(a) 82 (b) 328 (c) 164 (d) none

Q.3. Find the greatest number that divides 2149 and 1535 exactly.

(a) 704 (b) 37 (c) 714 (d) none

Q.4. Find the greatest number that divides 10997 and 14139 exactly.

(a) 1571 (b) 7570 (c) 571 (d) none

Q.5. Find the greatest number that divides 1241 and 949 without leaving any remainder.

(a) 146 (b) 73 (c) 173 (d) none

Q.6. The greatest number that divides 55250 and 8058 without leaving any reamainder is

(a) 38 (b) 40 (c) 34 (d) none

Q.7. Find the least number which is exactly divisible by 2, 3, 4, 5, 6 and 7.

(a) 840 (b) 420 (c) 210 (d) none

Q.8. Find the least number which is exactly divisible by 12, 15, 20 and 54

(a) 540 (b) 270 (c) 1080 (d) none

Q.9. Find the least number which when divided by 9, 12, 16 and 30 leaves in each case a remainder 3.

(a) 723 (b) 717 (c) 1443 (d) none

Q.10. Find the least number which when divided by 12,15,20 and 54 leaves in each case 4 as remainder.

(a) 540 (b) 544 (c) 536 (d) none

Q.11. Find the least number which when increased by 3 is exactly divisible by 2, 3, 4, 5, 6 and 7.

(a) 417 (b) 423 (c) 433 (d) none

Q.12. Find the least number which when increased by 3 is exactly divisible by 21, 25, 27 & 35

(a) 4728 (b) 4722 (c) 4725 (d) none

Q.13. Find the greatest number that divides 12288, 19139, & 28200 leaving the same remainder in each case.

(a) 133 (b) 422 (c) 221 (d) none

Q.14. **HCF** of two numbers is 101 and their sum is 1212. How many pairs of such numbers are possible.

(a) 3 (b) 1 (c) 2 (d) none

Q.15. **HCF** of two numbers is 21 and their product is 2646. How many pairs of such numbers are possible?

(a) 3 (b) 1 (c) 2 (d) none

Q.16. Find the **LCM** of 10/21, 20/63 & 55/56

(a) 7/220 (b) 110/7 (c) 7/110 (d) none

Q.17. Find the **HCF** of 2/5, 12/35 & 16/55

(a) 2/385 (b) 385/2 (c) 1/385 (d) none

Q.18. The least number which when divided by 8, 12, & 16 leaves 3 as remainder when divided by 7 leaves no remainder.

(a) 147 (b) 99 (c) 51 (d) none

Q.19. Which of the following pair is coprime ?

(a) (3,9) (b) (3,7) (c) (6,12) (d) (5, 25)

Q.20. A merchant has 3 kinds of wine. 117 litre of first kind, 130 litre of second kind & 143 litre of third kind. Find the least number of equal casks of equal size in which the wine can be stored without mixing.

(a) 30 (b) 32 (c) 34 (d) None

Q.21. Two numbers are in the ratio 15 : 11, their HCF is 13. The numbers are

(a) 75, 55 (b) 105, 77 (c) 15, 11 (d) 195 , 143

Q.22. Three numbers are in the ratio 1:2:3 and their HCF is 12. The numbers are

(a) 12, 24, 36 (b) 10, 20, 30 (c) 5, 10, 15 (d) 14, 8, 12

Q.23. 6 bells toll at a interval of 2, 4, 5, 7, 8 and 10 seconds. Beginning together after what interval of time will they toll again together.

(a) 540 (b) 240 (c) 280 (d) none

Q.24. Find the least number which when divided by 35 leaves remainder 30, when divided by 45 leaves remainder 40 & when divided by 55 leaves remainder 50.

(a) 3450 (b) 3455 (c) 3460 (d) none

Q.25. Find the least number of 4 digits which is exactly divisible by 2, 3, 4, 5, 6 and 7.

(a) 1260 (b) 1200 (c) 2520 (d) none

Q.26. Find the least number which when divided by 2, 3, 4, 5 & 6 leaves the remainders 1, 2, 3, 4 and 5 respectivly

(a) 100 (b) 617 (c) 62 (d) none

Q.27. Find the greatest number of four digits which is exactly divisible by 2, 3, 4, 5, 6 and 7.

(a) 9660 (b) 9888 (c) 9999 (d) none

Q.28. The **HCF** of two numbers is 11 and their **LCM** is 693 if one of the numbers is 77. Find the other number.

(a) 66 (b) 901 (c) 99 (d) none

SOLUTIONS

1. (c) 1001.

```
3003) 7007 (2
      6006
      1001) 3003 (3
            3003
             x
```

2. (c) 164.

```
1804) 2132 (1
      1804
       328) 1804 (5
            1640
             164) 328 (2
                  328
                   x
```

3. (d) none (307)

```
1535) 2149 (1
      1535
       614) 1535 (2
            1228
             307) 614 (2
                  614
                   x
```

4. (a) 1571

```
10997) 14139 (1
       10997
       -----
        3142 ) 10997 (3
                9426
               -----
                1571) 3142 (2
                      3142
                      ----
                        x
```

5. (b) 73.

```
949) 1241 (1
      949
      ---
      292) 949 (3
           876
           ---
            73) 292 (4
                292
                ---
                 x
```

6. (c) 34.

```
8058) 55250 (6
      48348
      -----
       6902 ) 8058 (1
              6902
              ----
              1156) 6902 (5
                    5780
                    ----
                    1122) 1156 (1
                          1122
                          ----
                            34) 1122 (13
                                102
                                ---
                                 102
                                 102
                                 ---
                                  x
```

7. (b)

2	2 – 3 – 4 – 5 – 6 – 7
3	1 –3 – 2 – 5 – 3 – 7
	1 – 1 – 2 – 5 – 1 – 7

LCM $= 2 \times 3 \times 2 \times 5 \times 7 =$ **420**

8. (a)

2	12 – 15 – 20 – 54
2	6 – 15 – 10 – 27
3	3 – 15 – 5 – 27
5	1 – 5 – 5 – 9
	1 – 1 – 1 – 9

∴ LCM $= 2 \times 2 \times 3 \times 5 \times 9 =$ **540**

9. (a)

2	9 – 12 – 16 – 30
2	9 – 6 – 8 – 15
3	9 – 3 – 4 – 15
	3 – 1 – 4 – 5

LCM $= 2 \times 2 \times 3 \times 3 \times 4 \times 5 = 720$

$$\text{Required No.} = \text{LCM} + 3 = 720 + 3 = \mathbf{723}$$

10. (b)
$$\text{LCM of 12, 15, 20 and 54} = 540$$
$$\text{Required No.} = 540 + 4 = \mathbf{544}$$

11. (a)
$$\text{Required No.} = \text{LCM} - 3 = 420 - 3 = \mathbf{417}$$

12. (b)

3	21 – 25 – 27 – 35
5	7 – 25 – 9 – 35
7	7 – 5 – 9 – 7
	1 – 5 – 9 – 1

$$\text{LCM} = 3 \times 5 \times 7 \times 5 \times 9 = 4725$$
$$\text{Required No.} = \text{LCM} - 3 = 4725 - 3 = \mathbf{4722}$$

13. (c) This question is based on **Example 2.**

So we proceed as follows

28200	28200	19139
–19139	–12288	–12288
9061	15912	6851

We have to find HCF of 9061, 15912 and 6851 which is equal to 221 So, **221** is the answer.

14. (c)
$$101x + 101y = 1212$$
$$101(x + y) = 1212$$
$$x + y = \frac{1212}{101} = 12$$

Possible values of x and y are

1	**11**
2	10
3	9
4	8
5	**7**
6	6

out of these only two pairs i.e., (1, 11) and (5, 7) are co-prime. Hence 2 pairs are possible.

15. (c)
$$21x \times 21y = 2646$$
$$xy = \frac{2646}{21 \times 21} = 6$$
$$xy = 6$$

possible value of xy are (1, 6) and (2, 3). Therefore, only **two** pairs are possible.

16. (d)
$$\text{LCM of } \frac{10}{21}, \frac{20}{63}, \frac{55}{56} = \frac{\text{LCM of } 10, 20, 55}{\text{HCF of } 21, 63, 56} = \mathbf{\frac{220}{7}}$$

17. (a)
$$\text{HCF of } \frac{2}{5}, \frac{12}{35}, \frac{16}{55} = \frac{\text{HCF of } 2, 12, 16}{\text{LCM of } 5, 35, 55} = \mathbf{\frac{2}{385}}$$

18. (a)

$$\begin{aligned} \text{LCM of 8, 12 and 16} &= 48 \\ 48 + 3 &= 51 \text{ not divisible by 7} \\ 48 \times 2 + 3 &= 99 \text{ not divisible by 7} \\ 48 \times 3 + 3 &= 147 \text{ divisible by 7} \end{aligned}$$

147 is the required no. which when divided by 8,12 and 16 leaves 3 as remainder and when divided by 7 leaves no remainder.

19. (b)

20. (a) Since the no. of cases required is least, therefore the size will be greatest.

$$\text{size of each case} = \text{HCF of 117, 130 and 143} = 13 \text{ litre}$$

$$\therefore \quad \text{Number of cases} = \frac{117}{13} + \frac{130}{13} + \frac{143}{13} = 9 + 10 + 11 = \mathbf{30}$$

21. (d)

22. (a)

23. (c)

$$\text{Required time} = \text{LCM of 2, 4, 5, 7, 8 and 10 sec}$$

$$\begin{array}{c|l} 2 & 2-4-5-7-8-10 \\ \hline 2 & 1-2-5-7-4-5 \\ \hline 5 & 1-1-5-7-2-5 \\ \hline & 1-1-1-7-2-1 \end{array}$$

$$\text{LCM} = 2 \times 2 \times 5 \times 7 \times 2 = 280$$

after **280 sec** they will toll again together

24. (c)

$$\begin{array}{ccc} 35 & 45 & 55 \\ -30 & -40 & -50 \\ \hline 5 & 5 & 5 \\ \hline \end{array}$$

$$\begin{array}{c|c} 5 & 35-45-55 \\ \hline & 7-9-11 \end{array}$$

$$\begin{aligned} \text{LCM} &= 5 \times 7 \times 9 \times 11 = 3465 \\ \text{Required No.} &= 3465 - 5 = \mathbf{3460} \end{aligned}$$

25. (a)

$$\begin{aligned} \text{LCM of 2, 3, 4, 5, 6 and 7} &= 420 \\ 420 \times 2 &= 840 \text{ which is of three digits} \\ 420 \times 3 &= \mathbf{1260} \text{ is of 4 digits and is required number.} \end{aligned}$$

26. (d)

$$\begin{array}{ccccc} 2 & 3 & 4 & 5 & 6 \\ -1 & -2 & -3 & -4 & -5 \\ \hline 1 & 1 & 1 & 1 & 1 \\ \hline \end{array}$$

$$\begin{aligned} \text{LCM} &= 2 \times 3 \times 2 \times 5 = 60 \\ \text{Required No.} &= 60 - 1 = \mathbf{59} \end{aligned}$$

27. (a) LCM of 2, 3, 4, 5, 6 and = 420

$$\text{Greatest No. of four digits} = 9999$$

Greatest no. of four digits exactly divisible by 420 is :

$$\begin{array}{r|l} 420 & 9999 \; (23 \\ & \underline{840} \\ & \;\;1599 \\ & \;\;\underline{1260} \\ & \;\;\;339 \end{array}$$

Required No. = 9999 – 339 = **9660**

28. (c) We know that HCF × LCM = Ist No. × IInd No.

Let IInd no. be x

$\therefore \quad 11 \times 693 = 77 \times x$

$$x = \frac{11 \times 693}{77} = \mathbf{99}$$

Day 6

Percentage

On **day 6** we will do the topic of **Percentage**. It is one of the most important topic of **Maths**. One of its most important use is of **Comparison**. It can give us the idea of performance of any organization or performance of any student in different exams. Let us take an **example**. A student in his **IX class** secures **300** marks out of **500** and in his **X class** secures **252** marks out of **400**. How can you compare his performance in the above given two classes? In which class he has done better? This can be done by taking his marks **out of 100** or **percentage of marks**. The method is given below :

$$\textbf{\% marks} = \frac{\textbf{marks obtained}}{\textbf{total marks}} \times \textbf{100}$$

∴ In **IX class** his score is $\frac{300}{500} \times 100 = \mathbf{60\%}$

In **X class** his score is $\frac{252}{400} \times 100 = \mathbf{63\%}$

In **IX class** he has secured **60%** and in **X class** he has secured **63%**. So his performance in **X class** is **better**.

In this topic first of all we will learn following two operation :

(1) To find % of a given number.

(2) To convert fraction into % equivalent.

To find % of given number

This is explained by following example

Example 1. 50% of 200 = ?

Sol : We are to find **50%** means **50/100**

$$\therefore \quad 50\ \% \text{ of } 200 = \frac{50}{100} \times 200$$

To convert fraction in %

Example 2. $\frac{1}{5} = ?\ \%$

Sol : $\frac{1}{5} = \frac{?}{100}$

$$\frac{1}{5} \times 100 = ?$$

$$20 = ?$$

∴ To convert a given fraction in % we are to simply multiply that fraction by 100

Now solve the following example:

Example 1. (a) $87\frac{1}{2}$ % of 800 = ? (b) $12\frac{1}{2}$ % of 400 = ?

(c) $37\frac{1}{2}$ % of ? = 300 (d) $83\frac{1}{3}$ % of 1212 = ?

Sol : Can you solve the above questions in **fraction of seconds**. **Yes,** It is possible. It is explained below :

(a) We are to find $87\frac{1}{2}$% of 800

$$\mathbf{87\frac{1}{2}\%} \text{ means } \mathbf{\frac{175}{2}\%} = \mathbf{\frac{175}{200}} = \mathbf{\frac{7}{8}}$$

So, simply we are to find **7/8 of 800** which is equal to **700**.

(b)

$$12\frac{1}{2}\% \text{ of } 400 = ?$$

$$25/200 \text{ of } 400 = ?$$

$$\mathbf{\frac{1}{8}} \text{ of } 400 = ?$$

$$50 = ?$$

∴ **50** is the answer

In this question if we knew $\mathbf{12\frac{1}{2}}$% = **1/8** then 1/8 of 400 = **50** could be found orally.

(c)

$$37\frac{1}{2}\% \text{ of } ? = 300$$

$$\frac{75}{200} \text{ of } ? = 300$$

$$\frac{3}{8} \text{ of } ? = 300$$

$$? = 300 \times \frac{8}{3}$$

$$= \mathbf{800}$$

(d)

$$83\frac{1}{3}\% \text{ of } 1212 = ?$$

$$\frac{250}{3}\% \text{ of } 1212 = ?$$

$$\frac{250}{300} \text{ of } 1212 = ?$$

$$\frac{5}{6} \text{ of } 1212 = ?$$

$$1010 = ?$$

∴ $? = \mathbf{1010}$

So in the above given examples we have seen that if we had known the fraction equivalent of **$87\frac{1}{2}$ %**, **$12\frac{1}{2}$ %**, **$37\frac{1}{2}$ %** and **$83\frac{1}{3}$ %** which are **7/8**, **1/8**, **3/8** and **5/6** respectively then we would have done these questions orally.

Following is the table which gives us the fractional equivalent of important percentages. Your first task is to learn these equivalent by heart and then solve the exercise given after this table.

%	Fractional Equivalent		% age	Fractional Equivalent
100%	1		$85\frac{1}{2}$%	$\frac{7}{8}$
80%	$\frac{4}{5}$		$62\frac{1}{2}$%	$\frac{5}{8}$
75%	$\frac{3}{4}$		$37\frac{1}{2}$%	$\frac{3}{8}$
60%	$\frac{3}{5}$		$12\frac{1}{2}$%	$\frac{1}{8}$
50%	$\frac{1}{2}$		$16\frac{2}{3}$%	$\frac{1}{6}$
40%	$\frac{2}{5}$		$33\frac{1}{3}$%	$\frac{1}{3}$
35%	$\frac{7}{20}$		$66\frac{2}{3}$%	$\frac{2}{3}$
30%	$\frac{3}{10}$		$83\frac{1}{3}$%	$\frac{5}{6}$
25%	$\frac{1}{4}$		$11\frac{1}{9}$%	$\frac{1}{9}$
20%	$\frac{1}{5}$		$22\frac{2}{9}$%	$\frac{2}{9}$
15%	$\frac{3}{20}$		$9\frac{1}{11}$%	$\frac{1}{11}$
5%	$\frac{1}{20}$		$14\frac{2}{7}$%	$\frac{1}{7}$

EXERCISE 1

Q.1. $66\frac{2}{3}\%$ of 3600 = ?

(a) 2400 (b) 1200 (c) 600 (d) none

Q.2. $33\frac{1}{3}\%$ of ? = 1200

(a) 1800 (b) 2400 (c) 3600 (d) none

Q.3. $66\frac{2}{3}\%$ of 36000 = ?

(a) 12000 (b) 18000 (c) 24000 (d) none

Q.4. $11\frac{1}{9}\%$ of 1818 = ?

(a) 22 (b) 202 (c) 212 (d) none

Q.5. 40% of 185 = ?

(a) 148 (b) 74 (c) 37 (d) none

Q.6. 80% of 85 = ?

(a) 68 (b) 64 (c) 36 (d) none

Q.7. 25% of 484 = ?

(*a*) 124 (*b*) 144 (*c*) 121 (*d*) none

Q.8. $62\frac{1}{2}\%$ of 1600 = ?

(a) 100 (b) 1000 (c) 10 (d) none

Q.9. $22\frac{2}{9}\%$ of 5454 = ?

(a) 1200 (b) 1202 (c) 1212 (d) none

Q.10. $14\frac{2}{7}\%$ of 1400 = ?

(a) 400 (b) 200 (c) 700 (d) none

Q.11. $87\frac{1}{2}\%$ 1600 = ?

(a) 1414 (b) 5600 (c) 700 (d) none

Q.12. $87\frac{1}{2}\%$ of 8008 = ?

(a) 707 (b) 717 (c) 7007 (d) none

Q.13. $33\frac{1}{3}\%$ of 3300 = ?

(a) 110 (b) 1100 (c) 1111 (d) none

Q.14. $33\frac{1}{3}\%$ of 5400 = ?

(a) 1800 (b) 180 (c) 1808 (d) none

Q.15. 75% of 400

(a) 300 (b) 100 (c) 303 (d) none

(Answers to above test are given in the Solutions at the end of this chapter)

Apart from the above types following are the types of questions which are frequently asked in exams.

1. Questions related to **Election**.
2. Questions related to **Marks in Examination**.
3. Questions related to **Population**.
4. Questions related to **Reduction in Consumption**.
5. Questions related to **Reduction in Price**.
6. Questions related to **Net % change**

Type I (Questions related to Elections)

Example 1. At an election a candidate secures 35% votes and is defeated by the other candidate by a margin of 450 votes. Find the total number of votes.

Sol :

$$\text{Let total votes} = x$$
$$\text{Loser gets} = 35\%x$$
$$\therefore \quad \text{Winner gets} = x - 35\%x = 65\%x$$
$$\text{Difference} = 65\%x - 35\%x = 30\%x$$

According to condition :

$$30\%x = 450$$
$$x = \frac{450}{30} \times 100 = \mathbf{1500}$$

Example 2. At an election a candidate secures 62% votes is elected with a majority of 144 votes. Find the total number of votes secured by successful candidate.

Sol :

$$\text{Let total votes} = x$$
$$\text{Winner gets} = 62\%\, x$$
$$\therefore \quad \text{Loser gets} = x - 62\%\, x = 38\%x$$
$$62\%\, x - 38\%\, x = 144$$
$$24\%\, x = 144$$
$$x = \frac{144}{24} \times 100 = 600$$

Votes secured by successfull candidate = $\frac{62}{100} \times 600$ = **372**

Example 3. At an election involving two candidates 135 votes were invalid. The winning candidate got 70% of the valid votes and won by 40000 votes. Find the total votes.

Sol : Let total valid votes $= x$

Winner gets $= 70\%x$

Loser gets $= x - 70\%x = 30\%x$

$\therefore$ $70\%x - 30\%x = 40000$

$40\%x = 40000$

$$x = \frac{40000}{40} \times 100 = 100000$$

Total valid votes $= 100000$

Invalid votes $= 135$

$\therefore$ Total votes $=$ **100135**

Type II (Questions on Marks in Examinations)

Example 4. Nilesh secured 360 marks in the examination. His % is 60%. Find the maximum marks.

Sol : Let total marks $= x$

$60\%x = 360$

$$x = \frac{360}{60\%}$$

$$= \frac{360}{60} \times 100 = \mathbf{600}$$

Example 5. A student had to secure 40% marks to pass. He got 440 marks and failed by 40 marks. Find the maximum marks.

Sol : Let maximum marks $= x$

$\therefore$ Pass marks $= 40\%x$

According to statement

pass marks $= 440 + 40 = 480$

$\therefore$ $40\%\, x = 480$

$$x = \frac{480}{40\%}$$

$$= \frac{480}{2} \times 5 = \mathbf{1200}$$

Example 6. A student secures 30% of total marks and fails by 30 marks. But another student secures 40% of marks and gets 30 marks more than the pass marks. Find the maximum marks.

Sol : Let total marks $= x$

Marks obtained by

Ist Student $=$ 30%x

$\therefore$ Pass marks $= 30\%x + 30$ **...(I)**

Marks obtained by

$$\textbf{IInd student} = \mathbf{40\%x}$$

$$\text{Pass marks} = 40\%x - 30 \quad \text{...(II)}$$

From I and II

$$30\% x + 30 = 40\%x - 30$$

$$40\% x - 30\%x = 30 + 30$$

$$10\% x = 60$$

$$x = \frac{60}{10} \times 100 = \mathbf{600}$$

Example7. Rahul gets 240 marks and fails by 24 marks. The pass % is 33%. Find the maximum marks.

Sol :

$$\text{Marks got by Rahul} = 240$$

$$\text{Marks by which he fails} = 24$$

$$\therefore \quad \text{Pass marks} = 240 + 24$$

$$= \mathbf{264} \quad \text{...(I)}$$

$$\text{Let maximum marks} = x$$

$$\therefore \quad \textbf{Pass marks} = \mathbf{33\%}\, x \quad \text{...(II)}$$

From I and II

$$33\% x = 264$$

$$x = \frac{264}{33} \times 100 = \mathbf{800}$$

Example 8. A candidate gets 25% marks and fails by 135 marks. Another candidate gets 56% marks and gets 144 more than the marks needed for passing. Find the maximum marks.

Sol :

$$\text{Let total marks} = x$$

$$25\% x + 135 = 56\%x - 144$$

$$56\% x - 25\% x = 135 + 144$$

$$31\% x = 279$$

$$x = \frac{279}{31} \times 100 = \mathbf{900}$$

Type III (Questions on Population)

Population Formula :

1. If the present population is P and $r\%$ is the rate of increase per annum then Population after n years is

$$= P\left(1 + \frac{r}{100}\right)^n$$

2. If $r\%$ is the rate of decrease per annum then population after n years is

$$= P\left(1 - \frac{r}{100}\right)^n$$

3. If the population increases at a rate of r_1% in the first year and decreases at a rate of r_2% in the second year then population after 2 years is

$$= \mathbf{P}\left(1+\frac{\mathbf{r_1}}{\mathbf{100}}\right)\left(1-\frac{\mathbf{r_2}}{\mathbf{100}}\right)$$

Example 9. The population of a town is 16000 and its annual increase is 5% what will be the population after 3 years ?

Sol :

$$\text{Population after } n \text{ year} = \mathbf{P}\left(1+\frac{\mathbf{r}}{\mathbf{100}}\right)^{\mathbf{n}}$$

Where P is present population and r is the rate %

$$\text{Population after three years} = 16000\left(1+\frac{5}{100}\right)^3$$

$$= 16000 \times \frac{21}{\mathbf{20}} \times \frac{21}{\mathbf{20}} \times \frac{21}{\mathbf{20}}$$

$$= 2 \times 9261 = \mathbf{18522}$$

Example 10. During one year the population of a village increases 5% and the next year it diminished by 5%. If at the end of second year the population was 1596 what was it at the beginning of the first year ?

Sol : Let P = Population at the beginning of 1st year.

$$\text{Population in 2 years} = \mathbf{P}\left(1+\frac{\mathbf{r_1}}{\mathbf{100}}\right)\left(1-\frac{\mathbf{r_2}}{\mathbf{100}}\right)$$

$$1596 = \text{P}\left(1+\frac{5}{100}\right)\left(1-\frac{5}{100}\right)$$

$$1596 = \text{P} \times \frac{21}{20} \times \frac{19}{20}$$

$$\therefore \quad \text{P} = 1596 \times \frac{20}{21} \times \frac{20}{19} = \mathbf{1600}$$

Type IV (Questions based on Reduction or Increase in Consumption)

There are **two ways** of asking the question in this type. In the **Ist way** the price of a **commodity** is **increased**, the **consumption** is to be **reduced** in such a way that **expenditure** remains the **same**. In the **second way** the price is **reduced** we are to **increase** the **consumption** in such a way that **expenditure** remains the **same**. Following **example** will explain it more **clearly**.

Example 11. Price of sugar is increased by 25%. How much percent must a house hold must reduce his consumption of sugar so as not to increase his expenditure ?

Sol : Let initial expenditure = 100 Rs.

Increase = 25%

$\therefore$ New Price of same quantity = **125 Rs.**

Now the house hold must reduce his consumption of sugar in such a way that expenditure remains 100. For that he should reduce the consumption by Rs. 25 **(Note : this is not 25%)**

On Rs. 125 he should reduce = 25 Rs.

On Rs. 1 he should reduce = $\frac{25}{125}$

On Rs. 100 he should reduce = $\frac{25}{125} \times 100$

$= \frac{1}{5} \times 100 = \mathbf{20\%}$

$\therefore$ % Reduction = $\frac{\textbf{increase}}{\textbf{100 + increase}} \times \mathbf{100}$

Example 12: Petrol prices are reduced by 10%. Find by how much % a consumer must increase his consumption of petrol so as not to decrease his expenditure.

Sol : Suppose the price of petrol consumed by the consumer is of Rs. 100

Reduction = 10 %

$\therefore$ New Price of same quantity = **Rs. 90**

He must increase his consumption in such a way that his expenditure remains the same. For that he has to increase his consumption by Rs. 10.

$\therefore$ % Increase = $\frac{10}{90} \times 100$

$= \frac{100}{9} = \mathbf{11\frac{1}{9}\%}$

$\therefore$ % Increase in consumption = $\frac{\text{increase}}{100 - \text{increase}} \times 100$

Type V (Reduction in Price)

Example 13. A reduction of 10% in the price of mangoes enables a person to buy 20 mangoes more for Rs. 27. What is the reduced price per dozen? Also find original price per dozen.

Sol : Due to reduction in price there is saving of 10% on Rs. 27

Saving in Rupees = 10% of 27

$= \frac{10}{100} \times 27 = \mathbf{2.70\ Rs.}$

With this Rs. 2.70 he can purchase 20 mangoes.

$\therefore$ Reduced price of 20 mangoes = 2.70 Rs.

Reduced price of 1 mango = $\frac{2.70}{20}$

$$\text{Reduced price of 12 mangoes} = \frac{2.70}{20} \times 12 = \textbf{1.62 Rs.}$$

Now reduced price is obtained after a reduction of 10% in original price. So it is 90% of original price.

$\therefore$ 90% of original price of 12 mangoes = 1.62 Rs.

$$\text{Original price of 12 mangoes} = \frac{1.62}{90\%}$$

$$= \frac{1.62}{90} \times 100 = \textbf{1.80 Rs.}$$

Type VI (Net % Change)

Example 14. Water tax is increased by 20 % and consumption is decreased by 20%. Find the % increase or decrease in the expenditure.

Sol : Let original Water Tax = 100 Rs/unit.

Original Consumption = 100 units

Therefore, original expenditure = 100 × 100 = 10000 Rs.

Increased tax = 120 Rs./unit

New consumption = 80 units

$\therefore$ New expenditure = 120 × 80

= 9600 Rs.

Decrease in expenditure = Original expenditure – new expenditure

= 10000 – 9600

= 400 Rs.

$$\therefore \quad \% \text{ Decrease} = \frac{400}{10000} \times 100 = \mathbf{4\%}$$

Shortcut : Whenever there are two % changes A and B then

$$\text{Net \% change} = A + B + \frac{AB}{100}$$

If there is an **increase** or **profit** etc. then the **% change** is taken as **+ve** and if there is a **decrease** or **loss** or **discount** etc. then it is taken as **–ve**.

Therefore,

Now in above case

$$A = +20$$

$$B = -20$$

$$\text{Therefore net \% change} = A + B + \frac{AB}{100}$$

$$= 20 - 20 + \frac{20 \times -20}{100}$$

$$= 0 - 4 = \mathbf{4\%}$$

–ve sign shows that there is a decrease. Therefore net result is 4% decrease

Now this shortcut is also helpful whenever there are three % changes, say **A**, **B** and **C**. In this we have to just find the resultant of **A** and **B**. Let it come out to be **D**.

Then net % change of **D** and **C** will be the final result. This shortcut is applicable in many topics like **Profit and Loss**, **Average, Mensuration** etc. This is explained whereever this is applicable.

Now we will do few more questions where this shortcut is applicable.

Example 15. The price of an article is increased by 20% but its sale is decreased by 10%. Find the % change in the revenue received by the dealer.

Sol :

$$\text{Net \% change} = A + B + \frac{AB}{100}$$

$$= 20 - 10 + \frac{20 \times -10}{100}$$

$$= 10 - 2 = \mathbf{8\%}$$

As the net % change is **+ve**, there is an **increase** of **8%** in revenue.

Example 16. On decreasing the price of computers by 20% its sale is increased by 30%. What is the effect on revenue received by the dealer ?

Sol :

$$\text{Net \% change} = A + B + \frac{AB}{100}$$

$$= -20 + 30 + \frac{-20 \times 30}{100}$$

$$= 10 - 6 = \mathbf{4\%}$$

EXERCISE 2

Q.1. (i) 50% of 88 = ...

(ii) 60% of 15 = ...

(iii) $87\frac{1}{2}$% of 6400 = ...

(iv) .....% of 1800 = 200

(v) $62\ \frac{1}{2}$% of 2400 = ...

Q.2. A student has to secure 35% marks to pass. He gets 650 marks and fails by 50 marks. Find the maximum marks.

(a) 1000 (b) 1500 (c) 2000 (d) none

Q.3. A student had to secure 40% marks to pass. He got 100 marks and secured 20 more than pass marks Find the maximum marks.

(a) 200 (b) 400 (c) 150 (d) none

Q.4. A student who gets 36% of marks, fails by 32 marks. Another candidate who gets 45% marks gets 40 marks more than are necessary for passing Find the maximum marks.

(a) 1600 (b) 800 (c) 400 (d) none

Q.5. At an election a candidate secures 40% votes and is defeated by the other candidate by a majority of 2020 votes. Find the total number of votes polled.

(a) 9900 (b) 10000 (c) 10100 (d) none

Q.6. At an election a candidate who secured 40% votes and is defeated by the other candidate by a margin of 300 votes. Find the total number of votes recorded.

(a) 3000 (b) 7500 (c) 1500 (d) none

Q.7. In an election a candidate got 35% of votes but was defeated by other candidate by 750 votes. Find the total number of votes polled.

(a) 2000 (b) 2500 (c) 3200 (d) none

Q.8. Rajesh is 40% older than Neeraj. Find how much % is Neeraj younger than Rajesh.

(a) $28\frac{4}{7}\%$ (b) $42\frac{6}{7}\%$ (c) 25% (d) none

Q.9. The population of a town increases at a rate of 4% per annum. At present it is 37500. What will it be after 2 years.

(a) 40000 (b) 40500 (c) 40200 (d) none

Q.10. During one year the population of a town increased by 5% and during the next year it decreased by 5%. If at the end of second year the population was 35112, what was it at the beginning of the first year?

(a) 35100 (b) 35200 (c) 35204 (d) none

Q.11. The price of milk has risen by 75%. Find by how much percent a house holder must reduce his consumption so as not to increase his expenditure.

(a) $28\frac{4}{7}\%$ (b) $16\frac{2}{3}\%$ (c) $42\frac{6}{7}\%$ (d) none

Q.12. Price of sugar has increased by 60%. How much % a house holder must reduce his consumption of sugar so as not to increase the expenditure?

(a) 25% (b) 30% (c) $37\frac{1}{2}\%$ (d) none

Q.13. Petrol prices are reduced by 20%. Find by how much percent a vehicle owner must increase his consumption of petrol so as not to decrease his expenditure of petrol.

(a) 20% (b) 25% (c) 42% (d) none

Q.14. A reduction of 20% in the price of rice enables a man to buy 10 kg more rice for Rs. 1600. Find the reduced price of rice per kg.

(a) 25 Rs. (b) 12 Rs. (c) 32 Rs. (d) none

Q.15. A reduction of 20% in the price of apples enables Sumeer to purchase 16 apples more for Rs.320. Find the reduced price of 10 apples.

(a) 40 Rs. (b) 26 Rs. (c) 33 Rs. (d) none

Q.16. Price of a car is first increased by 20%, then reduced by 20%. What is the net % change in the price of car?

(a) 4% increase (b) no change (c) 4% decrease (d) none

SOLUTIONS

EXERCISE 1

1. (a) $\frac{2}{3} \times 3600 = 2400$

2. (c) $\frac{1}{3}$ of ? $= 1200$

? $= 1200 \times 3 = 3600$

3. (c) $\frac{2}{3} \times 36000 = 24000$

4. (b) $\frac{1}{9} \times 1818 = 202$

5. (b) $\frac{2}{5} \times 185 = 74$

6. (a) $\frac{4}{5} \times 85 = 68$

7. (c) $\frac{1}{4} \times 484 = 121$

8. (b) $\frac{5}{8} \times 1600 = 1000$

9. (c) $\frac{2}{9} \times 5454 = 1212$

10. (b) $\frac{1}{7} \times 1400 = 200$

11. (d) $\frac{7}{8} \times 1600 = 1400$

12. (c) $\frac{7}{8} \times 8008 = 7007$

13. (b) $\frac{1}{3} \times 3300 = 1100$

14. (a) $\frac{1}{3} \times 5400 = 1800$

15. (a) $\frac{3}{4} \times 400 = 300$

EXERCISE 2

1. (i) $\frac{1}{2}$ of 88 = **44**

(ii) $\frac{3}{5} \times 15 = \mathbf{9}$

(iii) $\frac{7}{8} \times 6400 = \mathbf{5600}$

(iv)
$$x\% \text{ of } 1800 = 200$$
$$x\% = \frac{200}{1800} = \frac{1}{9}$$
$$x = \frac{100}{9} = 11\frac{1}{9}$$

(v) $\frac{5}{8} \times 2400 = \mathbf{1500}$

2. (c)
$$\text{Pass Marks} = 650 + 50 = 700$$
$$\text{Therefore } 35\%\, x = 700$$
$$x = \frac{700}{35} \times 100 = \mathbf{2000}$$

3. (a)
$$\text{Pass Marks} = 100 - 20 = 80$$
Therefore
$$40\%\, x = 80$$
$$x = \frac{80}{40} \times 100 = \mathbf{200}$$

4. (b)
$$\text{Let Total Marks} = x$$
$$\therefore \quad 36\%\, x + 32 = 45\%\, x - 40$$
$$9\%\, x = 72$$
$$x = \frac{72}{9} \times 100 = \mathbf{800}$$

5. (c)
$$\text{Let total votes} = x$$
$$\text{Loser gets} = 40\%\, x$$
$$\text{Therefore Winner gets} = x - 40\%\, x = 60\%\, x$$
$$\text{Therefore } 60\%x - 40\%x = 2020$$
$$20\%\, x = 2020$$
$$x = \frac{2020}{20} \times 100 = \mathbf{10100}$$

6. (c)
$$60\%\, x - 40\%\, x = 300$$
$$20\%\, x = 300$$
$$x = \frac{300}{20} \times 100 = \mathbf{1500}$$

7. (b)

$$65\%x - 35\%x = 750$$
$$30\%x = 750$$
$$x = \frac{750}{30}\times 100 = \mathbf{2500}$$

8. (a)

Let Neeraj's age = 100 yrs.

Let Rajesh's age = 140 yrs.

difference in age = 140 – 100 = 40 yrs.

∴ Neeraj is younger to Rajesh by $= \frac{40}{140}\times 100 = \frac{200}{7} = \mathbf{28\frac{4}{7}\%}$

9. (d) Population after 2 yrs.

$$= 37500\left(1+\frac{4}{100}\right)^2$$
$$= 37500\times\frac{676}{625} = \mathbf{40560}$$

10. (b) Let Population at the begining = P

$$\therefore\quad 35112 = P\left(1+\frac{5}{100}\right)\left(1-\frac{5}{100}\right)$$
$$35112 = P\times\frac{21}{20}\times\frac{19}{20}$$
$$P = 35112\times\frac{20}{21}\times\frac{20}{19} = 35200$$

11. (c)

$$\%\text{ Reduction} = \frac{75}{100}\times 100$$
$$= \frac{300}{7}\% = \mathbf{42\frac{6}{7}}$$

12. (c)

$$\%\text{ Reduction} = \frac{66}{160}\times 100$$
$$= \frac{75}{2}\% = \mathbf{37\frac{1}{2}\%}$$

13. (b) % Increase in petrol to maintain the same expenditure

$$= \frac{28}{80}\times 100 = 25\%$$

14. (c) ∴ Reduction in Rupees $= 1600\times\frac{20}{100} = 320$ Rs.

With this 320 Rs. he can purchase 10 Kg. rice now.

∴ reduced price Per kg. $= \frac{320}{10} = 32$ Rs.

15. (a) Reduction in Rupees $= 320\times\frac{20}{100} = 64$ Rs.

With this Rs. 64 sumeer can purchase 16 apples.

$\therefore$ Reduced price of 16 apples $=$ 64 Rs.

$\therefore$ Reduced price of 1 apples $= \dfrac{64}{16} = 4$ Rs.

Reduced price of 10 apples $= 4 \times 10 = 40$ Rs.

16. (c) Net % change $= 20 - 20 + \dfrac{20 \times -20}{100}$

$= -4$

$=$ **4% decrease**

Day 7

Profit and Loss - I

Today, we will do the topic of **Profit and Loss**. This is a very important topic as whole of the business activity revolves around this topic. Every **literate or illiterate, big** or **small** business unit makes the **practical use** of this topic. In business activity, the person having better understanding of this topic can analyse a situation better and can take fast and accurate decisions. Before going into details let us see how **profit** and **loss** are made.

Profit : Profit occurs when we **buy** an item or an article at a **cheaper** rate and **sell** it at a **higher** rate. If we buy a book at **Rs. 30** and **sell** it at **Rs. 51**. If it is asked how much profit one has made? You will quickly say **Rs. 21**.

Loss : Loss is made when we buy a certain item at higher rate and sell it at cheaper rate. If we buy an item for **Rs. 51** and **sell** it at **Rs. 30** then **loss** of **Rs. 21** is made.

Now, the **price at which an item is bought** is know as **Cost Price**. The **price at which it is sold** is known as **Selling Price.**

As already mentioned, if a thing is bought at **lower rate** and sold at **higher rate** means if **selling price is more than the cost price then Profit is made** and

Profit = Selling Price – Cost Price

Similarly if **Cost Price is more than the Selling Price then Loss is made**

Loss = Cost Price – Selling Price

Now it is hoped that you have understood.

Q.1. What is **Selling Price**?

Ans. Selling Price: **Price at which a thing is sold.**

Q.2. What is **Cost Price**?

Ans. Cost Price: **Price at which a thing is bought.**

Q.3. What is **Profit**?

Ans. Profit: **SP – CP** (It occurs when SP is greater than CP)

Q.4. What is **Loss**?

Ans. Loss: **CP – SP** (It occurs when CP is greater than SP)

Now fill the blanks in the following table :

Table 1

S. No. (C.P.)	*Cost Price (S.P.)*	*Selling Price*	*Whether Profit or Loss*	*Profit or Loss in Rs.*
1	188	208.00	--	--
2	200	240.00	--	--
3	716	825.00	--	--
4	8280	8080.00	--	--
5	317.50	322.50	--	--
6	440.00	407.00	--	--
7	1727	1832.00	--	--
8	525	327.00	--	--
9	10270	10198.00	--	--
10	118325	188292.00	--	--

(Answers to above table are given in the Solutions at the end of this chapter)

TO FIND PROFIT % OR LOSS %

Till now, we have learnt how to calculate Profit and Loss in terms of value. Now we will calculate them in Percentage. This is more clearly understood by the following example.

Example. A man buys a car for **Rs. 50000** and sells it for **Rs. 75000**. Find gain or loss%.

Sol : It is clear from the statement as **Selling Price** is greater than the **Cost Price**. Therefore there is a profit.

Profit in **Rupees** = 75000 – 50000 = **25000 Rs.**

Now question arises what is the **percentage of profit**?

To answer it we can proceed as follows:

By investing **Rs. 50000** he gets profit = 25000 Rs.

by investing **Rs. 1** he gets profit $= \frac{25000}{50000}$

by investing **Rs. 100** he gets profit $= \frac{25000}{50000} \times 100$

$= \frac{1}{2} \times 100 = \mathbf{50\%}$

Therefore **Profit** or **Loss Percentage** is always **calculated** on **Cost Price**.

Remember: Profit and Loss percentage are always **calculated in terms of Cost Price.**

$$\textbf{Profit \%} = \frac{\textbf{Profit}}{\textbf{C.P}} \times 100$$

and

$$\textbf{Loss \%} = \frac{\textbf{Loss}}{\textbf{C.P}} \times \textbf{100}$$

To make you well versed in this concept following table is made and you have to fill in the blanks.

Now fill in the blanks in the following Table:

Table 2

S. No.	*Cost Price (C.P.) in Rs.*	*Selling Price (S.P.) in Rs.*	*Profit or Loss in Rs.*	*Profit or Loss %*
1	80	100	--	--
2	100	120	--	--
3	260	286	--	--
4	480	516	--	--
5	2000	1800	--	--
6	10000	10800	--	--
7	600	400	--	--
8	8000	16000	--	--
9	16000	8000	--	--
10	50000	30000	--	--

(Answers to above table are given in the Solutions at the end of this chapter)

To find Selling Price or Cost Price when Profit or Loss Percentage is given.

Till now we have learnt how to find Profit, Loss, Profit % and Loss %. Now we will find C.P. and S.P. when Profit or Loss %age is given.

Example 1. An article is sold at **Rs. 240** and Profit is **20%.** Find the **Cost Price.**

Sol : This is solved as follows:

$$\text{Let the } \textbf{Cost Price} = 100 \text{ Rs.}$$

$$\text{Profit} = 20\%$$

$$\textbf{S.P.} = 120 \text{ Rs.}$$

$$\text{If } \textbf{S.P.} \text{ is Rs. 120 then } \textbf{C.P.} = 100 \text{ Rs.}$$

$$\text{If } \textbf{S.P.} \text{ is Rs. 1 then } \textbf{C.P.} = \frac{100}{200} \text{ Rs.}$$

$$\text{If } \textbf{S.P.} \text{ is Rs. 240 then } \textbf{C.P.} = \frac{100 \times 240}{120} = \textbf{200 Rs.}$$

This was the detailed method and direct formula for finding the Selling Price is

$$\textbf{S.P.} = \text{C.P.} \times \frac{100 + \text{Profit \%}}{100}$$ If there is a Loss

$$\text{Then } \textbf{S.P.} = \text{C.P.} \times \frac{100 - \text{Loss \%}}{100}$$

This formula can be written as

$$\text{S.P} = \text{C.P.} \times \frac{100 \pm \text{P/L}\,\%}{100}$$

SP is selling price CP is cost price P is profit % & L is loss % if there is a profit then sign is taken as positive, if there is a loss then sign is taken as negative

And similarly the formula for Cost Price is

$$\text{C.P.} = \text{S.P.} \times \frac{100}{100 \pm \text{P/L}\,\%}$$

Now as done earlier solve the following table:

Table 3

S. No.	*S.P.*	*C.P.*	*Profit%*	*Loss%*
1	110	--	10%	
2	90	--		10%
3	528	--	10%	*x*
4	432	--		10%
5	--	1535	25%	*x*
6	--	17120	5%	*x*
7	--	1210	10%	*x*
8	--	1210	*x*	10%
9	--	4500	30%	*x*
10	--	170	20%	*x*

(Answers to above table are given in the Solutions at the end of this chapter)

The purpose of this detailed exercise was to make your **concepts solid**. Now we will discuss about the types of questions asked in the exams. We have divided the questions in the following **eight types**. **Ist five** of these eight types we will do on **day 7** and rest on **day 8**.

Types of Questions

Type I. Simple Questions (in which CP or SP is to be found)

Type II. (Questions related to Mangoes, Oranges, Penciles etc.)

Type III. (When two or more quantities are mixed.)

Type IV. (Questions related to false weights.)

Type V. (When Profit & Loss given in term of CP & SP.)

Type VI. (When two percentage changes are given.)

Type VII. (Questions related to marked price & discount.)

Type VIII. Misc. Questions

Type I

Example 1. By selling a watch for Rs. 480 a man gains 20%. Find what would have been his gain or loss %. If he had sold it for 320.

Sol :

$$\textbf{S.P.}\text{ of watch} = 480 \text{ Rs.}$$

$$\text{Gain} = 20\%$$

$$\textbf{CP} = 480 \times \frac{100}{120} = 400 \text{ Rs.}$$

$$\text{IInd SP} = 320 \text{ Rs.}$$

$$\text{Loss} = \text{Rs. } 400 - 320 = 80 \text{ Rs.}$$

$$\text{Loss\%} = \frac{80}{400} \times 100 = \mathbf{20\%}$$

$$\left(\textbf{As loss \%} = \frac{\textbf{Loss}}{\textbf{C.P.}} \times \mathbf{100}\right)$$

Example 2. By selling a calculator for Rs. 858 a shopkeeper gain 10%. Find his gain or loss % when he sells it at Rs. 819.

Sol :

$$\text{SP} = 858 \text{ Rs.}$$

$$\text{Gain} = 10\%$$

$$\text{CP} = 858 \times \frac{100}{110} = \mathbf{780 \text{ Rs.}}$$

$$\text{IInd SP} = 819 \text{ Rs.}$$

Therefore Profit is made

$$\text{Profit in Rupees.} = 819 - 780 = 39 \text{ Rs.}$$

$$\text{Therefore Profit \%} = \frac{39}{780} \times 100 = \mathbf{5\%}$$

Example 3. By selling a computer at Rs. 23750 a shopkeeper loses 5%. Find his gain or Loss % when he sells it at Rs. 27500.

Sol :

$$\text{SP} = 23750 \text{ Rs.}$$

$$\text{Loss} = 5\%$$

$$\text{CP} = 23750 \times \frac{100}{95} = 25000 \text{ Rs.}$$

$$\text{IInd SP} = 27500 \text{ Rs.}$$

$$\text{Profit} = 25000 \text{ Rs.}$$

$$\text{Profit \%} = \frac{2500}{25000} \times 100 = \mathbf{10\%}$$

Example 4. By selling a Television for Rs. 12580 a dealer gains 25%. Find his gain or loss % when he sells it at a 11322.

Sol :

$$\text{SP} = 12580 \text{ Rs.}$$

$$\text{Gain} = 25\%$$

$$\text{CP} = 12580 \times \frac{100}{125} = 10064 \text{ Rs.}$$

$$\text{IInd SP} = 11322 \text{ Rs.}$$
$$\text{Profit} = 11322 - 10064 = 1258 \text{ Rs.}$$
$$\text{Profit \%} = \frac{1258}{10064} \times 100$$
$$= \frac{1}{8} \times 100$$
$$= \mathbf{12.5\%}$$

Example 5. By selling a T.V. Trolley for Rs. 1800 a carpentar loses 10%. Find his gain (in Rs.) when he sells it at Rs. 2500

Sol. :

$$\text{SP} = 1800 \text{ Rs.}$$
$$\text{Loss} = 10\%$$
$$\text{CP} = 1800 \times \frac{100}{90} = 2000 \text{ Rs.}$$
$$\text{IInd SP} = 2500 \text{ Rs.}$$
$$\text{Profit} = 2500 - 2000 = \mathbf{500 \text{ Rs.}}$$

Type II

Examples 6. By selling oranges 11 for a rupee a man loses 10%. How many for a rupee should he sell to gain 10%

Sol :

$$\text{SP of 11 oranges} = 1 \text{ Rs.}$$
$$\text{SP of 1 oranges} = \frac{1}{11} \text{ Rs.}$$
$$\text{Loss} = 10\%$$
$$\text{CP of 1 orange} = \text{S.P. of 1 orange} \times \frac{100}{100 - \text{loss \%}}$$
$$= \frac{1}{11} \times \frac{100}{90}$$
$$\text{IInd Profit} = 10\%$$
$$\text{IInd SP} = \frac{1}{11} \times \frac{100}{90} \times \frac{110}{100}$$
$$\text{SP of 1 orange} = \frac{1}{9} \text{ Rs. means that he should}$$

sell **9 oranges for a rupee.**

Example 7. A shopkeeper buys eggs at a rate of 7 for a rupee and sells then at a profit of 40%. How many does he sell for a rupee.

Sol :

$$\text{Let CP of 7 eggs} = 1 \text{ Rs.}$$
$$\text{CP of 1 egg} = \frac{1}{7}$$
$$\text{Profit} = 40\%$$

$$\text{SP of 1 egg} = \frac{1}{7} \times \frac{140}{100} = \text{Rs.}$$

Therefore, he should sell **5 eggs for a rupee.**

Example 8. By selling oranges at a rate of 12 for a rupee a man loses 20%. How many for a rupee should he sells to gain 20%.

Sol :

$$\text{SP of 12 oranges} = 1$$

$$\text{SP of 1 orange} = \frac{1}{12}$$

$$\text{Loss} = 20\%$$

$$\text{CP of 1 orange} = \frac{1}{12} \times \frac{100}{80}$$

$$\text{IInd Profit} = 20\%$$

$$\text{IInd SP} = \frac{1}{12} \times \frac{100}{80} \times \frac{120}{100}$$

$$\therefore \quad \text{SP of 1 orange} = \frac{1}{8} \text{ Rs.}$$

He should sell **8 oranges for a rupee.**

Example 9. By selling the oranges at a rate of 14 for a rupee, a fruit seller loses 4 %. How many oranges must he sell for a rupee to gain 12%.

Sol :

$$\text{Let SP of 14 oranges} = 1 \text{ Rs.}$$

$$\text{SP of 1 orange} = \frac{1}{14} \text{ Rs.}$$

$$\text{Loss} = 4\%$$

$$\text{CP of 1 orange} = \frac{1}{14} \times \frac{100}{96}$$

$$\text{Profit} = 12\%$$

$$\text{SP of 1 orange} = \frac{1}{14} \times \frac{100}{96} \times \frac{112}{100} = \frac{1}{12} \text{ Rs.}$$

He should sell **12 oranges for a rupee.**

Example 10. By selling 32 oranges for a rupee a man loses 40%. How many for a rupee must he sell to gain 20%.

Sol :

$$\text{SP of 32 oranges} = 1 \text{ Rs.}$$

$$\text{SP of 1 orange} = \frac{1}{32}$$

$$\text{Loss} = 40\%$$

$$\text{CP of 1 orange} = \frac{1}{32} \times \frac{100}{60}$$

Now he wants to gain 20%

$$\therefore \quad \text{SP of 1 orange} = \frac{1}{32} \times \frac{100}{60} \times \frac{120}{100}$$

$$= \frac{1}{16} \text{ Rs.}$$

∴ He should sell **16 orange for a rupee.**

Type III

In this type different quantity or different types are mixed.

Example 11. A fruit seller buys some mangoes at the rate of 20 for Rs. 3 and sells them at a rate of 25 for Rs.4. Find his gain or loss %.

Sol : Here cost price and selling price of different quantities are given. First we will find the cost price and selling price of some common quantities. For that purpose we take LCM of 20 and 25 which is equal to 100.

Now the question proceed as follows.

$$\text{CP of 20 Mangoes} = 3 \text{ Rs.}$$

$$\text{CP of 1 Mango} = \frac{3}{20} \text{ Rs.}$$

$$\text{CP fo 100 Mangoes} = \frac{3}{20} \times 100 = 15 \text{ Rs.}$$

$$\text{SP of 25 Mangoes} = 4 \text{ Rs.}$$

$$\text{SP of 1 Mango} = \frac{4}{25}$$

$$\text{SP of 100 Mangoes} = \frac{4}{25} \times 100 = 16 \text{ Rs.}$$

$$\therefore \quad \text{Profit} = 16 - 15 = 1 \text{ Rs.}$$

$$\text{Profit \%} = \frac{1}{15} \times 100$$

$$= \frac{20}{3}\% = \mathbf{6\frac{2}{3}\%}$$

Example 12. A person buys some apples at the rate of 6 for 90 p and sells them at a rate of 5 for 80 p. Find his gain percent.

Sol : LCM of 6 and 5 is 30 and so we assume that transaction of 30 apples is made. According to question

$$\text{CP of 6 apples} = 90 \text{ p.}$$

$$\text{CP of 1 apple} = \frac{90}{6}$$

$$\text{CP of 30 apples} = \frac{90}{6} \times 30 = 450 \text{ p.}$$

$$\text{SP of 5 apples} = 80 \text{ p.}$$

$$\text{SP of 1 apple} = \frac{80}{5}$$

$$\text{SP of 30 apples} = \frac{80}{5} \times 30 = 480 \text{ p.}$$

$$\text{Profit \%} = \frac{30}{450} \times 100 = \mathbf{\frac{20}{3}\%}$$

Type IV

Example 13. A merchant professes to sell his goods at cost price but uses a weight of 800 gms. instead of kilogram. Find his gain percent.

Sol : It means that he takes money for 1000 gms and gives goods worth 800 gm. Thus gains 200 gm.

$$\text{on 800 gm he gains} = 200 \text{ gm.}$$

$$\text{on 1 gm he gains} = \frac{200}{800}$$

$$\text{on 100 gm he gains} = \frac{200}{800} \times 100 = \mathbf{25\%}$$

Example 14. A merchant professes to sell his goods at cost price but uses a weight of 900 gm instead of Kg. Find his real gains percent.

Sol : For every 900 gm he gains 100 gms

$$\text{therefore Profit \%} = \frac{100}{900} \times 100$$

$$= \mathbf{\frac{100}{9}\%}$$

Example 15. A merchant professes to sell his goods at the cost price but still gains 25%. Find the weight he replaces for a Kg.

Sol : Let he uses x gm. less weight. then formula for

$$\textbf{Profit \%} = \mathbf{\frac{x}{1000 - x}}$$

$$\text{In this case profit} = 25\%$$

therefore

$$\frac{25}{100} = \frac{x}{1000 - x}$$

$$x = 200$$

Therefore he uses 1000 – 200 = **800 gms** in place of 1000 gms.

Type V

Example 16. If selling Price of 8 article is equal to cost price of 10 article. Find the loss or gain %.

Sol :

$$\text{Let CP of 1 article} = 1 \text{ Rs.}$$

$$\text{Let CP of 8 article} = 8 \text{ Rs.}$$

$$\text{Let CP of 10 article} = 10 \text{ Rs.}$$

$$\text{SP of 8 article} = \text{CP of 10 article means 10 Rs.}$$

$$\therefore \quad \text{CP of 8 articles} = 8 \text{ Rs.}$$

$$\text{SP of 8 articles} = 10 \text{ Rs.}$$
$$\text{therefore Profit} = 2 \text{ Rs.}$$
$$\text{Profit \%} = \frac{2}{8} \times 100 = \mathbf{25\%}$$

Short Method:

$$\text{SP of 8 articles} = \text{CP of 10 article}$$
$$= \text{CP of 8 art.} + \text{CP of 2 article}$$

Therefore **by selling 8 articles** the merchant **covers the cost of 8 articles** and further gets a **profit of 2 articles** and here profit is in terms of cost price.

$$\text{Therefore profit percentage is} = \frac{2}{8} \times 100 = \mathbf{25\%}$$

Note : Profit or loss should be in terms of cost price

Example 17. If Selling Price of 10 articles is equal to cost price of 8 articles. Find the profit or loss %.

Sol :

$$\text{SP of 10 article} = \text{CP of 8 article}$$
$$= \text{CP of 10} - \text{CP of 2}$$
$$\text{Therefore loss} = \frac{2}{8} \times 100 = \mathbf{20\%.}$$

Example 18. By selling 45m. of cloth a merchant gains the cost price of 15 m. Find the gain %.

Sol : The merchant gains the profit of cost price of 15 m. It means that he has first covered the cost price of 45 m.of cloth and then got the profit of cost price of 15 m of cloth.

$$\therefore \quad \text{SP of 45m} = \text{CP of 45m} + \text{CP of 15m}$$
$$\text{therefore Profit \%} = \frac{15}{45} \times 100$$
$$= \mathbf{\frac{100}{3}\%}$$

Example 19. By selling 33 m. of cloth a merchant gains the selling price of 11m. find the gains or loss%.

Sol : This question should be done very **carefully**. Because here profit is given **in terms of SP**. Here if profit had been **in term of CP** i.e. C.P. of 11 meter then profit percent would have been $\frac{11}{33} \times 100 = \frac{100}{3}$ percent. But here profit is **in terms of SP** i.e. SP of 11 meter. So we will proceed as follows:

$$\text{SP of 33 m} = \text{CP of 33 m} + \text{SP of 11 m}$$
$$\therefore \quad \text{SP of 33 m} - \text{SP of 11 m} = \text{CP of 33 m}$$
$$\text{SP of 22 m} = \text{CP of 33 m}$$
$$\text{SP of 22 m} = \text{CP of 22 m} + \text{CP of 11 m}$$
$$\therefore \quad \text{profit\%} = \frac{11}{22} \times 100 = 50\%$$

SOLUTIONS

Table 1

1. Profit Rs. 20
2. Profit Rs. 40
3. Profit Rs. 109
4. Loss Rs. 200
5. Profit Rs. 5.00
6. Loss Rs. 33
7. Profit Rs. 105
8. Loss Rs. 198
9. Loss Rs. 72
10. Profit Rs. 69967

Table 2

1. Profit Rs. 20 & 25%
2. Profit Rs. 20 & 20%
3. Profit Rs. 26 & 10%
4. Profit Rs. 36 & 7.5%
5. Loss Rs. 200 & 10%
6. Profit Rs. 800 & 8%
7. Loss Rs. 200 & 33.33%
8. Profit Rs. 8000 & 100%
9. Loss Rs. 8000 & 50%
10. Loss Rs. 20000 & 40%

Table 3

1. $$\text{C.P.} = \text{S.P.}\times\frac{100}{100+\text{P\%}} = 110\times\frac{100}{110} = \mathbf{100\ Rs.}$$

2. $$\text{C.P.} = \text{S.P.}\times\frac{100}{100-\text{L\%}} = 90\times\frac{100}{90} = \mathbf{100\ Rs.}$$

3. $$\text{C.P.} = \text{S.P.}\times\frac{100}{100+\text{P\%}} = 528\times\frac{100}{110} = \mathbf{480\ Rs.}$$

4. $$\text{C.P.} = \text{S.P.}\times\frac{100}{100-\text{L\%}} = 432\times\frac{100}{90} = \mathbf{480\ Rs.}$$

5. $$\text{S.P.} = \text{C.P.}\times\frac{100+\text{P\%}}{100} = 1535\times\frac{125}{100} = \mathbf{1918.75\ Rs.}$$

6. $$\text{S.P.} = \text{C.P.}\times\frac{100+\text{P\%}}{100} = 17120\times\frac{105}{100} = \mathbf{17976\ Rs.}$$

7. $$\text{S.P.} = \text{C.P.}\times\frac{100+\text{P\%}}{100} = 1210\times\frac{110}{100} = \mathbf{1331\ Rs.}$$

8. $$\text{S.P.} = \text{C.P.}\times\frac{100+\text{P\%}}{100} = 1210\times\frac{90}{100} = \mathbf{1089\ Rs.}$$

9. $$\text{S.P.} = \text{C.P.}\times\frac{100+\text{P\%}}{100} = 4500\times\frac{130}{100} = \mathbf{5850\ Rs.}$$

10. $$\text{S.P.} = \text{C.P.}\times\frac{100+\text{P\%}}{100} = 170\times\frac{120}{100} = \mathbf{204\ Rs.}$$

Day 8

Profit and Loss - II

On day 7 we have done basics of **profit and loss**. In addition to this **five types** of questions have been covered. On **day 8** we will cover questions of last **three types** followed by an exercise covering all the **eight types**.

Type VI : {When two percentage changes are given}

Type VII : {Questions related to marked price}

Type VIII : {Misc. Questions}

Type VI

Example 1. A man sold two watches for Rs. 20 each. On one he lost 10% and on another he gained10%. Find his total loss %.

Sol : Detailed Method

$$\text{Ist SP} = 20 \text{ Rs.}$$

$$\text{Gain} = 10\%$$

$$\therefore \quad \text{CP of Ist} = 20 \times \frac{100}{120} = \frac{200}{11} \text{ Rs.}$$

$$\text{IInd SP} = 20 \text{ Rs.}$$

$$\text{Loss} = 10\%$$

$$\therefore \quad \text{CP of IInd} = 20 \times \frac{100}{90} = \frac{200}{0} \text{ Rs.}$$

$$\text{Total SP} = 20 + 20 = \mathbf{40\ Rs.}$$

$$\text{Total CP} = \frac{200}{11} + \frac{200}{9} = \frac{\mathbf{4000}}{\mathbf{99}} \text{ Rs.}$$

As CP is greater than SP $\therefore$ there is a **loss**

$$\therefore \quad \text{loss} = \frac{4000}{99} - \frac{40}{1} = \frac{40}{99} \text{ Rs.}$$

$$\text{Loss Percentage} = \frac{\frac{40}{99}}{\frac{4000}{99}} \times 100$$

$$= \mathbf{1\%}$$

SHORTCUT

This is a very important shortcut illustrated as follows :

When ever there are two percent changes say A and B then net percentage is equal to

$$\mathbf{A + B + \frac{AB}{100}}$$

and If the change is **increasing** or there is a **profit** then it is taken as **positive**. If it is **decreasing** or there is a **loss** then it is taken as **negative**. This shortcut is applicable in lot of question of different types of Percentage, profit and loss area etc. Let us apply this shortcut to the previous question.

Here $A = +10\%$ (gain)

$B = -10\%$ (loss)

$$\text{Net percentage change} = 10 - 10 + \frac{10 \times -10}{100}$$

$$= 0 - 1 = -1\%$$

Negative sign shows that there is a loss

∴ Loss = **1%**

Example 2. A man sold two houses for Rs. 48000 each. On one he lost 20% and another he gained 20% . Find his total gain or loss in percent age as well as in rupees.

Sol : As discussed earlier this question is a combination of **shortcut and concept.**

$$\text{Net Percentage change} = -20 + 20 + \frac{-20 \times 20}{100}$$

$$= 0 - 4 = -4\%$$

∴ Net loss = 4%

Now total SP = 96000 which is 96% of CP

∴ 96% of CP = 96000 Rs.

$$CP = \frac{96000}{96} \times 100$$

$$= 100000 \text{ Rs.}$$

∴ Loss = 100000 – 96000 = **4000 Rs.**

Example 3. Two article are sold at 750 Rs. each. One at a profit of 30% another at a loss of 30%. Find his total gain or loss %.

Sol : By shortcut

$$\text{Net \% change} = 30 - 30 + \frac{30 \times -30}{100}$$

$$= 0 - 9 = -9\%$$

As Net change = – 9%

Therefore loss is 9%.

Note : This shortcut is applicable if selling Price of both the articles as well as profit % on first and loss % on second are same. If any of these are different then do the question as explained in detailed method of example 1.

Type VII

Example 4. A tradesman charges 20% above the cost price but allowed a discount of 10% for cash payment. What is gain percent.

Sol : Let cost price of item = 100 Rs.

Marked price = 120 Rs.

$$\text{Discount} = 120 \times \frac{10}{100} = 12 \text{ Rs.}$$

(Discount is always calculated on marked price and if no discount is given than the marked price and the selling price is same)

Net SP = 120 – 12 = 108 Rs.

$$\text{Profit \%} = \frac{8}{100} \times 100 = \mathbf{8\%}$$

Shortcut :

$$\text{Net \% change} = 20 - 10 + \frac{20 \times -10}{100}$$

$$= 10 - 2 = \mathbf{8\%}$$

Example 5. A tradesman allows a discount of 5% for cash payment. How much percent above the CP must he mark his goods to make a profit of 14%.

Sol : Let marked price = 100 Rs.

Discount = 5 Rs.

SP = 95

Profit = 14%

$$\text{CP} = 95 \times \frac{100}{114} = \frac{250}{3} \text{ Rs.}$$

$$\text{Price marked above} = 100 - \frac{250}{3}$$

$$= \frac{50}{3}$$

$$\text{\% marked above} = \frac{\frac{50}{3}}{\frac{250}{3}} \times 100$$

$$= \mathbf{20\%}$$

Shortcut :

$$\text{Net \% change} = A + B + \frac{AB}{100}$$

Net % change = 14% **(Given)**

Let he marks B% above C.P.

$$14 = -5 + B + \frac{-5 \times B}{100}$$

$$(14 + 5) = B - \frac{5B}{100}$$

$$19 = \frac{95B}{100}$$

$$B = \frac{100 \times 19}{95} = \mathbf{20\%}$$

Example 6. A merchant allows a discount of 10% for cash payment. How much% above the cost price must he mark his goods, so as to make a Profit of 17%.

Sol : $$\text{Net \% Change} = A + B + \frac{AB}{100}$$

Let mark price is B % above cost price

∴ $$A = -10\%$$

$$B = B\%$$

$$\text{Net \% change} = 17\%$$

$$17 = -10 + B + \frac{-10 \times B}{100}$$

$$17 + 10 = B - \frac{10B}{100}$$

$$27 = \frac{90B}{100}$$

$$B = 27 \times \frac{100}{90} = \mathbf{30\%}$$

Example 7. A tradesman allows a discount of 10%, what price should he mark on an article which costs 200 Rs. to gain 26%.

Sol : Let he mark B % above cost price

∴ $$26 = -10 + B + \frac{-10 \times B}{100}$$

$$36 = \frac{90B}{100}$$

$$B = \frac{100 \times 36}{90} = 40\%$$

$$40\% \text{ of } 200 = 80 \text{ Rs.}$$

∴ $$\text{Marked Price} = 200 + 80 = \mathbf{280\ Rs.}$$

Type VIII

Example 8 : I purchased two watches for Rs. 480. I sold one at a loss of 15% and second at a gain of 19%. The SP of both watches were same. Find the CP of each watch.

Sol : $$\text{Let CP of 1st watch} = x$$

$$\text{loss} = 15\%$$

∴ $$SP = x \times \frac{100 - 15}{100}$$

$$= \frac{85x}{100}$$

$$\text{Let CP of 2nd watch} = y$$
$$\text{Profit} = 10\%$$
$$\therefore \quad SP = y \times \frac{100+19}{100}$$
$$= \frac{119y}{100}$$

According to statement

$$\text{SP of Ist watch} = \text{SP of IInd watch}$$
$$\frac{85x}{100} = \frac{119y}{100}$$
$$\frac{x}{y} = \frac{119}{85} = \frac{7}{5}$$
$$\therefore \quad x : y = \mathbf{7 : 5}$$
$$\text{Cost price of 1st watch} = \frac{7}{12} \times 480$$
$$= \mathbf{280\ Rs.}$$
$$\text{Cost price of 2nd watch} = \frac{5}{12} \times 480 = 200$$
$$= \mathbf{200\ Rs.}$$

Example 9. I bought two horses for Rs. 1350. I sold one of them at a loss of 6% and another at a gain of $7\frac{1}{2}\%$. On the whole I neither gained nor lost. What did each horse cost?

Sol :

$$\text{Loss on one horse} = \text{gain on second}$$
$$6\% \text{ 1st cost} = 7\frac{1}{2}\% \text{ of 2nd cost price}$$
$$= \frac{7\frac{1}{2}\%}{6\%} = \frac{15}{2 \times 6} = \frac{15}{12}$$
$$= \frac{\mathbf{5}}{\mathbf{4}} = \mathbf{5:4}$$
$$\therefore \quad \text{CP of 1st horse} = \frac{5}{9} \times 1350 = \mathbf{750\ Rs.}$$
$$\text{CP of 2nd horse} = \frac{4}{9} \times 1350 = \mathbf{600\ Rs.}$$

Example 10. A sells an article at 10 percent profit. If he had bought it at 10% less and sold it for 60 Rs. less he would have gained 20%. Find the cost price of that article.

Sol :

$$\text{Let CP} = 100 \text{ Rs.}$$
$$\text{Then SP} = 110 \text{ Rs.}$$
$$\text{Second CP} = 90 \text{ Rs.}$$
$$\text{Profit} = 20\%$$

$$\text{Second SP} = 90 \times \frac{120}{100} = 108 \text{ Rs.}$$

$$\text{Difference} = 110 - 108 = 2 \text{ Rs.}$$

$$\text{If diff. is 2 Rs. than CP} = 100 \text{ Rs.}$$

$$\text{" " 1 " " " "} = \frac{100}{2}$$

$$\text{" " 60" " "} = \frac{100}{2} \times 60 = \mathbf{3000\ Rs.}$$

EXERCISE

Q 1. By selling a cycle of Rs. 240 a shopkeeper gain 25%, If he had sold it for Rs. 204, what would have been his gain percent

(a) $5\frac{5}{7}\%$ (b) $2\frac{2}{3}\%$ (c) $6\frac{2}{3}\%$ (d) none

Q.2. A man bought a car for Rs. 78000 and sold it at a profit of 10%. Find his selling price.

(a) 85800 Rs. (b) 85600 Rs. (c) 100 Rs. (d) none

Q.3. The SP of an article is 80 Rs. and loss is 20%. Find the Cost Price.

(a) 48 Rs (b) 120 Rs (c) 100 Rs. (d) none

Q.4. A man bought 5 oranges for a rupee. How many for a rupee should he sell to gain 25%

(a) 4 (b) 6 (c) 3 (d) none

Q.5. A man sold 8 apples for a rupee and lost 20%. How many for a rupee should he sell to gain 60%.

(a) 5 (b) 10 (c) 4 (d) none

Q.6. A man bought 9 bananas for a rupee. Due to bad weather he had to sell it at a loss of 10% How many for a rupee did he sell.

(a) 8 (b) 12 (c) 10 (d) none

Q.7. A shopkeeper professes to sell his goods at cost price, but uses a weight of 600 grams instead of kg. What is his real gain percent.

(a) 20% (b) 25% (c) 40% (d) none

Q.8. A dishonest dealer professes to sell his goods at cost price but still gains 100/3% on his out lay. What weight does he substitute for a kg.

(a) 800gm. (b) 750gm. (c) 850gm. (d) none

Q.9. A man bought 2 oranges for Rs. 5 and sold at a rate of 3 for Rs. 7. His real loss percent is.

(a) $6\frac{2}{3}\%$ (b) $7\frac{1}{7}\%$ (c) $7\frac{2}{7}\%$ (d) none

Q.10. Find the difference between a single discount of 40% and two successive discounts of 36% and 4% on Rs. 4000.

(a) 57.60 Rs. (b) 55.40 Rs. (c) 33.55 Rs. (d) none

Q.11. The manufacturer of a article makes a profit of 25%. The whole sale dealer makes a profit of 20% and retailer makes a profit of 28%. What is the cost of article retail for Rs. 48.

(a) 30 Rs. (b) 25 Rs. (c) 40 Rs. (d) none

Q.12. A manufacturer sells an article at a profit of 25% to wholesale dealer who sells it to a retailer at a profit of 5% and the retailer sells it for Rs. 756 and makes a profit fof 20%. What is the cost of article to manufacturer.

(a) 480 Rs. (b) 460 Rs. (c) 540 Rs. (d) none

Q.13. A manufacturer of an article sells his product to the whole sale dealer and makes a profit of 10%. The wholesale makes a profit of 20% and sells it to the retailer. The retailer sells the article for 41.25 and makes a profit of 25%. on his cost. What is the cost of production of the article.

(a) Rs. 25 (b) Rs. 30 (c) Rs. 28 (d) none

Q.14. A sells an article to B at a gain of 20%. B sells it to C at a gain of 15%, C gains $16\frac{2}{3}\%$. by selling it for Rs. 483. Find the cost price of A.

(a) 320 Rs. (b) 330 Rs. (c) 290 Rs. (d) none

Q.15. If selling price of 20 article is equal to cost price of 25 article. Find his real gain or loss%.

(a) gain 25% (b) loss 25% (c) gain 20% (d) none

Q.16. If cost price of 20 articles is equal to selling price of 25 article. Find his real gain or loss%.

(a) loss 20% (b) gain 20% (c) loss 25% (d) none

Q.17. By selling 60 metres cloth a merchant gain the cost price of 15 metres. Find his profit%.

(a) 25% (b) 20% (c) 60% (d) none

Q.18. By selling 60 metre of cloth a merchant gain the selling price of 15m. Find his real gain%.

(a) $\frac{100}{9}\%$ (b) $\frac{100}{3}\%$ (c) 25% (d) none

Q.19. A man sold two horses at Rs. 990 each. On one he gained 10% and on other he lost 10%. Find his total gain or loss.

(a) Rs. 25 profit (b) Rs. 100 loss (c) Rs. 20 loss (d) none

Q.20. A man sold two houses for Rs. 3840 each. On one he lost 20% and on other he gained 20%. Find his real gain or loss.

(a) Rs. 400 profit (b) Rs. 320 loss (c) Rs. 99 loss (d) none

Q.21. A dealer gives 10% discount on the marked price to his customers. How much percent should he mark his goods above cost price so as to gain 26 % after allowing the discount.

(a) 40% (b) 30% (c) 26% (d) none

Q.22. A man marked his radio at 25% above the CP and gave customers 12% discount on cash payment. In this way he gained Rs.55 Find the cost price of radio.

(a) Rs. 560 (b) Rs. 500 (c) Rs. 544 (d) none

Q.23. A dealer marked a radio at 25% above its cost price and gave purchaser15% discount on cash payment. In this way he gained Rs. 50. Find the cost price of radio.

(a) 1000 Rs. (b) 900 Rs. (c) 600 Rs. (d) none

Q. 24. I sold a watch for Rs. 50 and lost some thing. If I had sold for Rs. 65 my gain would have been double the former loss. Find the cost price of watch.

(a) 60 Rs. (b) 55 Rs. (c) 40 Rs. (d) none

Q.25. What profit percent is made by selling an article at a certain price If by selling 2/3 of that price there would be a loss of 10%.

(a) $33\frac{1}{3}\%$ (b) 35% (c) 40% (d) none

Q.26. A person sold a watch for 96 and got a percentage profit equal to cost price. Find the CP of watch.

(a) 60 Rs. (b) 50 Rs. (c) 40 Rs. (d) none

Q.27 A man bought two watches for Rs. 1210. He sold one at a gain of 8% and other at a loss of 10%. He found that SP of both watches was same. Find the cost price of each watch.

(a) Rs. 600, Rs. 510 (b) Rs. 560, Rs. 650 (c) Rs. 550, Rs. 660 (d) none

Q.28. A man bought two books for Rs. 2700. One he sold at a gain of 15% and another at a loss of 12% and on the whole he neither gained nor lost. What did each book cost?

(a) Rs.1400, Rs. 1300 (b) Rs. 1500, Rs. 1200 (c) Rs.1100, Rs.1600 (d) none

Q.29. A man bought goods for Rs. 800, sold half of them at a gain of 5%. At what gain percent must he sell the remaining so as to gain 20% on the whole .

(a) 25% (b) 30% (c) 40% (d) none

Q.30. What profit percent is made by selling an article at a certain price if by selling at 4/5 of that price there would have been a loss of 10%.

(a) $\frac{100}{3}\%$ (b) 40% (c) $\frac{100}{9}\%$ (d) none

SOLUTIONS

1. (d)

$$SP = 240 \text{ Rs.}$$
$$Gain = 25\%$$
$$CP = 240 \times \frac{100}{125} = 192 \text{ Rs.}$$
$$\text{2nd SP} = 204$$
$$\text{Profit\%} = \frac{12}{192} \times 100$$
$$= \frac{25}{4}\% = \mathbf{6\frac{1}{4}\%}$$

2. (a)

$$78000 \times \frac{110}{100} = \mathbf{85800 \text{ Rs.}}$$

3. (c)

$$80 \times \frac{100}{80} = \mathbf{100 \text{ Rs.}}$$

4. (a)

$$\text{CP of 1 orange} = \frac{1}{5} \text{ Rs.}$$
$$\text{Profit} = 25\%$$
$$\text{SP of 1 orange} = \frac{1}{5} \times \frac{125}{100} = \frac{1}{4} \text{ Rs.}$$

He should sell 4 oranges for a rupee.

5. (c)

$$\text{SP of 1 orange} = \frac{1}{8}\text{ Rs.}$$

$$\text{Loss} = 20\%$$

$$\text{CP} = \frac{1}{8}\times\frac{100}{80}\text{ Rs.}$$

$$\text{Gain} = 60\%$$

$$\text{SP of 1 orange} = \frac{1}{8}\times\frac{100}{80}\times\frac{160}{100} = \frac{1}{4}\text{ Rs.}$$

He should sell 4 oranges for a rupee.

6. (c)

$$\text{CP of one banana} = \frac{1}{9}\text{ Rs.}$$

$$\text{Loss} = 10\%$$

$$\text{SP of 9 banana} = \frac{1}{9}\times\frac{90}{100} = \frac{1}{10}\text{ Rs.}$$

So he should sell 10 bananas for a rupee.

7. (d)

$$\text{Profit} = \frac{400}{600}\times 100$$

$$= \frac{200}{3} = 66\frac{2}{3}\%$$

8. (b)

$$\frac{x}{1000-x} = \frac{1}{3}$$

$$3x = 1000 - x$$

$$4x = 1000$$

$$x = \mathbf{250\text{ gms.}}$$

$\therefore$ **he uses 1000 – 250 = 750 gms.**

9. (a) Let the no. of oranges be 6.

$$\text{CP of 2 oranges} = 5\text{ Rs.}$$

1 " " " " " $= \frac{5}{2}$ Rs.

" " " " 6 " $= \frac{5}{2}\times 6 = 15$ Rs.

SP of 3 " $= 7$ Rs.

1 " " " " $= \frac{7}{3}$ Rs.

" " " " " 6 " $= \frac{7}{3}\times 6 = 14$ Rs.

$$\text{Loss} = 15 - 14 = 1\text{ Rs.}$$

$$\text{Loss \%} = \frac{1}{15}\times 100 = \frac{20}{3}\% = \mathbf{6\frac{2}{3}\%}$$

10. (a) 1st case

$$\text{Discount} = 4000\times\frac{40}{100} = \mathbf{1600\text{ Rs.}}$$

2nd case

$$\text{1st Discount} = 4000 \times \frac{36}{100} = \mathbf{1440\,Rs.}$$

$$\text{Remaining Part} = 4000 - 1440 = 2560 \text{ Rs.}$$

$$\text{2nd discoun} = 2560 \times \frac{4}{100} = \mathbf{1022.40\ Rs.}$$

$$\text{Total discount in IInd case} = 1440 + 102.40 = 1542.40 \text{ Rs.}$$

$$\text{Difference} = 1600 - 1542.40 = \mathbf{57.60\ Rs.}$$

11. (b)

$$\text{SP of the retailer} = 48 \text{ Rs.}$$

$$\text{Profit} = 28\%$$

$$\text{CP of retailer} = 48 \times \frac{100}{128} = \frac{75}{2} \text{ Rs.}$$

$$\text{SP of wholesale dealer} = \frac{75}{2} \text{ Rs.}$$

$$\text{Profit} = 20\%$$

$$\text{CP of wholesale dealer} = \frac{75}{2} \times \frac{100}{120} = \frac{125}{4} \text{ Rs.}$$

$$\text{SP of manufacturer} = \frac{125}{4} \text{ Rs.}$$

$$\text{Gain} = 25\%$$

$$\text{His cost of production} = \frac{125}{4} \times \frac{100}{125} = \mathbf{25Rs.}$$

Shortcut :

$$\text{Cost of Production} = 48 \times \frac{100}{128} \times \frac{100}{120} \times \frac{100}{125} = \mathbf{25\ Rs.}$$

12. (a)

$$\text{CP} = 756 \times \frac{100}{120} \times \frac{100}{105} \times \frac{100}{125} = \mathbf{480\ Rs.}$$

13. (a)

$$\text{CP} = 41.25 \times \frac{100}{120} \times \frac{100}{105} \times \frac{100}{125} = \mathbf{25\ Rs.}$$

14. (d)

$$\text{CP} = 483 \times \frac{300}{350} \times \frac{100}{115} \times \frac{100}{120} = \mathbf{300\ Rs.}$$

15. (a)

$$\text{SP of 20 article} = \text{CP of 25 article}$$

$$= \text{CP of 20 art.} + \text{CP of 5 art.}$$

$$\text{Profit} = \frac{5}{20} \times 100 = 25\%$$

16. (a)

$$\text{SP of 25 article} = \text{CP of 20 article}$$

$$= \text{CP of 25 art.} - \text{CP of 5 art.}$$

$$\text{Loss \%} = \frac{5}{25} \times 100 = 20\%$$

17. (a)

$$\text{SP of 60 m} = \text{CP of 60m} + \text{CP of 15m}$$

$$\text{Profit \%} = \frac{15}{60} \times 100 = \mathbf{25\%}$$

18. (b)

$$\text{SP of 60m} = \text{CP of 60 m + SP of 15m}$$
$$\text{SP of 60m} - \text{SP of 15m} = \text{CP of 60}$$
$$\text{SP of 45m} = \text{CP of 60m}$$
$$= \text{CP of 45m + CP of 15m}$$
$$\text{Profit \%} = -\frac{15}{45}\times 100 = \frac{100}{3}\%$$

19. (c)

$$\text{\% Loss} = -10+10+\frac{-10\times 10}{100}$$
$$= -1\%$$
$$\therefore \quad \text{99\% of CP} = 1980$$
$$\text{CP} = \frac{1980}{99}\times 100$$
$$= \text{2000 Rs.}$$
$$\text{Loss} = 2000 - 1980 = \textbf{20 Rs.}$$

20. (b)

$$\text{Net loss} = \frac{-20\times 20}{100}$$
$$= -4\%$$
$$\text{96\% CP} = 7680$$
$$\text{CP} = \frac{7680}{96}\times 100$$
$$= \text{8000 Rs.}$$
$$\text{Net loss} = 8000 - 7680 = \textbf{320 Rs.}$$

21. (a)

$$\text{Let MP (mark price)} = 100$$
$$\text{Discount} = \text{10 Rs.}$$
$$\text{SP} = \text{90 Rs.}$$
$$\text{Profit} = 26\%$$
$$\text{CP} = 90\times\frac{100}{126}$$
$$= \frac{500}{7}\text{Rs.}$$
$$\text{Price marked above} = 100-\frac{500}{7} = \frac{200}{7}\text{Rs.}$$
$$\text{\% above} = \frac{\frac{200}{7}}{\frac{500}{7}}\times 100 = \textbf{40\%}$$

Shortcut :

$$\text{Net \% change} = A+B+\frac{AB}{100}$$
$$26 = -10+B+\frac{-10\times B}{100}$$

$$36 = \frac{90B}{-100}$$

$$B = \mathbf{40\%}$$

22. (d)

$$\text{Net \% Change} = 25 - 12 + \frac{25 \times -12}{100} = \mathbf{10\%}$$

$$10\% \text{ of CP} = 55$$

$$CP = \frac{55}{10} \times 100 = 550 \textbf{ Rs.}$$

23. (d)

$$\text{Net \% change} = 25 - 15 + \frac{25 \times 15}{100}$$

$$= 10 - \frac{15}{4} = \frac{25}{4}\%$$

$$\frac{25}{4}\% \text{ of CP} = 50 \text{ Rs.}$$

$$CP = \frac{50 \times 4 \times 100}{25} = \mathbf{800\,Rs.}$$

24. (b) Ist case

$$SP = 50 \text{ Rs.}$$

$$\text{Let loss} = x \text{ Rs.}$$

$$\therefore \quad \mathbf{CP} = SP + \text{loss}$$

$$= 50 + x \quad \textbf{... (1)}$$

IInd case

$$SP = 65 \text{ Rs.}$$

$$\text{Gain} = 2x \text{ (Double the former loss)}$$

$$\therefore \quad \mathbf{CP} = 65 - 2x \quad \textbf{... (2)}$$

From **(1)** & **(2)**

$$50 + x = 65 - 2x$$

$$3x = 15 \text{ Rs.}$$

$$x = 5 \text{ Rs.}$$

$$\therefore \quad CP = 65 - 10 = \mathbf{55\ Rs.}$$

25. (b) Let certain Price = x

$$SP = \frac{2x}{3}$$

$$\text{Loss} = 10\%$$

$$\therefore \quad CP = \frac{2x}{3} \times \frac{100}{90} = \frac{20x}{27}$$

Now we have to sell it at certain price i.e., at x.

$$\text{By selling at } x \text{ profit} = x - \frac{20x}{27} = \frac{7x}{27}$$

$$\text{Profit \%} = \frac{\text{Profit}}{\text{CP}} \times 100 = \frac{\frac{7x}{27}}{\frac{20x}{27}} \times 100 = \mathbf{35\%}$$

26. (a)

$$\text{SP} = 96$$

Let $\text{CP} = x$

$$\text{\% Profit} = x\%$$

$$\text{SP} = x \times \frac{100 + x}{100}$$

$$96 = \frac{100x + x^2}{100}$$

$$x^2 + 100\,x - 9600 = 0$$

$$x^2 + 160\,x - 60\,x - 9600 = 0$$

$$x\,(x + 160) - 60\,(x + 160) = 0$$

$$(x - 60)\,(x + 160) = 0$$

$$x = \mathbf{60\ Rs.}$$

NOTE : This question in objective type should be done by answer i.e., by cross check

$$60 \times \frac{160}{100} = \mathbf{96\ Rs.}$$

27. (c)

$$\text{Let CP of 1st} = x$$

$$\text{CP of 2nd} = y$$

Acc. to Question $108\,\%\,x = 90\%\text{ of } y$

$$\frac{x}{y} = \frac{90}{108} = \frac{5}{6}$$

$$\text{CP of 1st} = \frac{5}{11} \times 1210 = \mathbf{550\ Rs.}$$

$$\text{CP of 2nd} = \frac{6}{11} \times 1210 = \mathbf{660\ Rs.}$$

28. (b) In this question profit and loss are equal means 15% x = 12% y

$$\frac{x}{y} = \frac{12}{15} = \frac{4}{5}$$

$$\text{CP of 1st} = \frac{4}{9} \times 2700 = 1200\text{ Rs.}$$

$$\text{2nd} = \frac{5}{9} \times 2700 = \mathbf{1500\ Rs.}$$

29. (d)

$$\text{CP} = 800\text{ Rs.}$$

$$\text{Overall SP} = 800 \times \frac{120}{100} = 960\text{ Rs.}$$

$$\text{CP of Ist half of Art.} = \frac{800}{2} = 400\text{ Rs.}$$

$$\text{Profit} = 5\%$$

$$\therefore \quad \text{SP} = 400 \times \frac{105}{100} = \mathbf{420\ Rs.}$$

Remaining part i.e., $800 - 400 = 400$ Rs.

should be sold at $960 - 420 = 540$ Rs.

$$\therefore \quad \%\text{ profit in this half} = \frac{140}{400} \times 100 = \mathbf{35\%}$$

30. (d) Let certain price $= x$

$$\text{SP} = \frac{4x}{5}$$

$$\text{Loss} = 10\%$$

$$\text{CP} = \frac{4x}{5} \times \frac{100}{90} = \frac{8x}{9}$$

$$\text{On selling at } x \text{ profit} = x - \frac{8x}{9} = \frac{x}{9}$$

$$\therefore \quad \%\text{ profit} = \frac{\frac{x}{9}}{\frac{8x}{9}} \times 100 = \mathbf{12.5\%}$$

Section I

Test Yourself

1. Let x be the greatest number of 4 digits, which when divided by 15, 20 and 28 leaves in each case the remainder 2. The sum of digits of x is **(CLAT 2018)**
   (a) 19 (b) 21 (c) 23 (d) 25
2. A trader sells rice at a profit of 20% and uses weights which are 10% less than the correct weight. The total gain earned by him is: **(CLAT 2017)**
   (a) 35% (b) 30% (c) 33 1/3% (d) 22 2/9%
3. Suman saves 10% of her monthly salary. Now her expenditure increases by 30% and savings increases by 50%. The present increase in her salary is **(CLAT 2018)**
   (a) 25% (b) 30% (c) 32% (d) 40%
4. The marked price of an article is 20% more than the cost price. If the article is sold at a discount of 15% on its marked price, then the gain percent is **(CLAT 2018)**
   (a) 5% (b) 4.5% (c) 2.5% (d) 2%
5. a and b are inversely proportional to each other and are positive. If a increases by 100%, then b decreases by **(CLAT 2018)**
   (a) 50% (b) 75% (c) 100% (d) 200%
6. A mean spends 1/3 of his income on food, of the rest 1/4 on house rent and 1/5 on cloth. He still has ₹1,760 left with him. Find his income. **(NIFT UG 2017)**
   (a) ₹4,400 (b) ₹4,800 (c) ₹3,500 (d) ₹4,750
7. A number gets reduced to its one-third when 48 subtracted from it. What is two-third of that number? **(NIFT UG 2017)**
   (a) 22 (b) 76 (c) 36 (d) 48
8. Find the sum of all natural numbers between 100 and 175. **(NIFT UG 2017)**
   (a) 10450 (b) 10175 (c) 10170 (d) 10435
9. The sum of two numbers is 85 and their difference is 9. What is the difference of their squares? **(NIFT UG 2017)**
   (a) 765 (b) 845 (c) 565 (d) 645
10. Kiran, Kumar and kedar were watching cricket and eating groundnuts. They pop a groundnut in their mouth every 48 seconds, 72 seconds and 108 seconds respectively and munch it. If all of them pop a groundnut simultaneously at 8:20:00 hours, then at what time will they again pop again groundnuts in their mouth simultaneously? **(SNAP 2017)**
    (a) 8:22:20 hours (b) 8:27:12 hours (c) 8:30:12 hours (d) 8:25:15 hours
11. How many factors of $2^8 \times 36 \times 5^4 \times 10^5$ are multiples of 120? **(SNAP 2017)**
    (a) None of the mentioned options are correct (b) 660
    (c) 540 (d) 594

12. Two shopkeepers A and B sell machines at the same list price. A allows two successive discounts of 30% and 16% and B allows discounts of 20% and 26%. Which discount series is more advantageous to the purchaser? **(MAT Feb – 2016)**

(a) A's (b) B's (c) Both are equal (d) None of these

13. A person spends 40% of his salary on his educational expenses. He spends 60% of it in purchasing books and one half of the remaining in purchasing stationery items. If he saves ₹160 every month, which is one-fourth of the balance after spending over books and stationery items, what is his monthly salary? **(MAT Feb – 2010)**

(a) ₹8,000 (b) ₹4,800 (c) ₹9,600 (d) Data inadequate

14. An article is listed at ₹65. A customer bought this article for ₹56.16 with two successive discounts of which one is 10%. The other discount of this discount scheme that was allowed by the shopkeeper is ___________. **(MAT Feb – 2016)**

(a) 4% (b) 3% (c) 6% (d) 2.5%

15. Aman spends 1/3rd of his income on food, 1/4th of the rest on house rent and 1/5th of the rest on clothes. He still has ₹1,760 left with him. Find his income. **(RRB Non. Tec April-2016)**

(a) ₹8,800 (b) ₹6,160 (c) ₹4,400 (d) ₹2,640

16. On dividing 7865321 by a certain number, the quotient is 33612 and the remainder is 113. Find the divisor. **(RRB Non. Tec April-2016)**

(a) 243 (b) 234 (c) 423 (d) 432

17. During one year the population of a locality increased by 5% but during, the next year it decreased by 5%. If the total population at the end of the second year was 7,980, find the population at the beginning of the first year. **(RRB Non. Tec April- 2016)**

(a) 7,890 (b) 8,900 (c) 9,800 (d) 8,000

18. If the price of petrol increases by 50 percent and Rahul intends to spend only 20 percent more on petrol, by what percent should he reduce the quantity of petrol that he buys? **(ATMA 2010)**

(a) 10 (b) 15 (c) 20 (d) 25

19. An article is sold at a profit of 25%. If the selling price is doubled, the profit will be ________ **(RRB Non. Tec April- 2016)**

(a) 50% (b) 100% (c) 150% (d) 200%

20. Find the value of (1 – 1/3) (1 – 1/4) (1 – 1/5) ________ (1 – 1/100) **(RRB Non. Tec April-2016)**

(a) 1/50 (b) 2/5 (c) 1/20 (d) 1/60

21. If 50% of (x – y) = 30% of (x + y), then what per cent of x is y? **(RRB Non. Tec April-2016)**

(a) 25% (b) 30% (c) 35% (d) 40%

22. In a village of there are 4080 registered voters. 60% of the registered voters could cast their votes and none of the votes were invalid. If only two candidates were contesting the election and the respective ratio of the votes received by them is 12 : 5, what is the number of votes received the winning candidate? **(SBI Clerk– 2016)**

(a) 2304 (b) 1824 (c) 1584 (d) 1668

(e) 1728

23. 168 ÷? – 9 = 3 **(SBI Clerk – 2016)**

(a) 24 (b) 28 (c) 21 (d) 12

(e) 14

24. (204.75 + 275.25) ÷ 6 = ? **(SBI Clerk – 2016)**

(a) 60 (b) 120 (c) 80 (d) 0.25
(e) 90

25. 40% of 240 + ? = 120 **(MAT Feb – 2010)**

(a) 24 (b) 28 (c) 32 (d) 40
(e) 49

26. 11/5 –12/5 + 21/10 = ? **(SBI Clerk – 2016)**

(a) 19/10 (b) 7/10 (c) 8/10 (d) 5/10

27. $\left(\frac{1}{64}\right)^{-2/3} \times \frac{1}{16} \times 4 = ?$ **(SBI Clerk – 2016)**

(a) 16 (b) 8 (c) 6 (d) 4
(e) 1

28. $234 - \sqrt{?} - 23 = 14^2$ **(SBI Clerk – 2016)**

(a) 64 (b) 256 (c) 225 (d) 25
(e) 36

29. Which of the following fractions is less than $\frac{7}{8}$ and greater than $\frac{1}{3}$? **(NIFT – 2016)**

(a) $\frac{1}{4}$ (b) $\frac{23}{24}$ (c) $\frac{11}{12}$ (d) $\frac{17}{24}$

30. The sum of successive odd numbers from 1 to 15 is________. **(NIFT – 2016)**

(a) 81 (b) 64 (c) 49 (d) 36

31. A number when divided by 6 leaves a remainder 3. When the square of the number is divided by 6, the remainder is________. **(NCHMCT – 2016)**

(a) 0 (b) 1 (c) 2 (d) 3

32. Find the sum of all even natural numbers less than 75. **(NCHMCT – 2016)**

(a) 1406 (b) 1400 (c) 1369 (d) 16324

33. Present average age of A, B and C is 22 years. Three years ago, Average age of B and C is 18 years, then find A's age 9 years hence? **(SBI Preliminary 2018)**

(a) 24 years (b) 27 years (c) 30 years (d) 33 years
(e) 36 years

34. In place of 18% profit an article is sold at 42% profit and seller gets Rs.110.40 more. Find the selling price of article if it were sold at 25% profit? **(SBI Preliminary 2018)**

(a) Rs. 440 (b) Rs. 460 (c) Rs. 575 (d) Rs. 525

35. In a village three people contested for the post of village Pradhan. Due to their own interest, all the voters voted and no one vote was invalid. The losing candidates got 30% votes. What could be the minimum absolute margin of votes by which the winning candidates led by the nearest rival, if each candidate got an integral per cent of votes? **(SBI Clerk – 2016)**

(a) 4 (b) 2 (c) 1 (d) 3
(e) None of these

36. The price of an article is first increased by 20% and later on the price were decreased by 25% due to reduction in sales. Find the net percentage change in final price of Article.
(SBI Clerk – 2016)

(a) 20% (b) 18% (c) 38% (d) 10%
(e) None of these

37. K purchased a table for 11,000 and sold it for 13,500. What is his gain in percentage? **(RRB Non. Tec March – 2016)**

(a) 19.8% (b) 20.6% (c) 22.7% (d) 22%

38. The 1$ = ₹67.89. Money changer adds a margin of ₹1.11. What will be the cost of $ 150?? **(RRB Non. Tec March – 2016)**

(a) ₹10,183 (b) ₹10,350 (c) ₹10,330 (d) ₹10,450

39. A farmer purchased a piece of land for ₹18 lakh and spent ₹3 lakh for registration, fencing etc. He sold it for ₹24.57 lakh. Find his profit in percentage? **(RRB Non. Tec March – 2016)**

(a) 15% (b) 14.15% (c) 16.5% (d) 17%

40. Due to 20% hike in bus fare, a 10% fall in the passenger traffic was recorded. Yet the daily collection of a bus depot increased by ₹150. What was the daily collection of the bus depot after the fare hike? **(CMAT – 2016)**

(a) ₹2,100 (b) ₹2,250 (c) ₹2,025 (d) ₹1,875

41. Suezan offered a discount of 37.50% on MRP, and still makes a profit of 25%. Had she offered a discount of 40%, what would have been her profit or loss percentage? **(CMAT – 2016)**

(a) 13.33% (b) 20% (c) 37.50% (d) – 14.28%
(e) None of these

42. A car was sold at a loss of 18%. There would have been a profit of 15%, if the car would have been sold for ₹99,000 more. At what price the car should be sold to make a profit of 10%? **(CMAT – 2016)**

(a) ₹2,50,000 (b) ₹3,30,000 (c) ₹4,50,000 (d) ₹3,00,000

43. If p is 81% of q, but percentage of p is q ? **(CMAT – 2016)**

(a) 81% (b) 101.9% (c) 112.3% (d) 123.45%

44. $\frac{5}{9} \times (225.40 - 45.4) = ?^2$ **[SBI Clerk (2016)]**

(a) 15 (b) 5 (c) 100 (d) 25
(e) 10

45. $80 \div 24 \times ? = 60$ **[SBI Clerk (2016)]**

(a) 16 (b) 12 (c) 18 (d) 14
(e) 24

46. $\sqrt{784} \times 3 - ? = 58$ **[SBI Clerk (2016)]**

(a) 26 (b) 24 (c) 22 (d) 28
(e) 25

47. $\sqrt{\frac{160 \times 8}{5}} \times 8 = ?$ **[SBI Clerk (2016)]**

(a) 112 (b) 164 (c) 128 (d) 116
(e) 136

48. A number is mistakenly divided by 5 instead of being multiplied by 5. Find the percentage change in the result due to this mistake. **[SET Law (2016)]**

(a) 96% (b) 95% (c) 2400 (d) 200%

49. If 65% of x = 13% of y, then find the value of x if y = 2000 **[SET Law (2016)]**

(a) 200 (b) 300 (c) 400 (d) 500

50. A person spends 40% of his salary on his educational expenses. He spends 60% of it in purchasing books and one half of the remaining in purchasing stationary items. If he saves ₹160 every month, which is one-fourth of the balance after spending over books and stationery items, what is his monthly salary? **[MAT MAY 2016]**

(a) ₹4,800 (b) ₹8,000 (c) ₹9,600 (d) Data inadequate

51. Sudhir's uncle gave him ₹100 on his first birthday. On each birthday after that he doubled his previous gift. By the day after Sudhir's eighth birthday, what was the total amount that his uncle had given him? **[MAT MAY 2016]**

(a) ₹25,400 (b) ₹25,500 (c) ₹25,450 (d) ₹25,600

52. Find the smallest 4 digit number which is a perfect square. **[RRB Non. Tec April – 2016)]**

(a) 1000 (b) 1024 (c) 1081 (d) 1064

53. Ron gave 3/4th of 20% of her monthly salary to Simi. Out of the amount Simi received from Ron she gave 40% towards tuition fees and the remaining ₹2,700 Simi kept with herself. What was Ron's monthly salary? **[IBPS Clerical (2015)]**

(a) ₹30,000 (b) ₹35,250 (c) ₹35,000 (d) ₹27,000
(e) ₹24,000

54. When 2 is added to both the numerator and denominator of a possible fraction, the fraction thus obtained is 3/5. If the denominator is 6 more than the numerator, what is the original value of the denominator? **[IBPS Clerical (2015)]**

(a) 8 (b) 13 (c) 23 (d) 3
(e) 18

55. The square of a number x is 80 more than the square of another number, y. If the value of x is 1.5 times the value of y, what is the difference between the value of x and that of y?

[IBPS Clerical (2015)]

(a) 4 (b) 6 (c) 2 (d) 9
(e) 5

56. The sum of digits of a two-digit number is 12 and the difference between-the two-digits of the two-digit number is 6. What is the two-digit number? **[IBPS RRB Office Asst. Exam (2015)]**

(a) 39 (b) 84
(c) 93 (d) 75
(e) other than those given as options

57. In an examination, 'A' scored 25% of the total marks but failed by 56 marks and 'B' scored 50% of the total marks which was 144 marks more than the minimum passing marks. What was the minimum passing marks for the examination? **[RBI Assistant (2015)]**

(a) 256 (b) 196 (c) 284 (d) 180
(e) None of these

58. Abhay gave 30% of his money to Vijay. Vijay gave 2/3rd of what he received to his mother. Vijay's mother gave 5/8th of the money she received from Vijay, to the grocer. Vijay's mother is now left with ₹600. How much money did Abhay have initially?

[SBI PO Prelim Exam 2015)]

(a) ₹6,200/- (b) ₹8,000/- (c) ₹6,000/- (d) ₹8,200/-
(e) ₹10,200/-

59. A person spends 1/3 part of his income on food, ¼ on house rent and remaining ₹630 on other items. The house rent is: **[CLAT 2014]**

(a) ₹504 (b) ₹1512 (c) ₹378 (d) None of these

60. A farmer has some hens and some goats. If the total number of animal heads is 80 and the total number of animal feet is 200, what is the total number of goats? **[CLAT 2014]**

(a) 40 (b) 60

(c) 20 (d) Cannot be determined

61. A dealer marked his goods 20% above the cost price and allows a discount of 10%. Then the gain percent is: **[CLAT 2014]**

(a) 2% (b) 4% (c) 6% (d) 8%

62. Value of "A" in the expression, $5 + 12 \times 10 \div \frac{120}{240} = A \times 10$, is: **[CLAT 2014]**

(a) 11 (b) 24.5 (c) 34 (d) 6.5

63. The Least Common Multiple (L.C.M.) of 0.12, 9.60 and 0.60 is: **[CLAT 2014]**

(a) 9.60 (b) 0.12 (c) 0.6 (d) None of these

64. A man earns ₹20 on the first day and spends ₹15 on the next day. He again earns ₹20 on the third day and spends ₹15 on the fourth day. If he continues to save in this way, how soon will he have ₹60 in hand? **(CLAT 2016)**

(a) on 17th day (b) on 24th day (c) on 27th day (d) on 12th day

65. The number of 'three digit numbers' which are multiples of 9 are: **(CLAT 2016)**

(a) 99 (b) 100 (c) 101 (d) 98

66. A part of monthly hostel charges is fixed and the remaining depends on the number of days one has taken food in the mess. When a student A takes food for 20 days, she has to pay ₹1000 as hostel charges whereas a student B, who takes food for 26 days, pays ₹1180 as hostel charges. Find the fixed charges and the cost of food per day. **(CLAT 2016)**

(a) 400, 30 (b) 200, 20 (c) 300, 30 (d) 400, 40

67. The value of a machine depreciates every year at the rate of 10% on its value at the beginning of that year. If the present value of the machine is ₹729, its worth three years ago was: **(CLAT 2016)**

(a) ₹1,000 (b) ₹947.70 (c) ₹800 (d) ₹750.87

68. The traffic lights at three different signal points change after every 45 seconds, 75 seconds and 90 seconds respectively. If all change simultaneously at 7 : 20 : 15 hours, then they will change again simultaneously at **(CLAT 2016)**

(a) 7:28:00 hours (b) 7:27:30 hours (c) 7:27:50 hours (d) 7:27:45 hours

69. Raju bought 250 shirts at ₹150 each. He then spent ₹5000 on the packaging of the shirts. He marked the price of each shirts at ₹255 but while selling them he gave discount of 20% on each shirt. What is the percent profit earned by Raju, if he could sell all his shirts?

(IBPS Bank of Maharashtra 2016)

(a) 12% (b) 16% (c) 18% (d) 20%

70. The expression $2^{6n} - 4^{2n}$, where n is a natural number is always divisible by **(SSC CHSL 2016)**

(a) 15 (b) 18 (c) 36 (d) 48

71. In a shop, shirts are usually sold at 40% above the cost price. During a sale, the shopkeeper offers a discount of 10% off the usual selling price. If he manages to sell 72 shirts for ₹13,608, then his cost price per shirt, in ₹is **(SSC CHSL 2016)**

(a) 210 (b) 150 (c) 149 (d) 125

72. Shelf A has 4/5 of the number of books that shelf B has. If 25% of the books in A are transferred to B and then 25% of the books from B are transferred to A, then the percentage of the total number of books that A will have is **(SSC CHSL 2016)**

(a) 25% (b) 50% (c) 75% (d) 100%

73. A man has a certain number of small boxes to pack into parcels. If he packs 3, 4, 5 or 6 in a parcel, he is left with one over; if he packs 7 in a parcel, none is left over. What is the number of boxes he may have to pack? **(MAT Sept. 2016)**

(a) 301 (b) 106 (c) 309 (d) 400

74. Ms. Sharma's expenditure and savings are in the ratio of 3 : 2. Her income increases by 10%. Her expenditure also increases by 12%. How much percent does her savings increase? **(MAT Sept. 2016)**

(a) 13% (b) 6% (c) 7% (d) 11%

75. The sum of two numbers is 24 and their product is 143. The sum of their squares is **[NICL AO (2013)]**

(a) 296 (b) 295 (c) 290 (d) 298

76. The sum of two numbers is 45 and there product is 500. The G.C.F. number is **[NICL AO (2013)]**

(a) 5 (b) 9 (c) 10 (d) 15

77. The difference between a discount of 40% on ₹500 and two successive discounts of 36%, 4% on the same amount is **[NICL AO (2013)]**

(a) ₹0 (b) ₹2 (c) ₹1.93 (d) ₹7.20

78. 2/3 is a rational number whereas $\sqrt{2}/\sqrt{3}$ is **[CLAT (2013)]**

(a) Also a rational number (b) An irrational number

(c) Not a number (d) A natural periodic number

79. Greatest number which divides 926 and 2313, leaving 2 and 3 remainders, respectively, is **[CLAT (2013)]**

(a) 52 (b) 62 (c) 452 (d) 462

80. A single discount equivalent to a discount series 15% and 5% is **[CLAT (2013)]**

(a) 32% (b) 20% (c) 10% (d) 8.5%

SOLUTIONS

1. (c) LCM of 15,20 and 28 is 420. When greatest 4 digit number is divided by 420 the remainder is 339. Hence the required number is 9999 – 339 + 2 = 9662. So required sum of digits is 9 + 6 + 6 + 2 = 23.

2. (c) For 1000 gm he gets a price of 1200 gm while his costing is of 900 gm. hence

his profit % = $\frac{300}{900} \times 100$ = 33 1/3%

3. (c) Let salary = 100 Rs. Saving = 10

Expenditure = 100 – 10 = 90

New saving = 10 × 1.5 = 15

New expenditure = 1.3 × 90 = 117

Hence new salary = 117 + 15 = 132

Hence % increase = 32%

4. (d) Let CP = 100

So Marked price = 120

Selling price = 120 × 85 = 102

Hence profit% = 2%

5. (a) a = 1/b

ab = 1

so when a becomes 2a than b will become b/2 hence a decrease of 50%

6. (a) $\frac{2}{3}\times\frac{3}{4}\times\frac{4}{5}\times x = 1760$

$\Rightarrow x = 4400$

7. (d) $x - 48 = 1/3\ \times$

$\Rightarrow 2/3\ \text{x} = 48$

8. (b) Required series is 101 + 102 + 103 + 104 175

So sum is n/2 (a + l) = 74/2 (101 + 174) = 37 × 275 = 10175

9. (a) x + y = 85 and x – y = 9

Solving we get x = 47 and y = 38

So required difference = $47^2 - 38^2 = 2209 - 1444 = 765$

10. (b) LCM of 48, 72 and 108 is 432

∴ Required time = 8:20:00 + 432 second

= 8:20:00 + 7 min. 12 sec.

= 8:27:12

11. (a) $120 = 2^3 \times 3^1 \times 5^1$

$\therefore 2^8 \times 36 \times 5^4 \times 10^5$

$= 2^8 \times 2^2 \times 3^2 \times 5^4 \times 2^5 \times 5^5$

$= 2^{15} \times 3^2 \times 5^9$

$= 2^3 \times 3^1 \times 5^1\ (2^{12} \times 3^1 \times 5^8)$

∴ No of multiplies of 120 = (12 + 1) (1 + 1) (8 + 1) = 13 × 2 × 9 = 234

12. (a) A's net discount $= -30 - 16 + \frac{-30\times -16}{100}$

$= -46 + 4.8 = -41.2$

B's discount $= -20 - 26 + \frac{-20\times +26}{100}$

$= -46 + 5.2 = -40.8$

$= -40.8$

So for customer *A* is more advantages

13. (a) $(40\%\,x - 24\%x - 8\%\,x)\times\frac{1}{4} = 160$

$2\%\,x = 160$

$\Rightarrow\quad x = 8000$

14. (a) Discount% = $\frac{8.84}{65} \times 100 = 13.6\%$

$\therefore \quad -13.6 = -10 - B + \frac{-10 \times -B}{100}$

Solving we get $B = 4\%$

15. (c) $x \times \frac{2}{3} \times \frac{3}{4} \times \frac{4}{5} = 1760$

$\frac{2x}{5} = 1760$

$x = \frac{1760 \times 5}{2} = ₹4400$

16. (b) $7865321 = d \times 33612 + 113$

$\therefore \quad d = \frac{7865321 - 113}{33612} = 234$

17. (d) $P \times 1.05 \times .95 = 7980$

$P = \frac{7980}{1.05 \times .95} = 8000$

18. (c) $20 = 50 + B + \frac{50 \times B}{100}$

$-30 = \frac{150\,B}{100}$

$\Rightarrow \quad B = -20$ i.e., a decrease of 20%.

19. (c) Let C.P. = 100

S.P. = 125

New S.P. = 250

Profit % = $\frac{150}{100} \times 100 = 150\%$

20. (a) $\left(1 - \frac{1}{3}\right)\left(1 - \frac{1}{4}\right)\left(1 - \frac{1}{5}\right) \ldots\ldots \left(1 - \frac{1}{100}\right)$

$= \frac{2}{3} \times \frac{3}{4} \times \frac{4}{5} \ldots\ldots \frac{99}{100} = \frac{2}{100} = \frac{1}{50}$

21. (a) $\frac{x+y}{x-y} = \frac{5}{3}$

Applying compound and dividend do we get

$\frac{x}{y} = \frac{5+3}{5-3} = \frac{8}{2}$

$= \frac{4}{1}$

$\therefore \quad$ Required % = $\frac{1}{4} \times 100 = 25\%$

22. (e) Votes of winning candidate = $4080 \times .6 \times \frac{12}{17} = 1728$

23. (e) $168 \div x - 9 = 3$

$$\frac{168}{x} = 12$$
$$x = 14$$

24. (c) $480 \div 6 = 80$

25. (a) 40% of $240 + 2 = 120$

$$96 + x = 120$$
$$x = 120 - 96 = 24$$

26. (a) $\frac{11}{5} - \frac{12}{5} + \frac{21}{10} = \frac{22 - 24 + 21}{10} = \frac{19}{10}$

27. (d) $\left(\frac{1}{4^3}\right)^{\frac{-2}{3}} \times \frac{1}{16} \times 4$

$$= 4^{3 \times \frac{2}{3}} \times \frac{1}{16} \times 4$$

$$= 16 \times \frac{1}{16} \times 4 = 4$$

28. (c) $234 - \sqrt{x} - 23 = 196$

$$\sqrt{x} = 211 - 196$$
$$\sqrt{x} = 15$$
$$\Rightarrow \quad x = 225$$

29. (d) $\frac{7}{8} = .875$

$$\frac{1}{3} = .333 \text{ and } \frac{7}{24} = .708$$

30. (b) There will be in 8 odd number

Sum $= n^2 = 8^2 = 64$

31. (d) Remainder $= \frac{3^2}{6} = 3$

32. (a) There will be 37 even natural no less than 75

Sum $= 37 \times 38 = 1406$

33. (c) Total present age of A, B and C = 22 × 3 = 66 years

Present average age of B and C = 18 + 3 = 21 years

Total present age of B and C = 21 × 2 = 42 years

So present age of A = 66 – 42 = 24 years

A's age 9 years hence = 24 + 9 = 33 years

34. (c) Let CP of article = x

According to question

42% × –18% × = 110.40

24% × = 110.40

x = 460

Hence required selling price = 4601.25 = 575 Rs.

35. (a) Margin = 70% – 30%

= 40%x = .4x

So minimum value of margin = .4 × 10 = 4 votes

36. (d) Net% change

$$= +20 - 25 + \frac{20 \times -25}{100}$$

$$= -5 - 5 = 10\%$$

i.e., 10% change

37. (c) $\% \text{ profit} = \frac{2500}{11000} \times 100 = 22.77\%$

38. (b) Cost of $ (150) = 150 (67.89 + 1.11)

= 150 × 69 = ₹10350

39. (d) Total cost = 18 + 3 = 21 lakh

Total S.P. = 24.57

$\therefore \quad \text{Profit\%} = \frac{3.57}{24.57} \times 100 = 17\%.$

40. (d) $\text{Net\% charge} = 20 - 10 + \frac{20 \times -10}{100}$

= 10 – 2% = 8%.

$\therefore \quad 8\%\, x = 150$

$\Rightarrow \quad x = \frac{150}{8} \times 100 = ₹1875$

41. (b) Let CP = ₹100

∴ S.P. = ₹125

Now, $\text{MP} - \frac{3}{8}\text{MP} = 125$

$\frac{5}{8}\text{MP} = 125$

⇒ MP = ₹200

$\text{Discount} = 200 \times \frac{40}{100} = ₹80$

New SP = 200 – 80 = ₹120.

∴ New Profit = 120 – 100 = ₹20.

$\% \text{ profit} = \frac{20}{100} \times 100 = 20\%$

42. (b) 33% of CP = 99000

⇒ CP = 300000

∴ Required SP = 1.1 × 300000 = ₹3,33,000

43. (d) $p = 81\% \, q$

$q = \frac{100}{81}, \; q = 123.45\%$

44. (e) $\frac{5}{9} \times (225.40 - 45.4)$

$= \frac{5}{9} \times 180$

$\therefore \; x^2 = 100$

$\Rightarrow \; x = 10$

45. (c) $80 \div 24 \times x = 60$

$\frac{80}{24} \times x = 60$

$\Rightarrow \; x = \frac{60 \times 24}{80} = 18$

46. (a) $\sqrt{784} \times 3 - x = 58$

$28 \times 3 - x = 58$

$\Rightarrow x = 84 - 58 = 26$

47. (c) $\sqrt{\frac{160 \times 8}{5}} \times 8 = x$

$= \sqrt{256} \times 8 = x$

$\Rightarrow \; x = 128$

48. (a) Wrong result $= \frac{x}{5}$

Correct result $= 5x$

$\therefore \; \% \text{ Change} = \frac{5x - \frac{x}{5}}{5x} \times 100$

$= \frac{24x}{25x} \times 100 = 96\%$

49. (c) $65\% x = 13\% y$

$\therefore \; \frac{x}{y} = \frac{13\%}{65\%} = \frac{1}{5}$

$\therefore \; x : y = 1 : 5$

So if $x = k$

then $y = 5k$

According to question,

$5k = 2000$

$\therefore \; k = \frac{2000}{5} = 400 = x$

50. (b) Let Monthly Salary $= x$

Saving $= \frac{1}{4} (40\% x - 24\% x - 8\% x)$

$= 2\% x$

$\therefore \quad 2\%x = 160$

$\Rightarrow \quad x = ₹8000$

51. (b) It will be a GP with first term as 100 and common ratio as 2

$$\therefore \text{ Required Sum} = \frac{100 \times [2^8 - 1]}{2-1}$$

$$= 100 \times [256 - 1]$$

$$= 100 \times 255 = 25500$$

52. (b) It is $32^2 = 1024$

53. (a) Let Ron's salary = x

According to question

$$.6 \text{ of } 15\%\, x = 2700$$

$$\Rightarrow \quad x = 30000 \text{ Rs.}$$

54. (b) $\frac{x+2}{y+2} = \frac{3}{5} \Rightarrow \quad 5x - 3y = -4$

Also $\quad y - x = 6$

Solving, $x = 7, y = 13$

55. (a)

$$x = 1.5y$$
$$x^2 - y^2 = 80$$
$$(1.5y)^2 - y^2 = 80$$
$$2.25y^2 - y^2 = 80$$
$$1.25y^2 = 80$$
$$y^2 = \frac{80}{1.25}$$
$$\Rightarrow \quad y^2 = 64 \Rightarrow y = 8$$
$$x = 1.5 \times 8 = 12$$
$$x - y = 12 - 8 = 4$$

56. (e) Let the two digits be a and b.

$$a + b = 12 \quad \text{...(i)}$$

$$a - b = 6 \quad \text{...(ii)}$$

Adding and subtracting the two equations we get a = 9 and b = 3.

The number can be 93 as well as 39. So, can not be determined.

57. (a) Let the total marks are 'x'.

Hence 25% of x + 56 = 50% of x – 144

$$\Rightarrow \quad 25\% \text{ of } x = 200$$

$$\Rightarrow \quad x = \frac{200}{0.25} = 800.$$

Hence minimum passing marks = 25% of x + 56 = 25% of 800 + 56 = 256.

58. (b) Working backwards since Vijay's mother was left with ₹600 which must be 3/8th of what she received from Vijay as 5/8th of what she received was given to grocer, so total = 600 × 8/3

= 1600. Now it is 2/3rd of what Vijay received from Abhay so total = 1600 × 3/2 = 2400 which is 30% of total sum with Abhay so total will be 2400 × 100/30 = 8000.

59. (c) Let the total income be ₹ x

$$\Rightarrow \quad x - \left(\frac{x}{3} + \frac{x}{4}\right) = 630 \Rightarrow x = 1512$$

House rent = $\frac{1512}{4} = 378$

60. (c) Let the total number of goats be x and that of hens be y.

$\therefore \quad x + y = 80$ and $4x + 2y = 200$

Solving the above equation, we get $x = 20$.

61. (d) Let the CP of the goods be ₹100.

SP = 100 × 1.2 × 0.9 = 108.

Percentage profit = 8.

62. (b) $5 + 12 \times 10 \div \frac{120}{240} = A \times 10$

$$\Rightarrow \quad 5 + 12 \times 10 \times \frac{2}{1} \times = A \times 10$$

$$\Rightarrow \quad A = 24.5$$

63. (a) LCM of 0.12, 9.60 and 0.60

$$= \text{LCM of } \frac{12}{100}, \frac{960}{100} \text{ and } \frac{60}{100}$$

$$= \frac{\text{LCM of (12, 960 and 60)}}{\text{HCF of (100, 100 and 100)}} = \frac{960}{100} = 9.6$$

64. (a) A man saves ₹20 on the first day and spends ₹15 on the next day.

It means he saves ₹5 in 2 days.

So to save ₹40 he requires 16 days. Next ₹20 is saved in 1 day.

So total 17 days are required to earn ₹60.

65. (b) 1st 3 digit multiple of 9 is 108 and last 3 digit multiple is 999.

i.e., 9 × 12 to 9 × 111

$\therefore$ Number of multiples = 111 – 12 + 1 = 100

66. (a) Let the fixed charge per day be ₹X and variable charge per day be ₹ Y

For student A total charge, ₹1000 = $20Y + X$

For student B total charge, ₹1180 = $26Y + X$

after solving above equations we get $X = 400$ and $Y = 30$

67. (a) .9 × .9 × .9 of $x = 729$

$$\Rightarrow \quad x = 1000$$

68. (d) The traffic lights at three different signal points change after every 45 seconds, 75 seconds and 90 seconds respectively.

So their LCM = 450 seconds

Then they will change again simultaneously at 7 : 27 : 45 hours by adding 450 seconds in 7 : 20 : 15 hours

69. (d) Total cost of 250 shirts $= 250 \times 150 + 5000$

$= 37500 + 5000 = 42500$

Total selling price of 250 shirts $= 255 \times .8 \times 250 = 51000$

So profit % $= \dfrac{51000 - 42500}{42500} \times 100 = 20\%$

70. (d) Since 'n' is a natural number, so put $n = 1$ in $2^{6n} - 4^{2n}$, we get

$$2^{6n} - 4^{2n} = 2^6 - 4^2 = 64 - 16 = 48.$$

As it is divisible by 48, so the given expression is always divisible by 48.

71. (b) Let the cost price of 72 shirts is ₹ 'x'. We have,

$x \times 1.4 \times 0.9 = 13608 \Rightarrow x =$ ₹10800

Hence the cost price of each shirt $= \dfrac{10800}{72} =$ ₹150.

72. (b) Let the books on shelf B are 100 and on shelf A are 80 so that the total number of books is 180. When 25% of the books on shelf A are transferred to shelf B, then shelf B has 120 books and shelf A has 60 books. Again when 25% books of shelf B are transferred to shelf A, then the number of books on shelf A are $60 + 30 = 90$ which is 50% of the total books.

73. (a) Required number is 60k + 1 and should be divisible by 7.

$\Rightarrow$ 301 in the answer.

74. (c) Let the expenditure and saving of Ms. Sharma are ₹30 and ₹20 respectively. Hence her income = ₹50. As income increased by 10%, so the new income = ₹55. Her new expenditure $= 30 \times 1.12$ = ₹33.6

New saving $= 55 - 33.6 =$ ₹21.4.

Hence percentage increase in saving $= \dfrac{21.4 - 20}{20} \times 100 = \dfrac{1.4}{20} \times 100 = 7\%$

75. (c) $x^2 + y^2 = (x + y)^2 - 2xy \Rightarrow 24^2 - 2 \times 143 = 576 - 286 = 290$

76. (a) Let the numbers be x and y

Therefore, $(x - y)^2 = (x + y)^2 - 4xy \Rightarrow (x - y)^2 = 452 - 4 \times 500 = 25$

So $x - y = 5$ and $x + y = 45$

Solving the above equations, we get $x = 25$ and $y = 20$

GCF of 25 and 20 is 5 as 5 is the only highest common factor among these two numbers

77. (d) Difference $= \left(500 \times \dfrac{60}{100}\right) - \left(500 \times \dfrac{64}{100} \times \dfrac{96}{100}\right) = 7.20/-$

78. (b) $\dfrac{\sqrt{2}}{\sqrt{3}} = \sqrt{\dfrac{2}{3}} = \sqrt{0.666} ---$ = Irrational No.

79. (d) Required No. = GCD of (926 – 2, 2313 – 3)

= GCD of (924, 2310) = 462

80. (b) A single discount equivalent to a series of 15% and 5% is = 19.25% = 20%

SECTION-II

ARITHMETIC-II

Day 9

Time and Distance

TIME AND DISTANCE

A bus starts from **Delhi** at **8 am** and reaches **Chandigarh** at **1 pm** on same day. **Distance** between **Delhi** and **Chandigarh** is **250 Km**. So from **8 am to 1 pm** i.e. in **5 hours** the bus has covered **250 Km.**

So in 5 hours bus has covered distance = 250 Km.

1" " " " $= \frac{250}{5}$

= **50 km**

So bus covers on an average 50 km in one hour. This is known as **speed**. Speed is distance travelled by object per unit time

and $$\boxed{\text{Speed} = \frac{\text{Distance}}{\text{Time}}}$$

$$\boxed{S = \frac{D}{t}}$$

$$\boxed{t = \frac{D}{S}}$$

$$\boxed{D = S \times t}$$

where S is speed, D is distance and t is time taken.

Generally if the **distance** is in **Km**, we measure **time** in **hours** and **speed** in **Km per hour** and written as **km/h** and if **distance** is in **meter** then **time** is taken in **sec** and **speed** in **meter per second** and written as **m/sec.**

To Convert km/h to m/s and Vice Versa

Example. Convert 18 km/h to m/sec.

Sol. We know that 1 km = 1000 m

And 1 hour = 3600 sec

$\therefore$ $$18\frac{\text{km}}{\text{hour}} = 18 \times \frac{1000\text{ meter}}{3600\text{ sec}}$$

$$= 18 \times \frac{5}{18}\frac{m}{sec} = 5 \text{ m/sec}$$

So to convertkm/h to m/sec, we have to multiply by $\frac{1000}{3600}$ which in lowest form equals to $\frac{5}{18}$. So instead multiplying by $\frac{1000}{3600}$, we directly multiply by $\frac{5}{18}$. This way calculation becomes fast and can be done orally.

So **remember : To convert km/h to m/sec multiply by $\frac{5}{18}$.**

To Convert m/sec to km/h

Example. Convert 10 m/sec to km/h.

Sol : We know that $1 \text{ km} = 1000 \text{ m}$

Or $1000 \text{ m} = 1 \text{ km}$

$$1 \text{ m} = \frac{1}{1000} \text{ km}$$

And $1 \text{ hour} = 3600 \text{ sec}$

Or $3600 \text{ sec} = 1 \text{ hour}$

$$1 \text{ sec} = \frac{1}{3600} \text{ km}$$

$$\therefore \quad 10\frac{m}{sec} = \frac{10 \times \frac{1}{1000} \text{ km}}{\frac{1}{3600} \text{ hour}}$$

$$= 10 \times \frac{3600}{1000}\frac{km}{hour}$$

$$= 10 \times \frac{18}{5}\frac{km}{hour}$$

$$= 36\frac{km}{hour}$$

So to convert m/sec to km/h multiply by $\mathbf{\frac{18}{5}}$

So **remember:**

1. Speed $= \frac{\text{distance}}{\text{Time}}$

$$\mathbf{S} = \boxed{\frac{\mathbf{D}}{\mathbf{t}} \quad ...(1)}$$

2. Time $= \frac{\text{distance}}{\text{speed}}$

$$T = \boxed{\frac{D}{S} \quad ...(2)}$$

3. Distance = Speed × time

$$D = \boxed{S \times t} \quad ...(3)$$

4. **To convert speed in $\frac{\textbf{km}}{\textbf{hour}}$ to $\frac{\textbf{meter}}{\textbf{sec}}$ multiply by $\frac{\textbf{5}}{\textbf{18}}$**

5. **To convert speed in $\frac{\textbf{meter}}{\textbf{sec}}$ to $\frac{\textbf{km}}{\textbf{hour}}$ multiply by $\frac{\textbf{18}}{\textbf{5}}$**

Now based on above concepts we will solve some simple examples.

Example 1. A man covers 60 km in 5 hours. Find his speed in km/hour?

Sol :

$$S = \frac{D}{t}$$

$$S = \frac{60}{5} = \textbf{12 km/h}$$

Example 2. A bus covers a distance of 360 km in 5 hours. Find its speed in m/sec

Sol :

$$S = \frac{D}{t}$$

$$= \frac{360}{5} = 72 \text{ km/h}$$

$$72 \text{ km/h} = 72 \times \frac{5}{18} = \textbf{20 m/sec}$$

Example 3. A man reaches his destination 1 km far in 20 minutes . Find his speed in km/h.

Sol :

$$S = \frac{D}{t}$$

$$= \frac{1 \text{ Km}}{\frac{20 \text{ h}}{60}} = \textbf{3 km/h}$$

Example 4. Distance travelled by a person in 20 seconds at the rate of 300m/sec is ....km.

Sol :

$$\text{Distance} = \text{Speed} \times \text{time}$$

$$= 300 \times 20 = 6000 \text{ metre} = \textbf{6 km}$$

Example 5. A man covers 10 km in 2 hours, the distance covered by him in $7\frac{1}{2}$ hours is ....

Sol :

$$S = \frac{D}{t}$$

$$= \frac{10}{2} = 5 \text{ km/h}$$

$$\text{Now we have speed} = 5 \text{ km/h}$$

$$T = \frac{15}{2} \text{ hour}$$

$$\text{Distance} = \text{Speed} \times \text{time}$$

$$= 5 \times \frac{15}{2} = \mathbf{37.5 \text{ km}}$$

TYPES OF QUESTIONS

In addition to above discussed simple question we have divided this topic into following types and will go into details one by one.

Type I : Questions related to average speed

Type II : To find distance when speed is changed.

Type III : Question on usual time & distance

Type IV : Questions related to trains

Type V : Questions related to streams

Type VI : Misc.

Type I

Remember : $\text{Average Speed} = \dfrac{\text{Total Distance}}{\text{Total Time}}$

Example 6. A car covers first half of its journey at the rate of x km/h and second half at the rate of y km/h. Find its average speed over whole journey.

Sol : Let total distance $= 2k$

$$\text{Distance covered in 1st half} = \frac{2k}{2} = k$$

$$\text{Distance covered in 2nd half} = \frac{2k}{2} = k$$

$$\text{Time taken to complete 1st half } t_1 = \frac{k}{x}$$

$$\text{Time taken to complete 2nd half } t_2 = \frac{k}{y}$$

$$\text{Average speed} = \frac{\text{Total Distance}}{\text{Total Time}}$$

$$= \frac{k+k}{t_1+t_2} = \frac{2k}{\frac{k}{x}+\frac{k}{y}} = \frac{2k}{\frac{yk+kx}{xy}} = \frac{2k}{\frac{2(y+x)}{xy}}$$

$$= \frac{2xy}{x+y}$$

$$\therefore \quad \textbf{Average speed} = \frac{\mathbf{2xy}}{\mathbf{x+y}}$$

The result of above example can be **remembered** as a formula to find the **average speed** when **speeds** for **two halves** of journey are given.

Example 7. A cyclist completes first half of his journey at the rate of 5 km/h and second half at the rate of 3 km/h. Find his average speed of whole journey.

Sol :

$$\text{Average speed} = \frac{2xy}{x+y}$$

$$= \frac{2 \times 5 \times 3}{5+3} = \frac{2 \times 15}{8} = \frac{15}{4} = \mathbf{3\frac{3}{4}\ km/h}$$

Example 8. A car completes a certain journey in 8 hours. It covers first half of the distance at 40 km/h and rest at 60km/h. The length of journey is ....

Sol : Total length of journey [distance]

$$= \textbf{Average speed} \times \textbf{time}$$

$$= \frac{2xy}{x+y} \times t$$

$$= \frac{2 \times 40 \times 60}{40+60} \times 8 = \frac{2 \times 40 \times 60}{100} \times 8\ \mathbf{384\ km}$$

Note : If $\boldsymbol{a}$ is speed of the person in first half of the time and $\boldsymbol{b}$ is the speed in second half of the time then average speed is $= \frac{\boldsymbol{a+b}}{\mathbf{2}}$.

Type II

Before doing these types of questions understand the following concept. Suppose your college opens at **9.00 a.m.** in the morning. You walk at certain speed and reach **10 minutes late** i.e., at 9.10 am. On next day you start at the same time and walk a bit fast and reach college **15 minutes earlier** i.e., at **8.45 a.m.** Now the difference in timing is **9.10 – 8.45** i.e., **25 minutes**. In reaching **10 minute late** and **15 minutes earlier** the time difference is **10 + 15 = 25 minutes**.

This concept will be used in following questions :

Example 9. A person walks from his house at 4km/h and reaches his office 9 minutes late. If his speed had been 5km/h he would have reached the office 6 minutes earlier. How far is his office from house.

Sol :

$$\text{Let distance} = x \text{ km}$$

$$\text{Speed} = 4 \text{ km/h}$$

$$\text{Time taken by slower speed } t_1 = \frac{\text{Distance}}{\text{Speed}} = \frac{x}{4} \text{ hours}$$

$$\text{Time taken by faster speed } t_2 = \frac{x}{5} \text{ hours}$$

$$\text{Difference in timings} = 9 + 6 = 15 \text{ minutes (As discussed)}$$

$$t_1 - t_2 = \frac{15}{60}$$

$$\frac{x}{4} - \frac{x}{5} = \frac{15}{60}$$

$$\frac{15x - 12x}{60} = 15$$

$$3x = 15$$

$$x = \mathbf{5\ km}$$

Example 10. A student walks from his house at 3 km/h and reaches school 10 minutes late. If his speed had been 4 km/h he would have reached the school 10 minutes earlier. How far is his school from house.

Sol : Let distance $= x$ km

Difference in timing $= 10 + 10 = 20$ mnts. $= \frac{20}{60}$ hours

According to question

$$\frac{x}{3} - \frac{x}{4} = \frac{20}{60}$$

$$\frac{20x - 15x}{60} = 15$$

$$x = \mathbf{4\ kms}$$

Example 11. A car travels a distance of 300 km at a uniform speed. If the speed of the car is 5 km/h more, it takes two hours less to cover the same distance. The original speed of the car is ....

Sol : Let original speed $= x$

Increased speed $= x + 5$

According to question

$$\frac{300}{x} - \frac{300}{x+5} = \frac{2}{1}$$

$$\frac{300(x+5) - 300x}{x(x+5)} = \frac{2}{1}$$

$$\frac{300x + 1500 - 300x}{x(x+5)} = \frac{2}{1}$$

$$\frac{1500}{x(x+5)} = \frac{2}{1}$$

$$750 = x(x+5)$$

$$x^2 + 5x - 750 = 0$$

$$x^2 + 30x - 25x - 750 = 0$$

$$x(x+30) - 25(x+30) = 0$$

$$(x-25)(x+30) = 0$$

$$x = 25$$

$\therefore$ Original speed $=$ **25 km/h**

Type III

Before doing the questions let us understand an intresting concept.

Let distance between A and B is 30 km.

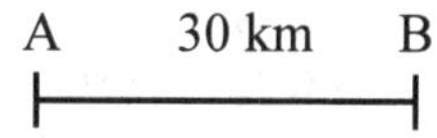

Suppose Rahul's usual speed is 30 km/h

So time taken by Rahul to reach B from A $= \frac{D}{S}$

$= \frac{30}{60}$

$= 1$ hour

So Rahul's usual time is 1 hour. ...(1)

Let he walks at $\frac{1}{2}$ of usual speed, i.e. at $\frac{1}{2} \times 30 = 15$ km/h

Then time taken $= \frac{30}{15}$

$=$ **2 hours** ...(2)

$=$ **1 × 2 = 2 times the usual time**

(1) and (2) implies

that by walking $\frac{1}{2}$ **of his usual speed**, a person takes $\frac{2}{1}$ **of his usual time**.

Similarly by walking $\frac{1}{3}$ **of usual speed**, a person takes $\frac{3}{1}$ **of usual time**.

Or by walking $\frac{4}{3}$ **of his usual speed**, a person takes $\frac{3}{4}$ **of usual time**.

This concept we will use in the following examples.

Example 12. By walking $\frac{3}{4}$ of his usual speed, a man reaches his office 20 minutes late. Find his usual time.

Sol : $\frac{3}{4}$ of usual speed means $\frac{4}{3}$ of usual time

He reaches 20 minutes late

$\therefore \quad \frac{4}{3}$ usual time – 20 min = usual time

$\frac{4}{3}$ ut – ut = 20 min (ut = usual time)

$\frac{1}{3}\, ut = 20$ min.

$ut =$ **60 min.**

Example 13. By walking $\frac{2}{3}$ of his usual speed a man reaches office 15 minutes late. Find his usual time.

Sol: $\frac{2}{3}$ of usual speed means $\frac{3}{2}$ of usual times

He reaches 15 minutes late.

$$\therefore \quad \frac{3}{2} \text{ of usual time} - 15 \text{ minute} = \text{usual time}$$

$$\frac{3}{2}ut - ut = 15 \text{ minute}$$

$$\frac{1}{2}ut = 15 \text{ minute}$$

$$ut = \mathbf{30\ minutes}$$

Example 14. By walking $\frac{5}{3}$ of usual speal a student reaches school 20 minutes earlier. Find his usual time.

Sol: $\frac{5}{3}$ usual speed means $\frac{3}{5}$ of usual time as he reaches earlier

$$\frac{3}{5}ut + 20 \text{ minutes} = ut$$

$$20 \text{ minutes} = ut - \frac{3}{5}ut$$

$$20 \text{ minutes} = \frac{2}{5}ut$$

$$ut = \frac{20 \times 5}{2} = \mathbf{50\ min.}$$

Type IV

Before doing the questions on this type note the following points:

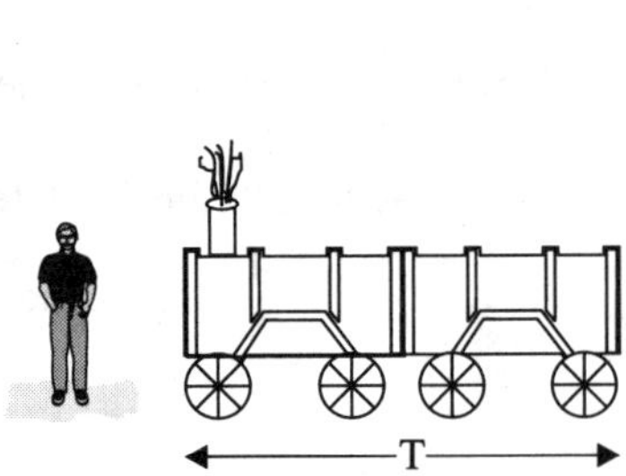

1. If a train has to cover a **pole** or **standing person** then train has to **cover** a **distance** equal to its **own length**. In adjacent figure the length of train is **T**. So to pass a standing person it has to cover a **distance** equal to **T**.
2. **If a train has to cover a bridge or tunnel or a platform etc.** then train has to cover a distance **equal to length of train + length of bridge or tunnel or platform**. In the adjacent figure length of train is **T** and platform is **P**. So train has to cover a distance **T + P**.

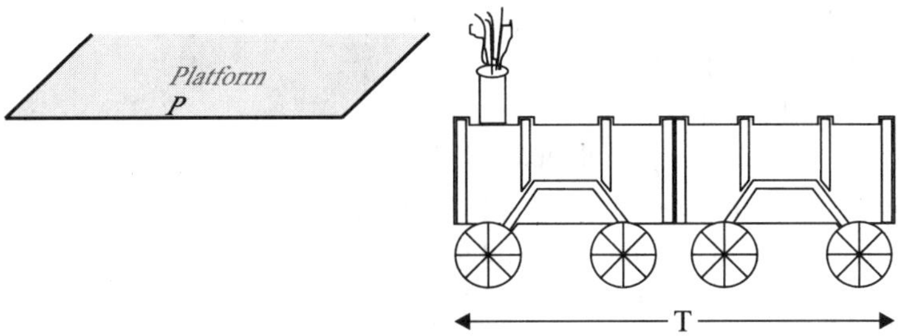

3. When two trains cross each other

(a) In opposite direction, (b) In same direction

(a) In opposite direction:

When two trains are in opposite direction their relative speed is sum of the speeds of two trains. Let the speed of first train is **x km/h** and that of second train is **y km/h**. Then speed at which they will cross each other will be equal to **(x + y)**. The distance will be equal to the sum of length of two trains i.e. $T_1 + T_2$.

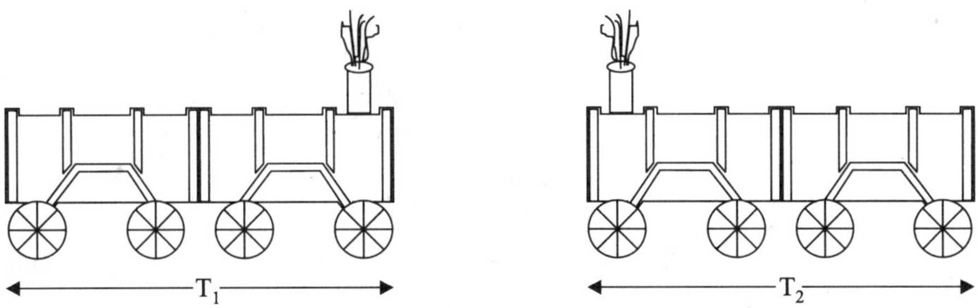

$$\text{So time taken} = \frac{\text{Distance}}{\text{Speed}} = \frac{\mathbf{T_1 + T_2}}{\mathbf{x + y}}$$

(b) In same direction:

Let two trains of length T_1 and T_2 travel with a speed of x and y and

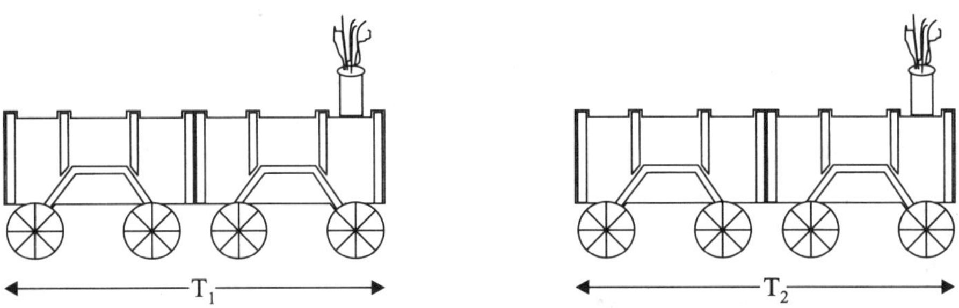

Time taken by faster train to cross slower train $= \dfrac{\mathbf{T_1 + T_2}}{x - y}$

Now solve the following examples:

Examples 15. A train whose length is 160 metre is running at the rate of 36 km/h. How long will it take to pass a pole.

Sol : Distance to be travelled = 160 m

Speed = 36 km/h

Now we have to find the time.

Now here the distance is in metre and speed is in km/h. To make both things compatible, we will convert 36 km/h to m/sec.

$$\text{Speed} = 36 \text{ km/h}$$

$$= 36 \times \frac{5}{18} = 10 \text{ seconds}$$

$$\text{Time} = \frac{\text{Distance}}{\text{Speed}} = \frac{160}{10} = \mathbf{16\ sec.}$$

Example 16. How long does a train 280 m long moving at the rate of 40 km/h take to pass a bridge 120 m long.

Sol : Distance to be travelled = length of train + length of bridge

= 280 + 120 = 400 m

$$\text{Speed} = 40 \times \frac{5}{18} = \frac{200}{18} \text{ m/s}$$

$$\text{Time} = \frac{\text{Distance}}{\text{Speed}}$$

$$= 400 \times \frac{18}{200} = \mathbf{36\ sec.}$$

Example 17. How long will 150 m long train running at the rate of 60km/h take to cross a bridge 300 m long?

Sol : Distance = 150 + 300 = 450 m

$$\text{Speed} = 60 \times \frac{5}{18} = \frac{300}{18} \text{ m/s}$$

$$\text{Time} = \frac{\text{Distance}}{\text{Speed}}$$

$$= 450 \times \frac{18}{300} = \mathbf{27\ sec.}$$

Example 18. Two trains of 150 m and 50 m long resp. are running on parallel tracks at the rate of 35 km/h and 30 km/h. In what time will they cross each other if they are running in same direction.

Sol : Distance = $T_1 + T_2 = 150 + 50 = 200$ m

Since they are running in the same direction.

Therefore their relative speed = $x - y$

= 35 – 30 = 5 km/h

$$= 5 \times \frac{5}{18} = \frac{25}{18} \text{ m/s}$$

$$\text{Time} = \frac{\text{Distance}}{\text{Speed}}$$

$$= 200 \times \frac{18}{25} = \mathbf{144\ sec.}$$

Example 19. Two trains of lengths 200 m and 100 m respectively are running on parallel tracks at the rate of 35 km/h and 25 km/h. In what time they cross each other if they are running in opposite direction.

Sol : Distance $= 200 + 100 = 300$ m

Relative speed $= 35 + 25 = 60$ km/h

$$\text{Time} = \frac{300 \times 18}{60 \times 5} = \mathbf{18\ sec}$$

Example 20. A train passes a platform of length of 150 m in 20 sec and a person standing on platform in 5 sec. Find the speed of train in km/h.

Sol : Train passes a platform means covers the **length of train + length of platform** in **20 sec**. and a person standing on platform means **its own length** in **5 sec**.

∴ Time taken by train to pass only the length of platform = 20 – 5 = 15 sec.

and length of bridge = 150 m

$$\therefore \quad \text{Speed} = \frac{150}{2}$$

$$= 10 \text{ m/s} = 10 \times \frac{18}{5} = \mathbf{36\ km/h}$$

Type V

Suppose a boat moves in **still water** at the rate of **5 km/h** and now let the **stream** start flowing at the rate of **3 km/h**. If the **direction of rowing** of boat is **same as that of flow of stream** then it is called **down stream** and **downstream speed** in this case is 5 + 3 = **8 km/h** and if the **direction of flow of stream** is **against the flow of rowing of boat** then it is called **upstream** and **upstream speed** in this case = 5 – 3 = **2 km/h**

Now generalising above concept

Let the speed of boat in still water $= x$

" " " " of stream $= y$

Downstream speed $= \boldsymbol{x+y}$ ...**(1)**

upstream speed $= \boldsymbol{x-y}$ ...**(2)**

We will use only equation **(1)** and **(2)** for solving the question on this topic.

Example 21. A man rows upstream 4 km and downstream 36 km in 2 hours each. Find his speed in still water and speed of stream.

Sol : Upstream speed $= \frac{4}{2} = 2$ km/h

Downstream speed $= \frac{36}{2} = 18$ km/h

Now,
$$x + y = 18$$
$$x - y = 2$$
$$2x = 20$$
$$2x = 20$$
$$x = 10 \text{ km/h}$$
$$y = 8 \text{ km/h}$$

$\therefore$ Speed of boat in still water = **10 km/h**

Speed of stream = **8 km/h**

Example 22. A man moves downstream at the rate of 10 km/h and upstream at the rate of 8 km/h. What is the speed of stream?

Sol :
$$x + y = 10$$
$$x - y = 8$$
$$- \quad + \quad\quad -$$
$$2y = 2$$
$$y = 1 \text{ km/h}$$

$\therefore$ Speed of stream = **1 km/h**

Example 23. A boat downstream at the rate of 1 km in 6 minutes and upstream at the rate of 1 km in 10 minutes. What is the velocity of boat in still water?

Sol : First we have to convert the speeds in km/h.

$$\text{Down stream speed} = \frac{\text{Distance}}{\text{Time}}$$
$$= \frac{1}{\frac{6}{60}} \text{ km/h} = 10 \text{ km/h}$$
$$\text{Upstream speed} = \frac{1}{\frac{10}{60}} \text{ km/h}$$
$$= 6 \text{ km/h}$$
$$x + y = 10$$
$$x - y = 6$$
$$2x = 16$$
$$x = \mathbf{8 \text{ km/h}}$$

Example 24. A boat moves down stream at the rate of 6km/h and it takes 20 minutes to move one km against stream. The speed of boat in still water is.........

Sol :
$$\text{Upstream speed} = \frac{1}{\frac{20}{60}} \text{ km/h} = 3 \text{ km/h}$$
$$\text{Downstream speed} = 6 \text{ km/h}$$

$$x + y = 6$$
$$x - y = 3$$
$$2x = 9$$
$$x = \mathbf{4.5\ km/h}$$

Example 25. The speed of boat in still water is 8 km/h . A boat goes 6 km and back to the starting point in 2 hours. The speed of stream is:

Sol : $x = 8$ km/h

Let $y = y$ km/h

upstream speed $= 8 - y$ km/h

down stream speed $= 8 + y$ km/h

Time taken to go 6 km upstream $T_1 = \dfrac{6}{8-y}$

Time taken to go 6 km downstream $T_2 = \dfrac{6}{8+y}$

According to question

$$T_1 + T_2 = 2 \text{ hours}$$
$$\frac{6}{8-y} + \frac{6}{8+y} = \frac{2}{1}$$
$$\frac{6(8+y)+6(8-y)}{(8+y)(8-y)} = \frac{2}{1}$$
$$\frac{48+6y+48-6y}{64-y^2} = \frac{2}{1}$$
$$48 = 64 - y^2$$
$$y^2 = 16$$
$$y = \mathbf{4\ km/h}$$

Example 26. A man can row $9\frac{1}{3}$ km/h in still water and he finds that it takes him thrice as much time to row up than to row down the same distance in river. The speed of stream is :

Sol : $x = \dfrac{28}{3}$ km/h

$y = y$ km/h

Let distance $= k$ km

Upstream speed $= \dfrac{28}{3} - y$

Downstream speed $= \dfrac{28}{3} + y$

Let T_1 be the time taken to row up

Then $T_1 = \dfrac{k}{\frac{28}{3} - y}$ hours

Let T_2 be the time taken to row down

Then $$T_2 = \frac{k}{\frac{28}{3}+y} \text{ hours}$$

According to question

$$T_1 = 3\,T_2$$

$$\frac{k}{\frac{28}{3}-y} = \frac{3k}{\frac{28}{3}+y}$$

$$\frac{28}{3}+y = 3\left[\frac{28}{3}-y\right]$$

$$\frac{28}{3}+y = 28-3y$$

$$4y = 28-\frac{28}{3}$$

$$4y = \frac{56}{3}$$

$$y = \frac{56}{3}\times\frac{1}{4} = \frac{14}{3} = \mathbf{4\frac{2}{3}\ km/h}$$

Type VI

Example 27. A man performs $\frac{1}{2}$ of the total journey by bus $\frac{1}{3}$ by rickshaw and the remaining 2 km on foot. Find his total journey.

Sol: Let total journey $= x$ km

Journey by bus $= \frac{x}{2}$ km

Journey by rickshaw $= \frac{x}{3}$

Remaining journey $= x-\frac{x}{2}-\frac{x}{3} = \frac{x}{6}$ km

According to statement $\frac{x}{6} = 2$ km

$\therefore$ $x =$ **12 km**

Example 28. A man takes 2 hours to cover a distance of 7 km . How long will it take to cover a distance of 52.5 km?

Sol: Speed $= \frac{7}{2}$ km/h

Now distance $= 52.5$ km

Time $= \frac{D}{S} = \frac{52.2\times 2}{7} =$ **15 hours**

Example 29. A thief steals a car at 2.00 pm. and drives at 60 km/h. The theft is discovered at 3.00 p.m. and the police starts chasing him at the rate of 80 km/h. At what time will the police catch the thief?

Sol : Between 2 pm. to 3 pm i.e., in one hour the thief will be 60 km away. Now police has to cover 60 km at a relative speed of (80 – 60) = 20 km/h.

So time taken by police to cover 60 km = $\frac{60}{20}$ = 3 hours

So police will catch the thief 3 hours after 3 pm i.e. at 6 pm.

Example 30. Excluding stoppage the speed of train is 60 km/h but including stoppage it is 48 km/h. How many minutes thus the train stope per hour?

Sol : Without stoppage train can travel 60 km in 1 hour. But with stoppage it can travel 48 km in an hour. So it will stop for the time in which it can travel 12 km @ 60 km/h i.e., $\frac{12}{60}$ hour = 12 minute.

EXERCISE

Q.1. Which is greater?

(i) 144 km/h or 30m/sec (ii) 70 km/h or 25m/sec
(iii) 20 m/sec or 72km/h (iv) 27.5 m/sec or 100km/h
(v) 25 m/sec or 90km/h

Q.2. A car covers first half of its journey at the rate of 15km/h and second half at the rate of 12km/h and total time taken to complete the journey is 9 hours. Find the length of journey.

(a) 60 km (b) 240km (c) 120 km (d) none

Q.3. A bus complete a cerain journey in 19 hours. It covers first half of the distance at a rate of 18 km/h and second half at the rate of 20km/h. The length of journey is ....

(a) 360 km (b) 240km (c) 120 km. (d) none

Q.4. A student walks from his house at 2 km/h and reaches school 5 minutes late and if his speed had been 3 km/h he would have reached the school 5 minutes earlier. How far is his school from house?

(a) 3km (b) 2.5 km (c) 2 km (d) none

Q.5. A person walks from his house at the rate of 6 km/h and reaches office 12 minutes late and if his speed had been 8 km/h he would have reached the office 8 minutes earlier. How far is his office from house?

(a) 6 km (b) 8 km (c) 10 km (d) none

Q.6. By walking $\frac{3}{2}$ of his usual speed a person reach his office 10 minutes earler . Find his usual times.

(a) 5 min (b) 10 min (c) 20 min (d) none

Q.7. By walking $\frac{2}{5}$ of his usual speed a student reach his school 30 minutes late. Find his usual time?

(a) 10 min (b) 20 min (c) 30 min (d) none

Q.8. A train whose length is 300m long is moving at the rate of 108 km/h. How long will it take to pass a pole?

(a) 10sec (b) 12 sec (c) 16 sec (d) none

Q.9. How long does a train 200 m long moving at the rate of 72 km/h take to pass a tunnel 40 m long?

(a) 6 sec (b) 12 sec (c) 18 sec (d) none

Q.10. How long does a train 600 m long moving at the rate of 144 km/h take to pass a bridge 200 m long?

(a) 20 sec (b) 30 sec (c) 40 sec (d) none

Q.11. Two trains of 400 m and 350 m long are running on parallel tracks at the rate of 180 km/h and 120 km/h respectively. How long will they take to cross each other if they are running in opposite direction?

(a) 5 sec (b) 9 sec (c) 12 sec (d) none

Q.12. Two trains 500 m and 200 m long are running on parallel tracks at the rate of 160 km/h and 200 km/h. How long the second train take to cross 1st train if they are running in same direction.

(a) 63 sec (b) 70 sec (c) 72 sec (d) none

Q.13. A train crosses a platform of length 350 m in 12 sec and a man standing on platform in 2 sec. Find the speed of train.

(a) 100 km/h (b) 126 km/h (c) 130 km/h (d) none

Q.14. A man moves downstream at a rate of 14 km/h and upstream at a rate of 6 km/h. Find his speed in still water.

(a) 4 km/h (b) 10 km/h (c) 16 km/h (d) none

Q.15. A man moves 20km down stream in 5 hours and 10km upstream in same time. Find the speed of stream.

(a) 4 km/h (b) 1 km/h (c) 1.6 km/h (d) none

Q.16. A boat moves down the stream at the rate of 20km/h and it takes 12 minutes for him to move one km against it. Find the speed of stream.

(a) 6 km/h (b) 7.5 km/h (c) 9 km/h (d) none

Q.17. A man can swim 4 km/h in still water. If the velocity of stream is 2 km/h, the time taken by him to swim to a place 12km upstream and back is:

(a) 6 hour (b) 7 hour (c) 8 hour (d) none

Q.18. A man can row three quarter of a km against the stream in $11\frac{1}{4}$ min. and return in $7\frac{1}{2}$ min. The speed of man in still water is.

(a) 5km/h (b) 6km/h (c) 8km/h (d) none

Q.19. A boat covers 6 km upstream and 12 km down stream in 10 hours while it covers 12 km upstream and 6 km down stream in 14 hours. The speed of boat in still water is .....

(a) 2 km/h (b) 1 km/h (c) .5 km/h (d) none

Q.20. The speed of a boat upstream is 10 km/h . The speed of stream is 5km/h. The speed of boat down stream is.

(a) 15 km/h (b) 20 km/h (c) 25 km/h (d) none

Q.21. A thief steals a car at 9 am and drive at the rate of 20 km/h. The theft is discovered and police starts chasing him at 9.30 am. in a jeep at the speed of 40 km/h.The thief will be caught at ....

(a) 10 am (b) 10.30 am. (c) 11 am. (d) none

Q.22. Excluding stoppage the speed of bus is 48 km/h and including stoppage it is 40 km/h. For how many minutes does the bus stop per hour?

(a) 8 min (b) 10 min (c) 12 min (d) none

Q.23. A and B are two stations 1000 km. A train start from A at 70 km/h and another train starts from B at same time at the rate of 30 km/h. How far from A will they cross each other?

(a) 700 km (b) 300 km (c) 70 km (d) none

Q.24. Two cyclist A and B starts from same place at the same time, one going to wards east at a rate of 20 km/h and another towards west at a rate of 10 km/h What time will they take to be 75 km apart.

(a) 3 hour (b) 2.5 hour (c) 2 hour (d) none

SOLUTIONS

1. (i) $144 \text{ km/h} = 144 \times \frac{5}{18} = 40 \text{ m/s}$ which is greater then 30 m/sec

Therefore, 144 km/h > 30 m/s

(ii) $25 \text{ m/sec} = 25 \times \frac{18}{5} = 90 \text{ km/h}$

Threrfore 25 m/sec > 70km/h

(iii) $72 \text{ km/h} = 72 \times \frac{5}{18} = 20 \text{ m/s}$

Therefore, both are equals.

(iv) $27.5 \text{ m/sec} = 27.5 \times \frac{18}{5} = 99.5 \text{ km/h}$

Therefore, 100 km/h > 27.5 m /sec

(v) $25 \text{ m/sec} = 25 \times \frac{18}{5} = 90 \text{ km/h}$ **both are equals.**

2. (c)
$$\text{Length of journey} = \frac{2xy}{x+y} \times \text{time}$$
$$= \frac{2 \times 15 \times 12}{27} \times 9 = \mathbf{120 \text{ km}}$$

3. (a)
$$\text{Length of journey} = \frac{2 \times 18 \times 20}{38} \times 19 = \mathbf{360 \text{ km}}$$

4. (d)
$$\frac{x}{2} - \frac{x}{3} = \frac{10}{60}$$
$$\frac{3x - 2x}{6} = \frac{1}{6}$$
$$\boldsymbol{x} = \mathbf{1 \text{ km}}$$

5. (b)

$$\frac{x}{6} - \frac{x}{8} = \frac{20}{60}$$

$$\frac{4x - 3x}{24} = \frac{1}{3}$$

$$\boldsymbol{x = 8 \text{ km}}$$

6. (d) $\frac{3}{2}$ usual speed means $\frac{2}{3}$ of usual time as he reaches earlier

$$\frac{2}{3}ut + 10 \text{ minutes} = ut$$

$$10 \text{ minutes} = ut - \frac{2}{3}ut$$

$$10 \text{ minutes} = \frac{1}{3}ut$$

$$\text{ut} = 10 \times 3 = \mathbf{30\ min.}$$

7. (b) $\frac{2}{5}$ of usual speed means $\frac{5}{2}$ of usual times

He reaches 30 minutes late.

$$\therefore \quad \frac{5}{2} \text{ of usual time} - 30 \text{ minute} = \text{usual time}$$

$$\frac{5}{2}ut - ut = 30 \text{ minute}$$

$$\frac{3}{2}ut = 30 \text{ minute}$$

$$ut = \mathbf{20\ minutes}$$

8. (a) To pass a pole train has to cover its own length.

$$\text{Distance} = 300 \text{ m}$$

$$\text{Speed} = 108 \text{ km/h}$$

$$= 108 \times \frac{5}{18} = 30 \text{ m/s}$$

$$\text{T} = \frac{\text{D}}{\text{S}} = \frac{300}{30} = \mathbf{10\ sec.}$$

9. (b)

$$\text{Total distance} = 200 + 40 = 240 \text{ m}$$

$$\text{Speed} = 72 \times \frac{5}{18} = 20 \text{ m/sec}$$

$$\text{T} = \frac{\text{D}}{\text{S}} = \frac{240}{20} = \mathbf{12\ sec}$$

10. (a)

$$\text{Total distance} = 800 \text{ m}$$

$$\text{Speed} = 144 \times \frac{5}{18} = 40 \text{ m/s}$$

$$\text{T} = \frac{\text{D}}{\text{S}} = \frac{800}{40} = \mathbf{20\ sec}$$

11. (b)

$$\text{Total distance} = 400 + 350 = 750 \text{ m}$$

$$\text{Relative speed} = 180 + 120 = 300 \text{ km/h}$$

$$= 300 \times \frac{5}{18} \text{ m/s}$$

$$\text{Time} = \frac{750 \times 18}{300 \times 5} = \mathbf{9\ sec}$$

12. (a)

$$\text{Total distance} = 500 + 200 = 700 \text{ m}$$

$$\text{Relative speed} = 200 - 160 = 40 \text{ km/h}$$

$$= 40 \times \frac{5}{18} = \frac{200}{18}$$

$$\text{Time} = \frac{700 \times 18}{200} = \mathbf{63\ sec}$$

13. (b) Same as example 20

14. (b)

$$x + y = 14$$

$$x - y = 6$$

$$2x = 20$$

$$x = \mathbf{10\ km/h}$$

15. (b)

$$\text{Downstream speed} = \frac{20}{5} = 4 \text{ km/h}$$

$$\text{Upstream speed} = \frac{10}{5}$$

$$= 2 \text{ km/h}$$

$$x + y = 4$$

$$x - y = 2$$

$$- \quad + \qquad -$$

$$2y = 2$$

$$\boldsymbol{y} = \mathbf{1\ km/h}$$

16. (b)

$$\text{Downstream speed} = 20 \text{ km/h}$$

$$\text{Upstream speed} = \frac{1}{\frac{12}{60}}$$

$$= 5 \text{ km/h}$$

$$x + y = 20$$

$$x - y = 5$$

$$- \quad + \qquad -$$

$$2y = 15$$

$$2y = 15$$

$$\boldsymbol{y} = \mathbf{7.5\ km/h}$$

17. (c)

$$x = 4 \text{ km/h}$$

$$y = 2 \text{ km/h}$$

$$\text{Downstream speed} = 4 + 2 = 6 \text{ km/h}$$

$$\text{Upstream speed} = 4 - 2 = 2 \text{ km/h}$$

$$\text{Distance} = 12 \text{ km/h}$$

$$\text{Total Time} = \frac{12}{6} + \frac{12}{2} = 2 + 6$$

$$= \textbf{8 hour}$$

18. (a)

$$\text{Upstream speed} = \frac{\frac{3}{4}}{\frac{45}{4 \times 60}} \text{ km/h}$$

$$= \textbf{4 km/h}$$

$$\text{Downstream speed} = \frac{\frac{3}{4}}{\frac{15}{2 \times 60}} \text{ km/h}$$

$$= \textbf{6 km/h}$$

$$x + y = 6$$

$$x - y = 4$$

$$2x = 10$$

$$x = \textbf{5 km/h}$$

19. (a)

$$\frac{6}{x-y} + \frac{12}{x+y} = 10$$

$$\frac{6(x+y)+12(x-y)}{x^2-y^2} = 10$$

$$6x + 6y + 12x - 12y = 10\,(x^2 - y^2) \quad \text{...(1)}$$

Also

$$\frac{12}{x-y} + \frac{6}{x+y} = 14$$

$$12\,(x+y) + 6\,(x-y) = 14\,(x^2 - y^2)$$

$$18x + 6y = 14\,(x^2 - y^2) \quad \text{...(2)}$$

Divide equation **(1)** by **(2)**

$$\frac{18x - 6y}{18x + 6y} = \frac{10}{14} = \frac{5}{7}$$

Cross multiplying, we get

$$126x - 42y = 90x + 30y$$

$$36x = 72y$$

$$\boldsymbol{x = 2y} \quad \text{...(3)}$$

By putting **(3)** in **(1)** we get

$$x = \textbf{2 km/h}$$

$$y = \textbf{1 km/h}$$

20. (b)

$$x - y = 10 \text{ km/h}$$

$$y = 5 \text{ km/h}$$

$$x - 5 = 10$$

$$x = 15 \text{ km./h}$$

$$\text{Downstream speed} = 15 + 5 = \mathbf{20\ km/h}$$

21. (a) In half an hour the thief will be

$$20 \times \frac{1}{2} = \mathbf{10\ km\ away.}$$

To catch thief police has to cover 10 km at a relative speed of

$$= 40 - 20 = 20\text{ km/h}$$

$$\text{Time taken} = \frac{10}{20} = \frac{1}{2}\text{ hour} = 30\text{ min.}$$

So thief will be caught at 9.30 + 30 min = **10 am**

22. (b) Due to stoppage it covers 8 km less per hour

$$\text{Time to cover 8 km} = \frac{8}{48} \times 60 = \mathbf{10\ min}$$

23. (a)

$$\text{Total distance} = 1000\text{ km}$$

$$\text{Relative speed} = 70 + 30 = 100\text{ km/h}$$

Time after which the train will meet

$$= \frac{1000}{100} = 10\text{ hour}$$

In 10 hour train A will cover distance

$$= 70 \times 10 = \mathbf{700\ km}$$

24. (b) Relative speed $= 20 + 10 = 30$ km/h

Distance to be covered $= 75$ km

Time taken to cover that distance

$$= \frac{75}{30} = \frac{5}{2}\text{ hour} = \mathbf{2.5\ hour}$$

Day 10

Average

On this day we will do a very simple yet important topic that is of **average**. Suppose the temp of 14th June is 40°C and on 15th June it is 44°C. If you are asked to find the average of these two days you will do it as follows:

$$\frac{40° + 44°}{2} = \frac{84°}{2} = \mathbf{42°}$$

So here there are two temperatures and to find the average of these two temperatures we had divided their sum by 2. If we had to find the average temp. of n days then we would have divided the sum of temperature by n days.

$$\textbf{Average of n observations} = \frac{\textbf{Sum of n observations}}{\textbf{n}}$$

Now we will do some more examples.

Example 1. A man spends Rs. 10 on 1st day, Rs. 15 on second day, Rs. 30 on third and Rs. 5 on 4th day. Find the average money spent by man in these four days.

Sol : Average money spent $= \frac{10+15+30+5}{4}$

$$= \frac{60}{4} = \mathbf{15\ Rs.}$$

Example 2. In a family there are five members. The age of father is 42 years mother is 38 years andsum of the 3 childern's age is 80 years. Find the average age of the family.

Sol : Average age $= \frac{42+38+80}{5} = \frac{160}{5}$

$$= \textbf{32 years}$$

Types of Questions

Type I : Questions related to temp.

Type II : Questions related to height, weight, age etc.

Type III : Questions on cricket score etc.

Type IV : Questions related to salary

Type V : Misc.

Type I

Example 3. The average temp of Monday, Tuesday and Wednesday was 40°C and Tuesday, Wednesday and Thursday is 45°C. The temperature of Monday is 30°C. Find the temperature of Thursday.

Sol :

Total of Monday + ~~Tuesday~~ + ~~Wednesday~~ = 40 × 3 = 120°

Total of ~~Tuesday~~ + ~~Wednesday~~ + Thursday = 45 × 3 = 135°

– – –

Monday	–	Thursday	=	–15°
Mon	–	Thursday	=	–15°
30°	–	Thursday	=	–15°
	–	Thursday	=	–15° –30°
	–	Thursday	=	–45°
		Thursday	=	**45° C**

Example 4. The average of Monday, Tuesday and Wednesday was 84°C and that of Tuesday Wednesday and Thursday was 87°C. If the temperature of Monday was 91°C. Find the temp of Thursday.

Sol :

Total of Monday + ~~Tuesday~~ + ~~Wednesday~~ = 84 × 3 =252°

Total of ~~Tuesday~~ + ~~Wednesday~~ + Thursday = 87 × 3 = 261°

– – –

Monday	–	Thursday	=	–9°
Mon	–	Thursday	=	–9°
91	–	Thursday	=	–9°
	–	Thursday	=	–9° –91°
	–	Thursday	=	–100°
		Thursday	=	**100°C**

Type II

Example 5. The average weight of a class of 20 students is 48 kg. If the weight of the teacher is included the average weight becomes 50 Kg. Find the weight of teacher.

Sol :

Total weight of 20 student = 48 × 20 = 960 kg.

Total weight of 21 person (20 student + 1 teacher) = 50 × 21 = 1050 kg.

Weight of teacher = 1050 – 960 = **90 kg.**

Direct Method:

Increase in average = 50 – 48 = 2 kg.

" " Total = 21 × 2 = 42 kg.

Teacher weight be such that it should maintain **previous average** 48 kg plus also **contribute** to increase 42 kg to increase the average by 2 kg.

$$\text{Weight of teacher} = 48 + 42 = \mathbf{90\ kg.}$$

Example 6. The average height of 30 students out of a class of 40 is 140 cm and that of remaining students is 120 cm. The average height of total class is..............?

Sol :

$$\text{Total height of 1st 30 students} = 30 \times 140 = 4200 \text{ cm}$$

$$\text{Total height of last 10 students} = 10 \times 120 = 1200 \text{ cm}$$

$$\text{Total height of 40 students} = 4200 + 1200 = 5400 \text{ cm.}$$

$$\text{Average height of 40 students} = \frac{5400}{40} = \mathbf{135\ cms}$$

Example 7. The average age of students in a class of 24 boys is 14.5 years. When a new boy join in the class average age decrease to 14.3 years. Find the age of new boy.

Sol :

$$\text{Average age of 24 boys} = 14.5 \text{ years}$$

$$\text{Total age of 24 boys} = 14.5 \times 24 = 348.0 \text{ years}$$

$$\text{Average age of 25 boys} = 14.3 \text{ years}$$

$$\text{Total age of 25 boys} = 14.3 \times 25 = 357.5 \text{ years}$$

$$\text{Age of new boy} = 357.5 - 348.0$$

$$= \mathbf{9.5\ years}$$

Example 8. The average age of the 5 childern is 8 years. If age of the father be included the average age is increased by 7 years. Find the age of father.

Sol :

$$\text{Total age of 5 children} = 5 \times 8 = 40 \text{ years}$$

$$\text{Total age of 6 persons} = 6 \times 15 = 90 \text{ years}$$

$$\text{Age of father} = 90 - 40 = \mathbf{50\ years}$$

Example 9. The average age of 8 oarsmen is increased by 2 years when one of the crew members whose age is 44 years, is replaced by new man. Find the age of new man.

Sol :

$$\text{Total increase} = 8 \times 2 = 16 \text{ years}$$

So in order to maintain the previous average the age of new man should be 44 years. But now he has to increase it by a total of 16 years.

$$\text{Age of new man} = 44 + 16 = \mathbf{60\ years}$$

Example 10. Three years ago the average age of a family of 6 members was 19 years. A baby having been born and the average age of the family is now the same as before. Find the present age of new born.

Sol :

Total age of members before 3 years = 6 × 19 = 114 years

Average after 3 years i.e., at present = 19 + 3 = 22 years

Total age of 6 members at present = 22 × 6 = 132 years

According to statement

$$\text{Present average of 7 members} = 19 \text{ years}$$

Total age of 7 members at present $= 7 \times 19 = 133$ years

Age of new born $= 133 - 132 =$ **1 years**

Example 11. There is a class of 40 girls whose average age is 15 years. The average age of 24 girls out of them is 14 years. Find the average age of rest.

Sol :

$$\begin{aligned}
\text{Total age of 40 girls} &= 15 \times 40 \\
&= 600 \text{ years} \\
\text{Total age of 24 girls} &= 24 \times 14 \\
&= 336 \text{ years} \\
\text{Total age of rest 16 girls} &= 600 - 336 \\
&= 264 \text{ years} \\
\text{Average age of rest 16 girls} &= \frac{264}{16} \\
&= \mathbf{16.5 \text{ years}}
\end{aligned}$$

Type III

Example 12. Score of a batsman in his five innings is 68, 72, 3, 42* and 26. Find his average of these innings. (* means not out)

Sol :

$$\begin{aligned}
\text{Total score of five innings} &= 68 + 72 + 3 + 42 + 26 \\
&= 211
\end{aligned}$$

But he has remained not out in one innings,

$$\therefore \quad \text{Total innings counted} = 4$$

$$\therefore \quad \text{Average Score} = \frac{211}{4} = \mathbf{52.75}$$

Note : In cricket score **not out innings** is not counted in total innings while its score is calculated in total score.

Example 13. A batsman makes a score of 85 in his 17th innings thereby increases his average by 3. What is his average after 17th innings? He has never been "not out".

Sol :

$$\begin{aligned}
\text{Let average after 16th innings} &= x \\
\text{Total after 16th innings} &= 16x \\
\text{Total after 17th innings} &= 16x + 85 \quad \ldots(1) \\
\text{Now, average after 17th innings} &= x + 3 \\
\therefore \quad \text{Total after 17th innings} &= 17(x + 3) \quad \ldots(2)
\end{aligned}$$

From (1) and (2)

$$\begin{aligned}
16x + 85 &= 17x + 51 \\
\therefore \quad x &= 34 \\
\text{So average after 16th innings} &= 34 \\
\text{And average after 17th innings} &= 34 + 3 = \mathbf{37}
\end{aligned}$$

Shortcut : Out of 85, **17 × 3 = 51** is used in increasing the average by 3 and **85 – 51 = 34** is used in maintaining the previous average.

Example 14. Average score of a cricketer in 2 matches is 28 and that in 3 others is 34. His average score in given matches is .... ?

Sol :

Total score in 2 matches = 2 × 28 = 56

Total in 3 matches = 3 × 34 = 102

Total of 5 matches = 56 + 102 = 158

Average of 5 matches = $\frac{158}{5}$

= **31.60**

Type IV

Example 15. A person spends Rs. 120 per month in his first seven months and Rs.150 for next five months. Find his monthly salary, If he saves Rs. 210 for whole year.

Sol :

Total expenditure of first seven months = 120 × 7 = 840 Rs.

Total of next five months = 150 × 5 = 750 Rs.

Saving = 210 Rs.

∴ Total income of twelve months = 840 + 750 + 210

= 1800 Rs.

Average income of twelve months = $\frac{1800}{12}$

= **150 Rs.**

Example 16. A man's monthly income is Rs. 1400. What should be his average monthly expenditure so as to save Rs. 7200 in a year.

Sol :

Total income = 1400 × 12 = 16800 Rs.

Saving = 7200 Rs.

Expenditure of Twelve months = 16800 – 7200 = 9600 Rs.

Average expenditure of twelve months should be = $\frac{9600}{12}$ Rs.

= **800 Rs.**

Example 17. The average salary of 30 workers in a factory is Rs. 5000 per month. If the salary of manager be included the average salary comes out to be Rs. 5500. Find the salary of manager.

Sol :

Total salary of 30 workers = 5000 × 30

= 150000 Rs.

Total salary of 31 (1 manager + 30 workers) = 31 × 5500

$$\begin{aligned} &= 170500 \text{ Rs.} \\ \text{Salary of manager} &= 170500 - 150000 \\ &= \mathbf{20500\ Rs.} \end{aligned}$$

Type V

This includes questions of average which can be easily solved by alligation and a few questions based on numbers.

Example 18. The average salary per head of the entire staff of an office including the officer and clerk is Rs. 60. The average salary per head of the officer is Rs. 400 and that of clerk is Rs. 56. If the number of officers is 12. Find the number of clerks.

Sol : This question can be done by two methods.

(i) By Algebra **(ii) By Alligation**

(i) By Algebra

$$\begin{aligned} \text{Let total staff} &= x \\ \text{Officers} &= 12 \\ \therefore \quad \text{Clerks} &= x - 12 \\ \text{Total salary of staff} &= 60x \\ \text{Total salary of officers} &= 12 \times 400 \\ &= 4800 \text{ Rs.} \\ \text{Total salary of clerks} &= (x - 12) \times 56 \\ \text{Total salary of staff} &= \text{Total salary of clerks + Total salary of officers} \\ \therefore \quad 60x &= 4800 + 56\,(x - 12) \\ 60x &= 4800 + 56x - 672 \\ 60x - 56x &= 4800 - 672 \\ 4x &= 4128 \\ x &= \frac{4128}{4} \\ &= 1032 \\ \therefore \quad \text{Total staff} &= 1032 \\ \text{No. of clerks} &= 1032 - 12 \\ &= \mathbf{1020} \end{aligned}$$

(ii) By Alligation

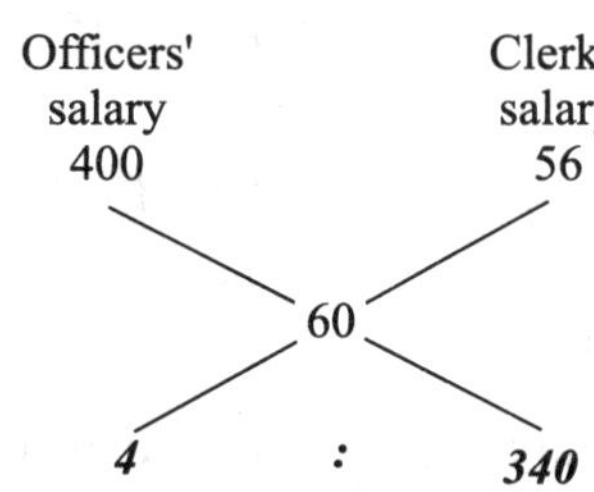

Ratio of **officer**	:	**clerk**
4	:	**340**
1	:	**85**

No. of officer $= 12$ ... (given)

$\therefore$ No. of clerks $= 12 \times 85$

$= \mathbf{1020}$

So alligation method is quite easy.

Example 19. The average of five numbers is 27. If one number is excluded, the average becomes 25. The excluded no. is .... ?

Sol : Total of five numbers $= 27 \times 5 = 135$

Total of four numbers $= 25 \times 4 = 100$

$\therefore$ Vth no. is $= 135 - 100 = \mathbf{35}$

Example 20. The average marks obtained by 8 students was 51 and by 9 other students was 68. The average marks of total 17 students was ....?

Sol : Total marks of 8 students $= 8 \times 51 = 408$

Total marks of 9 students $= 9 \times 68 = 612$

Toal marks of 17 students $= 408 + 612 = 1020$

$$\text{Average marks of 17 students} = \frac{1020}{17} = \mathbf{60}$$

Example 21. A company produces on an average 4000 items per month for its first three months. How many items it must produce over next 9 months to average 4375 item per month over the whole year?

Sol :

Total production of first 3 months $= 3 \times 4000 = 12000$

Total production of 12 months reqd. $= 12 \times 4375 = 52500$

Total production of 9 months reqd $= 52500 - 12000 = 40500$

$$\therefore \text{Average production of next 9 months} = \frac{40500}{9} = \mathbf{4500}$$

Example 22. What is the average of first 9 multiples of 5?

Sol : Sum of first 9 multiples of 5

$= 5 + 10 + 15 + 20 + 25 + 30 + 35 + 40 + 45$

$= 5\,[1 + 2 + 3 + 4 + 5 + 6 + 7 + 8 + 9]$

$= 5 \times [45] = 225$

$$\therefore \text{Average of first 9 multiples of 5} = \frac{225}{9} = \mathbf{25}$$

Example 23. *a*, *b*, *c*, *d*, *e* and f are six consecutive odd numbers. Find their average.

Sol :

$$\begin{aligned}
\text{First odd no.} &= a \\
\text{Second odd no.} &= a+2=b \\
\text{Similarly } c &= b+2=a+2+2=a+4 \\
d &= a+6 \\
e &= a+8 \\
f &= a+10
\end{aligned}$$

$$\begin{aligned}
\text{Their average} &= \frac{\text{Their sum}}{6} \\
&= \frac{a+a+2+a+4+a+6+a+8+a+10}{6} \\
&= \frac{6a+30}{6} = \frac{6(a+5)}{6} \\
&= \boldsymbol{a+5}
\end{aligned}$$

EXERCISE

Q. 1. The average temp. of Sunday, Monday, Tuesday was 96°F. The average temp of Monday, Tuesday and Wednesday was 99°F. If the temperature on Wednesday was 103°F. Find the temp on Sunday.

(a) 112°F (b 84°F (c) 94°F (d) none

Q.2. The average tempreture of Monday, Tuesday and Wednesday was 40°C. The average for Tuesday, Wednesday and Thursday be 41°C. If temperature of thursday was 45° C find the temp. of Monday.

(a) 43°C (b) 46°C (c) 48°C (d) none

Q.3. The average temp. on Ist four days of month is 40°C. The average temp. of 2nd, 3rd, 4th and 5th day is 38°C. The ratio of temp. of Ist day to 5th day is 4 : 3. Find the temp of first day

(a) 40°C (b) 32°C (c) 38°C (d) none

Q.4. There were 40 girls in a class. One of them weighing 50 kg. leaves the class. A new girl comes and increase the average weight of the class by 1/4 kg. Find the weight of new girl.

(a) 40 kg (b) 60 kg (c) 30kg (d) none

Q.5. There were 40 girls in a class. The average of the class is reduced by 1 year if 10 girls whose average age is 20 years leave the class and are replaced by 10 new girls. What is the average age of the new girls.

(a) 18 years (b) 16 years (c) 20 years (d) none

Q.6. The average weight of class of 40 students is 25 kg. If the teachers weight be included the average increasing by 1.5 kg. Find the weight of teacher.

(a) 86.5 kg. (b) 84.5 kg (c) 83.5 kg (d) none

Q.7. The average age of a class of 35 students is 18 years and if the teacher's age is included, the average becomes 19 years. Find the age of teacher.

(a) 58 years (b) 60 years (c) 52 years (d) none

Q.8. The average age of 3 childern in a family is 9 years. If the age of parents be included the average age increases by 10 years. Find the average age of parents.

(a) 58 years (b) 78 years (c) 68 years (d) none

Q.9. Three years ago, the average age of a family of five members was 16 years. A baby having been born and the average age of the family is now the same as before. Find the age of baby.

(a) 3 yrs (b) 2 yrs (c) 1 yrs (d) none

Q.10. The average age of 40 students in a class is 14 years 10 month . If by the admission of 5 more students, average age decrease by 2 months. Find average age of new students.

(a) 13.5 yrs (b) 13.25 yrs (c) 14 yrs (d) none

Q.11. A batsman makes a score of 50 in his fifth innings thereby increases his average by 2. What is his average after fifth innings. He has never been not out.

(a) 42 (b) 48 (c) 45 (d) none

Q.12. A person spends on an average Rs. 675 per months for first four months and Rs.800 per month for the next 8 months. Find his monthly income, if he saves Rs. 3872 during whole year.

(a) Rs. 1092 (b) Rs. 1081 (c) Rs.1084 (d) none

Q.13. The average expenditure of a man for first five months of a year is Rs. 350. If the average expenditure for next seven months is Rs. 300 and he saves Rs.950 at the end of the year. Find his monthly income.

(a) Rs.600 (b) Rs.300 (c) Rs.500 (d) none

Q.14. The average salary per head of the entire staff of an office including officer is Rs.80. The average salary of officers is Rs. 480 and of clerks is Rs. 60. Given that the number of officer is 15. Find the number of clerks.

(a) 400 (b) 20 (c) 120 (d) none

Q.15. The average age of 8 men is increased by 2 years if one of them whose age is 24 years is replaced by a fresh man. Find is the age of fresh man.

(a) 24 years (b) 26 years (c) 40 years (d) none

Q.16. There were 40 students in a hostel. If the number of students is increased by 10 the expense of the mess increased by Rs. 50. While the average diminished by Rs. 4. The original expenditure of the mess was .....?

(a) Rs. 25 (b) Rs. 1000 (c) Rs. 1040 (d) none

Q.17. Of the three numbers first is three times the third and second is twice the third. The average of three is 12.Find the first number.

(a) 18 (b) 12 (c) 6 (d) none

Q.18. Of three numbers second is 1/2 of the 1st and third is 1/2 of the second. if the average of three is 28. Find the numbers.

(a) 48, 32, 16 (b) 24,12, 6 (c) 48, 24,12 (d) none

Q.19. Average weight of A, B and C is 50 kg . Average weight of A and B is 42 Kg and that of B and C is 44 kg. Find the weight of B.

(a) 33 kg (b) 40 kg (c) 22 kg (d) none

Q.20. A shopkeeper earned Rs. 400 in 10 days . His average income for first four days is Rs. 40 a day. His average income for remaining days is......?

(a) Rs.60 (b) Rs.30 (c) Rs.45 (d) none

SOLUTIONS

1. (c)

Total of Sunday + Monday + Tuesday = 96 × 3 = 288°

Total of Monday + Tuesday + Wednesday = 99 × 3 = 297

– – –

Sunday – Wednesday = –9°

Sunday – 103° = –9°

Sunday = – 9° + 103° = **94°F**

2. (d)

Total of Monday + Tuesday + Wednesday = 40 × 3 = 120°

Total of Tuesday + Wednesday + Thursday = 41 × 3 = 123°

– – –

Monday – Thursday = –3°

Monday – 45° = –3°

Monday = –3° + 45° = **42°C**

3. (b)

I + II + III + IV = 40 x 4 = 160

II + III + IV + V = 38 x 4 = 152

– – – –

I –V = 8

Ist – Vth = 8

$4x$ – $3x$ = 8

x = 8

∴ Temp on first day = 4 × 8 = **32°C**

4. (b) Total increase = $40 \times \frac{1}{4}$ = 10 kg

weight of new girl = 50 + 10 = **60 kg.**

5. (b) Total decrease = 40 × 1 = 40 years.

Total age of 10 girls [leaving] = 10 × 20 = 200 years

Total age of new coming girls = 200 – 40 = 160 years

Average age of 10 new coming girls = $\frac{160}{10}$ = **16 years**

6. (a) Total wt. of 40 students = 40 × 25 = 1000 kg.

Total wt. of 41

[40 students + 1 teacher] = 41 × 26.5 = 1086.5 years

Weight of teacher = 1086.5 – 1000 **= 86.5 kg.**

7. (d) Teacher's age = 36 × 19 – 35 × 18

= 684 – 630 = **54 years**

8. (d) Total age of childern $= 3 \times 9 = 27$ years

Total age of family $= 5 \times 19 = 95$ years

Total age of parents $= 95 - 27 = 68$ years

Average age of parents $= \dfrac{68}{2} =$ **34 years**

9. (c) Total age of five members 3 years ago $= 5 \times 16 = 80$ years

Total age of five members at present $= 5 \times 19 = 95$ years

Total age of 6 members at present $= 6 \times 16 = 96$ years

Age of baby $= 96 - 95 =$ **1 year**

10. (d) Total age of 40 students $= 40 \times 178$ months $= 7120$ months

Total age of 45 students $= 45 \times 176 = 7920$ months

Total age of 5 students $= 7920 - 7120 = 800$ months

Average of five $= \dfrac{800}{5} = 160$ months

$=$ **13 years 4 months**

11. (a) Let average after 4 innings $= x$

$$\therefore \quad 4x + 50 = 5(x + 2)$$

$$4x + 50 = 5x + 10$$

$$x = 40$$

Average after 5th innings $= 40 + 2 = 42$

12. (b) Total income $= 675 \times 4 + 800 \times 8 + 3872$

$=$ Rs. 2700 + Rs. 6400 + Rs. 3872

$=$ Rs. 12972

Average $= \dfrac{12972}{12} =$ **1081 Rs.**

13. (d) Total income $= 350 \times 5 + 300 \times 7 + 950$

$= 1750 + 2100 + 950 = 4800$ Rs.

Average $= \dfrac{4800}{12} =$ **400 Rs.**

14. (d)

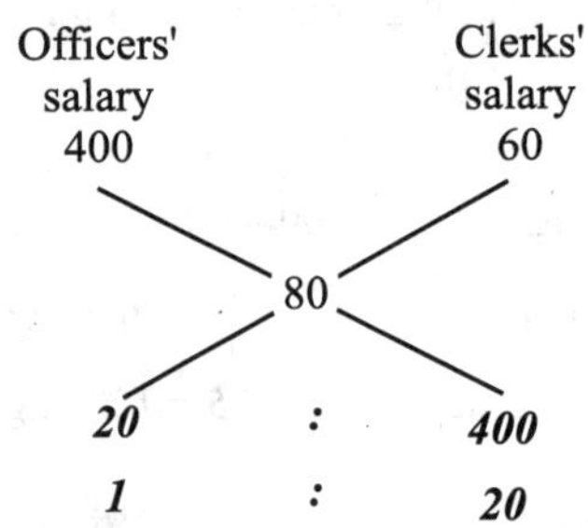

$\therefore$ Ratio of officers to clerks $= 1 : 20$

Number of officers $= 15$ **... (Given)**

$\therefore$ Number of clerks $= 20 \times 15 =$ **300**

15. (c)

16. (b) Let average expense $= x$

Total expense $= 40x$

Total increase in expense $= 50$

Total new expense $= 40x + 50$ ...(i)

New average $= x - 4$

Total new expense $= 50\,(x - 4)$ ...(ii)

From **(i)** and **(ii)**

$40x + 50 = 50\,x - 200$

$10x = 250$

$x = 25$

Original expense $= 40x = 40 \times 25 =$ **1000 Rs.**

17. (a) Let third no. $= x$

$\therefore$ 1st no. $= 3x$

And IInd no $= 2x$

According to statement

$$\frac{x + 3x + 2x}{3} = 12$$

$$6x = 36$$

$$x = 6$$

$\therefore$ 1st no. $= 3 \times 6$ $=$ **18**

18. (c) Let 1st number $= x$

$\therefore$ 2nd number $= \frac{x}{2}$

3rd number $= \frac{x}{2} \times \frac{1}{2} = \frac{x}{4}$

According to statement

$$\frac{x + \frac{x}{2} + \frac{x}{4}}{3} = 28$$

$$\frac{7x}{4} = 84$$

$$x = \frac{84 \times 4}{7} = 48$$

$\therefore$ 1st no. $=$ **48,** 2nd no. $=$ **24**

3rd no. $=$ **12**

19. (c) Weight of

$A + B + C = 150$ kg ...(i)

$A + B = 84$ kg ...(ii)

$B + C = 88$ kg ...(iii)

Add **(ii)** and **(iii)**

$A + 2B + C = 172$ kg ...(iv)

Subtract **(i)** from **(iv)** we get B $=$ **22 kg**.

20. (d) Total income of last 6 days $= 400 - 4 \times 40 = 400 - 160 = 240$ Rs.

$\therefore$ Average income of last 6 days $= \frac{240}{6}$ **= 40 Rs.**

Day 11

Partnership

On day 11 we will do the topic of **Partnership**. Partnership is of two types.

1. **Simple**
2. **Compound**

SIMPLE PARTNERSHIP

In Simple Partnership two or more persons join hands to do a business such that the capital of all the partners is used for same time. The profit is invested in the ratio of their investments. So in simple partnership

Investment ratio = Profit ratio

COMPOUND PARTNERSHIP

In compound partnership like simple partnership two or more persons join hands to do a business but their capital are used for different times.

The basic rule of **Investment Ratio = Profit ratio** is followed, But here **Investment is taken as Capital × Time**

So in compound partnership

Investment ratio = Profit Ratio

and **Investment** = **Capital × Time**

Types of Questions

Type I : Questions on Simple Partnership

Type II : Questions on Compound Partnership

Type III : Working and Sleeping Partners

Type IV : Questions related to Rent etc.

Type V : Misc.

Type I [Simple Partnership]

Example 1. Three men A, B and C invest Rs. 2000, Rs. 3000 and Rs. 4000. How should they divide a profit of Rs. 2700?

Sol : Investment Ratio = 2000 : 3000 : 4000 = **2 : 3 : 4**

$$\text{Profit of A} = \frac{2}{9} \times 2700 = \mathbf{600\ Rs.}$$

$$\text{Profit of B} = \frac{3}{9} \times 2700 = \textbf{900 Rs.}$$

$$\text{Profit of C} = \frac{4}{9} \times 2700 = \textbf{1200 Rs.}$$

Example 2. A and B enter a business with A investing Rs. 2000 and B investing Rs. 3000. If they receive Rs. 5000 as profit after one year, what is the share of B?

Sol :

$$\text{Investment Ratio} = 2000 : 3000 = \mathbf{2 : 3}$$

$$\text{Share of B} = \frac{3}{5} \times 5000 = \textbf{3000 Rs.}$$

Example 3. A and B enter into a partnership A invests Rs. 4000. How much should B invest so that Profit after one year may be divided in the ratio 2 : 7.

Sol :

$$\text{Let B invest} = \text{Rs. } x$$

$$\text{Investment ratio} = \text{Profit Ratio}$$

$$\frac{4000}{x} = \frac{2}{7}$$

$$x = \frac{4000 \times 7}{2} = \textbf{14000 Rs.}$$

Example 4. Three Investers A, B and C invest Rs. 4700. A invests Rs. 500 more than B and B invest Rs. 300 more than C. How should they divide the profit of Rs. 940.

Sol :

$$\text{Let investment of C} = x \text{ Rs.}$$

$$\text{Then investment of B} = x + 300$$

$$\text{And investment of A} = x + 300 + 500$$

According to question

$$x + x + 300 + x + 800 = 4700 \text{ Rs.}$$

$$3x = 3600 \text{ Rs.}$$

$$x = 1200 \text{ Rs.}$$

$$\therefore \quad \text{Investment of C} = 1200 \text{ Rs.}$$

$$\text{And investment of B} = 1500 \text{ Rs.}$$

$$\text{And investment A} = 2000 \text{ Rs.}$$

$$\text{Investment Ratio} = 2000 : 1500 : 1200 = \mathbf{20 : 15 : 12}$$

$$\text{Profit of A} = \frac{20}{47} \times 940 = \textbf{400 Rs.}$$

$$\text{Profit of B} = \frac{15}{47} \times 940 = \textbf{300 Rs.}$$

$$\text{Profit of C} = \frac{12}{47} \times 940 = \textbf{240 Rs.}$$

Type II (Compound Partnership)

Rule:

$$\text{Investment Ratio} = \text{Profit Ratio}$$

And

$$\textbf{Investment} = \textbf{Capital} \times \textbf{Time}$$

Example 5. A and B start a business. A invests Rs.1000 for 8 months and B invests Rs.1200 for 7 months. If they receive profit of Rs. 4100 after one year, find the profit of A.

Sol : Investment of A $= 1000 \times 8 = 8000$ Rs.

Investment of B $= 1200 \times 7 = 8400$ Rs.

$\therefore$ Investment Ratio $= 8000 : 8400 = \mathbf{20 : 21}$

Profit of A $= \frac{\mathbf{20}}{41} \times 4100 = \mathbf{2000\ Rs.}$

And Profit of B $= \frac{21}{41} \times 4100 = \mathbf{2100\ Rs.}$

Example 6. A, B and C start a business together. A invests Rs. 1200 for 6 months, B invests Rs.800 for 7 months and C Rs. 600 for 8 Months. How should they divide a profit of Rs. 396?

Sol : Investment of A $= 1200 \times 6 = 7200$ Rs.

Investment of B $= 800 \times 7 = 5600$ Rs.

Investment of C $= 600 \times 8 = 4800$ Rs.

Investment Ratio $= 7200 : 5600 : 4800$

$= 18 : 14 : 12 = \mathbf{9 : 7 : 6}$

Profit of A $= \frac{9}{22} \times 396 = \mathbf{162\ Rs.}$

Profit of B $= \frac{7}{22} \times 396 = \mathbf{126\ Rs.}$

Profit of C $= \frac{6}{22} \times 396 = \mathbf{108\ Rs.}$

Example 7. There are three partners in a business. A invests Rs. 6000 for 3 months, B invests Rs.5000 for 5 months and C Rs. 1000 for 2 months. Total gain is Rs. 4500. How should they divide the profit?

Sol : Investment of A $= 6000 \times 3 = 18000$ Rs.

Investment of B $= 5000 \times 5 = 25000$ Rs.

Investment of C $= 1000 \times 2 = 2000$ Rs.

Ratio of Investments $= 18000 : 25000 : 2000 = \mathbf{18 : 25 : 2}$

Profit of A $= \frac{18}{15} \times 4500 = \mathbf{1800\ Rs.}$

Profit of B $= \frac{25}{45} \times 4500 = \mathbf{2500\ Rs.}$

Profit of C $= \frac{2}{45} \times 4500 = \mathbf{200\ Rs.}$

Example 8. A and B start a business. A invests Rs. 1200 for 8 months and B invests Rs. 2000 for 5 months. If the profit received at the end of the year was Rs. 4900, find the Profit of A.

Sol : Investment of A $= 1200 \times 8 = 9600$ Rs.

Investment of B $= 2000 \times 5 = 10000$ Rs.

$$\text{Investment Ratio} = 9600 : 10000 = \mathbf{24 : 25}$$

$$\text{A's Share} = \frac{29}{49} \times 4900 = \mathbf{2400\ Rs.}$$

Example 9. A and B enter into a buisness with their capitals in the ratio 1/2 : 1/3. After 5 months A withdraws half of his capital and after 7 months more a profit of Rs. 1650 is divided. Find the profit of A.

Sol :

$$\text{Investment ratio} = \frac{1}{2}.\frac{1}{3} = \frac{3:2}{6} = \mathbf{3 : 2}$$

$$\text{Let A's Capital} = 300\ \text{Rs.}$$

$$\text{Let B's Capital} = 200\ \text{Rs.}$$

According to question

$$\text{A's investment} = 300 \times 5 + 150 \times 7 = 1500 + 1050 = \mathbf{2550\ Rs.}$$

$$\text{B's Investment} = 200 \times 12 = \mathbf{2400\ Rs.}$$

$$\text{Investment Ratio} = 2550 : 2400 = \mathbf{17 : 16}$$

$$\text{As share} = \frac{17}{33} \times 1650 = \mathbf{850\ Rs.}$$

Type III [Working and Sleeping Partner]

Example 10. A is working and B is sleeping Partner in a business. A puts in Rs. 5000 and B Rs. 6000. A receives $12\frac{1}{2}$% for managing. The rest being divided propotional to their capitals. What does each receive out of the total profit of Rs. 880?

Sol :

$$\text{Total Profit} = 880$$

$$\text{A's salary} = \frac{1}{8} \times 880 = \mathbf{110\ Rs.}$$

$$\text{Remaining Profit} = 880 - 110 = \mathbf{770\ Rs.}$$

$$\text{Ratio of Investment} = \mathbf{5 : 6}$$

$$\text{A's Profit} = \frac{5}{11} \times 770 + 110 = 350 + 110 = \mathbf{460\ Rs.}$$

$$\text{B's profit} = \frac{6}{11} \times 770 = \mathbf{420\ Rs.}$$

Type IV

Example 11. A, B and C rent a room for two years at Rs. 360 per annum. A remains in it for whole time, B for 16 months and C for 4 months during the occupancy of B. How much must each pay out of rent?

Sol :

$$\text{Rent for 1 year} = 360\ \text{Rs.}$$

$$\text{Rent for 1 month} = \frac{360}{12} = 30\ \text{Rs.}$$

From question it is clear that for 4 months A, B and C have lived together and for 12 months A and B have lived together and rest 8 months A has lived alone. So rent is to he shared accordingly.

$$\text{Rent of 4 months} = 4 \times 30 = 120 \text{ Rs.}$$

It is divided equally among A, B and C.

1. $$\text{Each's share} = \frac{1}{3} \times 120 = 40 \text{ Rs.}$$

2. $$\text{Now total rent for 12 months} = 12 \times 30 = 360 \text{ Rs.}$$

It is divided among A and B

$$\text{So each has to pay} = \frac{1}{2} \times 360 = 180 \text{ Rs.}$$

3. $$\text{Rent for 8 months} = 8 \times 30 = 240 \text{ Rs.}$$

$$\text{Total rent paid by A} = 40 + 180 + 240 = \mathbf{460 \text{ Rs.}}$$

$$\text{Total rent paid by B} = 40 + 180 = \mathbf{220 \text{ Rs.}}$$

$$\text{Total rent paid by C} = \mathbf{40 \text{ Rs.}}$$

Type V

Example 12. A starts a business with Rs. 4200 and later on B joined him with Rs. 7200. After one year profits were divided equally. Find when did B joined A.

Sol :

Let time for which B's money used $= x$

$$\therefore \quad \frac{4200 \times 12}{7200 \times x} = \frac{1}{1}$$

$$4200 \times 12 = 7200 \times x$$

$$x = \frac{4200 \times 12}{7200} = \mathbf{7 \text{ months}}$$

$$\therefore \quad \text{B joined after } \mathbf{12 - 7} = \mathbf{5 \text{ months}}$$

Example 13. A and B enter into a business. B who invests 1600 for 8 months claim 5/8 of the profit. How much did A contribute If his money remains in the business for 4 months.

Sol : $$\text{B's profit} = \frac{5}{8}$$

$$\text{A's Profit} = 1 - \frac{5}{8} = \frac{3}{8}$$

$$\text{Profit Ratio} = \frac{3}{8} : \frac{5}{8} = \mathbf{3 : 5}$$

$$\frac{x \times 4}{1600 \times 8} = \frac{3}{5}$$

$$x = \frac{3 \times 1600 \times 8}{4 \times 5} = \mathbf{1920 \text{ Rs.}}$$

Example 14. Vivek and Jatin start a business . Vivek invests Rs. 400 for 6 months. Jatin invests Rs. 600 for whole year. If at the end of the year Jatin gets Rs. 720 as profit, what was the profit of Vivek?

Sol : Let Profit of Vivek = x

$$\therefore \quad \frac{400 \times 6}{600 \times 12} = \frac{x}{720}$$

$$x = \frac{400 \times 6 \times 720}{600 \times 12} = \mathbf{240\ Rs.}$$

Example 15. A started a business with Rs. 4500 and after some time B joined A with Rs. 5400. The ratio of their profit at the end of the year is 2 :1. When did B join?

Sol : Let B join for x months

$$\frac{4500 \times 12}{5400 \times x} = \frac{2}{1}$$

$$x = \frac{4500 \times 12}{5400 \times 12} = 5 \text{ months}$$

$\therefore$ B joined after 12 – 5 = **7 months.**

EXERCISE

Q.1. Rajesh, Sumeer and Vikas invest Rs. 4000, Rs. 5000 and Rs. 8000 respectively in a business. How should they divide a profit of Rs. 8500?

(a) Rs. 2500, Rs. 4000, Rs. 2000
(b) Rs. 3500, Rs. 4000, Rs. 1000
(c) Rs. 2000, Rs. 2500, Rs. 4000
(d) None

Q.2. A, B and C invest Rs. 5000 in the ratio of 3 : 8 :14. What should be the profit of B if the total profit is Rs. 2500.

(a) Rs. 800 (b) Rs. 1400 (c) Rs. 400 (d) None

Q.3. A, B,C and D enter into a partnership. A invests 1/3 of the capital, B 1/4, C 1/5 and D the rest. If the total profit is Rs. 840 then what is the profit of D?

(a) Rs. 210 (b) Rs. 182 (c) Rs. 282 (d) None

Q.4. Three persons A, B and C rent a pasture. A puts in 12 cows for 4 months, B 16 cows for 5 months and C 24 cows for 5 months. If A pays Rs. 480, how much should B pay?

(a) Rs. 1200 (b) Rs. 4800 (c) Rs. 400 (d) None

Q.5. A, B and C enter into a partnership. A invests Rs. 1200, B Rs.1500 and C Rs. 2000. After 2 months C withdraws Rs. 200 and after 5 months, B invests Rs. 300 more. Total profit after a year is Rs.1130. Find the profit of B.

(a) Rs. 402 (b) Rs. 440 (c) Rs. 288 (d) None

Q.6. A invests Rs.1500 and B Rs. 900 in a business. A is a sleeping partner and B is working partner. B gets 10% of the total profit as salary and rest of the profit is divided in the ratio of their capitals. If total profit is Rs.1600. How should they divide it?

(a) Rs. 900, Rs. 700 (b) Rs. 800, Rs. 800 (c) Rs. 1000, Rs. 600 (d) none

Q.7. A and B enter into a business with investments in the ratio 1/3 : 1/5. A with draws half of his capital at the end of 15 months and after 15 months more a profit of Rs. 5400 is divided. Find the share of A.

(a) Rs. 3000 (b) Rs. 2400 (c) Rs. 2900 (d) none

Q.8. Gagan started a business with Rs. 37500. After some time Ashu joined with him Rs. 50000. After one year they shared the profit equally. Find after how many months Ashu joined the business.

(a) 9 months (b) 3 months (c) 8 months (d) none

Q.9. A, B and C start a business. A invest half of the capital for 1/3 of the time, B 1/3 of the capital for 1/4 of the time and C the rest of capital for the whole year. How should they divide a profit of Rs. 2000.

(a) Rs. 4800, Rs.400, Rs.800 (b) Rs.800, Rs.400, Rs.800

(c) Rs.800, Rs.600, Rs.600 (d) none

Q.10. A and B started a joint firm. A's investment was four times the investment of B and time period of B is twice the time period of A. If A got Rs. 2000 as profit then their total profit is:

(a) Rs. 1000 (b) Rs. 3500 (c) Rs. 4000 (d) none.

Q. 11. A, B and C accepted a work for Rs.770. Together B and C do 5/7 of the work and rest is done by A. The share of A should be:

(a) Rs. 500 (b) Rs. 550 (c) Rs. 220 (d) none

SOLUTIONS

1. (c) Investment Ratio of Rajesh, Sumeer and Vikas = **4 : 5 : 8**

$$\text{Profit of Rajesh} = \frac{4}{7} \times 8500 = \mathbf{2000\ Rs.}$$

$$\text{Profit of Sameer} = \frac{5}{17} \times 8500 = \mathbf{2500\ Rs.}$$

$$\text{Profit of VIkas} = \frac{8}{17} \times 8500 = \mathbf{4000\ Rs.}$$

2. (a)

$$\text{Profit of B} = \frac{8}{25} \times 2500 = \mathbf{800\ Rs.}$$

3. (b)

$$\text{D's Share} = 1 - \frac{1}{3} - \frac{1}{4} - \frac{1}{5}$$

$$= \frac{60 - 20 - 15 - 12}{60} = \mathbf{\frac{13}{60}}$$

$$\therefore \quad \text{Investment Ratio} = \frac{1}{3} : \frac{1}{4} : \frac{1}{5} : \frac{13}{60}$$

$$= \frac{20:15:12:13}{60} = \mathbf{20 : 15 : 12 : 13}$$

$$\therefore \quad \text{Profit of D} = \frac{13}{60} \times 840 = \mathbf{182\ Rs.}$$

4. (d)

$$\text{Ratio of rent to be paid} = 12 \times 4 : 16 \times 5 : 24 \times 5$$

$$= 48 : 80 : 120 = \mathbf{6 : 10 : 15}$$

$6x = 480$ ... **(Given)**

$x = 80$ Rs.

∴ B's rent $= 10 \times 80 =$ **800 Rs.**

5. (a) Investment Ratio of A, B and C

$= \mathbf{1200 \times 12 : 1500 \times 5 + 1800 \times 7 : 2000 \times 2 + 1800 \times 10}$

$= \mathbf{14400 : 7500 + 12600 : 4000 + 18000}$

$= \mathbf{14400 : 20100 : 22000}$

$= \mathbf{144 : 201 : 220}$

∴ Profit of B $= \frac{201}{565} \times 1130 =$ **402 Rs.**

6. (a) B's salary $= \frac{1600 \times 10}{100} = 160$ Rs.

Remaining Profit $= 1600 - 160 = 1440$ Rs.

Investment Ratio $= 1500 : 900$

$= 15 : 9 =$ **5 : 3**

Profit of A $= \frac{5}{8} \times 1440 =$ **900 Rs.**

" " B $= \frac{3}{8} \times 1440 + 160 =$ **700 Rs.**

7. (a) Investment Ratio $= \frac{1}{3} : \frac{1}{5} = \frac{5:3}{15} =$ **5 : 3**

Let A invests = 500 Rs.

B invest = 300 Rs.

A investment $= 500 \times 15 + 250 \times 15$

$= 7500 + 3750 = 11250$ Rs.

B's Investment $= 300 \times 30 = 9000$ Rs.

Investment Ratio $= 11250 : 9000 = 5 : 4$

A's share $= \frac{5}{9} \times 5400 =$ **3000 Rs.**

8. (b) Let Ashu join for x months

∴ $\frac{37500 \times 12}{50000 \times x} = \frac{1}{1}$

$x = \frac{37500 \times 12}{50000} = 9$ months

∴ Ashu joined after $= 12 - 9 =$ **3 months**

9. (b) C's Capital $= 1 - \frac{1}{2} - \frac{1}{3}$

$\frac{6-3-2}{6} = \frac{1}{6}$

Investment of A $= \frac{1}{2} \times 4 =$ **2 Rs.**

$$\text{Investment of B} = \frac{1}{3} \times 3 = \underline{\mathbf{1\ Rs.}}$$

$$\text{Investment of C} = \frac{1}{6} \times 12 = \mathbf{2\ Rs.}$$

So profit should be in the ratio **2 : 1 : 2**

$$\text{A's Profit} = \frac{2}{5} \times 2000 = \mathbf{800\ Rs.}$$

$$\text{B's Profit} = \frac{1}{5} \times 2000 = \mathbf{400\ Rs.}$$

$$\text{C's Profit} = \frac{2}{5} \times 2000 = \mathbf{800\ Rs.}$$

10. (d)

$$\frac{4 \times 1}{1 \times 2} = \frac{2000}{K}$$

$$K = 1000 \text{ Rs.}$$

$$\therefore \quad \text{Total Profit} = 2000 + 1000 = \mathbf{3000\ Rs.}$$

11. (c)

$$\text{Work done by A} = 1 - \frac{5}{7}. = \frac{2}{7}$$

$$\therefore \quad \text{Profit of A} = \frac{2}{7} \times 770 = \mathbf{220\ Rs.}$$

Day 12

Ratio & Proportion and Unitary

RATIO

Let us start this topic with a simple example. Ram and Shyam's daily wages are **Rs. 200** and **Rs.100** res.

$$\text{Ram's daily wages} = \text{Rs. } 200 \qquad ...(1)$$

$$\text{Shyam's daily wages} = \text{Rs. } 100 \qquad ...(2)$$

Divide **(1)** by **(2)**

$$\frac{\text{Ram's daily wages}}{\text{Shyam's daily wages}} = \frac{200}{100} = \frac{2}{1}$$

So $\frac{2}{1}$ is the ratio of **Ram's daily wages** to **Shyam's daily wages**. It is written as **2 : 1** and pronounced as **two is to one**. While finding the ratio of two quantites you must keep in mind that **units** of these quantities should be same.

So ratio of two quantities is a fraction that one quantity is of other.

In ratio first term is called **antecedent** and the second term is **consequent.**

So in **2 :1**, **2** is called **antecedent** and **1** is **consequent.**

Proportion

As discussed above ratio of Ram's salary and Shyam's salary is 2 : 1. Now Mohan's daily salary is Rs. 400 and Hari's salary is Rs. 200

$$\frac{\text{Mohan's salary}}{\text{Hari's salary}} = \frac{400}{200} = \frac{2}{1}$$

Ratio of **Mohan's salary to Hari's salary** is 2 : 1 **which is same as** the ratio of **Ram's salary to Shyam's salary**. These ratios are called in **proportion**. So **when two ratios are equal they are called in proportion.**

So 200 : 100 and 400 : 200 are equal and are called in proportion

$$200 : 100 = 400 : 200$$

It is pronounced as ratio of **200 is to 100** is same as ratio of **400 is to 200** and is written as

$$200 : 100 :: 400 : 200$$

So generalising

If $a : b$ is same as $c : d$ then these are called in **proportion** and is written as

$$a : b :: c : d$$

Now *a* is called **First Proportion**

b is called **Second Proportion**

c is called **Third Proportion**

d is called **Fourth Proportion**

$$a : b = c : d$$

$$\frac{a}{b} = \frac{c}{d}$$

Or $$ad = bc$$

Ist proportion × IV proportion = IInd proportion × IIIrd proportion is remembered as shown

$$a : b :: c : d$$

$$ad = bc$$

Continued Proprotion : If **second** and **third** proportion are same then the proportion are said to be continued proportion.

200 : 400 :: 400 : 800 is an example of continued proportion.

Generlising

a* : *b* :: *b* : *c is continued proportion and ***a*** is known as **first proportion**, ***b*** as **mean** and *c* as **third proportion**

$$a : b = b : c$$

$$\frac{a}{b} = \frac{b}{c}$$

$$b^2 = ac$$

$$b = \sqrt{ac}$$

Or $$\text{Mean prop.} = \sqrt{\text{Ist proportion} \times \text{IIIrd proportion}}$$

Example. Find the mean proportion of 3 and 75.

Sol : This question is written as follows

$$3 : b :: b : 75$$

$$b^2 = 3 \times 75$$

$$b^2 = 225$$

$$b = \sqrt{225} = \mathbf{15}$$

Now we hope that you have understood the basic concepts of the topic and we will now do the questions related to this topic. We have divided the questions of this topic into following types

Types of Questions

Type I : Simple Question

Type II : Questions Related to coins

Type III: Proportion

Type IV: Related to Ages and number

Example 1. Divide Rs. 400 in the ratio 1 : 3.

Type I

Sol :

$$\text{Sum of ratio's} = 1 + 3 = 4$$

$$\text{Ist part} = \frac{1}{4} \times 400 = 100 \text{ Rs.}$$

$$\text{IInd part} = \frac{3}{4} \times 100 = \mathbf{300 \text{ Rs.}}$$

Example 2. Divide Rs. 600 in the ratio 1 : 2 : 3.

Sol :

$$\text{Sum of ratio's} = 1 + 2 + 3 = 6$$

$$\text{Ist part} = \frac{1}{6} \times 600 = 100 \text{ Rs.}$$

$$\text{IInd part} = \frac{2}{6} \times 600 = 200 \text{ Rs.}$$

$$\text{IIIrd part} = \frac{3}{6} \times 600 = 300 \text{ Rs.}$$

Example 3. Divide Rs. 8080 in ratio 1 : 2 : 5

Sol.

$$\text{Ist part} = \frac{1}{8} \times 8080 = \mathbf{1010 \text{ Rs.}}$$

$$\text{IInd part} = \frac{2}{8} \times 8080 = \mathbf{2020 \text{ Rs.}}$$

$$\text{IIIrd part} = \frac{5}{8} \times 8080 = \mathbf{5050 \text{ Rs.}}$$

Type II Questions Related to Coins

Example 4. A bag contains100 paise 50 paise and 25 paise coins in the ratio 5/2 : 3 : 4. If the total amount is Rs. 110. Find the no. of coins of each type.

Sol : Total Sum = **110 Rs.**

Now the **ratio of coins** is given in **numbers** whereas **total sum** is given in terms of **value** i.e., **Rs. 110**. Therefore we have to first convert it into value shown below.

	100 p		50 p		25 p
	$\frac{5}{2}$	:	3	:	4
Ratio in value	$100 \times \frac{5}{2}$	:	50×3	:	25×4
	250	:	150	:	100
	5	**:**	**3**	**:**	**2**

$$\therefore \quad \text{Value of rupee coins} = \frac{5}{10} \times 110$$

$$= 55 \text{ Rs.}$$

In 55 Rs. rupee coins are **55**.

$$\text{Value of 50 P coins} = \frac{3}{10} \times 110 = 33 \text{ Rs.}$$

$$\text{The number of 50 p. coins in 33 Rs. is} = 33 \times 2 = \mathbf{66}$$

$$\text{The value of 25 p coins} = \frac{2}{10} \times 110$$

$$= 22 \text{ Rs.}$$

$$\text{The number of 25 p coin in 22 Rs.} = 22 \times 4 = \mathbf{88}$$

Example 5. A bag contains 25 p, 10 p and 5 p coins in the ratio 1:2:3. if their total value is Rs. 24 the number of 10 p coin is ....

Sol : Total Sum = 24 Rs.

Ratio of 25 p, 10 p and 5p coin is = 1 : 2 : 3

Ratio of these coins in terms of value = $25 \times 1 : 10 \times 2 : 5 \times 3$

25 p : 20 p : 15 p

5 : 4 : 3

$$\text{Value of 10 p coins} = \frac{4}{12} \times 24 = \mathbf{8 \text{ Rs.}}$$

No. of 10 paise coins in 8 Rs. = $8 \times 10 =$ **80 coins**

Type III (Proportion)

Example 6. If a:b = 2:3 and b:c = 4:5. Find a:b:c

Sol : We will do it by N formula shown below

a : b : c

2 : 3

4 : 5

$2 \times 4 : 4 \times 3 : 3 \times 5$

$\therefore$ $a : b : c$ = **8 : 12 : 15**

It is hoped that you have undetstood the N method.

Example 7. If a : b = 1 : 2 and b : c = 1 : 5. Find a : b : c

Sol :

a : b : c

1 : 2

: 5

$1 \times 1 : 2 \times 1 : 2 \times 5$

$\therefore$ $a : b : c$ = **1 : 2 : 10**

Example 8. Divide Rs. 350 among A, B and C such that A's share : B's share = 2 : 3 and B's share : C's share = 4 : 5

Sol : Total sum = 350 Rs.

A : B : C

2 : 3

4 : 5

8 : 12 : 15

$$\text{A's share} = \frac{8}{35}\times 350 = \mathbf{80\ Rs.}$$

$$\text{B's share} = \frac{12}{35}\times 350 = \mathbf{120\ Rs.}$$

$$\text{C's share} = \frac{15}{35}\times 350 = \mathbf{150\ Rs.}$$

Example 9. Divide Rs. 81 into 3 parts such that 1/2 of the Ist is equal to 1/3 of the IInd and 1/4 of the IIIrd.

Sol :

$$\text{Total sum} = 81\text{ Rs.}$$
$$\text{Let Ist part} = x$$
$$\text{IInd part} = y$$
$$\text{IIIrd part} = z$$

$$\therefore \quad \frac{1}{2}x = \frac{1}{3}y = \frac{1}{4}z = k \text{ (suppose)}$$

$$\frac{1}{2}x = k$$
$$x = 2k$$

Similarly

$$y = 3k$$
$$z = 4k$$
$$x : y : z = 2k : 3k : 4k$$
$$= 2 : 3 : 4$$

$$\text{Ist part} = \frac{2}{9}\times 81 = \mathbf{18\ Rs.}$$

$$\text{IInd part} = \frac{3}{9}\times 81 = \mathbf{27\ Rs.}$$

$$\text{IIIrd part} = \frac{4}{9}\times 81 = \mathbf{36\ Rs.}$$

Example 10. Divide Rs. 53 among A, B and C such that A gets Rs. 7 more than B and B Rs. 8 more than C.

Sol : Let C gets Rs. x

$\therefore$ B will get Rs. $x + 8$

and C will get Rs. $x + 8 + 7 = x + 15$

$$\therefore \quad x + x + 8 + x + 15 = 53$$
$$3x + 23 = 53$$
$$3x = 30$$
$$x = 10$$

$$\text{C's share} = \mathbf{10\ Rs.}$$
$$\text{B's share} = 10 + 8 = \mathbf{18\ Rs.}$$
$$\text{A's share} = 18 + 7 = \mathbf{25\ Rs.}$$

Example 11. The fourth proportion of 4, 8 and 12 is

Sol : Let fourth proportion is x

$$\therefore \quad 4 : 8 :: 12 : x$$
$$4x = 8 \times 12$$
$$x = \frac{8 \times 12}{4} = \mathbf{24}$$

Example 12. The third Proportion of 7 and 14 is

Sol : This is clearly a case of continued Proportion.

Here 7 is First Proportion and 14 is mean.

Let x is third Proportion.

$$\therefore \quad 7 : 14 :: 14 : x$$
$$7x = 14 \times 14$$
$$x = \frac{14 \times 14}{7} = \mathbf{28}$$

Example 13. Find the mean proportion of 6 and 96

Sol : Let x be mean proportion of 6 and 96.

$$\therefore \quad 6 : x :: x : 96$$
$$x^2 = 6 \times 96$$
$$x^2 = 576$$
$$x = \mathbf{24}$$

Example 14. What number should be added to 5,11,14 and 29 to make it proportionate.

Sol : Let x be added to each to make it proportionate

$$\therefore \quad 5 + x : 11 + x :: 14 + x : 29 + x$$
$$\frac{5+x}{11+x} = \frac{14+x}{29+x}$$
$$(5 + x)(29 + x) = (11 + x)(14 + x)$$
$$145 + 5x + 29x + x^2 = 154 + 11x + 14x + x^2$$
$$145 + 34x = 154 + 25x$$
$$34x - 25x = 154 - 145$$
$$9x = 9$$
$$x = \mathbf{1}$$

Type IV

(Questions related to Numbers and ages)

Example 15. Ratio of present ages of Japneet and Manjot is 1 : 2. If sum of their ages is 18 year. What is the present age of Manjot?

Sol : Let Japneet's age $= x$

Since their ages are in ratio $= 1 : 2$

$\therefore$ Manjot's Age $= 2x$

According to statement

$$x + 2x = 18 \text{ years}$$
$$3x = 18 \text{ years}$$
$$x = 6 \text{ years}$$
$$\text{Manjot's age} = 2x = 2 \times 6 = \mathbf{12\ years}$$

Example 16. If ratio of Gagan's present age to Vishal's present age is 1 : 2 and sum of their ages is 36 years. Find the present age of Vishal.

Sol. Let present age of Gagan $= x$

$$\text{Present age of Vishal} = 2x$$
$$x + 2x = 36$$
$$3x = 36$$
$$x = 12 \text{ years}$$
$$\therefore \quad \text{Present age of Vishal} = 2 \times 12 = \mathbf{24\ years}$$

Example 17. Shammi's present age to Pawan's present age is 7 : 9. Three years ago their ages were in the ratio 3 : 4. Find their present ages?

Sol. Present age of Shammi $= 7x$

$$\text{Present age of Pawan} = 9x$$

Before 3 years

$$\text{Shammi's age} = 7x - 3$$
$$\text{Pawan's age} = 9x - 3$$

According to statement

$$\frac{7x-3}{9x-3} = \frac{3}{4}$$
$$28x - 12 = 27x - 9$$
$$28x - 27x = 12 - 9 = 3$$
$$x = 3 \text{ years}$$
$$\therefore \quad \text{Shami's age} = 7 \times 3 = \mathbf{21\ years}$$
$$\text{Pawan's age} = 9 \times 3 = \mathbf{27\ years}$$

Example 18. A father is three times as old as son. After 10 years, Father will be twice old as his son. Find the present age of son.

Sol : Let present age of son $= x$ years

" " " " father $= 3x$ years

After 10 years

$$\text{Age of son} = x + 10 \text{ years}$$
$$\text{Age of father} = 3x + 10 \text{ years}$$

According to statement

$$3x + 10 = 2(x + 10)$$
$$3x + 10 = 2x + 20$$
$$3x - 2x = 20 - 10$$
$$x = \mathbf{10\ years}$$

Example 19. Six years ago a man was three times as old as his son. Six years hence, he will be twice as old as his son then will be. Find the present age of father.

Sol : Let six years ago

$$\text{Age of man} = x \text{ years}$$
$$\text{" " father} = 3x \text{ years}$$
$$\text{Present age of the son} = x + 6$$
$$\text{" " " " father} = 3x + 6$$

After six years

$$\text{Age of the son} = x + 6 + 6 = x + 12$$
$$\text{" " " father} = 3x + 6 + 6 = 3x + 12$$

According to condition

$$3x + 12 = 2(x + 12)$$
$$3x + 12 = 2x + 24$$
$$3x - 2x = 24 - 12$$
$$x = 12 \text{ years}$$

Therefore Present age of father = $3 \times 12 + 6$ = **42 years**

Example 20. Two numbers whose sum is 52 are such that $\frac{1}{5}$ of one of them is equal to $\frac{1}{8}$ th of the second. Find them.

Sol :

$$\text{Let 1st number} = x$$
$$\text{And 2nd number} = y$$
$$\frac{1}{5}x = \frac{1}{8}y$$
$$\frac{x}{y} = \frac{5}{8}$$
$$x : y = 5 : 8$$
$$\therefore \quad \text{1st number} = \frac{5}{13} \times 52 = \mathbf{20}$$
$$\text{2nd number} = \frac{8}{13} \times 52 = \mathbf{32}$$

Example 21. A number consists of two digits whose sum is 5. If 9 be added to the number, the digits are reversed. Find the number.

Sol :

$$\text{Let digit at unit place} = x$$
$$\text{" " " ten's place} = y$$
$$x + y = 5 \qquad \textbf{...(1)}$$
$$\text{Number} = 10y + x$$

Acoording to statement

$$10y + x + 9 = 10x + y$$
$$9x - 9y = 9$$
$$x - y = 1 \qquad \textbf{...(2)}$$

Solving **(1)** and **(2)** we get

$$x = \mathbf{3} \quad \text{and} \quad y = \mathbf{2}$$

$$\therefore \quad \text{Number} = 10y + x = 10 \times 2 + 3 = \mathbf{23}$$

Unitary Method

Unitary word arises from **Unit** means **1**. In this we bring question to the **single quantity** then **proceed** as explained in following example.

Example 22. If the cost of 12 pencils is Rs. 24. Find the cost price of 1 pencil.

Sol : Cost of 12 pencils = 24 Rs.

Cost of 1 pencil obviously will be less than cost of 12 pencils.

$$\therefore \quad \text{cost of 1 pencil} = \frac{24}{12} = \mathbf{2\ Rs.}$$

Example 23. If 8 men can do a piece of work in 6 days. In how many days will 1 man take to do the work?

Sol : This is a very simple question but understand one **basic** clearly. In these types of questions first see in the statement that **what is to be found**? In above question from statement "**In how many days**" implies that **days** are to be **found**. So the quantity which is to be found is to be put after **equal to** (=) or **on right hand side** as shown below.

Time taken by 8 men to do the work = 6 days

Time taken by 1 man to do the work = **....** days

The above blank can be filled by two ways, either as **6/8** or as **6 × 8**. Now at this point you have to put your common sense to use. Now think, if **8 men** can do the work in **6 days**, how many days **1 man** takes to do the work, **less or more**? 1 man will take **more days** because **capacity** of 1 man will be less than 8 men. So the **required answer** will be **6 × 8** not **6/8**.

Example 24. If 10 mangoes cost Rs. 40. What is the cost of 1 mango?

Sol : Cost of 10 mangoes = 40 Rs.

Cost of 1 mango = ... Rs.

Now cost of 10 mangoes = 40 Rs. So to buy 1 mango, obviously you have to pay less than Rs. 40. So the answer is $\frac{40}{10}$ = **4 Rs** not 40 × 10.

Now it is hoped that you have understood the basics. Now let us solve following questions.

Example 25. If 10 pens cost Rs. 30. Then find the cost of 15 pens.

Sol : Cost of 10 pens = 30 Rs.

$$\text{Cost of 1 pen} = \frac{30}{10}\ \text{Rs.}$$

$$\text{Cost of 15 pens} = \frac{30}{10} \times 15 = \mathbf{45\ Rs.}$$

Example 26. If 12 men can do a piece of work in 25 days. In how many days 5 men take to do the same work?

Sol : Time taken by 12 men to do the work = 25 days

Time taken by 1 man to do the work = 25×12

Time taken by 5 men to do the work = $\frac{25 \times 12}{5}$ = **60 days**

Now the same method is applicable when more than one quantity is given and we have to find a single quantity. Let us solve few of these types.

Example 27. If 15 men can harvest a field measuring 24 hectares in 20 days. In how many days 9 men will harvest a field measuring 18 hectares?

Sol : Now 1st step is to see what is to be found. In above question the underlined term "<u>In how many days</u>" means we have to find the days. ∴ We have to keep the days on right hand side and we will proceed as follows.

Time taken by 15 men to reap 24 hectares = 20 days

Time taken by 1 man to reap 24 hectare = 20×15

Time taken by 1 man to reap 1 hectare = $\frac{20 \times 15}{24}$

Time taken by 9 men to reap 1 hectare = $\frac{20 \times 15}{24 \times 9}$

Time taken by 9 men to reap 18 hectare = $\frac{20 \times 15 \times 18}{24 \times 9}$ **= 25 days**

Example 28. A garrison of 600 men has provision for 20 weeks. If at the end of 2 weeks they are reinforced by 300 men. How long will now the provision last?

Sol : Provision left = 20 – 2 = 18 weeks

For 600 men provision is for = 18 weeks

For 1 man provision is for = 18×600

For (600 + 300) = 900 men provision is for = $\frac{18 \times 600}{900}$

= **12 weeks**

Example 29. 8 men or 12 women can do a piece of work in 40 days. In how many days the work can be done by 6 men and 6 women working together?

Sol. 8 men = 12 women

1 man = $\frac{12}{8}$ women

6 men = $\frac{12}{8} \times 6$ women

= 9 women

So 12 women can do the work = 40 days

1 woman can do the work = 40×12

(6 men + 6 women) = 9 W + 6 W = 15 women can do the work

= $\frac{40 \times 12}{15}$ = **32 days**

EXERCISE

Q.1. Rs. 800 are divided into two parts in the ratio 3 : 5. Find the second part.

(a) 300 Rs. (b) 500 Rs. (c) 600 Rs. (d) none

Q.2. Rs. 4840 are divided into three parts in the ratio 2 : 3 : 6. Find the second part.

(a) 440 Rs. (b) 2640 Rs. (c) 880 Rs. (d) none

Q.3. On a scale of a map 8 cm represents 8.8 km. If the distance between two points on the map is 80.5 cms the distance between two points is approximately ......

(a) 90 km (b) 885 km (c) 900 km (d) none

Q.4. A bag contains 50 paise, 25 paise and 10 paise coins in the proportion of 3 : 4 : 5. The total value of these coins is Rs. 12. Find the number of 25 paise coins.

(a) 8 (b) 16 (c) 24 (d) none

Q. 5. A bag contains one rupee coins, 25 paise and 10 paise coins. Their values are in proprotion of 8 : 5 : 4. The total number of coins in the bag is 102. Find the number of coins of 10 paise.

(a) 30 (b) 60 (c) 120 (d) none

Q.6. If 4A = 5B and 6B = 7C, Find A : B : C.

(a) 28 : 30 : 35 (b) 35 : 30 : 28 (c) 35 : 28 : 24 (d) none

Q.7. If 20% x = 40% y, then find x : y

(a) 1 : 4 (b) 1 : 2 (c) 2 : 1 (d) none

Q.8. If A : B = 2 : 3, B : C = 4 : 5 and C : D = 6 : 7, find A : D

(a) 35 : 20 (b) 35 : 16 (c) 16 : 35 (d) none

Q.9. In a ratio which is 4 : 5 the antecedent is 24 then consequent is ....

(a) 30 (b) 25 (c) 28 (d) none

Q.10. Divide Rs. 1162 into three parts such that 4 times the first may be equal to the 5 times the second and 7 times the third. The first part is ....

(a) 490 Rs. (b) 480 Rs. (c) 440 Rs. (d) none

Q.11. Find the mean proportion of 3 and 108

(a) 18 (b) 24 (c) 30 (d) none

Q.12. The third proportion of 9 and 18 is

(a) 34 (b) 36 (c) 44 (d) none

Q.13. Two numbers are in the ratio 2 : 3. If 10 be subtracted from each of them then they are in to ratio 4 : 7. Find the second number.

(a) 10 (b) 20 (c) 30 (d) none

Q.14. What number should be added to 5, 12, 10 and 22 to make it proportionate?

(a) 1 (b) 2 (c) 3 (d) none

Q.15. Ratio of present ages of Gurjyot and Simarpreet is 5 : 9. Ratio of their ages after 3 years will be 2 : 3. Find the present age of Gurjyot.

(a) 9 year (b) 10 years (c) 15 years (d) none

Q.16. Ratio of present ages of Gagan and Ashu is 2 : 1. Ratio of their ages after 30 years will be 7 : 6. Find the present age of Gagan.

(a) 6 years (b) 12 years (c) 10 years (d) none

Q.17. Two numbers whose sum is 144 are such that $\frac{1}{4}$ of 1st = $\frac{1}{5}$ of 2nd. Find the second number.

(a) 80 (b) 64 (c) 120 (d) none

Q.18. Sum of digits of a number is 6. If 18 be subtracted from the number the digits reverse their position. The number is ....

(a) 24 (b) 22 (c) 32 (d) none

Q.19. Some men promised to do a job in 18 days but 6 of them became absent and remaining did the job in 20 days. What is original number of men?

(a) 30 (b) 40 (c) 60 (d) none

Q.20. A garrison of 1500 soldiers have provision for 25 days. At the end of 5 days, a reinforcement arrives and it is found that now provision will last only for 12 days more. What is strength of the reinforcement?

(a) 2500 (b) 1000 (c) 1500 (d) none

Q.21. A fort is provisioned for 44 days. After 14 days a reinforcement of 200 men arrives and food will now last only for 24 days. How many men were there in the fort?

(a) 600 (b) 800 (c) 1200 (d) none

Q.22. If 8 men can reap 80 hectare in 12 days. How many hectares 36 men reap in the same field in 36 days?

(a) 540 hec (b) 1080 hec (c) 270 hec (d) none

Q.23. A garrison of 700 men has provision for 20 weeks. If at the end of 4 weeks they are reinforced by 700 men. How long will the provision last?

(a) 8 weeks (b) 16 weeks (c) 24 weeks (d) none

Q.24. A man undertakes to do a piece of work in 150 days. He employs 200 men. He finds that only a quarter of work is done in 50 days. How many additional men should be em ploy so that work may be finished in time?

(a) 300 men (b) 100 men (c) 80 men (d) none

SOLUTIONS

1. (b) $\frac{5}{8} \times 800 = \mathbf{500\ Rs.}$

2. (d) $\frac{3}{11} \times 4840 = \mathbf{1320\ Rs.}$

3. (d)

$$8 \text{ cm} = 8.8 \text{ km}$$

$$1 \text{ cm} = \frac{8.8}{8}$$

$$80.5 \text{ cm} = \frac{8.8}{8} \times 80.5 = \mathbf{88.5\ km}$$

4. (b) Ratio of no of 50 p, 25 p and 10 p coins $= 3 : 4 : 5$

Ratio of value of 50 p, 25 p and 10 p coins $= 3 \times 50 : 4 \times 25 : 5 \times 10$

$= 150 : 100 : 50 = 3 : 2 : 1$

$\therefore$ Value of 25 p coins $= \frac{2}{6} \times 12 = 4$ Rs.

In 4 Rs. numbers of 25 p coins $= 4 \times 4 = \mathbf{16}$

5. (b) Ratio of value of one rupee, 25 p and 10 p coins $= 8 : 5 : 4$

Ratio of number of one rupee, 25 p and 10 p coin

$= 8 \times 1, 5 \times 4 : 4 \times 10 = 8 : 20 : 40$

$\therefore$ numbers of 10 p coins $= \frac{40}{68} \times 102 = 60$

6. (c)

$$4A = 5B$$

$$\frac{A}{B} = \frac{5}{4} \text{ or, } A : B = 5 : 4 \quad \text{or, } 6B = 7C$$

$$\frac{B}{C} = \frac{7}{6} \text{ or, } B : C = 7 : 6$$

A : B : C

5 : 4

7 : 6

35 : 28 : 24

7. (c)

$$20\% x = 40\% y$$

$$\frac{x}{y} = \frac{40\%}{20\%}$$

$$\frac{x}{y} = \frac{2}{1} = \mathbf{2 : 1}$$

8. (c)

$$\frac{A}{B} = \frac{2}{3}, \frac{B}{C} = \frac{4}{5}, \frac{C}{D} = \frac{6}{7}$$

$$\frac{A}{B} \times \frac{B}{C} \times \frac{C}{D} = \frac{2}{3} \times \frac{4}{5} \times \frac{6}{7}$$

$$\frac{A}{D} = \frac{16}{35} = \mathbf{16 : 35}$$

9. (a)

$$4x = 24$$

$$x = 6$$

$\therefore$ consequence $= 5x = 5 \times 6 = \mathbf{30}$

10. (a)

$$4x = 5y = 7z = k$$

$$x = \frac{k}{4} \text{ or, } y = \frac{k}{5} \text{ or, } z = \frac{k}{7}$$

$$x : y : z = \frac{k}{4} : \frac{k}{5} : \frac{k}{7}$$

$$= \frac{1}{4} : \frac{1}{5} : \frac{1}{7} = \frac{35:28:20}{140}$$

$$1\text{st part} = \frac{35}{83} \times 1162 = \textbf{Rs.490}$$

11. (a) Let x be the mean proportion

$$\therefore \quad 3 : x :: x : 108$$
$$x^2 = 3 \times 108 = 324$$
$$x = 18$$

12. (b) Let x be the third proportion.

$$\therefore \quad 9 : 18 :: 18 : x$$
$$9x = 18 \times 18$$
$$x = \frac{18 \times 18}{9} = \mathbf{36}$$

13. (d)

$$\frac{2x - 10}{3x - 10} = \frac{4}{7}$$
$$14x - 70 = 12x - 40$$
$$2x = 30$$
$$x = 15$$
$$\therefore \quad \text{Second number} = \mathbf{45}$$

14. (b) Let x be the required number.

$$5 + x : 12 + x :: 10 + x : 22 + x$$

$$(5 + x)(22 + x) = (12 + x)(10 + x)$$
$$110 + 5x + 22x + x^2 = 120 + 12x + 10x + x^2$$
$$27x + 110 = 22x + 120$$
$$5x = 10$$
$$x = 2$$

15. (d) Present age of Gurjyot = $5x$

Present age of Simerpreet = $9x$

After 3 years

$$\text{Gurjyot's age} = 5x + 3$$
$$\text{Simerpreet's age} = 9x + 3$$
$$\frac{5x + 3}{9x + 3} = \frac{2}{3}$$
$$15x + 9 = 18x + 6$$
$$3x = 3$$
$$x = 1$$
$$\therefore \quad \text{Present age of } \textbf{Gurjyot} = 5x = 5 \times 1 = \textbf{5 years}$$

16. (b)

$$\text{Let present age of Gagan} = 2x$$
$$\therefore \quad \text{present age of Ashu} = x$$

After 30 years

$$\text{Age of Gagan} = 2x + 30$$
$$\text{Age of Ashu} = x + 30$$

$$\frac{2x+30}{x+30} = \frac{7}{6}$$
$$12x + 180 = 7x + 210$$
$$5x = 30$$
$$x = 6$$

$\therefore$ Present age of **Gagan** $= 2x = 2 \times 6 =$ **12 years**

17. (a)
$$\frac{1}{4}x = \frac{1}{5}y$$
$$5x = 4y$$
$$\frac{x}{y} = \frac{4}{5}$$
$$x : y = 4 : 5$$
$$y = \frac{5}{9} \times 144 = \mathbf{80}$$

18. (d) Let digit at unit place $= x$

ten's place $= y$

$\therefore$ number $= 10y + x$

Number obtained by reversing the digit $= 10x + y$

According to statement

$$10y + x - 18 = 10x + y$$
$$9y - 9x = 18 \quad \text{...(1)}$$
$$y + x = 6 \quad \text{...(2)}$$

Multiply **(2)** by 9 we get

$$9y + 9x = 54 \quad \text{...(3)}$$

By solving **(1)** and **(3)** we get

$$y = \mathbf{4}$$
$$x = \mathbf{2}$$

$\therefore$ Number $= 10 \times 4 + 2 =$ **42**

19. (c) Let original number of men $= x$

x men can do the job in $=$ 18 days

1 man can do the job in $= 18x$ days

$x - 6$ men can do the job $= \frac{18x}{x-6}$ days

According to question $\frac{18x}{x-6} = 20$

$$18x = 20x - 120$$
$$2x = 120$$
$$x = 60$$

20. (b) Remaining days $= 25 - 5 = 20$ days

For 20 days, provision is sufficient for = 1500 soldiers

For 1 day provision is sufficient for = 1500×20 soldiers

For 12 days provisions are sufficient

for $= \frac{1500 \times 20}{12}$

$= 2500$ soldiers

$\therefore$ strength of reinforcement $= 2500 - 1500 = \mathbf{1000}$

21. (b) Remaining days $= 44 - 14 = 30$ days

Let x be original number of men

$\therefore$ For x men food lasts for $=$ 30 days

for 1 man food last for $=$ 30x days

for $x + 200$ men food lasts for $= \dfrac{30x}{x+200}$

According to statement

$$\frac{30x}{x+200} = 24$$

$$30x = 24x + 4800$$

$$6x = 4800$$

$$\boldsymbol{x = 800}$$

22. (b) Here we have to find "How many hectare". So we will put hectare on right hand side.

8 men in 12 days reap $=$ 80 hectares

8 men in 1 day reap $= \dfrac{80}{12}$ hectares

1 man in 1 day reaps $= \dfrac{80}{12 \times 8}$

36 men in 1 day reap $= \dfrac{80}{12 \times 8} \times 36$

36 men in 36 days reap $= \dfrac{80}{12 \times 8} \times 36 \times 36 =$ **1080 hectares**

23. (a) Remaining provision $= 20 - 4 = 16$ week

$\therefore$ for 700 men provision lasts $=$ 16 weeks

for 1 man provision last $= 16 \times 700$

for $(700 + 700) = 1400$ men provision last $= \dfrac{16 \times 700}{1400} =$ **8 weeks**

24. (b) Work done $= \dfrac{1}{4}$

Remaining work $= 1 - \dfrac{1}{4} = \dfrac{3}{4}$

Remaining days $= 150 - 50 = 100$ days

In 50 days 1/4 work is done by $=$ 200 men

in 50 days 1 work is done by $= 200 \times \dfrac{4}{1}$

In 1 day 1 work is done by $= 200 \times 4 \times 50$

In 100 days 1 work is done by $= \dfrac{200 \times 4 \times 50}{100}$

In 100 days 3/4 work is done by $= \dfrac{200 \times 4 \times 50}{100} \times \dfrac{3}{4} = 300$ men

So additional men required $= 300 - 200 =$ **100 men**

Day 13

Time and Work

Before starting this topic let us understand the **basics** of this **topic**. Let A can complete a work in 2 days. So on **day 1** A will do **half of the work** and on **day 2** he will do the **remaining half of the work**. Now suppose B too, alone can do the piece of work in 2 days. So B will do **half of the work on Ist day** and **half the work on 2nd day**. Now let A and B **start together**. On **Ist day A** will do **half** of the work and **B** also does **half** of the work. So **half + half = full work** is completed by **A and B** on the **day Ist** itself. This concept is done Mathematically as follows :

$$\text{A's one day work} = \frac{1}{2}$$

$$\text{B's one day work} = \frac{1}{2}$$

(A + B)'s one day work $= \frac{1}{2} + \frac{1}{2} = 1$. So they will do the work in one day. Now let us do it in another way. If A can do work in 2 days. So he will do $\frac{1}{2}$ of the work in one day. In other words If A's one day work is $\frac{1}{2}$ then he will do the work in $\frac{2}{1}$ days and "if A's one day work is $\frac{1}{3}$ then he will do the work in $\frac{3}{1}$ days." **(If A's one day work is $\frac{1}{n}$ then he will do the work in $\frac{n}{1}$ days.)** Now let us solve one more example "If A can do a piece of work in 10 days and B in 15 days. In how many days will they complete the work, working together?"

Sol :

$$\text{A's one day work} = \frac{1}{10}$$

$$\text{B's one day work} = \frac{1}{15}$$

$$\text{(A + B)'s one day work} = \frac{1}{10} + \frac{1}{15}$$

$$= \frac{3+2}{30} = \frac{1}{6}$$

$$\text{So (A + B)'s one day work} = \frac{1}{6}$$

So, as discussed, they will do the work in **6 days.**

The following are the types of Questions in this topic.

Type I : (Simple questions)

Type II : (where someone joins or leaves in between)

Type III: (Related to pipes and cistern)

Type IV: Misc.

Type I (Simple)

Example 1. If A can do a job in 12 days and B alone can do it in 8 days. Working together in how many days they will finish it.

Sol :

$$\text{A's One day work} = \frac{1}{12}$$

$$\text{B's one day work} = \frac{1}{8}$$

$$\text{(A + B)'s one day work} = \frac{1}{12} + \frac{1}{8}$$

$$= \frac{2+3}{24} = \frac{5}{24}$$

$$\text{Now (A + B)'s one day work} = \frac{5}{24}$$

They will finish the work in $\frac{24}{5}$ days.

Example 2. A and B can together do a work in 8 days. B alone can do it in 24 days. In how many days A alone will do the work?

Sol :

$$\text{(A + B)'s one day work} = \frac{1}{8}$$

$$\text{B's one day work} = \frac{1}{24}$$

A's one day work = (A + B)'s one day work – B's one day work

$$= \frac{1}{8} - \frac{1}{24} = \frac{3-1}{24} = \frac{2}{24} = \frac{1}{12}$$

So A will do the work in **12 days.**

Example 3. A and B can do a piece of work in 4 days. B and C in 6 days and C and A in 2 days. In what time A, B and C will complete the work working together.

Sol :

$$\text{(A + B)'s one day work} = \frac{1}{4} \quad \text{...(1)}$$

$$\text{(B + C)'s one day work} = \frac{1}{6} \quad \text{...(2)}$$

$$\text{(C + A)'s one day work} = \frac{1}{2} \quad \text{...(3)}$$

Adding **(1)**, **(2)** and **(3)**, we get

$$2\,(A + B + C)\text{'s one day} = \frac{1}{4} + \frac{1}{6} + \frac{1}{2} = \frac{3+2+6}{12} = \frac{11}{12}$$

$$2\,(A + B + C)\text{'s one day work} = \frac{11}{12}$$

$$(A + B + C)\text{'s one day work} = \frac{11}{12} \times \frac{1}{2} = \frac{11}{24}$$

A, B and C together can do the work in $\mathbf{\frac{24}{11}}$ **days.**

Example 4. A and B can do a piece of work in 8 days, B and C can do the same work in 10 days and C and A can do the same work in 15 days. In how many days A alone can do the work ?

Sol :

$$(A + B)\text{'s one day work} = \frac{1}{8} \quad \text{...(1)}$$

$$(B + C)\text{'s one day work} = \frac{1}{10} \quad \text{...(2)}$$

$$(C + A)\text{'s one day work} = \frac{1}{15} \quad \text{...(3)}$$

Adding **(1)**, **(2)** and **(3)**, we get

$$2\,(A + B + C)\text{'s one day work} = \frac{1}{8} + \frac{1}{10} + \frac{1}{15} = \frac{15+12+8}{120} = \frac{35}{120} = \frac{7}{24}$$

$$2\,(A + B + C)\text{'s one day work} = \frac{7}{24}$$

$$(A + B + C)\text{'s one day work} = \frac{7}{24} \times \frac{1}{2} = \frac{7}{48}$$

$$(B + C)\text{'s one day work} = \frac{1}{10}$$

A's one day work = (A + B + C)'s one day work – (B + C)'s one day work

$$= \frac{7}{48} - \frac{1}{10} = \frac{35-24}{240} = \frac{11}{240}$$

So, A alone can do the work in $\mathbf{\frac{240}{11}}$ **days.**

Example 5. If A can do 7/10 of a piece of work in 14 days and with the help of B, he is able to finish the work in 10 days, what time will B take to finish the work ?

Sol : A can do $\frac{7}{10}$ of work in 14 days

A can do 1 one work in $14 \times \frac{10}{7} = 20$ days

$$A\text{'s one day work} = \frac{1}{20}$$

$$(A + B)\text{'s one day work} = \frac{1}{10}$$

B's one day work = (A + B)'s 1 day work – A's 1 day work

$$= \frac{1}{10} - \frac{1}{20} = \frac{2-1}{20} = \frac{1}{20}$$

A will finish the work in **20 days.**

Type II

Example 6. A can do the piece of work in 10 days and B in 15 days. They started working together, but after 3 days A left the work. what time will B take to finish the work ?

Sol : A's one day work $= \frac{1}{10}$

B's one day work $= \frac{1}{15}$

(A + B)'s one day work $= \frac{1}{10} + \frac{1}{15} = \frac{3+2}{30} = \frac{5}{30} = \frac{1}{6}$

Since A and B have worked together for 3 days.

(A + B)'s 3 day work $= \frac{3}{6} = \frac{1}{2}$

Remaining work $= 1 - \frac{1}{2} = \frac{1}{2}$

Now this work has to be done by B. B can do one work in 15 days

B can do the $\frac{1}{2}$ of the work in $15 \times \frac{1}{2} = \frac{15}{2} = \mathbf{7\frac{1}{2}}$ **days**

Exaample 7. A can do a piece of work in 20 days. He worked for 5 days. After this B did the remaining work in 3 days. How many days A and B will together take to finish the whole work.

Sol : A's one day work $= \frac{1}{20}$

A's five day work $= \frac{5}{20} = \frac{1}{4}$

Remaining work $= 1 - \frac{1}{4} = \frac{3}{4}$

B can do $\frac{3}{4}$ of work in 3 days

B can do 1 work in $3 \times \frac{4}{3}$ = **4 days**

B's one day work $= \frac{1}{4}$

A's one day work $= \frac{1}{20}$

(A + B)'s one day work $= \frac{1}{4} + \frac{1}{20} = \frac{5+1}{20} = \frac{6}{20} = \frac{3}{10}$

So, A and B together can do the work in $\mathbf{\frac{10}{3}}$ **days.**

Example 8. A and B can do a piece of work in 10 days, B and C in 12 days and C and A in 15 days. If B alone works for 15 days and then joined by A and C, In how many days will the work be finished.

Sol :

$$(A + B)\text{'s one day work} = \frac{1}{10}$$

$$(B + C)\text{'s one day work} = \frac{1}{12}$$

$$(C + A)\text{'s one day work} = \frac{1}{15}$$

$$2\,(A + B + C)\text{'s one day work} = \frac{1}{10} + \frac{1}{12} + \frac{1}{15} = \frac{6+5+4}{60} = \frac{15}{60} = \frac{1}{4}$$

$$A + B + C \text{ one day} = \frac{1}{8}$$

B's 1 day work = (A + B + C)'s 1 day work – (C + A)'s 1 day work

$$= \frac{1}{8} - \frac{1}{15} = \frac{15-8}{120} = \mathbf{\frac{7}{120}}$$

$$\text{B's 15 day work} = \frac{7}{120} \times 15 = \frac{7}{8}$$

$$\text{Remaining work} = 1 - \frac{7}{8} = \frac{1}{8}$$

This remaining work is to be done by A + B + C, A + B + C does I work in 8 days

A + B + C does $\frac{1}{8}$ of work in $8 \times \frac{1}{8}$ days = **1 day**

Total time taken to finish the work = 15 + 1 = **16 days.**

Type III (Pipes and Cistern)

The method of doing the **questions** on **pipes and cistern** is **almost same** as that of doing the **questions** of **Time and Work**, but with a **bit difference** as **negative work** comes into play due to **leakage** in the **questions** of **pipes and cistern**. This is clearly explained below.

Let pipe **A** can **fill** a tank in **10 hours** then its **one hour work** will be $\mathbf{\frac{1}{10}}$ and another **leak** can **empty** it in **15** hours then its **one hour work** is equal to $-\mathbf{\frac{1}{15}}$. Suppose **both** start working together then their work is $\mathbf{\frac{1}{10} - \frac{1}{15} = \frac{1}{30}}$. Net work is **+ve.** ∴ Tank will be **filled** in **30 hours**. Now we will do some more examples.

Example1. Pipes A and B can fill a tank in 30 minutes and 60 minutes resp. Both together can fill it in .... minutes.

Sol : A's one minute work $= \frac{1}{30}$

B's one minute work $= \frac{1}{60}$

(A + B)'s one minute work $= \frac{1}{30} + \frac{1}{60} = \frac{2+1}{60} = \frac{3}{60} = \frac{1}{20}$

∴ Together they wil fill it in **20 minutes.**

Example 2. A tap can fill a cistern in 3 hours and another can empty it in 6 hours. If both taps are opened together, the time to fill the tank is ....

Sol : Net filling in 1 hour $= \frac{1}{3} - \frac{1}{6} = \frac{2-1}{6} = \frac{1}{6}$

∴ Time taken to fill the tank is **6 hours**

Example 3. A tank can be filled by two pipes in 2 hours and 4 hours res. When full the tank can be emptied by third pipe C in 6 hours. If all the taps are opened together, the tank will be full in .... hour.

Sol : (A + B + C)'s one hour work $= \frac{1}{2} + \frac{1}{4} - \frac{1}{6} = \frac{6+3-2}{12} = \frac{1}{6}$

∴ Tank will be filled in $\frac{12}{7}$ **hours**= $1\frac{5}{7}$ hours

Example 4. A cistern has two taps which fill it in 12 minutes and 15 minutes res. There is also a waste pipe in the cistern. When all the pipes are opened, the empty cistern is full in 20 minutes. How long will the waste pipe take to empty full cistern.

Sol : Let waste pipe take x minutes to empty the full cistern

Waste pipe's one minute work $= \frac{1}{x}$

$$\therefore \quad \frac{1}{12} + \frac{1}{15} - \frac{1}{x} = \frac{1}{20}$$

$$\frac{1}{12} + \frac{1}{15} - \frac{1}{20} = \frac{1}{x}$$

$$\frac{10+8-6}{120} = \frac{1}{x}$$

$$\frac{12}{120} = \frac{1}{x}$$

x = **10 minutes.**

∴ Waste pipe takes 10 minutes to empty a full cistern.

Example 5. A tank can be filled by one tap in 20 minutes and by another in 25 minutes. Both the taps are kept open for 10 minutes and then second is turned off. In how minutes more is the tank completely filled.

Sol : 1 minute work of both taps $= \frac{1}{20}+\frac{1}{25} = \frac{5+4}{100} = \frac{9}{100}$

10 minute's work of both taps $= \frac{9}{100}\times 10 = \frac{9}{10}$

Remaining work $= 1-\frac{9}{10} = \mathbf{\frac{1}{10}}$

Now, this $\frac{1}{10}$ of the tank is to be filled by Ist tap. It fills full tank in 20 minutes.

$\therefore$ It will fill $\frac{1}{10}$ of tank in = **2 minutes.**

Example 6. A pipe can fill a cistern in 12 minutes and another fill it in 15 minutes, but a third pipe can empty it in 6 minutes. The first two are kept open for 6 minutes in the beginning and then the third pipe is also opened, In what time is the cistern is emptied?

Sol : First two pipe's one minute work $= \frac{1}{12}+\frac{1}{15} = \frac{5+4}{60} = \frac{9}{60}$

First two pipe's 6 minutes work $= \frac{9}{60}\times 6 = \mathbf{\frac{9}{10}}$

Now one minute work of three pipes $= \frac{1}{12}+\frac{1}{15}-\frac{1}{6} = -\mathbf{\frac{1}{60}}$

–ve sign shows that if all the pipes are opened together they will empty the full cistern in 60 minutes. Now we have $\frac{9}{10}$ of cistern filled with water.

$\therefore$ Time taken to empty $\frac{9}{10}$ of water in cistern $= \frac{9}{10}\times 60 =$ **54 minute.**

Type IV (Misc.)

A and B working separately can do a piece of work in 9 and 12 days respectively. If they work for a day alternately, A beginning, in how many days will the work be completed?

Sol: A's one day work $= \frac{1}{9}$

B's one day work $= \frac{1}{12}$

After two days the work completed $= \frac{1}{9}+\frac{1}{12} = \frac{7}{36}$

After 4 days the work completed $= \frac{7}{36}\times 2 = \frac{14}{36}$

After 6 days the work completed $= \frac{7}{36}\times 3 = \frac{21}{36}$

$$\text{After 8 days the work completed} = \frac{7}{36} \times 4 = \frac{28}{36}$$

$$\text{After 10 days the work completed} = \frac{7}{36} \times 5 = \frac{35}{36}$$

$$\text{After 12 days the work completed} = \frac{7}{36} \times 6 = \frac{42}{36} \text{ But } \mathbf{\frac{42}{36}} \text{ is } \textbf{more than 1 work}$$

∴ work is done between **10th** and **12th** day

$$\text{Now work done after 10 days} = \frac{35}{36}$$

$$\text{Remaining work} = 1 - \frac{35}{36} = \mathbf{\frac{1}{36}}$$

Now it is the turn of A Time taken by A to complete 1 work = 9 days

∴ Time taken by A to $\frac{1}{36}$ of the work = $9 \times \frac{1}{36}$ days = $\mathbf{\frac{1}{4}}$ **days**

∴ Total time taken to completer the work = $10 + \frac{1}{4} = \mathbf{10\frac{1}{4}}$ **days**

Example 7. A can do a work in 2 days and B in 3 days. If they got Rs.100 for that work what is A's share.

Sol :

$$\text{A's one day work} = \frac{1}{2}$$

$$\text{B's one day work} = \frac{1}{3}$$

$$\text{Ratio of wages of A : B} = \frac{1}{2} : \frac{1}{3} = \frac{3:2}{6} = \mathbf{3:2}$$

$$\therefore \quad \text{A's share} = \frac{3}{5} \times 100 = \textbf{60 Rs.}$$

Example 8. 2 Men and 3 boys can do a piece of work in 16 days, 5 men and 6 boys can do it in 7 days. How long will 4 men and 8 boys take to do the same work?

Sol : 2 men + 3 boys one day work = $\frac{1}{16}$ **...(1)**

5 men + 6 boys one day work = $\frac{1}{7}$ **...(2)**

Multiply **(1)** by 2 and subtract from **(2)**

$$\text{5m + 6b one day work} = \frac{1}{7}$$

$$\text{4m + 6b one day work} = \frac{1}{8}$$

$$- \qquad -$$

$$\text{1 men's one day work} = \frac{1}{56} \qquad \textbf{...(3)}$$

Put **(3)** in **(1)**

$$2 \times \frac{1}{56} + 3 \text{ boys 1 day work} = \frac{1}{16}$$

$$3 \text{ boy's day work} = \frac{1}{16} - \frac{1}{28} = \frac{3}{112}$$

$$1 \text{ boy's one day work} = \frac{3}{112} \times \frac{1}{3} = \frac{1}{112}$$

Now we have to find that in how many days 4 men and 8 days take to do the work.

$$4 \text{ men's} + 8 \text{ boys one day work} = 4 \times \frac{1}{56} + 8 \times \frac{1}{112}$$

$$= \frac{1}{14} + \frac{1}{14} = \mathbf{\frac{1}{7}}$$

∴ They will take **7 days** to complete the work.

Shortcut : $(2\,M + 3\,B) \times 16 = (5\,M + 6\,B) \times 7$

$$32\,M + 48\,B = 35\,M + 42\,B$$

∴ $3\,M = 6\,B$

or $1\,M = 2\,B$

Now we can put the above value in any of the condition and solve it as follows:

$$2\,M + 3\,B = 4\,B + 3\,B = 7\,B$$

and $4\,M + 8\,B = 8\,B + 8\,B = 16\,B$ $(\because 1\,M = 2\,B)$

Now question becomes 'if 7 boys can do the work in 16 days in how many days 16 boys will take to do the same work? So they will do it in $\frac{7 \times 16}{16} = 7$ ***days'.***

Example 9. A and B promise to do a work for Rs. 75. A alone can do it in 20 days and B in 30 days, with the help of C they are able to finish it in 8 days. How will they distribute the wages?

Sol : Every body will get the wages according to their labour.

$$\text{A's one day work} = \frac{1}{20}$$

$$\text{A's 8 days work} = \frac{8}{30} = \frac{2}{5}$$

$$\text{B's one day work} = \frac{1}{30}$$

$$\text{B's 8 day work} = \frac{8}{30} = \frac{4}{15}$$

Remaining work $= 1 - \frac{2}{5} - \frac{4}{15} = \mathbf{\frac{1}{3}}$ C did the $\frac{1}{3}$ of work in 8 days.

∴ wages are divided in the ratio of $\frac{2}{5} : \frac{4}{15} : \frac{1}{3} = \frac{6:4:5}{15} = \mathbf{6 : 4 : 5}$

$$\text{A's share} = \frac{6}{15} \times 75 = \mathbf{30\ Rs.}$$

$$\text{B's share} = \frac{4}{15} \times 75 = \mathbf{20\ Rs.}$$

$$\text{C's share} = \frac{5}{15} \times 75 = \mathbf{25\ Rs.}$$

Example 10. A can do as much of work in 2 days as B in 3 days and B as much in 4 days as C in 5 days. In what time A, B and C together do the work If A can do it in 22 days ?

Sol.

$$\text{A's one day work} = \frac{1}{22}$$

$$\text{3 day work of B} = \text{2 days work of A} = 2 \times \frac{1}{22} = \frac{1}{11}$$

$$\text{B's one day work} = \frac{1}{11} \times \frac{1}{3} = \frac{1}{33}$$

$$\text{C's 5 days work} = \text{B's 4 day work} = \frac{4}{33}$$

$$\text{C's one day work} = \frac{4}{33} \times \frac{1}{5} = \frac{4}{165}$$

$$\text{(A + B + C)'s one day work} = \frac{1}{22} + \frac{1}{33} + \frac{4}{165} = \frac{15+10+8}{330} = \frac{33}{330} = \mathbf{\frac{1}{10}}$$

$\therefore$ They will do the work in **10 days**

EXERCISE

Q.1. A can do a piece of work in 25 days and B in 75 days. Working together, In how many days will they complete the work.

(a) $16\frac{2}{3}$ days (b) $18\frac{3}{4}$ days (c) $15\frac{1}{6}$ days (d) none

Q.2. Ram can do a piece of work in 10 days and Hari alone can do the same work in 20 days. They started working together. In how many days will they complete the work.

(a) $6\frac{1}{4}$ days (b) $5\frac{1}{7}$ days (c) $6\frac{2}{3}$ days (d) none

Q.3. A, B, and C can do a piece of work in 24 days. A and B can do it in 40 days, B and C can do it in 30 days. How many days will A and C take to finish work?

(a) 20 days (b) 30 days (c) 40 days (d) none.

Q.4. A and B can do a piece of work in 4 days, B and C in 6 days and A, B and C all together in 3 days, what time will A and C take to do it?

(a) 4 days (b) 8 days (c) 12 days (d) none.

Q.5. A and B can do a piece of work in 6 days. A alone can do it in 8 days. A worked for 5 days and then left it. In what time will B finish the rest of the work?

(a) 7 days (b) 9 days (c) 12 days (d) none

Q.6. A and B can do a work in 8 days B and C in 12 days, C and A in 16 days. All worked for 2 days. How many days will A take to finish the work.

(a) 14 days (b) 7 days (c) $3\frac{1}{2}$ days (d) none.

Q.7. A can do a piece of work in 12 hours and with the help of his son, he is able to do it in 8 hours. How many hours will the son alone take to finish the work?

(a) 20 days (b) 24 days (c) 4 days (d) none

Q.8. A pipe can fill a tank in y hours and another can empty it in x hours. They can together fill it in .... (x > y).

(a) $\frac{1}{y}+\frac{1}{x}$ hrs (b) $\frac{1}{y}-\frac{1}{x}$ hrs (c) $\frac{1}{\frac{1}{y}-\frac{1}{x}}$ hrs (d) none

Q.9. A tank is filled by a pipe A in 32 min. and pipe B in 36 min. When filled, it can be emptied by pipe C in 20 minute. If all the three pipes are opened simultaneously 1/3rd of tank will be filled in

(a) $55\frac{1}{13}$ min (b) $36\frac{12}{13}$ min (c) $35\frac{2}{13}$ min (d) none

Q.10. Two pipes x and y can fill a cistern in 24 minute and 32 min. res. If both the pipes are opened together, then after how much time should y be closed so that tank is full in 18 minutes.

(a) 10 min (b) 18 min (c) 8 min (d) none

Q.11. A cistern is filled in 8 hours but takes two hours longer to fill because of a leak in its bottom. If the cistern is fill the leak will empty it in :

(a) 40 hours (b) 20 hours (c) 10 hours (d) none

Q.12. Two taps can separately fill a cistern in 10 minutes and 15 min. resp. and when waste pipe is open, they can together fill it in 18 minutes. The waste pipe can empty the full cistern in

(a) 9 min (b) 12 min (c) 15 min (d) none

Q.13. A cistern has a leak which should empty it in 8 hours. A tap is turned on which admits 6 litre a minute into the cistern and it is now emptied in 12 hours. How many litres does cistern hold?

(a) 8860 litres (b) 8640 litres (c) 8840 litres (d) none

Q.14. Two pipes A and B can fill a tank in 15 hours and 20 hours resp. While a third pipe can empty the full tank in 25 hours. All the three pipes are opened in the beginning. After 10 hours C is closed. The time taken to fill the tank is :

(a) 24 hours (b) 12 hours (c) 2 hours (d) none

Q.15. Pipe A can fill a tank in 20 hours while pipe B alone can fill it in 30 hours and pipe C can empty the full tank in 40 hours. If all the pipes are opened together half the tank will be filled in :

(a) 120/7 hrs (b) 7 hrs (c) 40/7 hrs (d) none

Q.16. A and B can do a piece of work in 10 and 12 days res. If they work for a day alternately. In how many days the work be completed.

(a) 11 days (b) $10\frac{1}{4}$ days (c) $10\frac{5}{6}$ days (d) none

Q.17. A and B can do a piece of work in 30 and 40 days res. They began the work together, but A left after some days and B did the remaining work in 12 days. After how many days did A leave?

(a) 12 days (b) 10 days (c) 24 days (d) none

Q.18. A and B can do a piece of work in 3 days, B and C in 4 days and A and C in 6 days. A, B and C can together complete the work in .... days?

(a) $6\frac{2}{3}$ days (b) $2\frac{2}{3}$ days (c) $2\frac{1}{3}$ days (d) none

Q.19. If 2 men and 2 boys can do a piece of work in 10 days and 3 men and 1 boy can do it in 8 days, how long will 3 men and 7 boys take to do it?

(a) 5 days (b) 10 days (c) 7 days (d) none

SOLUTIONS

1. (b) (A + B)'s one day work $= \frac{1}{25} + \frac{1}{75} = \frac{3+1}{75} = \frac{4}{75}$

$\therefore$ They will finish the work in $\frac{74}{4}$ days i.e $\mathbf{18\frac{3}{4}}$ **days**

2. (c) Ram + Hari's one day work $= \frac{1}{10} + \frac{1}{20} = \frac{3}{20}$

$\therefore$ working days $\mathbf{\frac{20}{3}}$ **days.**

3. (c) A + B + C's one day work $= \frac{1}{24}$

2 (A + B + C)'s one day work $= 2 \times \frac{1}{24} = \frac{1}{12}$ ... **(1)**

(A + B)'s one day work $= \frac{1}{40}$ ... **(2)**

B + C one day work $= \frac{1}{30}$ ... **(3)**

Add **(2)** and **(3)**

(A + 2B + C)'s one day work $= \frac{1}{30} + \frac{1}{40} = \frac{7}{120}$ ...**(4)**

Subtract **(4)** from **(1)**

2A + 2B + 2C one day work $= \frac{1}{12}$

A + 2B + C one day work $= \frac{7}{120}$

– – – –

(A + C)'S one day work $= \frac{1}{40}$

$\therefore$ A and C together can do the work in **40 days**

4. (a)

$$A + B + C\text{'s one day work} = \frac{1}{3}$$

$$2\,(A + B + C)\text{'s one day work} = 2 \times \frac{1}{3} = \frac{2}{3} \quad ...(1)$$

$$(A + B)\text{'s one day work} = \quad ...(2)$$

$$B + C \text{ one day work} = \frac{1}{6} \quad ...(3)$$

Add **(2)** and **(3)**

$$(A + 2B + C)\text{'s one day work} = \frac{1}{4} + \frac{1}{6} = \frac{5}{12} \quad ...(4)$$

Subtract **(4)** from **(1)**

$$2A + 2B + 2C \text{ one day work} = \frac{2}{3}$$

$$A + 2B + C \text{ one day work} = \frac{5}{12}$$

$$- \quad - \quad - \quad -$$

$$(A + C)\text{'s one day work} = \frac{1}{4}$$

$\therefore$ A and C together can do the work in **4 days**

5. (b)

$$(A + B)\text{'s one day work} = \frac{1}{6}$$

$$A\text{'s one day work} = \frac{1}{8}$$

$$B\text{'s one day work} = \frac{1}{6} - \frac{1}{8} = \frac{1}{24}$$

$$A\text{'s 5 day work} = \frac{5}{8}$$

$$\text{Remaining work} = 1 - \frac{5}{8} = \frac{3}{8}$$

This work is to be done by B. B does 1 work in 24 days

So B does $\frac{3}{8}$ work $24 \times \frac{3}{8}$ in days = **9 days**

6. (a)

$$2(A + B + C)\text{'s one day work} = \frac{1}{8} + \frac{1}{12} + \frac{1}{16} = \frac{13}{48}$$

$$A + B + C \text{ one day work} = \frac{13}{48} \times \frac{1}{2} = \frac{13}{96}$$

$$A\text{'s one day work} = \frac{13}{96} - \frac{1}{12} = \frac{5}{96}$$

Since A, B, C together have done work for 2 days.

$$\therefore \quad \text{Their 2 days work} = \frac{13}{96} \times 2 = \frac{13}{48}$$

$$\text{Remaining work} = 1-\frac{13}{48} = \frac{35}{48}$$

$$\text{A does one work in} = \frac{96}{5} \text{ days}$$

$$\text{SO A does } \frac{35}{48} \text{ work in} = \frac{96}{5}\times\frac{35}{48} = \mathbf{14\ days}$$

7. (b) $$\text{Son's one day work} = \frac{1}{8}-\frac{1}{12}=\frac{1}{24}$$

∴ They will do it in **24 days**.

8. (c) $$\text{Both's one hour work} = \frac{1}{y}-\frac{1}{x}$$

∴ They will fill the tank in $\dfrac{1}{\frac{1}{y}-\frac{1}{x}}$ hrs

9. (b) One minute work of three pipes $= \dfrac{1}{32}+\dfrac{1}{36}-\dfrac{1}{20}$

$$= \frac{40+40-72}{1440} = \frac{13}{1440}$$

$$\therefore \text{ They will fill the full tank in} = \frac{1440}{13} \text{ min}$$

$$\text{They will fill the } \frac{1}{3} \text{ tank in} = \frac{1440}{13}\times\frac{1}{3}$$

$$= \mathbf{\frac{480}{13}\ min = 36\frac{12}{13}\ min}$$

10. (c) Very important question.

Let y be closed after k minutes. $(18-k)$ minute will be the time for which only x will opreate and equation will become as follows.

$$\left(\frac{1}{24}+\frac{1}{32}\right)k+(18-k)\frac{1}{24} = 1$$

$$\left(\frac{4+3}{96}\right)k+\frac{18}{24}-\frac{k}{24} = 1$$

$$\frac{7}{96}k-\frac{k}{24} = 1-\frac{18}{24}\ \frac{7k-4k=96-72}{96}$$

$$3k = 24$$

$$k = \mathbf{8\ min}$$

∴ y be closed after **8 minutes**

11. (a) Let leak will empty it in x hours

$$\therefore \quad \frac{1}{8}-\frac{1}{x} = \frac{1}{10}$$

$$\frac{1}{8}-\frac{1}{10} = \frac{1}{x}$$

$$\frac{1}{40} = \frac{1}{x}$$

$$x = \textbf{40 hours}$$

12. (a) Let waste pipe can empty the full cistern in x min

$$\therefore \quad \frac{1}{10} + \frac{1}{15} - \frac{1}{x} = \frac{1}{18}$$

$$\frac{1}{10} + \frac{1}{15} - \frac{1}{18} = \frac{1}{x}$$

$$\frac{18 + 12 - 10}{180} = \frac{1}{x}$$

$$\frac{20}{180} = \frac{1}{x}$$

$$\frac{1}{9} = \frac{1}{x}$$

$$x = \textbf{9 hours}$$

13. (b)

$$\text{One hour work of leak} = -\frac{1}{8}$$

$$\text{Let one hour work of tap} = \frac{1}{x}$$

$$\therefore \quad \frac{1}{x} - \frac{1}{8} = -\frac{1}{12}$$

$$\therefore \quad \frac{1}{x} = \frac{1}{8} - \frac{1}{12} = \frac{1}{24}$$

$\therefore$ Tap will fill the cistern in 24 hours

$$\therefore \quad \text{Capacity of cistern} = 24 \times 60 \times 6 = \textbf{8640 litre}$$

14. (b)

$$\text{10 hours work of three pipes} = \left(\frac{1}{15} + \frac{1}{20} - \frac{1}{25}\right)10 = \frac{23}{30}$$

$$\text{Remaining work} = 1 - \frac{23}{30} = \frac{7}{30}$$

$$\text{(A + B)'s one hours work} = \frac{1}{15} + \frac{1}{20} = \frac{7}{60}$$

$\therefore$ They will do one work in $\frac{60}{7}$ hrs. They will do $\frac{7}{30}$ work in $\frac{60}{7} \times \frac{7}{30}$ = **2 hours**

$$\therefore \quad \text{Total time} = 10 + 2 = \textbf{12 hours}$$

15. (d)

$$\text{(A + B + C)'s one hour work} = \frac{1}{20} + \frac{1}{30} - \frac{1}{40} = \frac{7}{120}$$

$$\therefore \quad \text{They will fill full tank in} = \frac{120}{7} \text{ hrs}$$

$$\text{and half tank in} = \frac{1}{2} \times \frac{120}{7} = \frac{60}{7} \textbf{ hrs}$$

16. (c)

$$\text{A's one day work} = \frac{1}{10}$$

$$\text{B's one day work} = \frac{1}{12}$$

$$\text{Work done after 2 days} = \frac{1}{10} + \frac{1}{12} = \frac{11}{60}$$

$$\text{Work done after 10 days} = \frac{11}{60} \times 5 = \frac{55}{60} = \frac{11}{12}$$

$$\text{Remaining work} = 1 - \frac{11}{12} = \frac{1}{12}$$

This $\frac{1}{12}$ work will be done by A. Now A does one work in 10 days

$\therefore$ A does $\frac{1}{12}$ work in $10 \times \frac{1}{12} = \mathbf{\frac{5}{6}}$ **days**

Total time $\mathbf{10 + \frac{5}{6}} = \mathbf{10\frac{5}{6}}$ **days**

17. (a) Work done by B in last 12 days $= \frac{12}{40} = \frac{3}{10}$

$$\text{Remaining work} = 1 - \frac{3}{10} = \frac{7}{10}$$

This $\frac{7}{10}$ work has been done by (A + B)

$$\text{(A + B)'s one day work} = \frac{1}{30} + \frac{1}{40} = \frac{7}{120}$$

$\therefore$ (A + B) can do one work in $\frac{120}{7}$ days

$\therefore$ (A + B) will do $\frac{7}{10}$ work in $\frac{120}{7} \times \frac{7}{10} =$ **12 days**

$\therefore$ A left after **12 days**

18. (b) 2 (A + B + C)'s one day work $= \frac{1}{3} + \frac{1}{4} + \frac{1}{6} = \frac{9}{12} = \frac{3}{4}$

$\therefore$ (A + B + C)'s one day work $= \frac{3}{4} \times \frac{1}{2} = \frac{3}{8}$

$\therefore$ They will do the work in $\frac{8}{3}$ **days** Or $\mathbf{2\frac{2}{3}}$ **days**

19. (a) (2M + 2B)'s one day work $= \frac{1}{10}$ **...(1)**

(3M + 1B)'s " " $= \frac{1}{8}$ **...(2)**

Multiply (2) by 2, **we get**

(6M + 2B)'s " " " $= \frac{1}{4}$ **...(3)**

Subtracting (1) from (3), we get

$$4M\text{'s one day work} = \frac{1}{4} - \frac{1}{10} = \frac{3}{20}$$

$$1\ M\text{'s one day work} = \frac{3}{20} \times \frac{1}{4}$$

$$1\ M\text{'s one day work} = \frac{3}{80} \quad \text{...(4)}$$

Putting (4) in (2), we get

$$3 \times \frac{3}{80} + 1\ B\text{'s one day work} = \frac{1}{8}$$

$$1B\text{'s one day work} = \frac{1}{8} - \frac{9}{80} = \frac{1}{80}$$

$$\therefore \quad (3M + 7B)\text{'s one day work} = \frac{9}{80} + \frac{7}{80} = \frac{16}{80} = \mathbf{\frac{1}{5}}$$

$\therefore$ They will do the work in **5 days**.

Shortcut : $(2\ M + 2\ B) \times 10 = (3\ M + 1\ B) \times 8$

$$20\ M + 20\ B = 24\ M + 8\ B$$

$$\therefore \quad 4\ M = 12\ B$$

or $$1\ M = 3\ B$$

Now we can put the above value in any of the condition and solve it as follows :

$$2\ M + 2\ B = 6\ B + 2\ B = 8\ B$$

and $$3\ M + 7\ B = 9\ B + 7\ B = 16\ B \qquad (\because 1\ M = 3\ B)$$

Now question becomes 'if 8 boys can do the work in 10 days in how many days 16 boys will take to do the same work? So they will do it in $\frac{8 \times 16}{10}$ = ***5 days****'.*

Day 14

Alligation or Mixture

ALLIGATION OR MIXTURE

Today we will do **alligation or mixture**.

ALLIGATION OR MIXTURE

Suppose we have two quantities of rice. One priced at Rs. **10 a kg** and other **Rs. 20 a kg.** We mixed **one kg** of each of them.

Then the total cost of (1kg +1kg)

= 2 kg will be $= 10 + 20 =$ **30 Rs.**

1 kg will be $= \frac{30}{2} =$ **15 Rs.**

Rs. 15 is exactly midway between Rs. 10 and 20 as shown below.

10 11 12 13 14 15 16 17 18 19 20

So when we mix same quantity of two types of mixture i.e., in the **ratio 1 : 1** the value comes out to be **exactly** in **middle**. The value of mixture **can't be lower** than the **cheaper value** and **cannot be higher** than the **dearer**, In whatever ratio we may mix each of them. Now suppose we have to make the value of mixture at Rs. 11, which is nearer to cheaper one then we have to mix more of Rs.10 and less of Rs. 20. *But what will be the exact ratio*? For that following method is adopted:

1st 2nd
10 20

11

(20 – 11) (11 – 10)
9 : 1

As shown above we have written **1st value** on **left hand side**, **second** on **right hand side** and **desired in between**. Then **subtracted lower value** from **higher one** and written it diagonaly as shown. So the ratio comes out to be 9:1. Now we will cross-check the result means we will see that by mixing **9 kg. of Ist type** with **1 kg. of second type** whether we get a mixture worth Rs. 11 a kg. or not.

Now according to result if we mix 9 kg. of 1st type and 1 kg of 2nd

Then value of 9 kg of 1st type $= 9 \times 10$

$= 90$ Rs.

" " 1 kg " 2nd " $= 1 \times 20$

$= 20$ Rs.

Total value of $(9 + 1) = 10$ kg $= 90 + 20$

$= 110$ Rs

Value of 1 kg. of mixture $= \frac{110}{10} = \mathbf{11\ Rs}$

Now we will solve some more **examples.**

Example 1. In what ratio must rice worth Rs. 40 a kg be mixed worth rice Rs. 60 a kg. So as to produce rice Rs. 45 a kg.

Sol :

1st 40   2nd 40

45

(60 - 45)   (45 - 40)

15 : 5

3 : 1

NOTE : If the question had been "In what ratio must rice worth Rs. 60 a kg be mixed worth rice Rs. 40 a kg. So as to produce rice Rs. 45 a kg." then we would have solved it as follows:

Sol :

1st 60   2nd 40

45

(45 - 40)   (60 - 45)

5 : *15*

1 : *3*

So we have to solve in the same sequence as asked in the question.

Now following five types of questions are asked in the exams:

Type I : (Simple Questions)

Type II : (In which profits are given)

Type III : (Related to milk and water)

Type IV : (Milk is sold at cost price)

Type V : (Repetitive Mixing)

Type VI : (Misc.)

We will do each type of questions one by one.

TYPE I

These types of questions we have already done in the example. Now we will do some more examples:

Example 1. Find the ratio in which tea worth Rs. 10 per kg. be mixed with tea worth Rs. 5 per kg., so as to produce tea worth Rs. 7 per kg.

Sol :

1st 10   2nd 5

7

(7 - 5)   (10 - 7)

2 : 3

Example 2. Find the ratio in which rice at Rs. 4.20 per kg be mixed at rice worth Rs. 5.40 per kg. So as to produced mixture worth Rs. 5.00 per kg.

Sol :

1st 4.20   2nd 5.40

5.00

.40   .80

1 : 2

TYPE II

Example 3. In what ratio must two kinds of tea worth Rs. 27 and Rs. 31 per kg be mixed so that by selling the mixture at Rs. 35 per kg 25% may be gained.

Sol : In this type of questions first we have to find the cost price of the mixture.

$$SP = 35 \text{ Rs.}$$

$$\text{Profit} = 25\%$$

$$CP = 35 \times \frac{110}{10} = \mathbf{28 \text{ Rs.}}$$

Now we have to mix these two quantities in such a way that value of mixture may be Rs. 28 a kg. So we will proceed as follows

1st 27   2nd 31

28

3 : 1

Therefore required ratio = **3 : 1**

Example 4. In what ratio must two kinds of sugar at Rs. 1.15 and Rs. 1.24 per kg be mixed so that by selling at Rs. 1.50 per kg, 25% may be gained.

Sol :

$$SP = 1.50 \text{ Rs.}$$

$$\text{Profit} = 25\%$$

$$CP = 1.50 \times \frac{110}{10} = \mathbf{1.20\ Rs.}$$

1st 115   2nd 124

120

4 : 5

Therefore required ratio = **4 : 5**

Type III

Example 5. In what proportion must water be mixed with milk worth Rs. 5.00 per litre, so that the value of mixture may be Rs. 4.00 per litre.

Sol : Value of water is taken as zero.

Milk 5.00   Water 0

4.00

4 : *1*

Therefore required ratio = **4 : 1**

Example 6. How much water be mixed in 38 litre of milk worth Rs. 5.00 per litre, so that value of mixture is Rs. 3.80 per litre.

Sol :

Milk 5.00   Water 0

3.80

3.80   ***1.20***

19   ***6***

In 19 litre milk water added = 6 litre

" 1 " " " $= \frac{6}{19}$

" 38 " " " " $= \frac{6}{19} \times 38 =$ **12 litres**

Type IV

Example 7. In what ratio must water be mixed with milk worth Rs. 6.00 per litre. So as to gain 25% by selling the mixture at cost price.

Sol : This question can be done by simple logic. We want to sell the mixture at cost price and still gain 25%. What is the way out? Simply add 25% of water

$$25\% \text{ water means } = \frac{25}{100}$$

i.e., in the ratio **1 : 4**

Example 8. How much water be added to 25 litres of milk worth Rs. 12 a litre so that by selling the milk at cost price 20% may be gained.

Sol : $20\% \text{ means } = \frac{1}{5}$ means in the ratio 1 : 5

Means 1 litre of water be added in 5 litre of milk or in other way

$$\text{In 5 litre milk water added} = 1 \text{ litre}$$

$$1 \text{ " " " "} = \frac{1}{5}$$

$$25 \text{ " " " "} = \frac{1}{5} \times 25 = \mathbf{5\ litres}$$

Example 9. A milk man buys milk at Rs. 15 per litre. He mixed the water with it and sell the mixture at cost price. In what proportion is water mixed with milk if his profits were $6\frac{1}{4}\%$

Sol : $\text{Simply } 6\frac{1}{4}\% = \frac{25}{4}\% = \frac{1}{16}$

$$\text{Ratio of water to milk} = \mathbf{1 : 16}$$

Type V

Example 10. Two equal vessels A and B contains mixture of milk and water. In vessel A the ratio of milk to water is 5 : 3 and in vessel B the ratio of milk to water is 4 : 1. Then two vessels are poured in to the third vessel C. Now find the ratio of milk to water in vessel C.

Sol : $\text{In vessel A milk} = \frac{5}{8} \text{ part}$

$$\text{And water} = \frac{3}{8} \text{ part}$$

$$\text{In vessel B milk} = \frac{4}{5} \text{ part}$$

$$\text{water} = \frac{1}{5} \text{ part}$$

$$\text{Total milk} = \frac{5}{8} + \frac{4}{5} = \frac{57}{40}$$

$$\text{Total water} = \frac{3}{8} + \frac{1}{5} = \frac{23}{40}$$

$$\therefore \qquad \text{milk : water} = \frac{57}{40} : \frac{23}{40} = \mathbf{57 : 23}$$

Direct Method : We can avoid fractions if we take quantities in each vessels as LCM of (5 + 3 = 8) and (4 + 1 = 5) that is i.e.,, LCM of 8 and 5 which is 40.

$$\text{Now milk vessel A} = \frac{5}{8} \times 40 = 25 \text{ litres}$$

$$\text{water vessel A} = \frac{3}{8} \times 40 = 15 \text{ litres}$$

$$\text{Now milk vessel B} = \frac{4}{5} \times 40 = 32 \text{ litres}$$

$$\text{water vessel B} = \frac{1}{5} \times 40 = 8 \text{ litres}$$

Therefore ratio of milk : Water = (25 + 32) : (15 + 8) = **57 : 23**.

Example 11. Two equal vessels A and B contain milk and water in the ratio 1 : 1 and 3:1. In what ratio these mixed so that ratio of milk to water in new mixture is 3 : 2.

Sol : $$\text{Milk in vessel A} = \frac{1}{2}$$

$$\text{" " " B} = \frac{3}{4}$$

$$\text{" " mixture} = \frac{3}{5}$$

So ratio is found as followed

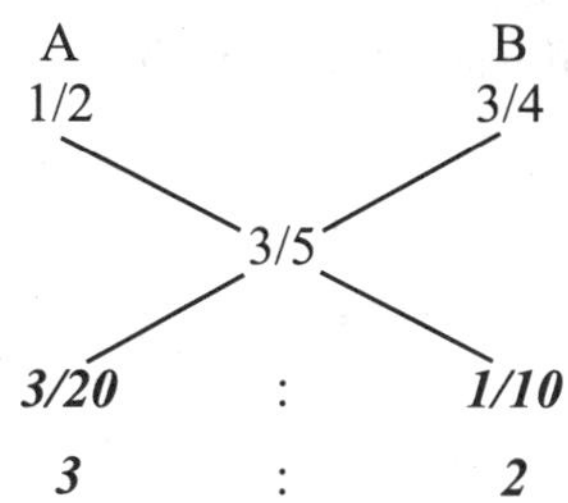

Two quantities are mixed in the ratio **3 : 2.**

Type VI

Let a vessel contain A litre of milk, B of milk is taken out of it and replaced by the same quantity of water and the process is repeated n times.Then

$$\frac{\textbf{Milk left in container after n operations}}{\textbf{Original quantity of milk in container}} = \left(\frac{\mathbf{A - B}}{\mathbf{A}}\right)^n$$

This is very important formula. Now solve the following examples based on this formula.

Example 12. A container is full of milk. One third of the milk is taken out of it and replaced by the same quantity of water. Then again one third of the mixture is taken out and replaced by same quantity of water. This process is repeated 3 times. If capacity of container 54 litres. Find the milk left in the container after the end of 3rd operation.

Sol :

$$\frac{\textbf{Milk left in container after n operations}}{\textbf{Original quantity of milk in container}} = \left(\frac{\mathbf{A - B}}{\mathbf{A}}\right)^n$$

Here $n = 3 \ \& \ \text{B} = \text{A}/3$

$$\frac{\textbf{Milk left in container after n operations}}{\textbf{Original quantity of milk in container}} = \left(\frac{\mathbf{A - B}}{\mathbf{A}}\right)^n$$

$$= \left(\frac{\frac{2\text{A}}{3}}{\text{A}}\right)^3$$

$$= \left(\frac{2}{3}\right)^3 = \frac{8}{27}$$

$$\text{Milk Left after 3 operations} = \frac{8}{27} \times \text{original quantity of milk}$$

$$= \frac{8}{27} \times 54 = \textbf{16 litre}$$

Example 13. A container has capacity of 20 litres and is full of milk. 10 Litre of milk is taken out of it and replaced by same quantity of water again 10 litres of mixture is taken out and replaced by equal quantity of water. How much milk is left in the mixture.

Sol : Here

$$n = 2$$

$$\text{A} = 20$$

$$\text{B} = 10$$

$$\text{Milk left after 2 operations} = \left(\frac{20-10}{20}\right)^2 \times 20$$

$$= \left(\frac{10}{20}\right)^2 \times 20$$

$$= \frac{1}{4} \times 20 = \textbf{5 litre}$$

Type VI

Alligation is used in solving the questions of various other topics e.g. percentage, profit and loss, simple interest, average etc. A few of the example are given below.

Example 14. In a zoo there are rabbits and pigeons. If their heads are counted they are 90 while their legs are 224. Find the Number of pigeons in the zoo.

Sol : This question can be done by 2 methods.

1. by Algebra
2. by Alligation

By Algebra :

Let pigeons in the zoo $= x$

" rabbits " " $= y$

Head 90 means $x + y = 90$ ... **1** {as each one has one head}

As a rabbit has four legs and pigeon has 2 legs

$\therefore \quad 4x + 2y = 224$ ... **2**

Multiply equation 1 by 4 & subtract equation 2 from it.

$$4x + 4y = 360$$
$$4x + 2y = 224$$
$$- \quad - \quad -$$
$$2y = 136$$
$$y = \frac{136}{2} = 68$$

Pigeon in zoo = **68**

By Alligation

Let all are rabbits then number of legs $= 90 \times 4 = 360$

Let all are pigeon then number of legs $= 90 \times 2 = 180$

Now we have to mix them in such a way that the mixture of them have 224 legs which is done by Alligation as follows:

Rabbit 360   Pigeon 180

224

44 : ***136***

11 : ***34***

$\therefore$ ratio of rabbits to pigeons = **11 : 34**

$\therefore$ Number of pigeons $= \frac{34}{45} \times 90 = \mathbf{68}$

Example 15. In an examination out of 1000 students 70% boys and 80% girls are passed. It total pass percentage was 76%. Find the Number of girls.

Sol : This question we will solve by Alligation

Boys 70%   Girls 80%

76%

4 : 6

2 : 3

$\therefore$ Ratio of Boys to Girls = **2 : 3**

$\therefore$ Girls = $\frac{3}{5} \times 1000$ = **600**

Example 16. Divide Rs. 85 among 100 students such that each boy got a rupee and each girl 50 P. Find the number of girls.

Sol : Let all are boys then

total sum = $100 \times 100 = 10000$ P.

Let all are girls then total sum = $100 \times 50 = 5000$ P.

Boys 10000   Girls 5000

8500

3500 : 1500

7 : 3

$\therefore$ Ratio of Boys to Girls = 7 : 3

Number of girls = $\frac{3}{5} \times 1000$

= **30**

So it is obvious that Alligation method is quiet simple. Further there are some question in average SI,CI which can be done by alligation which we will do while doing those topics.

Now solve the following exercise.

EXERCISE

Q.1. In what ratio must two types of wheat at Rs. 445 per quintal and Rs. 670 per quintal be mixed so as to produce a mixture worth Rs. 620 per quintal.

(a) 7 : 2   (b) 1 : 6   (c) 2 : 7   (d) none

Q.2. In what ratio must two kinds of sugar worth 132.25 per quintal be mixed with 182.25 per quintal so as to produce a mixture worth Rs. 152.25 per quintal.

(a) 3 : 2   (b) 2 : 3   (c) 3 : 1   (d) none

Q.3. In what ratio must rice at Rs. 8 per kg be mixed with rice at 12 per kg so as to produce a mixture worth Rs. 9 per kg.

(a) 1 : 3 (b) 3 : 1 (c) 2 :1 (d) none

Q.4. In what ratio must oil worth Rs. 70.50 per litre be mixed with Rs. 82.50 per litre so as to produce oil worth Rs. 77.50 per litre.

(a) 7 : 5 (b) 9 : 3 (c) 5 : 7 (d) none

Q.5. In what ratio must two kinds of tea worth Rs. 7.50 per kg and Rs. 12.00 per kg so that by selling the mixture at Rs. 12.50 per kg 25% may be gained.

(a) 4 : 5 (b) 5 : 4 (c) 2 : 7 (d) none

Q.6. In what ratio must pickle worth Rs. 25 per kg be mixed worth Rs. 40 per kg. so that by selling it at Rs. 42 a kg a profit of 40% is made.

(a) 1 : 2 (b) 1:3 (c) 2 : 1 (d) none

Q.7. In what ratio rice worth Rs. 6 per kg.be mixed with rice worth Rs. 3 per kg so that by selling the mixture at Rs. 6.40 per kg. a profit of 60% is made.

(a) 1 : 2 (b) 4 : 1 (c) 3 : 2 (d) none

Q.8 In what ratio must tea worth Rs. 9 per kg be mixed with tea worth Rs. 14 per kg so that by selling the mixture at Rs. 11 per kg a profit of 10% is made.

(a) 1 : 4 (b) 4 : 1 (c) 3 : 2 (d) none

Q.9. In what ratio must water be mixed with milk worth Rs. 12 per litre so as to produce a mixture worth Rs. 10 per litre.

(a) 1 : 4 (b) 5 : 1 (c) 2 : 3 (d) 5 : 6

Q.10. In what ratio must water be mixed with milk worth Rs. 5.20 per litre so as to gain 20% by selling the mixture at Rs. 4.80 per litre.

(a) 12 : 1 (b) 10 : 3 (c) 3 : 12 (d) none

Q.11. In what ratio must water be mixed with milk so as to gain 30% by selling the mixture at cost price.

(a) 3 : 7 (b) 7 : 3 (c) 3 : 10 (d) 5 : 6

Q.12. In what ratio must water be mixed with milk so as to gain 35 % by selling the mixture at cost price

(a) 7 : 20 (b) 20 : 7 (c) 7 :13 (d) none

Q.13. Three equal vessel contain mixture of milk and water . The proportion of milk and water in them is respectively 4:1, 5:2, and 6:1. If these mixtures are mixed in a single vessel, what will be the proportion of milk to water in new mixture.

(a) 83 : 11 (b) 22 : 83 (c) 83 : 22 (d) none

Q.14. Three equal glasses are filled with milk and water. the ratio of milk to water in them is respectively 3:4, 4:5 and 6:1. If these three glasses are mixed in a single container then what is the ratio of milk to water in new vessel?

(a) 80 : 109 (b) 109 : 80 (c) 109 : 63 (d) none

Q.15. A bucket contain 36 liter of milk 1/6 of milk drawn off and replaced by water. Again 1/6 of mixture were drawn off and replaced by water. What will be the quantity of milk in new mixture.

(a) 25 litre (b) 30 litre (c) 32 litre (d) none

Q.16. A barrel contains 128 litres of wine 1/4 of the wine is taken out of barrel and replaced by water. This process is repeated 3 times. What is the quantity of wine in new mixture.

(a) 27 litre (b) 54 litre (c) 33 litre (d) none

Q.17. A contained container 50 liter of milk 1/5 of milk is taken out and replaced by equal quantity of water. Now 1/5 of mixture is again taken out and replaced by water. This process is repeated 3 times. What is the ratio of milk left to the original milk.

(a) 64 : 125 (b) 4 : 5 (c) 5 : 4 (d) none

Q.18. A container contained certain amount of milk. 1/2 of the milk is taken out and replaced by water. Now 1/2 of mixture is again taken out and replaced by water. This process is repeated 3 times and the quantity of milk left in the new mixture is 3 litre. What was the original quantity of milk.

(a) 24 litre (b) 27 litre (c) 8 litre (d) none

Q.19. In a zoo there are rabbits and pigeon. If their heads are counted they are 35 while their legs are 120. Find the number of rabbits in the zoo.

(a) 25 (b) 15 (c) 10 (d) none

Q.20. In a class there are 150 students. 60% of boys are passed while 45% of girls are passed. The pass percentage of class is 55%. Find the number of girls in the class.

(a) 50 (b) 100 (c) 115 (d) none

Q.21. Divide Rs. 49.25 among 150 childern such that each boy gets 50 paise and each girl 25 paise. The number of boys in class.

(a) 103 (b) 47 (c) 50 (d) none

SOLUTIONS

1. (c)

First 445   Second 670

620

50 : ***175***

2 : ***7***

2. (a)

Ist 132.25   IInd 182.25

152.25

30 : ***20***

3 : ***2***

3. (b)

Ist IInd

8 12

9

3 : *1*

4. (c)

Ist IInd

70.50 82.50

77.50

5 : *7*

5. (a)

$$SP = 12.50$$

$$\text{profit} = 25\%$$

$$CP = 12.50 \times \frac{100}{125} = \mathbf{10\ Rs.}$$

It means cost price of mixture be 10 Rs. we have to mix two varities of Rs. 7.50 per kg. and Rs. 12 per kg. so as to produce mixture worth Rs. 10 per kg.

Ist IInd

7.50 12.00

10

2* : *2.50

4* : *5

6. (c)

$$SP = 42$$

$$\text{Profit} = 40\%$$

$$CP = 42 \times \frac{100}{140} = \mathbf{30\ Rs.}$$

Ist IInd

25 40

30

10* : *5

2* : *1

7. (a)

$$SP = 6.40\ Rs.$$

$$\text{Profit} = 60\%$$

$$CP = 6.40 \times \frac{100}{160} = \mathbf{4\ Rs.}$$

Ist 6   IInd 3

4

1 **:** ***2***

8. (b)

$$CP = 11 \times \frac{100}{110} = \mathbf{10\ Rs.}$$

Ist 9   IInd 14

10

4 **:** ***1***

9. (b)

Milk 12   Water 0

10

10 **:** ***2***

5 **:** ***1***

10. (b)

$$CP = 4.80 \times \frac{100}{120} = \mathbf{4\ Rs.}$$

Milk 5.20   Water 0

4.00

4.00 : 1.20

400 : 120

10 : 3

11. (d) 3 : 10

12. (a)

13. (c) Milk in all the three vessels $= \frac{4}{5} + \frac{5}{7} + \frac{6}{7} = \frac{83}{35}$

water in the three vessels $= \frac{1}{5} + \frac{2}{7} + \frac{1}{7} = \frac{22}{35}$

Ratio of milk to water = **83 : 22**

14. (b) Milk in the new mixture $= \frac{3}{7} + \frac{4}{9} + \frac{6}{7} = \frac{109}{63}$

$$\text{Water in the new mixture} = \frac{4}{7}+\frac{5}{9}+\frac{1}{7}=\frac{80}{63}$$

$$\text{Ratio of milk to water} = \frac{109}{63}:\frac{80}{63}$$

$$= \mathbf{109 : 80}$$

15. (a)

$$\frac{\text{Milk aftern operations}}{\text{Original milk}} = \left(\frac{A-B}{A}\right)^n$$

$$\frac{\text{Milk after 2 operation}}{\text{Original Milk}} = \left(\frac{A-\frac{A}{6}}{A}\right)^2 = \left(\frac{\frac{5A}{6}}{A}\right)^2 = \left(\frac{5}{6}\right)^2 = \frac{25}{36}$$

$$\text{Milk left} = \frac{25}{36}x \text{ original quantity}$$

$$\text{of milk} = \frac{25}{36}\times 36 = \mathbf{25\ litre}$$

16. (b)

$$\text{Wine left} = \left(\frac{A-B}{A}\right)^3 \times 128$$

$$= \left(\frac{A-\frac{A}{4}}{A}\right)^3 \times 128$$

$$= \left(\frac{\frac{3A}{4}}{A}\right)^3 \times 128 = \left(\frac{3}{4}\right)^3 \times 128$$

$$= \frac{27}{64}\times 128 = \mathbf{54\ litre}$$

17. (a)

$$\frac{\text{Milk left after 3 operations}}{\text{Original quatity of milk}} = \left(\frac{A-B}{A}\right)^3$$

$$= \left(\frac{A-\frac{A}{5}}{A}\right)^3 = \left(\frac{\frac{4A}{5}}{A}\right)^3$$

$$= \left(\frac{4}{5}\right)^3 = \frac{64}{125} = \mathbf{64:125}$$

18. (a)

$$\frac{\text{Milk left}}{\text{Original milk}} = \left(\frac{A-B}{A}\right)^3$$

$$= \left(\frac{A-\frac{A}{2}}{A}\right)^3 = \left(\frac{\frac{A}{2}}{A}\right)^3$$

$$\frac{\text{3 litre}}{\text{original milk}} = \left(\frac{1}{2}\right)^3$$

$$\text{Original quantity of milk} = 3\times 8 = 24 \text{ ltr}$$

19. (a) Let all be rabbits

$\therefore$ their legs $= 35 \times 4 = 140$

Let all be pigeons

$\therefore$ their legs $= 35 \times 2 = \mathbf{70}$

Rabbits 140   Pigeons 70

120

50 : *20*

5 : *2*

Therefore rabbits $= \frac{5}{7} \times 35 = \mathbf{25}$

20. (a)

Boys 60%   Girls 45%

55%

10 : *5*

2 : *1*

$\therefore$ number of girls $= \frac{1}{3} \times 150 = \mathbf{50}$

21. (b) Let all are boys

Then total money $= 150 \times 50$

$= 7500$ p.

Let all are girls

Then total money $= 150 \times 25$

$= 3750$ p.

Boys 7500   Girls 3750

4925

1175 **:** ***2575***

47 **:** ***103***

Therefore number of boys $= \frac{47}{150} \times 150 = \mathbf{47}$

Day 15

Simple Interest

Today we will do the topic of **SIMPLE INTEREST**. Suppose **Ram** lent **Rs. 200** to **Shyam** for **1 year** on the condition that **after 1 year** Shyam will pay Rs. 220. So the **additional amount i.e., 220 – 200 = 20 Rs.** Shyam pays is known as **Interest**. **Interest calculated per hundred** is known as rate percent.

In above case on Rs. 200 Interest = 20 Rs.

$$\text{Rs. } 1 \text{ "} = \frac{20}{200}$$

$$\text{Rs. } 100 \text{ "} = \frac{20}{200} \times 100 = 10 \text{ Rs.}$$

∴ **Rate percent** is **10%** and it is for **one year**. So rate percent is **10% per annum**.

The money lent is known as **principal** and **Interest** if calculated on basic money only is known as **Simple Interest.** for example every year the Interest in above given example will be calculated on Rs. 200 and generlized formula for Simple Interest is:

$$\textbf{Simple Interest} = \frac{\mathbf{P \times R \times T}}{\mathbf{100}} \quad \text{...(1)}$$

Where P = Principal

R = Rate % per annum.

From formula 1 we can deduct the formula of **time**, **rate** and **principal**.

$$\mathbf{R} = \frac{\mathbf{S.I \times 100}}{\mathbf{P \times T}} \quad \text{...(2)}$$

and

$$\mathbf{P} = \frac{\mathbf{S.I \times 100}}{\mathbf{R \times T}} \quad \text{...(3)}$$

and

$$\mathbf{T} = \frac{\mathbf{S.I \times 100}}{\mathbf{P \times R}} \quad \text{...(4)}$$

We have divided questions in following types :

Type I : When Simple Interest is to be calculated.

Type II : When rate, time or principal is to be found.

Type III : When amount is to be found.

Type IV : Double Questions.

Type V : Question on partial lending.

Type VI : Misc. Questions.

Type I

Example 1. Find the simple Interest on Rs. 800 for 3 years at 5% per annum.

Sol :

$$SI = \frac{P \times R \times T}{100}$$

$$= \frac{800 \times 3 \times 5}{100} = \textbf{120 Rs.}$$

Example 2. Find the Simple Interest on Rs.1800 for 5 years at 2% per annum.

Sol :

$$SI = \frac{1800 \times 5 \times 2}{100} = \textbf{180 Rs.}$$

Example 3. Find the Simple Interest on Rs.700 from 2nd March 1999 to 19 Dec. 1999 at $4\frac{1}{2}\%$ per annum.

Sol : First we will calculate the time. You have to keep in mind that **Interest is not charged for the day** on which **money is borrowed** but it is charged **for the day it is returned.**

Time

March = 29 Days
April = 30 Days
May = 31 Days
June = 30 Days
July = 31 Days
Aug = 31 Days
Sep = 30 Days
Oct = 31 Days
Nov = 30 Days
Dec = 19 Days

292 Days

$$\text{Time in years} = \frac{292}{365} = \frac{4}{5} \text{ years}$$

$$SI = \frac{P \times R \times T}{100}$$

$$= \frac{700 \times 9 \times 4}{100 \times 2 \times 5} = \textbf{25.20 Rs.}$$

Example 4. Find the Simple Interest on Rs. 200 for 7 months at 5 paise per rupee per month.

Sol : 5 Paise per rupee per month means 5% per month.

As rate is given per month and time also is in months

$$\therefore \quad SI = \frac{200 \times 7 \times 5}{100} = \textbf{70 Rs.}$$

Example 5. Find the Simple Interest on Rs. 2400 for 9 months at 6% per annum.

Sol : As the rate % is given per annum So we have to convert 9 months into years

$$9 \text{ months} = \frac{9}{12} \text{ years}$$

$$= \frac{3}{4} \text{ years}$$

$$SI = \frac{P \times R \times T}{100}$$

$$= \frac{2400 \times 6 \times 3}{100 \times 4} = \textbf{108 Rs.}$$

Type II

Example 6. In what time will Rs. 5000 amount to Rs. 6000 at 5% per annum Simple Interest.

Sol :

$$SI = A - P$$

$$= 6000 - 5000 = 1000 \text{ Rs}$$

$$\text{Rate} = 5\%$$

$$T = \frac{\text{S.I.} \times 100}{P \times R}$$

$$= \frac{1000 \times 100}{5000 \times 5} = \textbf{4 years}$$

Example 7. At what rate percent per annum Simple Interest will a sum of money double itself in 10 years.

Sol :

$$\text{Let sum} = 100 \text{ Rs.}$$

$$\text{Amount} = 200 \text{ Rs.}$$

$$SI = 200 - 100 = 100 \text{ Rs.}$$

$$T = 10 \text{ years}$$

$$\text{Rate} = \frac{\text{S.I.} \times 100}{P \times T} = \frac{100 \times 100}{100 \times 10} = \textbf{10\%}$$

Example 8. On what sum the Simple interest for 10 years at 5% per annum equals to Rs. 3600?

Sol :

$$P = \frac{\text{S.I.} \times 100}{R \times T}$$

$$= \frac{3600 \times 100}{5 \times 10} = \textbf{7200 Rs.}$$

Example 9. A lent Rs. 100 to B for 4 years and Rs. 160 to C for 5 years. The rate of Interest was same in both the cases. He got from both Rs. 240 as Interest. Find the rate percent.

Sol : Let r be the rate percent

$$\therefore \quad \frac{100 \times r \times 4}{100} + \frac{160 \times r \times 5}{100} = 240 \text{ Rs.}$$

$$4r + 8r = 240$$
$$12r = 240$$
$$r = \mathbf{20\%}$$

Type III

Example 10. What sum will amount to Rs. 9900 in 2 years at 5% Simple Interest.

Sol : Method I

Let
$$P = 100$$
$$R = 5\%$$
$$T = 2 \text{ years}$$
$$SI = \frac{100 \times 5 \times 2}{100} = 10 \text{ Rs.}$$
$$A = 100 + 10 = 110 \text{ Rs.}$$

If A is Rs.110 then $P = 100$

" " 1 " " $= \frac{100}{110}$

" " 9900 " $= \frac{100 \times 9900}{110} = \mathbf{9000\ Rs.}$

Direct Formula

$$\mathbf{P = \frac{100 \times A}{100 + RT}}$$
$$= \frac{100 \times 9900}{100 + 10} = \frac{100 \times 9900}{110} = \mathbf{9000\ Rs.}$$

Type IV

Example 11. Rs. 300 becomes Rs. 420 in 2 years at simple Interest .What will Rs. 620 amount to 3½ years the rate being same in both the cases.

Sol : 1st case

$$A = P + SI = 420 \text{ Rs.}$$
$$P = 300 \text{ Rs.}$$
$$SI = 420 - 300 = 120 \text{ Rs.}$$
$$\text{Time} = 2 \text{ years}$$
$$\text{Rate} = \frac{120 \times 100}{300 \times 2} = \mathbf{20\%}$$

IInd case

$$P = 620 \text{ Rs.}$$
$$R = 20\% \text{ [From 1st case]}$$
$$T = \frac{7}{2} \text{ years}$$
$$SI = \frac{620 \times 20 \times 7}{100 \times 2} = 434 \text{ Rs.}$$

$$\text{Amount} = 620 + 434 = \mathbf{1054\ Rs.}$$

Example 12. A sum of money amounts to Rs. 1326 in 6 years at 5% per annum Simple Interest. In what time will it amount to Rs. 1530.

Sol :

$$A = 1326$$

$$P = \frac{100 \times A}{100 + RT}$$

$$= \frac{100 \times 1326}{130} = 1020 \text{ Rs.}$$

$$A = 1530 \text{ Rs.}$$

$$SI = \text{Rs. } 1530 - \text{Rs. } 1020 = \text{Rs. } 510$$

$$R = 5\%$$

$$T = \frac{S.I. \times 100}{P \times R} = \frac{510 \times 100}{1020 \times 5} = \mathbf{10\ years}$$

Example 13. A sum of money amounts to Rs.1760 in 2 years and Rs. 2000 in 5 years at simple interest. Find the sum and rate percent.

Sol :

$$P + 5 \text{ years SI} = 2000$$

$$P + 2 \text{ years SI} = 1760$$

$$3 \text{ Years SI} = 240$$

$$3 \text{ years SI} = 240$$

$$1 \text{ years SI} = \frac{240}{3} = 80 \text{ Rs.}$$

$$2 \text{ Years SI} = 2 \times 80 = 160 \text{ Rs.}$$

$$\text{Principal} = \text{Rs.}1760 - \text{Rs.}160 = 1600 \text{ Rs.}$$

$$R = \frac{S.I. \times 100}{P \times T} = \frac{160 \times 100}{1600 \times 2} = \mathbf{5\%}$$

Example 14. In what time will Rs. 400 at 3% per annum produce the same income as Rs. 600 in 4 years at 5% Simple Interest.

Sol : Let T be the required time

According to question :

$$\frac{400 \times 3 \times T}{100} = \frac{600 \times 4 \times 5}{100}$$

$$12\,T = 120$$

$$T = \mathbf{10\ Years}$$

Type V

Example 15. A man had Rs.1000 a part of which he lent at 4% and left over money at 5%. After 1 years he received Rs. 44 as total Interest. The money he lent at 5% was ....?

Sol : Ist Method

Let he lent Rs. x at 4%

$$\therefore \quad \text{Money lent at 5\%} = (1000 - x)$$

$$\frac{x \times 4 \times 1}{100} + \frac{(1000 - x) \times 5 \times 1}{100} = 44$$

$$\frac{4x + 5000 - 5x}{100} = 44$$

$$-x = 4400 - 5000 = -600$$

$$-x = -600$$

$$x = 600 \text{ Rs.}$$

Money lent at 4% = **600 Rs.**

And Money lent at 5 % = **400 Rs.**

2nd Method : By Alligation

Let he lent all the money at 4 %

$$\text{then Interest} = \frac{1000 \times 4 \times 1}{100} = \mathbf{40\ Rs.}$$

Let he lent all the money at 5%

$$\text{then Interest} = \frac{1000 \times 5 \times 1}{100} = \mathbf{50\ Rs.}$$

By Alligation

Interest when Money lent at 4%   Interest when Money lent at 5%

40   50

44

6 : ***4***

3 : ***2***

$$\text{Money at 5\%} = \frac{2}{5} \times 1000 = \mathbf{400\ Rs.}$$

Type VI

Example 16. A man buys a land and gives for it 10 times the annual rent. Find the rate of Interest he gets for his money.

Sol : This is a tricky question and has been asked several times in different competitions

Let annual rent = 1 Rs.

He buys it at Rs. 1 × 10 = 10 Rs.

Now by investing Rs.10 he will get Rs.1 as rent, if he lent it to some body. In other words by investing Rs. 10 he will get Rs.1 as interest after 1 year.

Let

$$P = 10 \text{ Rs.}$$
$$SI = 1 \text{ Rs.}$$
$$T = 1 \text{ year}$$
$$\text{Rate \%} = \frac{1 \times 100}{10 \times 1} = \mathbf{10\%}$$

Example 17. Rs. 1500 be placed at Interest at a rate of 10% and Interest be added to the principal after every 5 years. In how many years will it amount to Rs. 2500.

Sol : First we will find the Simple Interest after 5 years

$$SI = \frac{1500 \times 5 \times 10}{100} = 750 \text{ Rs.}$$

Now it is added to principal

$$\therefore \quad \text{New principal} = 1500 + 750 = 2250 \text{ Rs}$$
$$A = 2500 \text{ Rs.}$$
$$SI = 2500 - 2250 = 250 \text{ Rs.}$$
$$T = \frac{250 \times 100}{2250 \times 10} = \frac{10}{9} \text{ years}$$
$$\text{Total time} = 5 + \frac{10}{9} = \frac{55}{9} = \text{years} = \mathbf{6\frac{1}{9}} \text{ years}$$

EXERCISE

Q.1. Find the simple Interest on Rs. 1020 at 5% per annum in 4 years.

(a) Rs. 408 (b) Rs. 210 (c) Rs. 200 (d) none

Q.2. Find the simple Interest on Rs. 5000 at 6% per annum from 15 May, 2000 to 20th Dec, 2000.

(a) 360 Rs. (b) 180 Rs. (c) 140 Rs. (d) none

Q.3. On what sum the simple Interest for 3 years at 5% per annum amount to Rs. 150.

(a) Rs. 1000 (b) Rs. 1500 (c) Rs. 1050 (d) none

Q.4. At what rate percent per annum will a sum of money double itself in 15 years.

(a) $6\frac{2}{3}\%$ (b) $6\frac{1}{4}\%$ (c) $5\frac{2}{3}\%$ (d) none

Q.5. In what time Rs. 2600 amount to Rs. 3055 at the rate of 3 ½% per annum.

(a) 2½ years (b) 5 years (c) 3 years (d) none

Q.6. Find the sum of money when increases 1/10 of itself every year amount to Rs. 600 in 5 years.

(a) 380 Rs. (b) 280 Rs. (c) 480 Rs. (d) none

Q.7. A sum of money amounts to Rs.1100 in 2 years and Rs. 1400 in 6 years at simple Interest. Find the rate percent.

(a) $7\frac{17}{19}\%$ (b) $3\frac{1}{13}\%$ (c) $4\frac{2}{3}\%$ (d) none

Q.8. If a sum of money doubles itself in 6 years. It becomes 5 times in how many years.?

(a) 12 years (b) 24 years (c) 10 years (d) none

Q.9. In what time Rs. 540 at 5 percent per annum will produce the same Interest as Rs.1800 in 5 years at 6 percent per annum.

(a) 10 years (b) 30 years (c) 25 years (d) none

Q.10. A lent 180 to B for 10 years and 200 to C for 2 years at Simple interest, the rate of interest being same in both the cases. He received Rs. 220 as total interest. Find the rate%.

(a) 5% (b) 10% (c) 8 % (d) none

Q.11. A man had Rs. 3000 a part of which he lent at 4% and rest at 5%. The whole annual interest received was Rs. 130. the money he lent at 4% was....?

(a) Rs 2000 (b) Rs 1000 (c) Rs 800 (d) none

Q.12. A man had Rs. 5800 a part of which he lent at 4% and rest at 6%. The whole annual interest received was Rs. 292. The money lent at 4% was....?

(a) Rs 3000 (b) Rs 2600 (c) Rs 2700 (d) none

Q.13. A man buys a land and gives for it 20 times the annual rent. Find the rate of Interest he gets for his money.

(a) 10% (b) 3% (c) 4% (d) none

Q.14. At what rate percent per annum simple Interest will a sum of money be 1.1 times in 2 years?

(a) 2% (b) 3% (c) 4% (d) none

Q.15. Find the rate of Simple Interest when a sum of money treble itself in 25 years.

(a) 4% (b) 10% (c) 8% (d) done

SOLUTIONS

1. (d)

$$SI = \frac{1020 \times 5 \times 4}{100} = \mathbf{204\ Rs.}$$

2. (b) Time

May = 16 days
Jun = 30 days
Jul = 31 days
Aug = 31 days
Sep = 30 days
Oct = 31 days
Nov = 30 days
Dec = 20 days

Total = 219

Total = 219

$$\text{Time in years} = \frac{219}{365} = \frac{3}{5} \text{ years}$$

$$SI = \frac{5000 \times 6 \times 3}{100 \times 5} = \mathbf{180\ Rs.}$$

3. (a)

$$P = \frac{S.I. \times 100}{R \times T}$$

$$= \frac{150 \times 100}{5 \times 3} = \mathbf{1000\ Rs.}$$

4. (a) Let

$$P = 100 \text{ Rs.}$$

$$A = 200 \text{ Rs.}$$

$$SI = 200 - 100 = 100 \text{ Rs.}$$

$$T = 15 \text{ years}$$

$$R = \frac{100 \times 100}{100 \times 15} = \frac{20}{3} = 6\frac{2}{3}\%$$

5. (b)

$$\text{Amount} = 3055 \text{ Rs.}$$

$$\text{Principal} = 2600 \text{ Rs.}$$

$$SI = 3055 - 2600$$

$$= 455 \text{ Rs.}$$

$$R = \frac{7}{2}\%.$$

$$T = \frac{455 \times 100 \times 2}{2600 \times 7} = \mathbf{5\ years}$$

6. (d)

$$\frac{1}{10} \text{ means rate } \frac{1}{10} \times 100 = 10\%$$

$$P = \frac{100 \times A}{100 + RT} = \frac{100 \times 600}{100 + 50}$$

$$= \frac{600 \times 100}{150} = \mathbf{400\ Rs.}$$

7. (a)

$$P + 6 \text{ years SI} = 1400 \text{ Rs.}$$

$$P + 2 \text{ years SI} = 1100 \text{ Rs.}$$

$$- \quad - \qquad\qquad -$$

$$4 \text{ years SI} = 300$$

$$1 \text{ year SI} = \frac{300}{4} = 75 \text{ Rs.}$$

$$2 \text{ years SI} = 150$$

$$P = 1100 - 150 = 950$$

$$\text{Rate} = \frac{150 \times 100}{950 \times 2} = 7\frac{17}{19}\%$$

8. (b) Let

$$P = 100 \text{ Rs.}$$

$$A = 200 \text{ Rs.}$$

$$SI = 100 \text{ Rs.}$$

$$T = 6 \text{ years}$$

$$R = \frac{S.I. \times 100}{P \times T}$$

$$= \frac{100 \times 100}{100 \times 6} = \frac{\mathbf{50}}{\mathbf{3}}\%$$

2nd case

Let

$$P = 100 \text{ Rs.}$$
$$A = 500 \text{ Rs.}$$
$$SI = 400 \text{ Rs.}$$
$$R = \frac{\mathbf{50}}{\mathbf{3}}\% \text{ [from case 1]}$$
$$T = \frac{400 \times 100 \times 3}{100 \times 50} = \mathbf{24\ years}$$

Shortcut : Let $\left.\begin{matrix} P = 100 \\ A = 200 \end{matrix}\right\} SI = 100$

2nd case

Let $\left.\begin{matrix} P = 100 \\ A = 500 \end{matrix}\right\} SI = 400$

SI of 100 Rs. is obtained in 6 years

SI of 400 which is 4 times the earlier SI should be obtained in

$$6 \times 4 = 24 \text{years}$$

(In both the cases principal should be same)

9. (d) Let T be the required time.

$$\frac{540 \times 5 \times T}{100} = \frac{1800 \times 6 \times 5}{100}$$
$$T = \mathbf{20\ Years}$$

10. (b) Let r be the rate

$$\frac{18 \times r \times 10}{100} + \frac{200 \times r \times 2}{100} = 220$$
$$18\,r + 4\,r = 220$$
$$22r = 220$$
$$r = 10\%$$

11. (a) We will do it by alligation as in example.

Let whole money be given at 4%

$$\text{Interest} = \frac{3000 \times 4 \times 1}{100} = 120 \text{ Rs.}$$

Let whole money be given at 5%

$$\text{then Interest} = \frac{3000 \times 5 \times 1}{100}$$
$$= 150 \text{ Rs.}$$

By alligation

Interest when Money lent at 4% — 120   Interest when Money lest at 5% — 150

130

20 : **10**

2 : **1**

Money lent at 4% $= \frac{2}{3} \times 3000$

$=$ **2000 Rs.**

12. (d) same as above

Interest when Money lent at 4% — 232   Interest when Money lest at 5% — 348

292

56 : **60**

14 : **15**

Money lent at 4% $= \frac{14}{29} \times 5800$

$=$ **2800 Rs.**

13. (c) Same as example 16

Rate $= \frac{1 \times 100}{20 \times 1} =$ **5%**

14. (d) Let

P = 100 Rs.

Amount = 100 × 1.1 = 110 Rs.

SI = 10 Rs.

Time = 2 years

Rate $= \frac{10 \times 100}{100 \times 2} =$ **5%**

15. (c) Let

P = 100 Rs.

A = 300 Rs.

SI = 200 Rs.

T = 25 Years

R $= \frac{200 \times 100}{100 \times 25} =$ **8%**

Compound Interest

On day 15 we have done the topic of simple interest. It is hoped that you have clearly understood the topic. **Today** we will do the topic of **compound interest**. Let us under stand it more clearly:

Suppose you have lent **Rs. 2000** at Interest of **5%** per annum for **2 years**. First we know that Simple Interest for **1st year** and **IInd** year will be same i.e., $= \frac{2000 \times 5 \times 1}{100}$ **= 100 Rs.**

Total **simple Interest** for **2 years** will be = 100 + 100 = **200 Rs.**

In C.I. the first year Interest will be same as of SI i.e., **Rs. 100.** But **IInd year interest** is calculated on **P + SI of Ist year** i.e., on Rs. 2000 + Rs. 100 = **2100 Rs.** So, **IInd year** interest in **CI** becomes $= \frac{2100 \times 5 \times 1}{100}$ = **105 Rs.** So after two years total CI = 100 + 105 = **205 Rs.**

So it is **Rs. 5 more** than the **simple Interest** . *This increase is due to the fact that* ***SI is added to the principal*** *and this* ***Rs.105*** *is also added in the* ***principal*** *if we have to find the* ***compound interest*** *after* ***3 years****. Now it is hoped that you are clear that why CI is more than SI when time is more than one year. This concept will help us to solve a lot of questions directly. Now we will discuss* ***direct formula*** *for* ***amount*** *in case of compound interest.*

The formula is given below :

$$A = P\left(1+\frac{r}{100}\right)^t$$

Where A = Amount

P = Principal

r = Rate % per annum.

t = Time

A = P + CI

$$P\left(1+\frac{r}{100}\right)^t = P + CI$$

$$CI = P\left(1+\frac{r}{100}\right)^t - P$$

$$CI = P\left[\left(1+\frac{r}{100}\right)^t - 1\right]$$

TYPES OF QUESTIONS

Type I : To find CI and Amount
Type II : To find rate, Principal or time
Type III : When difference between CI and SI is given
Type IV : When interest is calculated half yearly or quarterly etc.
Type V : When both rate and principal have to be found

Type I

Example 1. Find the amount of Rs. 800 in 2 years at 5% per annum.

Sol :

$$A = P\left(1+\frac{r}{100}\right)^t$$

$$A = 800\left(1+\frac{5}{100}\right)^2$$

$$= 800\left(\frac{21}{20}\right)^2 = 800 \times \frac{441}{400} = \mathbf{882\ Rs.}$$

Example 2. Find the amount of Rs. 1000 in 2 years at 10% per annum compound interest.

Sol :

$$A = P\left(1+\frac{r}{100}\right)^t$$

$$= 1000\left(1+\frac{10}{100}\right)^2 = 1000 \times \frac{121}{100} = \mathbf{1210\ Rs.}$$

Example 3. Find the amount of Rs. 6250 in 2 years at 4% per annum compound interest.

Sol :

$$A = P\left(1+\frac{r}{100}\right)^t$$

$$= 6250\left(1+\frac{4}{100}\right)^2 = 6250 \times \frac{676}{625} = \mathbf{6760\ RS.}$$

Example 4. Find the amount of Rs. 16000 at 5% per annum compound interest in 3 years.

Sol :

$$A = P\left(1+\frac{r}{100}\right)^t$$

$$= 16000\left(1+\frac{5}{100}\right)^3 = 16000 \times \frac{9261}{8000} = \mathbf{18522\ Rs.}$$

Example 5. Find the amount of Rs. 20000 at 10% compound interest in 3 years.

Sol :

$$A = 20000\left(1+\frac{10}{100}\right)^3$$

$$= 20000\left(1+\frac{10}{100}\right)^3$$

$$= 20000\left(\frac{11}{10}\right)^3 = 20000\times\frac{1331}{1000} = \textbf{26620 Rs.}$$

Example 6. What will be compound interest on Rs. 20000 at the rate of 5% per annum for 2 years.

Sol :

$$CI = P\left[\left(1+\frac{r}{100}\right)^t - 1\right]$$

$$= 20000\left[\left(1+\frac{5}{100}\right)^2 - 1\right]$$

$$= 20000\left[\frac{441}{400} - 1\right] = 20000\times\frac{41}{400} = \textbf{2050 Rs.}$$

Example 7. What will be the compound interest on Rs. 31250 at a rate of 4% per annum for 2 years.

Sol :

$$CI = P\left[\left(1+\frac{r}{100}\right)^t - 1\right]$$

$$= 31250\left[\left(1+\frac{4}{100}\right)^2 - 1\right]$$

$$= 31250\left[\frac{676}{625} - 1\right] = 31250\times\frac{51}{625} = \textbf{2550 Rs.}$$

Example 8. Find the compound interest on Rs. 24000 at 5% per annum for 3 years.

Sol :

$$CI = P\left[\left(1+\frac{r}{100}\right)^t - 1\right]$$

$$= 24000\left[\left(1+\frac{5}{100}\right)^3 - 1\right]$$

$$= 24000\left[\frac{9261}{8000} - 1\right] = 24000\times\frac{1261}{8000} = \textbf{3783 Rs.}$$

Example 9. A sum amounts to Rs. 24200 in 2 years at 10% per annum compound interest. The sum is

Sol :

$$A = P\left(1+\frac{r}{100}\right)^t$$

$$24200 = P\left(1+\frac{10}{100}\right)^2 = P\left(\frac{11}{10}\right)^2$$

$$P = 24200 \times \frac{100}{121} = \mathbf{20000\ Rs.}$$

Example 10. A sum amounts to 4056 in 2 years at 4% Compound Interest per annum. The sum is ....

Sol :

$$A = P\left(1+\frac{r}{100}\right)^t$$

$$4056 = P\left(1+\frac{4}{100}\right)^2 = P\left(\frac{26}{25}\right)^2$$

$$P = 4056 \times \frac{625}{676} = \mathbf{3750\ Rs.}$$

Example 11. In what time Rs. 1200 will amount to Rs.1323 at 5% compound Interest?

Sol :

$$A = P\left(1+\frac{r}{100}\right)^t$$

$$1323 = 1200\left(1+\frac{5}{100}\right)^t$$

$$\frac{1323}{1200} = \left(\frac{21}{20}\right)^t$$

$$\left(\frac{21}{20}\right)^t = \left(\frac{21}{20}\right)^2$$

$$t = \mathbf{2\ years}$$

Example 12. The time in which Rs. 15625 will amount to Rs. 17576 at 4% compound interest is?

Sol :

$$A = P\left(1+\frac{r}{100}\right)^t$$

$$17576 = 15625\left(1+\frac{4}{100}\right)^t$$

$$\frac{17576}{15625} = \left(\frac{26}{25}\right)^t$$

$$\left(\frac{26}{25}\right)^t = \left(\frac{26}{25}\right)^3$$

$$t = \mathbf{3\ years}$$

Example 13. The rate percent if compound Interest of Rs. 15625 for 3 years is Rs. 1951.

Sol :

$$A = P + CI = 15625 + 1951 = 17576\ \text{Rs.}$$

$$A = P\left(1+\frac{r}{100}\right)^t$$

$$17576 = 15625\left(1+\frac{r}{100}\right)^3$$

$$\frac{17576}{15625} = \left(1+\frac{r}{100}\right)^3$$

$$\left(\frac{26}{25}\right)^3 = \left(1+\frac{r}{100}\right)^3$$

$$\frac{26}{25} = 1+\frac{r}{100}$$

$$\frac{26}{25}-1 = \frac{r}{100}$$

$$\frac{1}{25} = \frac{r}{100}$$

$$r = \mathbf{4\%}$$

Type II

Example 14. The difference between SI and CI on a certain sum of money for 2 years at r% per annum is D. The sum is ....?

Sol : Let P = 100

$$r = r\%$$

$$t = 2 \text{ years}$$

$$\text{SI} = \frac{100\times r\times 2}{100} = \mathbf{2r}$$

$$\text{CI} = 100\left[\left(1+\frac{r}{100}\right)^2-1\right]$$

$$= 100\left[\left(\frac{100+r}{100}\right)^2-1\right]$$

$$= 100\left[\frac{(100+r)^2}{100^2}-1\right]$$

$$= 100\left[\frac{100^2+r^2+200r-100^2}{100^2}\right]$$

$$= 100\left[\frac{r^2+200r}{100^2}\right]$$

$$= \frac{r^2+200r}{100}$$

$$\textbf{CI} - \textbf{SI} = \frac{\textbf{r}^2 + \textbf{200r}}{\textbf{100}} - \textbf{2r}$$

$$= \frac{r^2 + 200r - 200r}{100} = \frac{\textbf{r}^2}{\textbf{100}}$$

If difference is $\frac{r^2}{100}$ then prinicipal $= 100$ Rs.

If difference is 1 then principal $= \frac{100 \times 100}{r^2}$

If difference is D then principal $= \left(\frac{100}{r}\right)^2 \times D$

$$\boxed{\textbf{Principal} = \left(\frac{\textbf{100}}{\textbf{r}}\right)^2 \times \textbf{D}} \quad ...(1)$$

Where D is difference between CI and SI for 2 years. We will take 1 as the general formula to find principal of CI and SI for 2 years is given.

Example 15. The difference between SI and CI on a certain sum of money for 2 years at 5% per annum is Rs. 312. Find the sum.

Sol : From the last question $\quad \textbf{P} = \left(\frac{\textbf{100}}{\textbf{r}}\right)^2 \times \textbf{D}$

Where r is rate and D is difference between SI and CI

$$= \left(\frac{100}{5}\right)^2 \times 312 = 400 \times 312 = \textbf{124800 Rs.}$$

Example 16. The difference between SI and CI for certain sum of money for 2 years at 10% per annum is Rs. 80. The sum is.

Sol :

$$P = 80\left(\frac{100}{10}\right)^2 = 80 \times 100 = \textbf{8000 Rs.}$$

Example 17. If the difference between SI and CI for certain sum of money for **3 years** at **r%** per annum is **D. The sum is .... .**

Sol : Let

$$P = 100 \text{ Rs.}$$

$$r = r\%$$

$$t = 3 \text{ years}$$

$$\text{SI} = \frac{100 \times r \times 3}{100} = \mathbf{3r}$$

$$\text{CI} = 100\left[\left(1 + \frac{r}{100}\right)^3 - 1\right]$$

$$= 100\left[\left(\frac{100+r}{100}\right)^3 - 1\right]$$

$$= 100\left[\frac{(100+r)^3}{100^3} - 1\right]$$

$$= 100\left[\frac{100^3 + r^3 + 300r(100+r)}{100^3} - 1\right]$$

$$= 100\left[\frac{100^3 + r^3 + 300r(100+r)}{100^3} - 1\right]$$

$$= \frac{r^3 + 30000r + 300r^2}{100^2}$$

$$\text{CI} - \text{SI} = \frac{r^3 + 30000r + 300r^2}{100^2} - 3\text{r}$$

$$= \frac{r^3 + 30000r + 300r^2 - 30000r}{100^2}$$

$$= \frac{r^3 + 300r^2}{100^2}$$

If difference is $\frac{r^3 + 300r^2}{100^2}$ then principal = 100 Rs.

If difference is 1 Rs. then principal $= \frac{100 \times 100^2}{r^3 + 300r^2}$

If difference is D then principal $= \frac{100 \times 100^2}{r^3 + 300r^2} \times \text{D}$

$$= \frac{\text{D} \times 100^3}{r^2(r+300)}$$

$$\mathbf{P} = \frac{\mathbf{D \times 100^3}}{\mathbf{r^2 (r + 300)}} \qquad \textbf{...(2)}$$

Where D is the difference between CI and SI for 3 years and r is rate%

So (2) is taken as general formula when difference between SI and CI for 3 years is given.

Example 18. The difference between CI and SI on a certain sum of money at 5 p.a. is Rs. 122. The sum is .....

Sol : From last question $$P = \frac{D \times 100^3}{r^2(r+300)}$$

$$= \frac{122 \times 100 \times 100 \times 100}{25 \times 305} = \mathbf{16000\ Rs.}$$

Example 19. If difference between CI and SI for a Certain sum of money for 3 years at 4% per annum is Rs. 1064. Find the sum.

Sol : $$P = \frac{D \times 100^3}{r^2(r+300)}$$

$$= \frac{1064 \times 100 \times 100 \times 100}{16 \times 304} = \mathbf{218750\ Rs.}$$

Type III

1. Remember

When interest is compounded half yearly

Then $$\textbf{Amount} = P\left(\frac{1+\frac{R}{2}}{100}\right)^{2t}$$

i.e., in half yearly compound interest rate is halved and time is doubled.

2. When interest is compounded quarterly then rate is made 1/4 and time is made 4 times.

Then $$A = P\left(\frac{1+\frac{R}{4}}{100}\right)^{4t}$$

3. When rate of interest is R_1%, R_2% and R_3% for 1st, 2nd and 3rd year respectively.

Then $$A = \mathbf{P\left(1+\frac{R_1}{100}\right)\left(1+\frac{R_2}{100}\right)\left(1+\frac{R_3}{100}\right)}$$

Example 20. Find the compound Interest on Rs. 5000 at 20% per annum for 1.5 year compound half yearly

Sol : When interest is compounded half yearly

then $$A = P\left(\frac{1+\frac{R}{2}}{100}\right)^{2t}$$

$$A = 5000\left(1+\frac{\frac{20}{2}}{100}\right)^{\frac{3}{2}\times 2}$$

$$= 5000\left(1+\frac{10}{100}\right)^3$$

$$= 5000 \times \frac{1331}{1000} = 6655 \text{ Rs.}$$

$$\text{CI} = 6655 - 5000 = \mathbf{1655\ Rs.}$$

Example 21. Find compound Interest on Rs. 47145 at 12% per annum for 6 months, Compounded quarterly.

Sol : As interest is compounded quarterly

$$\therefore \quad A = P\left(\frac{1+\frac{R}{4}}{100}\right)^{4t}$$

$$A = 47145\left(1+\frac{\frac{12}{4}}{100}\right)^{\frac{1}{2}\times 4}$$

$$= 47145\left(1+\frac{3}{100}\right)^2$$

$$= 47145 \times \frac{103}{100} \times \frac{103}{100} = 50016.13 \text{ Rs.}$$

$$\therefore \quad \text{CI} = 50016.13 - 47145 = \mathbf{2871.13\ Rs.}$$

Example 22. Find the compound Interest on Rs. 18750 for 2 years when the rate of interest for 1st year is 4% and for 2nd year 8%

Sol :

$$A = P\left(1+\frac{R_1}{100}\right)\left(1+\frac{R_2}{100}\right)$$

$$= 18750 \times \frac{104}{100} \times \frac{108}{100} = 21060 \text{ Rs.}$$

$$\therefore \quad \text{CI} = 21060 - 18750 = \mathbf{2310\ Rs.}$$

Example 23. Find the compound interest on Rs. 8000 for $2\frac{1}{2}$ year at 8% per annum.

Sol :

$$P = 8000 \text{ Rs.}$$

$$t = 2\frac{1}{2} \text{ years}$$

$$r = 8\% \text{ for 2 years } 4\% \text{ half yearly}$$

$$\text{CI} = 8000\left(1+\frac{8}{100}\right)^2\left(1+\frac{4}{100}\right) - 8000$$

$$= 8000\times\frac{108}{100}\times\frac{108}{100}\times\frac{104}{100}-8000$$

$$= 9704.44 - 8000 = \mathbf{1704.44\ Rs.}$$

Type IV

Example 24. The compound interest on a certain sum for two years is Rs. 52 and simple Interest for the same period at same rate is Rs. 50. Find the sum and the rate.

Sol : We will do this question by **basic concept.** As discussed in the beginning of this chapter we know that **Simple interest** is same every year and there is no difference between **SI and CI** for 1st year. The difference arises in the 2nd year because interest of 1st year is added in **principal** and **interest** is now charged on **principal + Simple interest of 1st year.**

So in this question

$$\text{2 years SI} = 50 \text{ Rs.}$$

$$1'' = 25 \text{ Rs.}$$

Now CI for 1st year is same as SI for 1st year

i.e., **= 25 Rs.**

$\therefore$ CI for 2nd year $= 52 - 25 =$ **27 Rs.**

This additional interest **27– 25 = 2 Rs.** is due to the fact that 1st year SI i.e., Rs. 25 is added in principal. It means that **additional 2 Rs.** interest is charged on **Rs. 25**

$$\text{Rate\%} = \frac{2}{25}\times 100 = \mathbf{8\%}$$

Shortcut :

$$\text{Rate\%} = \frac{\text{CI}-\text{SI}}{\frac{\text{SI}}{2}}\times 100$$

$$= \frac{2}{\frac{50}{2}}\times 100$$

$$= \frac{2}{25}\times 100 = \mathbf{8\%}$$

$$P = \frac{\text{SI}\times 100}{\text{R}\times\text{T}} = \frac{50\times 100}{8\times 2} = \mathbf{312.50\ Rs.}$$

Example 25. The simple interest on a sum of money for 2 years is Rs. 600 and the compound interest on the same sum at same rate for 2 years is Rs. 615. Find the sum and rate %.

Sol : This example is same as **Example 24.**

Therefore

$$\text{Rate\%} = \frac{\text{CI}-\text{SI}}{\frac{\text{SI}}{2}}\times 100$$

$$= \frac{15}{300}\times 100 = \mathbf{5\%}$$

Example 26. A sum of money lent CI amounts in 2 year to Rs. 8820 and in 3 years to Rs. 9261. Find the sum and rate%.

Sol : First method

Amount after 3 years = 9261 Rs.

" " 2 " = 8820 Rs.

– –

By subtracting last year's interest = 441 Rs.

It is clear that this **Rs. 441** is **SI** on **Rs. 8820** from 2nd to 3rd year i.e., for 1 year.

$$\text{Rate\%} = \frac{441 \times 100}{8820 \times 1} = \mathbf{5\%}$$

Also

$$A = P\left(1 + \frac{r}{100}\right)^t$$

$$8820 = P\left(1 + \frac{5}{100}\right)^2$$

$$= P\left(\frac{21}{20}\right)^2$$

$$P = 8820 \times \frac{400}{441} = \mathbf{8000\ Rs.}$$

Second Method :

$$A = P\left(1 + \frac{r}{100}\right)^t$$

$$9261 = P\left(1 + \frac{r}{100}\right)^3 \quad \text{...(1)}$$

$$8820 = P\left(1 + \frac{r}{100}\right)^2 \quad \text{...(2)}$$

Divide (1) by (2)

$$\frac{9261}{8820} = \frac{\left(1 + \frac{r}{100}\right)^3}{\left(1 + \frac{r}{100}\right)^2}$$

$$\frac{9261}{8820} = 1 + \frac{r}{100}$$

$$\frac{9261}{8820} - 1 = \frac{r}{100}$$

$$\frac{441}{8820} = \frac{r}{100}$$

$$r = 5\%$$

Now
$$8820 = P\left(1+\frac{5}{100}\right)^2$$

$$8820 = P\left(\frac{21}{20}\right)^2$$

$$P = \frac{8820 \times 400}{441}$$

$$= \mathbf{8000\ Rs.}$$

EXERCISE

Q. 1 – 10 Fill in the blanks in following table

Q.No.	Principal (Rs.)	Rate% (compounded)	Time (years)	Amount (Rs.)	CI (Rs.)
1.	5500	10	3	—	—
2.	10800	$16\frac{2}{3}$	3	—	—
3.	1200	5	2	—	—
4.	2760	4	2	—	—
5.	18200	5	—	20065.50	—
6.	1500	—	2	1815	315
7.	3200	—	2	3528	328
8.	4000	—	3	5324	1324
9.	—	5	3	9261	—
10.	—	4	2	676	—

Q. 11 – 20 Fill in the blanks

Q.No.	Difference between CI and SI (Rs.)	Rate%	Time (years)	Principal (Rs.)
11.	428	4	2	—
12.	183	5	3	—
13.	760	4	3	—
14.	180	6	2	—
15.	1292	4	3	—
16.	642.50	10	2	—
17.	3750	4	2	—
18.	4509	6	2	—
19.	540	6	2	—
20.	4650	10	3	—

Hint : For Q. 11 – 20

USE SHORTCUT FOR 2 YEARS

$$\text{Principal} = \left(\frac{100}{r}\right)^2 \times D$$

And USE SHORTCUT FOR 3 YEARS

$$\text{Principal} = \frac{D \times 100^3}{r^2(r+300)}$$

Q. 21. Find the compound Interest on Rs. 12500 at 8% per annum for 9 months compounded quarterly.

(a) 1020 Rs. (b) 1428 Rs. (c) 510 Rs. (d) none

Q. 22. Find the compound Interest on Rs. 32000 at 20% per annum for 1 year, compounded half yearly.

(a) 6320 Rs. (b) 6720 Rs. (c) 6400 Rs. (d) none

Q. 23. The simple interest on a sum of money for two years is Rs. 80 and compounded Interest on same sum at same rate percent per annum for same time is Rs. 84. Find the rate %.

(a) 5% (b) 4% (c) 6% (d) none

Q. 24. The simple interest on a certain sum of money for two years is Rs. 40 and the CI on same sum at same rate for same time is Rs. 40.50. Find the sum.

(a) Rs. 1600 (b) Rs. 800 (c) Rs. 600 (d) none

Q. 25. Invested a sum of money at CI. It amounted to Rs. 2420 in 2 years and 2662 in 3 years. The rate percent is...... .

(a) 10% (b) 12% (c) 14% (d) none

SOLUTIONS

1.

$$A = P\left(1+\frac{r}{100}\right)^t$$

$$= 5500 \times \frac{1331}{1000} = 7320.50$$

$$CI = 7320.50 - 5500 = \mathbf{1820.50}$$

2.

$$A = 10800\left(1+\frac{50}{3\times 100}\right)^3$$

$$= 10800\left(\frac{7}{6}\right)^3 = \mathbf{17150\ Rs}$$

$$CI = 17150 - 10800 = \mathbf{6350\ Rs.}$$

3.

$$A = 1200\left(1+\frac{5}{100}\right)^2$$

$$= 1200 \times \frac{441}{400} = \mathbf{1323\ Rs.}$$

$$CI = 1323 - 1200 = \mathbf{123\ Rs.}$$

4.
$$A = 2760\left(1+\frac{4}{100}\right)^2$$
$$= 2760 \times \left(\frac{26}{25}\right)^2 = 2760 \times \frac{676}{625} = \mathbf{2985.21\ \ Rs.}$$
$$CI = 2985.21 - 2760 = \mathbf{225.21\ Rs.}$$

5.
$$20065.50 = 18200\left(1+\frac{5}{100}\right)^t$$
$$\frac{20065.50}{18200} = \left(\frac{21}{20}\right)^t$$
$$\frac{441}{400} = \left(\frac{21}{20}\right)^t$$
$$\left(\frac{21}{20}\right)^2 = \left(\frac{21}{20}\right)^t$$
$$\boldsymbol{t} = \mathbf{2\ years}$$

6.
$$1815 = 1500\left(1+\frac{r}{100}\right)^2$$
$$\frac{1815}{1500} = \left(1+\frac{r}{100}\right)^2$$
$$\frac{121}{100} = \left(1+\frac{r}{100}\right)^2$$
$$\left(\frac{11}{10}\right)^2 = \left(1+\frac{r}{100}\right)^2$$
$$\frac{11}{10} = 1+\frac{r}{100}$$
$$\frac{11}{10} - 1 = \frac{r}{100}$$
$$\frac{1}{10} = \frac{r}{100}$$
$$\boldsymbol{r} = \mathbf{10\%}$$

7.
$$3528 = 3200\left(1+\frac{r}{100}\right)^2$$
$$\frac{3528}{3200} = \left(1+\frac{r}{100}\right)^2$$
$$\frac{441}{400} = \left(1+\frac{r}{100}\right)^2$$

$$\left(\frac{21}{20}\right)^2 = \left(1+\frac{r}{100}\right)^2$$

$$\frac{21}{20} = 1+\frac{r}{100}$$

$$\frac{21}{20}-1 = \frac{r}{100}$$

$$\frac{1}{20} = \frac{r}{100}$$

$$\boldsymbol{r = 5\%}$$

8.

$$5324 = 4000\left(1+\frac{r}{100}\right)^3$$

$$\frac{5324}{4000} = \left(1+\frac{r}{100}\right)^3$$

$$\frac{1331}{1000} = \left(1+\frac{r}{100}\right)^3$$

$$\left(\frac{11}{10}\right)^3 = \left(1+\frac{r}{100}\right)^3$$

$$\frac{11}{10}-1 = \frac{r}{100}$$

$$\boldsymbol{r = 10\%}$$

9.

$$9261 = P\left(1+\frac{5}{100}\right)^3$$

$$9261 = P\left(\frac{21}{20}\right)^3$$

$$9261 = P\times\frac{9261}{8000}$$

$$\mathbf{P = 8000\ Rs.}$$

10.

$$676 = P\left(1+\frac{4}{100}\right)^2$$

$$676 = P\left(\frac{26}{25}\right)^2$$

$$676 = P\times\frac{676}{625}$$

$$\mathbf{P = 625\ Rs.}$$

11. Since time is 2 years

Therefore $P = D\times\left(\frac{100}{r}\right)^2$

$$= 428 \times \left(\frac{100}{4}\right)^2$$

$$= 428 \times 625 = \mathbf{267500\ Rs.}$$

12. Here time is 3 years

Therefore $$P = \frac{D \times 100^3}{r^2 (r + 300)}$$

$$= \frac{183 \times 100^3}{5^2 (5 + 300)}$$

$$= \frac{183 \times 100 \times 100 \times 100}{25 \times 305} = \mathbf{24000\ Rs.}$$

13. Here time is 3 years

Therefore $$P = \frac{D \times 100^3}{r^2 (r + 300)}$$

$$= \frac{760 \times 100^3}{4^2 (4 + 300)}$$

$$= \frac{760 \times 100 \times 100 \times 100}{16 \times 304} = \mathbf{156250\ Rs.}$$

14. Since time is 2 years

Therefore $$P = D \times \left(\frac{100}{r}\right)^2$$

$$= 180 \times \left(\frac{100}{6}\right)^2$$

$$= 180 \times \frac{2500}{9} = \mathbf{50000\ Rs.}$$

15. Here time is 3 years

Therefore $$P = \frac{D \times 100^3}{r^2 (r + 300)}$$

$$= \frac{1292 \times 100^3}{4^2 (4 + 300)}$$

$$= \frac{1292 \times 100 \times 100 \times 100}{16 \times 304} = \mathbf{265625\ Rs.}$$

16. Since time is 2 years

Therefore $$P = D \times \left(\frac{100}{r}\right)^2$$

$$= 242.50 \times \left(\frac{100}{10}\right)^2$$

$$= 242.50 \times 100 = \mathbf{24250\ Rs.}$$

17. Since time is 2 years

Therefore $$P = D\times\left(\frac{100}{r}\right)^2$$

$$= 3750\times\left(\frac{100}{4}\right)^2$$

$$= 3750 \times 625 = \mathbf{2543750\ Rs.}$$

18. Since time is 2 years

Therefore $$P = D\times\left(\frac{100}{r}\right)^2$$

$$= 4509\times\left(\frac{100}{6}\right)^2$$

$$= 4509\times\frac{2500}{9} = \mathbf{1252500\ Rs.}$$

19. Since time is 2 years

Therefore $$P = D\times\left(\frac{100}{r}\right)^2$$

$$= 540\times\left(\frac{100}{6}\right)^2$$

$$= 540\times\frac{2500}{9} = \mathbf{150000\ Rs.}$$

20. Here time is 3 years

Therefore $$P = \frac{D\times 100^3}{r^2(r+300)}$$

$$= \frac{4650\times 100^3}{10^2(10+300)}$$

$$= \frac{4650\times 100\times 100\times 100}{100\times 310} = \mathbf{150000\ Rs.}$$

21. **(d)** As interest is compounded quarterly

$$\therefore \quad A = P\left(\frac{1+\frac{R}{4}}{100}\right)^{4t}$$

$$A = 12500\left(1+\frac{\frac{8}{4}}{100}\right)^{\frac{9}{12}\times 4}$$

$$= 12500\left(1+\frac{2}{100}\right)^3$$

$$= 12500 \times \frac{102}{100} \times \frac{102}{100} \times \frac{102}{100}$$

$$= 12500 \times \frac{132651}{125000} = \textbf{13265.10 Rs.}$$

$$\therefore \quad CI = 13265.10 - 12500 = \textbf{765.10 Rs.}$$

22. (b) When interest is compounded half yearly

then

$$A = P\left(\frac{1+\frac{R}{2}}{100}\right)^{2t}$$

$$A = 32000\left(1+\frac{\frac{20}{2}}{100}\right)^{1\times 2}$$

$$= 32000\left(1+\frac{10}{100}\right)^{2}$$

$$= 32000 \times \frac{121}{100} = 38720 \text{ Rs.}$$

$$CI = 38720 - 32000 = \textbf{6720 Rs.}$$

23. (d)

$$\text{Rate\%} = \frac{CI - SI}{\frac{SI}{2}} \times 100$$

$$= \frac{84-80}{\frac{80}{2}} \times 100 = \frac{4}{40} \times 100 = \textbf{10\%}$$

24. (b)

$$R = \frac{50}{20} \times 100 = 2.5\%$$

$$P = \frac{40 \times 100}{2.5 \times 2} = \textbf{800 Rs.}$$

25. (a) This question is same as example 26.

$$\text{Last year interest} = 2662 - 2420 = 242 \text{ Rs.}$$

Therefore

$$\text{rate\%} = \frac{242 \times 100}{2420 \times 1} = \textbf{10\%}$$

Section II

Test Yourself

1. Twelve men and 5 women can complete a work in 2 days whereas 4 men and 3 women can complete the same work in 5 days. In how many days can 8 men complete the work? **(CLAT 2018)**

   (a) 3 (b) 4 (c) 5 (d) 6

2. Four pipes A, B, C and D can fill a tank with water in 15, 20, 30 and 60 hours respectively. Pipe A is opened at 4 a.m., B at 5 a.m., C at 6 a.m. and D at 7 a.m. When is the tank filled up completely? **(CLAT 2018)**

   (a) 9:30 a.m. (b) 10:00 a.m. (c) 10:30 a.m. (d) 11:00 a.m.

3. A vessel contains a mixture of milk and water in the ratio of 5:3 respectively. How much of the mixture must be siphoned off and replaced with water, so that the mixture may be half milk and half water? **(CLAT 2017)**

   (a) 1/7 (b) 1/3 (c) 1/4 (d) 1/5

4. The difference between simple interest and compound interest at the same rate for rupees 5,000 for two years is rupees 98. The rate of interest is: **(CLAT 2017)**

   (a) 12% (b) 10% (c) 10½% (d) 14%

5. Gold and copper are as heavy as water by 19 and 9 times respectively. The ratio in which these two metals be mixed so that the mixture is 17 times as heavy as water is: **(CLAT 2017)**

   (a) 2:3 (b) 4:1 (c) 3:2 (d) 3:4

6. A train 'X' leaves station 'A' at 3 p.m. and reaches station 'B' at 4.30 p.m., while another train 'Y' leaves station 'B' at 3.00 p.m. and reaches station 'A' at 4.00 p.m. These two trains cross each other at: **(CLAT 2017)**

   (a) 3.36 p.m. (b) 3.30 p.m. (c) 3.20 p.m. (d) 3.40 p.m.

7. Three partners A, B, C shared their' profit in the ratio 4:5:6. If A partnered for 10 months, B partnered for 15 months and C partnered for 16 months, what is the ratio of their investments? **(NIFT UG 2017)**

   (a) 6:7:9 (b) 27:20:15 (c) 48:40:45 (d) 40:75:96

8. 16 persons can do a work in 10 days working 9 hours a day. In how many days can 12 persons do this work while working 8 hours a day? **(NIFT UG 2017)**

   (a) 10 days (b) 18 days (c) 12 days (d) 15 days

9. Naresh puts Rs. 10,000 at interest compounded manually for 2 years. Had he invested it at 1% more rate of interest compounded annually, he would have earned Rs. 209 more. What is the interest rate? **(NIFT UG 2017)**

   (a) 7% (b) 6% (c) 5% (d) 4%

10. After covering five-eighth of my journey, I find that I have travelled 60 km. How much journey is left? **(NIFT UG 2017)**

(a) 35 km (b) 40 km (c) 44 km (d) 36 km

11. A train travelling at 48 kmph completely crosses another train having half of its length and tavelling in opposite direction at 42 kmph in 12 seconds. It also passes a railway platform in 45 seconds. The length of the platform is ________. **(MAT Feb 2016)**

(a) 400 m (b) 450 m (c) 560 m (d) 600 m

12. A sum of ₹ 7,700 is to be divided among three brothers Sunil, Sumant and Surat in such a way that simple interest on each part at 5% per annum after 1,2 and 3 years respectively remains equal. The share of Sunil is more than that of Surat by _________. **(Mat 2016)**

(a) ₹ 2,800 (b) ₹ 2,500 (c) ₹ 3,000 (d) ₹ 2,700

13. Three numbers A, B and C are in the ratio of 12 : 15 : 25. If sum of these numbers is 312, the ratio between the difference of B and A and the difference of C and B is _________.

(Mat 2016)

(a) 3 : 7 (b) 10 : 3 (c) 3 : 10 (d) 5 : 1

14. Mixture of milk and water has been kept in two separate containers. Ratio of milk to water in one of the containers is 5 : 1 and that in the other container is 7 : 2. In what ratio the mixtures of these two containers should be added together so that the quantity of milk in the new mixture may become 80%? **(Mat 2016)**

(a) 2 : 3 (b) 3 : 2 (c) 4 : 5 (d) None of these

15. Walking at $5/7^{th}$ of his usual speed, a boy reaches school 6 minutes late. Find his usual time to reach school. **(RRB Non. Tec 2016)**

(a) 16 minutes (b) 15 minutes (c) 14 minutes (d) 12 minutes

16. The average of 5 quanities is 6. The average of 3 of them is 8. What is the average of the remaining two numbers. **(RBI Non. Tec April 2016)**

(a) 4 (b) 5 (c) 3 (d) 3.5

17. What quantity of watrer should be added to reduce 9 litres of 50% acidic liquid to 30% acidic liquid? **(RRB Non. Tec April 2016)**

(a) 4.5 litres (b) 3 litres (c) 6 litres (d) 7.5 litres

18. A and B can do a piece of work in 12 day, B and C in 15 days, C and A in 20 days. In how many days will the job be finished, if A, B and C work together? **(RRB Non. Tec April 2016)**

(a) 10 days (b) 15 days (c) 8 days (d) 16 days

19. The simple interest on ₹ 170 increases by ₹ 17 when the time increases by 5 years. Find the rate per cent per annum. **(RRB Non. Tec April 2016)**

(a) 1.7% (b) 10% (c) 5% (d) 2%

20. If A : B = 3 : 4 and B : C = 5 : 6 then A : B : C = ? **(RRB Non. Tec April 2016)**

(a) 10 : 15 : 24 (b) 15 : 25 : 24 (c) 15 : 16 : 24 (d) 15 : 20 : 24

21. Jar A contains 'X' ml of mixture of milk and water in the respective ration of 3 : 2. Jar B contains 'X' ml mixture of milk and water in the respective ratio of 4 : 1, which is added to Jar A. If the resultant quantity of milk in Jar A is 84 ml, what is the value of 'X'? **(SBI Clerk 2016)**

(a) 120 (b) 80 (c) 72 (d) 81

(e) 60

22. The average weight of 26 students in a class was 41.5 kg. After 10 new students joined the class, its average weight increased by 2 kg, what is the average of 10 new students who joined the class? (in kg) **(SBI Clerk 2016)**

(a) 49.3 (b) 47.6 (c) 48.3 (d) 47.9

(e) 48.7

23. A train, 218 m long, travelling at 63kmph can cross a platform in 32 seconds. If a man can cross the same platform in 3 minutes, what is the speed of the man (in m/sec)? **(SBI Clerk 2016)**
(a) 1.8 (b) 1.9 (c) 2.3 (d)

24. Ira and John start a business together. The capital invested by Ira was twice of that invested by John. At the end of six months from the start of the business, John doubled his investment. If at the end of the year, the total profit earned was ₹ 8400, what was John's share from the profit? **(SBI Clerk 2016)**
(a) ₹ 3600 (b) ₹ 3500 (c) ₹ 4900 (d) ₹ 4800
(e) ₹ 4200

25. When a certain sum of money is invested for 2 years @10 p.c.p.a. (compound interest compounded annually), it amounts to ₹ 2904. Had the rate of interest been 20%, what would have been the amount received after 2 years? **(SBI Clerk 2016)**
(a) ₹ 3586 (b) ₹ 3482 (c) ₹ 3260 (d) ₹ 3624
(e) ₹ 3456

26. The distance travelled downstream by a boat in 4 hours is 16 km more than the distance travelled upstream by the boat in the same time. What is the speed of the current? (in kmph) **(SBI Clerk 2016)**
(a) 2.5 (b) 3 (c) 3.5 (d) 2

27. The ratio between (X + 8) and Y is 5 : 4 respectively. If 36 is 125% more than Y, what is the value of X? **(SBI Clerk 2016)**
(a) 12 (b) 16 (c) 24 (d) 18
(e) 20

28. 8 men can finish a piece of work in 21 days. 14 men started working and after 3 days were replaced by 9 women. These 9 women finished the remaining work in 24 days. In how many days 9 women can finish the whole work? **(LIC AAO March 2016)**
(a) 18 (b) 26 (c) 36 (d) 32

29. There are 5 consecutive odd numbers. If the difference between the square of the average of the first two odd numbers and square of the average of the last two odd numbers is 492, what is the smallest odd number? **(LIC AAO March 2016)**
(a) 37 (b) 42 (c) 41 (d) 35
(e) 39

30. A boat can travel 12.8 km downsream in 32 minutes. If the speed of the water current is 1/5th of the speed of the boat in still water, what distance (in km) the boat can travel in 33 minutes in still water? **(LIC AAO March 2016)**
(a) 11 (b) 10.2 (c) 8.5 (d) 9
(e) 6.5

31. Forty five men can complete a work in 16 days. Six days after they started working, 30 more men joined them. How many days will they now take to complete the remaining work? **(NIFT 2016)**
(a) 16 days (b) 12 days (c) 6 days (d) 10 days

32. A man takes 3 hours 45 minutes to row a boat 15km downstgream of a river and 2 hours 30 minutes to cover a distance of 5 km upstream. Find the speed of the river current in km/hr. **(NIFT 2016)**
(a) 1.0 km/hr (b) 1.5 km/hr (c) 1.2 km/hr (d) 1.3 km/hr

33. Find the compound interest on ₹ 48,000 for one year at 8% per annum when compounded half yearly? **(NIFT 2016)**

(a) ₹ 3,916.20 (b) ₹ 3,812.10 (c) ₹ 3,916.80 (d) ₹ 3,716.80

34. To gain 25% after allowing a discount of 10%, the shopkeeper must mark the price of the article which costs him ₹ 360 as ________. **(NIFT 2016)**

(a) ₹ 500 (b) ₹ 450 (c) ₹ 460 (d) ₹ 486

35. A pond was dug by four men: Ram, Rahim, Rahum and Rohan. They were paid a total of ₹ 1500 and the money is to be shared in the ratio of 1 : 2 : 2 : 1, respectively. What is the share of Ram and Rohan together? **(NIFT 2016)**

(a) ₹ 1000 (b) ₹ 250 (c) ₹ 500 (d) None of these

36. In a mixture of 120 litres of milk and water, the amount of water is 40 litres and the remaining is milk? What is the ratio of milk to water? **(NIFT 2016)**

(a) 1 : 2 (b) 1 : 4 (c) 4 : 1 (d) 2 : 1

37. The cost of five chairs and three tables is ₹ 3,110. Cost of one chair is ₹ 210 less than cost of one table. What is the cost of two tables and two chairs? **(NIFT 2016)**

(a) ₹1,660 (b) ₹ 1,860
(c) ₹ 2,600 (d) Cannot be determined

38. A library has an average of 510 visitors on Sundays and 240 on other days. The average number of visitors per day in a month of 30 days beginning with a Sunday is ________? **(NCHMCT 2016)**

(a) 250 (b) 276 (c) 280 (d) 285

39. Theratio between the present ages of P and Q is 6 : 7. If Q is 4 years older than P, what will be the ratio of the ages of P and Q after 4 years? **(NCHMCT 2016)**

(a) 3 : 4 (b) 3 : 5 (c) 4 : 3 (d) None of these

40. A and B started a business with initial investments in the ratio 5 : 7. If after one year their profits were in the ratio 1 : 2 and the period for A's investment was 7 months, B invested the money for **(SBI Clerk 2016)**

(a) 6 months (b) 2½ months (c) 10 months (d) 4 months
(e) 7 months

41. An army lost 10% its men in war, 10% of the remaining due to diseases and 10% of the rest were disabled. Thus, the strength was reduced to 729000 active men. Find the original strength? **(SBI Clerk 2016)**

(a) 1000000 (b) 1200000 (c) 1500000 (d) 1800000
(e) None of these

42. What is the difference between the compound interests on ₹ 5000 for 1 years at 4% per annum compounded yearly and half-yearly? **(SBI Clerk 2016)**

(a) 2 (b) 3 (c) 4 (d) 8
(e) None of these

43. A person carries ₹165/- in the form of currency notes of ₹ 5, ₹ 10 and ₹ 20 in the ratio of 3 : 2 : 1. What is the value of currency notes of ₹ 20 denomination? **(RRB Non. Tec March 2016)**

(a) 60 (b) 100 (c) 40 (d) 80

44. What is the 4th proportional of 3, 8, 12? **(RRB Non. Tec March 2016)**

(a) 36 (b) 26 (c) 32 (d) 16

45. Vessel A contains a mixture of milk and water in the respective quantities of 20 litres and 4 litres. Vessel B contains a mixture of milk and water in the respective ratio of 3 : 1. Mixtures from vessel A and B are both mixed together in Vessel C. If the resultant percentage of water in Vessel C was 20%, what was the initial quantity of mixture in vessel B? (inlitres)?

(NABARD Asst. Manager 2016)

(a) 6 (b) 20 (c) 12 (d) 28

46. Aman invests a certain sum –₹ 'X' for two years in scheme A offering compound interest at 20% p.a. (compounded annually). He also invests ₹ 'X –1600' in scheme B for three years offering simple interest at 10% p.a. The interest he received from scheme A was twice the interest he received from scheme B. What is the sum of money that Aman had invested in scheme B?

(NABARD Asst. Manager 2016)

(a) ₹ 4400 (b) ₹ 4800 (c) ₹ 4200 (d) ₹ 4600
(e) ₹ 4000

47. Shyam bought two articles – A and B, the cost prices of which were same. Shyam labelled both the articles at 40% more than its cost price. He then sold article a at 30% discount and article B at ₹ 35 more than the selling price of article A. If in the whole deal he earned a profit of 5%, what was the cost price of article A? **(NABARD Asst. Manager 2016)**

(a) ₹ 250 (b) ₹ 100 (c) ₹ 300 (d) ₹ 360
(e) ₹ 150

48. The distance between point X and point Y is 80 km. Boat A could travel from point X to Y upstream and point Y to X downstream in total 30 hours. Boat B could travel from point X to Y upstream and point Y to X downstream in total 9 hours. If the speed of boat B in still water is thrice of boat A in still water, what is the speed of Boat B? (Please note: speed of the current is constant for both **(NABARD Asst. Manager 2016)**

(a) 24 kmph (b) 27 kmph (c) 12 kmph (d) 18 kmph
(e) 21 kmph

49. The average runs of a cricketer in a tournament, in which he played 14 matches, are 47. His average runs in the first seven matches are 57 and that in the last five matches are 44. If the runs made by him in 8th match are 15, how many runs did he made in 9th match?

(SBI Clerk 2016)

(a) 24 (b) 32 (c) 26 (d) 22
(e) 28

50. An equal amount of sum, P is invested in scheme A and scheme B. Both the schemes A and B offer simply interest at the rate of 12% and 9% respectively. If at the end of two years total amount received from both the schemes together was ₹ 21,780, what is the value of P?

(SBI Clerk 2016)

(a) 9000 (b) 9600 (c) 12000 (d) 8400

51. A, B and C, each working alone, can finish a piece of work in 27, 33 and 45 days respectively. A starts by working alone for 12 days, then B takes over from A and works for 11 days. At this stage Ctakes over from B and completes the remaining work. In how many days the whole work was completed? **(SBI Clerk 2016)**

(a) 33 (b) 31 (c) 39 (d) 35
(e) 37

52. Train A, travelling at 'S' m/sec,. can cross a platform double its length in 21 seconds. The same train, travelling at 'S + 5' m/sec, can cross the same platform in 18 seconds. What is the value of 'S'? **(SBI Clerk (2016)**

(a) 27.5 (b) 32.5 (c) 30 (d) 35
(e) 25

53. Five years ago average age of A and B was 15 years. Now the average age of A, B and C is 20 years. What would be the age of C after 10 years? **(SET Law 2016)**
(a) 35 years (b) 40 years (c) 30 years (d) 50 years

54. A tank can be filled by a tap A in 20 minutes and by tap B in 60 minutes. Both the taps are kept open for 5 minutes and then tap A is closed. After this, in how much time the tank will be completely filled? **(SET Law 2016)**
(a) 20 minutes (b) 30 minutes (c) 45 minutes (d) 40 minutes

55. A train is 150 meter long, and has a speed of 36 km/hr. Find the time it will take to cross a person standing on the platform? **(SET Law 2016)**
(a) 15 seconds (b) 12 seconds (c) 10 seconds (d) 18 seconds

56. The simple interest on ₹ 5,000 at 8 percent/annum for the period of 73 days will be **(SET Law 2016)**
(a) ₹ 90 (b) ₹ 80 (c) ₹ 75 (d) ₹ 70

57. The average monthly expenditure of a family was ₹ 2750 for the first 3 months. ₹ 3150 for for the next three months and ₹ 6750 for next three months. Find the average income of he family for the 9 months, if they save ₹ 650 per month? **(SET Law 2016)**
(a) 4866.66 (b) 5123.33 (c) 4666.66 (d) 4216.66

58. The average age of 8 persons in a committee is increased by 2 years when two men aged 35 years and 45 years are substituted by two women. Find the average age of the two women. **(SET Law 2016)**
(a) 48 (b) 45 (c) 51 (d) 42

59. The average temperature for Wednesday, Thursday and Friday was 40°C. The average for Thursday, Fridayand Saturday was 41°C. If the temperature on Saturday was 42°C, what was the temperature on Wednesday? **(SET Law 2016)**
(a) 39°C (b) 44°C (c) 38°C (d) 41°C

60. In a mixture of 80 litres of milk and water, 25% of the mixture is milk. How much water should be added to the mixture so that milk becomes 20% of the mixture? **(SET Law 2016)**
(a) 20 litres (b) 15 litres (c) 25 litres (d) 24 litres

61. By selling bouquets for ₹ 63, florist gains 5%. At what price should he sell the bouquets to gain 10% on the cost price? **(SET Law 2016)**
(a) ₹ 66 (b) ₹ 69 (c) ₹ 72 (d) ₹ 72.50

62. A shopkeeper bought 240 chocolates at ₹ 9 per dozen. If he sold all of them at ₹ 1 each, what was his profit per cent? **(SET Law 2016)**
(a) 30% (b) 33.33% (c) 24% (d) 27%

63. A feeding bottle is sold for 120. Sales tax accounts for one-fifth of this and profit one-third of the remainder. find the cost price of the feed bottlle? **(SET Law 2016)**
(a) 64 (b) 72 (c) 68 (d) 76

64. ₹1694 is repaid after two years at compound interest. Which of the following is the value of the principal and the rate? **(SET Law 2016)**
(a) ₹1200,20% (b) ₹1300,15% (c) ₹1400,10% (d) ₹1500,12%

65. The students present in an auditorium are asked to stand in rows. If there were 4 students extra in a row, there would be 4 less rows. However, if 2 students were less in a row, there would be 4 more rows. The number of students present in the autditorium is _______. **(MAT May 2016)**

(a) 96 (b) 80 (c) 100 (d) 128

66. One fourth of Nitin's money is equal to one-sixth of Yash's money. If both together have ₹ 600, what is the difference between their amounts? **(MAT May 2016)**

(a) ₹ 240 (b) ₹ 160 (c) ₹ 200 (d) ₹ 120

67. If 5 students utilize 18 pencils in 9 days, how long at the same rate will 66 pencils last for 15 students? **(IBSAT 2008)**

(a) 12 days (b) 10 days (c) 11 days (d) None of these

68. Three friends Anu, Bina and Charu divide ₹ 1,105 amongst them in such a way that if 10, 20, and 15 are removed from the sums that Anu, Bina and Charu received respectively, then the share of the sums that they got will be in the ratio of 11 : 18 : 24. How much did Charu receive? **(MAT May 2016)**

(a) ₹ 510 (b) ₹ 495 (c) ₹ 480 (d) ₹ 375

69. I have to reach a certain place at a certain time and I find that I shall be 15 minutes too late, if I walk at 4 km an hour, and 10 minutes too soon, if I walk at 6 km an hour. How far have I to walk. **(MAT May 2016)**

(a) 5 km (b) 25 km (c) 10 km (d) None of these

70. A train travels a distance of 300 km at a constant speed. If the speed of the train is increased by 5 kman hour the journey would have taken 2 hours less. The original speed of the train was **(MAT May 2016)**

(a) 20 km/hr (b) 25 km/hr (c) 28 km/hr (d) 30 km/hr

71. By selling 20 chairs, a shopkeeper earns a profit equal to the selling price of 4 chairs. What is his gain%? **(RRB Non. Tec April 2016)**

(a) 20% (b) 25% (c) 33.33% (d) >15%

72. Two numbers are in the ratio 2 : 3. If 12 is subtracted from both the numbers the ratio becomes 5 : 8. Find the numbers. **(RRB Non. Tec April 2016)**

(a) 16 and 24 (b) 35 and 56 (c) 72 and 108 (d) 20 and 48

73. A train of length 210 m runs at a speed of 108 km/hr. How long will it take for the train to cross a platform of length 150 m? **(RRB Non. Tec April 2016)**

(a) 8 seconds (b) 9 seconds (c) 10 seconds (d) 12 seconds

74. If 4 men can paint a wall of length 96 m in 2 days, how, how long will it take 6 men to paint a wall of 72 m? **(IRMA 2008)**

(a) 1 day (b) 2 days (c) 3 days (d) 4 days

75. If the difference between the compound interest and simple interest at 5% for 2 years is ₹ 225, find the principal. **(RRB Non. Tec April 2016)**

(a) ₹ 75,000 (b) ₹ 80,000 (c) ₹ 85,000 (d) ₹ 90,000

76. The selling price of a book is 19/15 times its cost price. Find the profit %. **(RRB Non. Tec April 2016)**

(a) 20% (b) 22% (c) 26.67% (d) 30%

77. In an election, 1/6th of the votes were invalid, if candidate 'A' got 8/15th of the valid votes and candidates 'B' got 4900 votes, find the total number of votes. **(RRB Non. Tec April 2016)**

(a) 10,500 (b) 11,800 (c) 12,600 (d) 14,200

78. There are two varieties of rice- A and B, which are mixed in the respective ratio of 5 : 3. Rice A is worth ₹ 16 per kg and rice B is worth ₹ 28 per kg. What was the worth (in per kg) of the new mixture of rice (of varieties –A and B)? **(IBPS Clerical 2015)**

(a) ₹ 18.75/- (b) ₹ 22.50/- (c) ₹ 17.75/- (d) ₹ 25.25/-

(e) None of these

79. A can finish a piece of work in 30 days. He worked alone for 6 days and B joined him. Together they could finish the remaining work in 9 days. In how many days B alone can finish the same piece of work? **(IBPS Clerical 2015)**

(a) 12 (b) 18 (c) 15 (d) 14

(e) 10

80. Two cars (A & B) started from the same point at the same time in opposite directions (one towards north and other towards south.) Car A travelled a distance of 122 km in 5 hours. If the speed of the car B is 8.5 m/s, what is the distance between them after 3 hours? **(IBPS Clerical 2015)**

(a) 158 km (b) 144 km (c) 185 km (d) 165 km

(e) 195 km

SOLUTIONS

1. (b) **$(12m + 5w) \times 2 = (4m + 3w) \times 5$**

$24m + 10w = 20m + 15w$

$\Rightarrow \quad 4m = 5w$

$12m = 4m \times 3 = 5w \times 3 = 15w$

And $\quad 8m = 4m \times 2 = 5w \times 2 = 10w$

As **$12m + 5w = 15w + 5w = 20w$ can do the work in 2 days**

$\Rightarrow$ 8m = 10w will take double time i.e. 4 days

2. (d) Let total capacity of tank is 60 liters

So A will fill 60/15 = 4 liters per hour

B will fill 60/20 = 3 liters per hour

C will fill 60/30 = 2 liters per hour

D will fill 60/60 = 1 liters per hour

till 7 a.m. tank filled is $3 \times 4 + 2 \times 3 + 1 \times 2 = 20$ liters

So further time taken = (60 – 20)/(4 + 3 + 2 + 1) = 40/10 = 4 hours.

So tank will be filled up completely at**11:00 a.m.**

3. (d) Milk = 5/8

Water = 3/8

Let x liter of mixture is withdrawn

So according to question

5/8 – 5/8 x = 1/2

Solving we get x = 1/5

4. (d) $P = D(100/r)^2$

$5000 = 98(100/r)^2$

$5000/98 = (100/r)^2$

$2500/49 = (100/r)^2$

$(50/7)^2 = (100/r)^2$

$50/7 = 100/r$

So $r = 14\%$

5. (b)

6. (a) Let distance is 3 km

So speed of X = 3/1.5 = 2 km/h

speed Y = 3/1 = 3 km/h

So they will meet after 3/(3 + 2) = 3/5 hours i.e. 36 min. Hence meeting time is 3:36 pm

7. (d) Required ratio = 4 × 10 : 5 × 15 : 6 × 16 = **40 : 75 : 96**

8. (d) Required number of days = $10 \times \frac{16}{12} \times \frac{9}{8} = 15$ days

9. (d) By options

$10000\ [1.05^2 - 1.04^2] = 10000 \times 0.00209 = 209$

Hence original rate = 4%

10. (d) Let total journey = x

So 5/8 x = 60

x = 96

remaining journey = 3/8 of 96 = 36 km

11. (a) L_1 = Length of Ist train

L_2 = Length of IInd train

$$L_1 + L_2 = (48 + 42) \times \frac{5}{18} \times 12$$

$$= 90 \times \frac{5}{18} \times 12$$

$$= 300 \text{ m}$$

$$L_1 : L_2 = 2 : 1$$

$$\therefore \quad L_1 = 200 \text{ m}$$

$$L_2 = 100 \text{ m}$$

$$L_1 + P = 48 \times \frac{5}{18} \times 45$$

$$= 600$$

$$\therefore \quad P = 600 - L_1 = 600 - 200 = 400 \text{ m}$$

12. (a) $5\%\, x = 10\%\, y = 15\%\, z$

$$x = 2y = 3z$$

$$\Rightarrow \quad x : y : z = 6 : 3 : 2$$

$$\therefore \quad \text{Required difference} = \frac{4}{11} \times 7700 = ₹\,2800$$

13. (c) Required difference = (15 – 12) : (25 – 15) = 3 : 10

14. (a) Milk and water in first container = 5 : 1

Milk and water in second container = 7 : 2

Milk and water in mixture = 80% : 20% = 4 : 1

By Allgation (milk)

First $\frac{5}{6}$ Second $\frac{7}{9}$

$\frac{4}{5}$

$\frac{1}{30}$: $\frac{1}{45}$

3 : 2

15. (b) $\frac{7}{5}$ of ut = ut + 6

$\frac{7}{5}$ ut – ut = 6

$\frac{2}{5}$ ut = 6

ut = $\frac{6 \times 5}{2}$ = 15 min

16. (c) Required average = $\frac{5 \times 6 - 3 \times 8}{2} = \frac{30 - 24}{2} = \frac{6}{2} = 3$

17. (c) Let x litre of water added

So, $\frac{4.5 + x}{9 + x} = \frac{7}{10}$

$\Rightarrow \quad x = 6$

18. (a) $2(A + B + C)$ one day work $= \frac{1}{12} + \frac{1}{15} + \frac{1}{20}$

$= \left(\frac{5 + 4 + 3}{60}\right) = \frac{12}{60} = \frac{1}{5}$

$\therefore$ $(A + B + C)$'s one day work $= \frac{1}{5} \times \frac{1}{2} = \frac{1}{10}$

So they will done work in 10 days.

19. (d) % Increase = $\frac{17}{170} \times 100 = 10\%$

$r \times 5 = 10\%$

$r = \frac{10\%}{5} = 2\%$

20. (d)

A	:	*B*	:	*C*
3	:	4		
		5	:	6
15	:	**20**	:	**24**

21. (e) $\frac{3}{5}x + \frac{4}{5}x = 84$

$\frac{7}{5}x = 84$

$\Rightarrow \quad x = 60$

22. (e) Increase in weight $= 36 \times 2 = 72$ kg

$\therefore$ Average weight of new students $= 41.5 + \frac{72}{10}$

$= 41.5 + 7.2 = 48.7$ kg

23. (b) $218 + P = 63 \times \frac{5}{18} \times 32$

$218 + P = 560$

$\therefore$ P(Length of Platform) $= 560 - 218$

$= 342$ m

So speed of man $= \frac{342}{180} = 1.9$ m/sec

24. (a) Ratio of profit = Ratio of Investment

$$\frac{\text{John Profit}}{\text{Ira Profit}} = \frac{x \times 6 + 2x \times 6}{12 \times \mathbf{2x}}$$

$$= \frac{18x}{24x} = \frac{3}{4}$$

$\therefore$ John Profit $= \frac{3}{7} \times 8400 =$ ₹ 3600

25. (e) $P \times 1.21 = 2904$

$$P = \frac{2904}{121} = 2400$$

Required Amount $= 2400 \times (1.2)^2$

$= 2400 \times 1.44$

$= 3456$

26. (d) $(x + y) \times 4 - (x - y) \times 4 = 16$

$x + y - x + y = 4$

$2y = 4$

$y = 2$ km/h

27. (a) $\frac{x+8}{y} = \frac{5}{4}$

$4x + 32 = 5y$

Also $y + \frac{5}{4}y = 36$

$\frac{9}{4}y = 36 \Rightarrow y = 16$

Putting the value of y we get $4x + 32 = 5 \times 16$

$4x = 80 - 32 = 48$

$\therefore$ $x = \frac{48}{4} = 12$

28. (a) Total work = 168 unit

So 1 man can do 1 unit per day

14 men in 3 day do 42 unit

$\therefore$ Remaining work = 168 – 42 = 126

Now 9 w can do 126 units in 24 D

They will do 168 units in $\dfrac{24 \times 126}{168} = 18$ Days

29. (a) By options

Numbers are 37, 39, 41, 43, 45

30. (a)

$$x + y = \frac{12.8}{32} \times 60$$
$$= 24 \text{ k/m}$$

Also $$y = \frac{x}{5}$$

$$x + \frac{x}{5} = 24$$

$\therefore$ $$x = 20 \text{ km/h}$$

$$\text{Distance travelled} = 20 \times \frac{33}{60}$$
$$= 11 \text{ km}$$

31. (c) Let total work = 45 × 16 = 720 units

$\therefore$ 1 man can do 1 unit/day

$\therefore$ in 6 days 45 men will do 45 × 6 = 270 unit

Remaining work = 720 – 270

= 450 units

75 men would do in $\dfrac{450}{75} = 6$ day

32. (a)

$$x + y = \frac{15}{3\frac{3}{4}} = \frac{15 \times 4}{15}$$
$$= 4 \text{ k/m}$$
$$x - y = \frac{5}{2\frac{1}{2}} = 2 \text{ k/m}$$

$\therefore$ $$x = 3, \quad y = 1$$

33. (c) $\text{CI} = 48000 \left[\left(1 + \dfrac{4}{100}\right)^2 - 1\right]$

$$= 48000 \times \left[\frac{676}{625} - 1\right]$$

$$= 48000 \times \frac{51}{625} = 3916.8$$

34. (a) Required marked price $= 360 \times \frac{1.25}{.9} =$ ₹ 500

35. (c) Required share

$$= \frac{1+1}{1+2+2+1} \times 1500$$

$$= \frac{2}{6} \times 1500$$

$$=$$ ₹ 500

36. (d) Required ratio = 80 : 40

= 2 : 1

37. (a)

$$3T + 5C = 3110 \quad ...(i)$$

$$T - C = 210 \quad ...(ii)$$

Adding (i) and (ii) $4T + 4C = 3320$

$\Rightarrow$ $2T + 2C = \frac{3320}{2} =$ ₹ 1660

38. (d) Required Average $= \frac{5 \times 510 + 25 \times 240}{30}$

39. (d) Ages = 24 : 28

After 4 years = 28 : 32 = 7 : 8

40. (c) Investment Ratio = Profit Ratio

$$\frac{5x \times 7}{7x \times k} = \frac{1}{2}$$

Solving we get,

$$k = 10$$

41. (a) Let original strength = x

According to Question,

$$x \times .9 \times .9 \times .9 = 729000$$

$$x \times .729 = 729000$$

$\Rightarrow$ $x = 1000\ 000$

42. (a) Required difference

$$5000\left(1+\frac{2}{100}\right)^2 - 5000\left(1+\frac{4}{100}\right)^1$$

5000 [(1.02)2 – 1.04]

= 5000 [1.0404 – 1.04]

= 5000 × .0004 = 2

43. (a) Ratio of numbers 3 : 2 : 1.

Ratio of values = 3 × 5 : 2 × 10 : 1 × 20

= 15 : 20 : 20

= 3 : 4 : 4

$\therefore$ values of ₹ 20 $= \frac{4}{11} \times 165$

$= ₹\ 60$

44. (c) $3 : 8 :: 12 : d$

$$3d = 8 \times 12$$

$$d = \frac{8 \times 12}{3} = 32$$

45. Milk and water in $A = 20 : 4 = 5 : 1$

Milk and water in $B = 3 : 1$

Milk and water in $C = 80\%, 20\% = 4 : 1$

By Allgation (water)

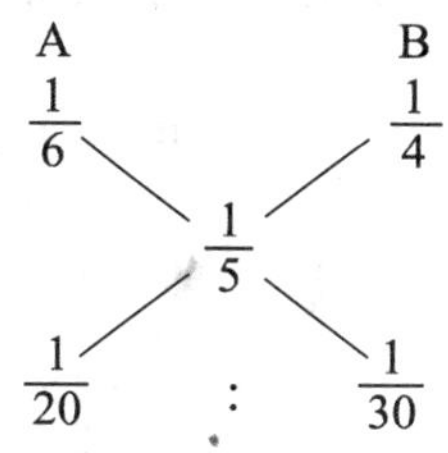

$= 3 : 2$

$= 3x = 24$

$\Rightarrow \quad x = 8$

$\therefore$ Quantity in $B = 2 \times 8 = 16$ litre.

46. (a) CI for 2 years $= x\,[(1.20)^2 - 1]$

$= x \times .44$

$= .44x$

SI for 3 years $= \dfrac{(x - 1600) \times 10 \times 3}{100}$

$= .3\,(x - 1600)$

$\therefore \quad .44x = 2 \times .3\,(x - 1600)$

$= .60x - 960$

$.16x = 960$

$\Rightarrow \quad x = 6000$

$\therefore$ Money invested at $B = 6000 - 1600 = ₹\ 4400$.

47. (a) Let CP $= x$

$1.4x \times .7 + 1.4x \times .7 + 35 = 1.05 \times 2x$

$35 = 2.1x - 1.96x$

$\Rightarrow \quad .14x = 35$

$x = \frac{35}{14} = ₹\ 250.$

48. (d) $\frac{80}{x+y} + \frac{80}{x-y} = \frac{30}{1}$

$\frac{80}{3x+y} + \frac{80}{3x-y} = \frac{9}{1}$

Solving (1) and (2) we get

$x = 6$ and $y = 2$

$\therefore$ Speed of boat B in still water $= 6 \times 3 = 18$ k/h.

49. (a) Required run $= 14 \times 47 - 7 \times 57 - 5 \times 44 - 15 = 24$

50. (a) $1.24x + 1.18x = 21780$

$$x = \frac{21780}{2.42} = 9000$$

51. (a) Let total work = LCM of 27, 33, 45 = 1455 unit

$$\text{Work done} \quad \text{A in one day} = \frac{1485}{27} = 55 \text{ unit}$$

$$\text{B in one day} = \frac{1485}{33} = 45 \text{ unit}$$

$$\text{C in one day} = \frac{1485}{40} = 33 \text{ units}$$

$$\text{Work done by C} = 1485 - 12 \times 55 - 11 \times 45$$
$$= 1485 - 660 - 485$$
$$= 1485 - 1155$$
$$= 330$$

$\therefore$ C will take $\frac{330}{33} = 10$ more days

$\therefore$ Total no of days = 12 + 11 + 10 = 33

52. (c) $3\text{ L} = \text{S} \times 21$

$3\text{ L} = (\text{S} + 5) \times 18$

$\therefore$ $21\text{ S} = 18\text{ S} + 90$

$\text{S} = 30$ m/sec.

53. (c) Present Average age of A and $B = 15 + 5 = 20$ years

$\therefore$ Present total age of A and $B = 20 \times 2 = 40$ years

Present total age of A, B and $C = 20 \times 3 = 60$ years

$\therefore$ Present age of $C = 60 - 40 = 20$ years

Hence age of C after 10 years = 20 + 10 = 30 years

54. (d) Let capacity of tank = 60 litres

$\therefore$ Tank filled by tap A in 1 minute $= \frac{60}{20} = 3$ litres

and tank filled by tap B in 1 min $= \frac{60}{20} = 1$ litres

So tank filled by both tabs

A and B in 5 minutes $= (3 + 1) \times 5$

$= 4 \times 5 = 20$ litres

Tank to be filled $= 60 - 20$

$= 40$ litres

$\therefore$ B will take $\frac{40}{1} = 40$ minutes more to fill the tank.

55. (a) To cross a person train has to cross a distance equal to own length.

$$\text{Time taken to cross the person} = \frac{150}{36 \times \frac{5}{18}} = 15 \text{ seconds}$$

56. (b) Required SI $= \dfrac{5000 \times 8 \times 73/365}{100} = ₹\ 80$

57. (a) Average Income $= \dfrac{2750 \times 3 + 3150 \times 3 + 6750 \times 3 + 650 \times 9}{9}$

$$= \frac{8250 + 9450 + 20250 + 5850}{9} = \frac{43800}{9} = 4866.66$$

58. (a) Average age of two women $= \dfrac{35 + 45 + 8 \times 2}{2}$

$$= \frac{80 + 16}{2} = \frac{96}{2} = 48 \text{ years}$$

59. (a) As the difference in Average temperature in two cases is 41 – 40 = 1°

So total difference (for 3 days) in two cases in 1 × 3 = 3°.

This increase in temperature is due to difference of Saturday temp and Wednesday temp. as Thursday and Friday are two so common days in both the cases and will create no difference.

∴ Saturday – Wednesday = 3

42 – Wednesday = 3

Wednesday = 42 – 3 = 39°

60. (a) Milk in original mixture = 25% of 80

= 20 litres

Water in original mixture = 80 – 20 = 60 litres

Let x litre of water added to make milk 20% as water 80%

$$\frac{60 + x}{80 + x} = 80\% = \frac{4}{5}$$

Solving we get x = 20 litres

61. (a) Required SP $= 63 \times \dfrac{100}{105} \times \dfrac{110}{100} = ₹\ 66$

62. (b) SP of dozen = 1 × 12 = ₹ 12

CP of dozen = ₹ 9

∴ Profit% $= \dfrac{12 - 9}{9} \times 100 = \dfrac{3}{9} \times 100 = 33.33\%$

63. (a) SP = ₹ 120

Sales tax $= 120 \times \dfrac{1}{5} = ₹\ 24$

∴ Reminder = 120 – 24 = ₹ 96

Profit $= 96 \times \dfrac{1}{3} = ₹\ 32$

∴ CP of bottle = 96 – 32 = ₹ 64

64. (e) By checking options one by one

option (c) is correct as $1400 \times \left(1+\frac{10}{100}\right)^2 = 1400 \times \frac{125}{100} =$ ₹ 1694

65. (a) Let no. of students in each row = x

No. of rows = y

No. of students = $x \times y = xy$

According to Ist condition

$$(x+4)(y-4) = xy$$
$$xy - 4x + 4y - 16 = xy$$
$$4x - 4y = -16$$
$$x - y = -4 \quad ...(1)$$

According to second condition

$$(x-2)(y+4) = xy$$
$$xy + 4x - 2y - 8 = xy$$
$$4x - 2y = 8$$
$$2x - y = 4 \quad ...(2)$$

Solving 1 and 2

We get $x = 8$ and $y = 12$

$\therefore$ No. of students = $8 \times 12 = 96$

66. (d) Let Nitin Money = x

Yash Money = y

$$\frac{1}{4}x = \frac{1}{6}y$$
$$\therefore \quad \frac{x}{y} = \frac{4}{6} = \frac{2}{3}$$
$$x : y = 2 : 3$$

if
$$x = 2k$$
$$y = 3k$$

According to question,

$$2k + 3k = 600$$
$$5k = 600$$
$$k = \frac{600}{5} = 120$$

$\therefore$ Difference = $3k - 2k = k$

= ₹ 120

67. (c) Required no. of days = $9 \times \frac{5}{15} \times \frac{66}{18} = 11$ days

68. (b) Total share after removing the respective share = 1105 – 10 – 20 – 15

= 1105 – 45 = 1060

$\therefore$ Charu's Share $= \dfrac{24}{11+18+24} \times 1060 + 15$

$= \dfrac{24}{53} \times 1060 + 15$

$= 480 + 15 =$ ₹ 495

69. (a) Let Distance $= x$

$$\frac{x}{4} - \frac{x}{6} = \frac{25}{60}$$

Solving we get $x = 5$ km

70. (b) Let original speed of train $= x$ km/h

$$\therefore \quad \frac{300}{x} - \frac{300}{x+5} = \frac{2}{1}$$

Solving we get $x = 25$ km/h

71. (b) SP of 20 = CP of 20 + SP of 4

$\Rightarrow$ SP of 20 – SP of 4 = CP of 20

SP of 16 = CP of 20

SP of 16 = CP of 16 + CP of 4

$\therefore$ Profit% $= \dfrac{4}{16} \times 100 = 25\%$

72. (c) $\dfrac{2x-12}{3x-12} = \dfrac{5}{8}$

Solving we get $x = 36$

$\therefore$ Numbers are 72 and 108.

73. (a) $T = \dfrac{210+150}{108 \times \dfrac{5}{18}} = \dfrac{360}{30} = 12$ seconds

74. (a) No. of days $= 2 \times \dfrac{72}{96} \times \dfrac{4}{6} = 1$ day

75. (d) $P = D \times \left(\dfrac{100}{r}\right)^2$

$= 225 \times \left(\dfrac{100}{5}\right)^2 = 225 \times 400 =$ ₹ 90000

76. (c) Let CP = ₹ 15

$\therefore$ SP = ₹ 19

$\therefore$ Profit% $= \dfrac{4}{15} \times 100 = 26.67\%$

77. (c) Let total votes $= x$

Invalid votes = $\frac{x}{6}$

Valid votes = $x - \frac{x}{6} = \frac{5x}{6}$

∴ Winners vote = $\frac{8}{15} \times \frac{5x}{6} = \frac{4x}{9}$

Now $\frac{4x}{9} + 490 + \frac{x}{6} = x$

Solving this we get $x = 12600$

78. (e) $\frac{16 \times 5 + 3 \times 28}{8} = \frac{80 + 84}{8} = 20.5$ /kg

79. (b) Let the total work be 30 units;

A does the work in 30 days, therefore his per day work is 1 unit

In six days he does 6 units. Hence, remaining work is 24 units.

As A and B do the remaining work in 9 days

(A + B)'s one day's work = 24/9

B's one days work = $\frac{24}{9} - 1 = \frac{5}{3}$

Days required by B to do the work = $\frac{30 \times 3}{5} = 18$ days

80. (d) Speed of car A = $\frac{122}{5} = 24.4$ km/hr

Speed of car B = $8.5 \times \frac{18}{5} = 30.6$ km/hr

In one hour the distance between them is 24.4 + 30.6 = 55 km

So distance between them after 3 hours = 55 × 3 = 165 km

SECTION-III

MENSURATION AND GEOMETRY

Day 17

Mensuration (2D)

Today we will do the topic of mensuration (2 D) i.e., the two dimensional figures like rectangle, square, triangle, circle, parallelogram, rhombus and trapezium. In the mensuration it is very important that you should know and understand the basic formulae.

So, We will first see the basic formulae.

Rectangle

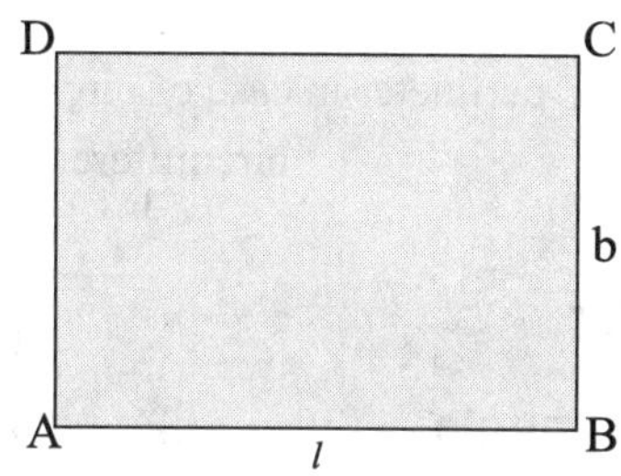

Adjacent figure ABCD is a rectangle. In rectangle all the angles are of 90° and has two pairs of equal and opposite sides. First we will find **area.** Now, what is **area**?

The shaded portion in the figure is area. Area is the product of length and breadth or we say the product of two lengths. That is why its units are always in m^2 or cm^2 etc. and **perimeter** is the total length of boundry. In this figure AB + BC + CD + DA i.e., is $2l + 2b$ is **perimeter** and diagonal is the longest side in any figure. Here **AC** and **BD** are **diagonals.**

So in **rectangle** :

$$\text{Area} = lb$$

$$\boxed{\mathbf{A = lb}}$$

$$\text{Perimeter} = 2(\text{length} + \text{breadth})$$

$$\boxed{\mathbf{P = 2(l + b)}}$$

$$\text{Diagonal} = \sqrt{l^2 + b^2}$$

$$\boxed{\mathbf{D} = \sqrt{l^2 + b^2}}$$

Square

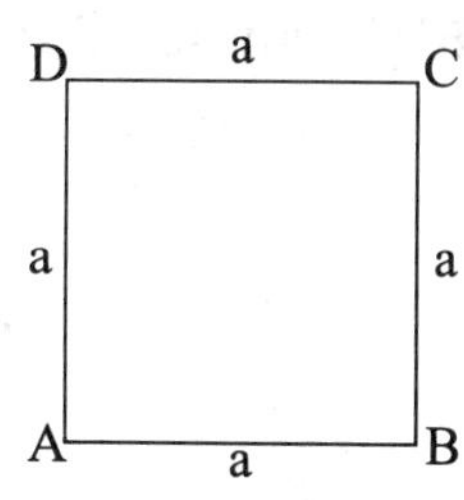

In square, all sides are equal and all angles are of 90°.

$$\text{Area} = (\text{side})^2$$

$$\boxed{\mathbf{A = a^2}}$$

$$\textbf{Perimeter} = \mathbf{4 \times side}$$
$$= 4a$$

$$\boxed{P = 4a}$$

$$\text{Diagonal} = \sqrt{a^2 + a^2}$$
$$= \sqrt{2a^2}$$
$$= \sqrt{2}\,a$$

$$\boxed{D = \sqrt{2}\,a}$$

Circle

Let r be the radius of circle

then Area $= \pi r^2$

where $\pi = \frac{22}{7}$

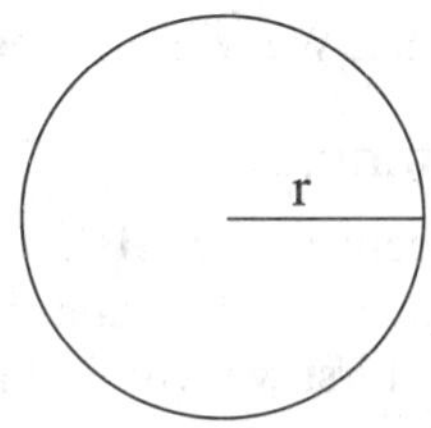

$$\boxed{A = \pi r^2}$$

and perimeter in case of circle is called **circumference.**

$$\text{circumference of circle} = 2\pi \times \text{radius}$$
$$= 2\pi r$$

$$\boxed{C = 2\pi r}$$

Triangle

$$\text{Area} = \frac{1}{2} \times \text{base} \times \text{height}$$

$$\boxed{A = \frac{1}{2} \times b \times h}$$

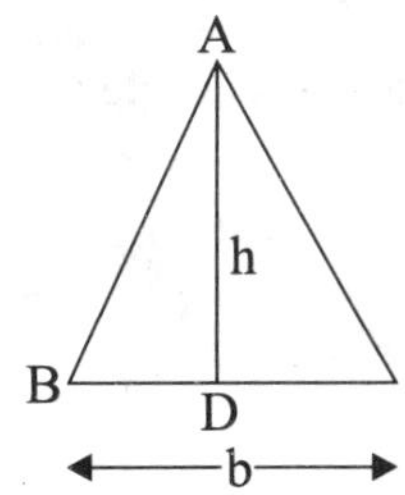

(Triangle having all sides equal)

$$\text{Area of equilateral triangle} = \frac{\sqrt{3}}{4}(\text{side})^2$$

$$\boxed{A = \frac{\sqrt{3}}{4} \times a^2}$$

Right Angled Triangle

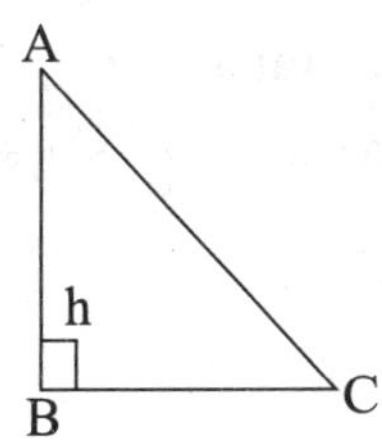

In **right angled triangle** one angle is **90°** and the side opposite to right angle is **longest** and is called **hypotenuse.**

Also $(\text{hypotenuse})^2 = (\text{base})^2 + (\text{height})^2$

In given figure $\angle B = 90°$

$\therefore \quad AC^2 = AB^2 + BC^2$

$$\text{Area of right angled triangle} = \frac{1}{2} \times \text{base} \times \text{height}$$

$$= \frac{1}{2} \times BC \times AB$$

Hero's Formula

This formula is used to find the area of triangle when length of three sides is given. By this formula

$$\textbf{Area of triangle} = \sqrt{S(S-a)(S-b)(S-c)}$$

where a, b, c are lengths of three sides resp.

And $$S = \frac{a+b+c}{2}$$

llgm: In parallelogram opposite sides are equal and parallel.

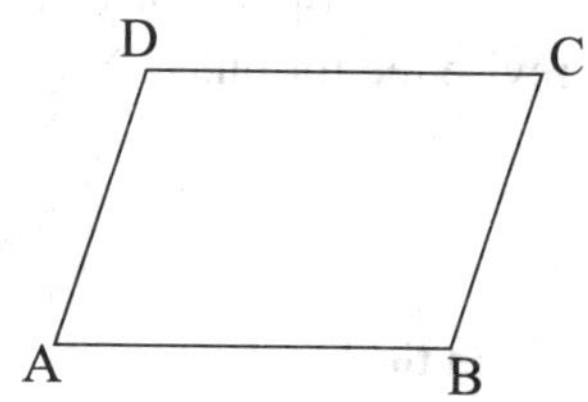

$$\text{Area of llgm} = \text{base} \times \text{height}$$

$$\boxed{A = b \times h}$$

Rhombus

In **rhombus** all sides are equal and diagonal bisect each other at 90° i.e., in figure

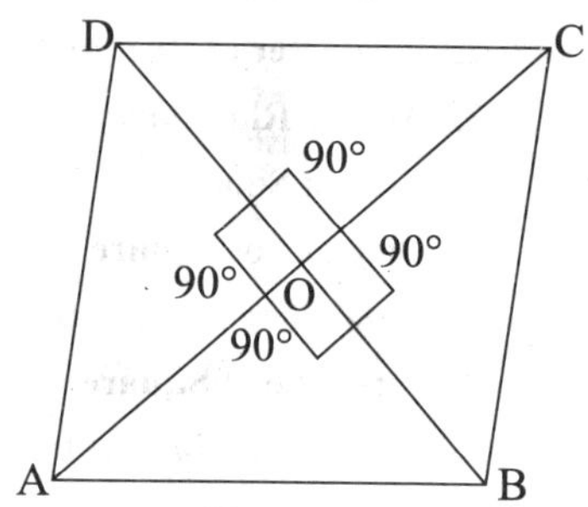

$$AO = OC$$

$$BO = OD$$

and

$$\angle AOB = \angle BOC = \angle COD = \angle DOA = 90°$$

$$\text{Area of rhombus} = \frac{1}{2} \times \text{length of 1st diagonal} \times \text{length of IInd diagonal}$$

$$\boxed{A = \frac{1}{2} \times d_1 \times d_2}$$

Trapezium

In trapezium one pair of opposite side is parallel and one non-parallel as shown in figure:

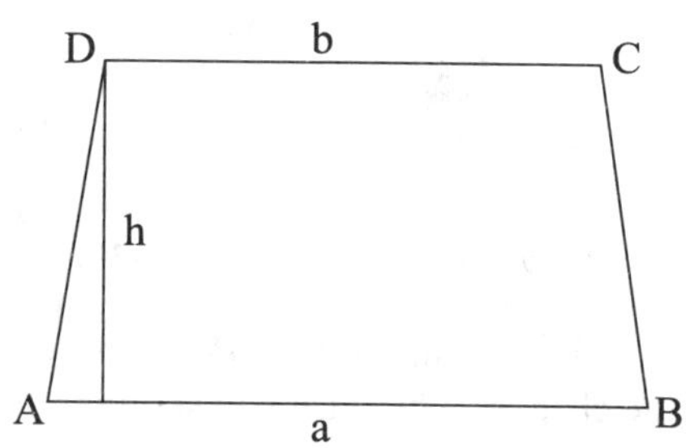

$$\text{Area} = \frac{1}{2} \times \text{sum of parallel sides} \times \text{height}$$

$$\boxed{A = \frac{1}{2} \times (a+b) \times h}$$

So as stated eariler you should have all the above formulae on your fingertips. So it is advised that before starting the question you should learn these formulae by heart and should devote at least 10 – 15 minutes for formulae just now.

:

:

:

:

Break for formulae revision.

:

:

:

:

It is hoped that you have gone through all the formulae. Now let us have a simple test to revise the formulae. Now solve the following exercise :

EXERCISE

Fill in the blanks

1. Area of **Circle** = ...
2. **Perimeter of Rectangle** = ...
3. Area of **Rhombus** = ...
4. Area of **Square** = ...
5. Diagonal of **Square** = ...
6. Circumference of **Circle** = ...
7. Perimetre of **Square** = ...
8. Area of **Equilateral triangle** = ...
9. Area of **llgm** = ...
10. Area of **Trapezium** = ...

SOLUTIONS

1. πr^2 **2.** $2(l+b)$ **3.** $\frac{1}{2}\times d_1\times d_2$ **4.** a^2

5. $\sqrt{2}\,a$ **6.** $2\pi r$ **7.** $4a$ **8.** $\frac{\sqrt{3}}{4}\times a^2$

9. $b\times h$ **10.** $\frac{1}{2}\times(a+b)\times h$

If your all the **answers** in above **exercise** are **correct** then you should do the next part of the chapter, otherwise you should revise the **formulae** again.

Types of Questions

Type I : Simple Question [where **direct formulae** are used]
Type II : Allied Questions [where **cost** is to be **calculated** using formulae]
Type III : Question related to **area of path** etc.
Type IV : Question related to **% change in area**
Type V : **Misc**

Type I

Example 1. Area of a rectangle is 128 cm^2. Its breadth is 8 cm. Find its length.

Sol :

$$A = 128 \text{ cm}^2$$
$$b = 8 \text{ cm.}$$
$$A = lb$$
$$128 = 8l$$
$$l = \frac{128}{8} = \mathbf{16\ cm}$$

Example 2. Diagonal of a rectangle is 10 cm. Its length is 8 cm. Find its breadth.

Sol :

$$\sqrt{l^2 + b^2} = 10 \text{ cm}$$

Squaring both sides

$$l^2 + b^2 = 100 \text{ cm}^2$$
$$64 + b^2 = 100 \text{ cm}^2$$
$$b^2 = 36 \text{ cm}^2$$
$$b = \mathbf{6\ cm.}$$

Example 3. Diagonal of a square park is 3910 √2 m. Find its Area.

Sol :

$$\sqrt{2}a = 3910\sqrt{2} \text{ m}$$
$$a = 3910 \text{ m}$$
$$A = (3910)^2 = \mathbf{15288100\ m^2}$$

Example 4. The length and breadth of a rectangle are in the ratio 4 : 3. Its area is 1452 cm^2. Find its perimeter.

Sol :

$$\text{Let length} = 4x$$
$$\text{then breadth} = 3x$$
$$4x \times 3x = 1452 \text{ cm}^2$$
$$12x^2 = 1452 \text{ cm}^2$$
$$x^2 = \frac{1452}{12} = 121 \text{ cm}^2$$
$$x^2 = 121 \text{ cm}^2$$
$$x = 11 \text{ cm}$$

$$\therefore \quad \text{length} = 11 \times 4 = 44\text{cm.}$$

and breadth $= 11 \times 3 = 33$ cm

$$\text{perimeter} = 2(44 + 33)$$
$$= 2(77) = \mathbf{154\ cm^2.}$$

Example 5. Area of a square field is 6050 m^2. Find the length of the diagonal.

Sol :
$$A = 6050\ m^2$$
$$a^2 = 6050\ m^2$$
$$a = \sqrt{6050}\ m$$
$$D = \sqrt{2}\, a$$
$$= \sqrt{2}\sqrt{6050}$$
$$= \sqrt{12100} = \mathbf{110\ m.}$$

Example 6. Find the area and circumference of a circle where radius is 7 cm.

Sol :
$$\text{Area} = \pi r^2$$
$$= \frac{22}{7} \times 7 \times 7 = 154\ cm^2$$
$$\text{Circumference} = 2\pi r$$
$$= 2 \times \frac{22}{7} \times 7 = \mathbf{44\ cm.}$$

Example 7. Area of circle is 616 cm^2. Find its circumference.

Sol :
$$\pi r^2 = 616\ cm^2$$
$$r^2 = \frac{616}{\pi}$$
$$= \frac{616}{22} \times 7$$
$$= 196\ cm^2$$
$$r^2 = 196\ cm^2.$$
$$r = \mathbf{14\ cm.}$$
$$\text{Circumference} = 2\pi r$$
$$= 2 \times \frac{22}{7} \times 14 = \mathbf{88\ cm.}$$

Example 8. Area of two circular fields is in the ratio 16 : 49. If the radius of later is 14 m. What is the radius of former.

Sol : Let r_1 and r_2 be the radii of Ist & IInd circle resp.

$$\therefore \quad \frac{\pi r_1^2}{\pi r_2^2} = \frac{16}{49}$$
$$\frac{r_1^2}{r_2^2} = \frac{16}{49}$$

$$\frac{r_1}{r_2} = \frac{4}{7}$$

Let $r_1 = 4x$

$\therefore$ $r_2 = 7x$

$7x = 14$ m ... **(Given)**

$$x = \frac{14}{7} = 2$$

$\therefore$ radius of Ist circle $= 4 \times 2 =$ **8 m**

Example 9. Area of Rhombus is 120 cm^2. If one of the diagonal is 24cm. Find the perimeter of rhombus.

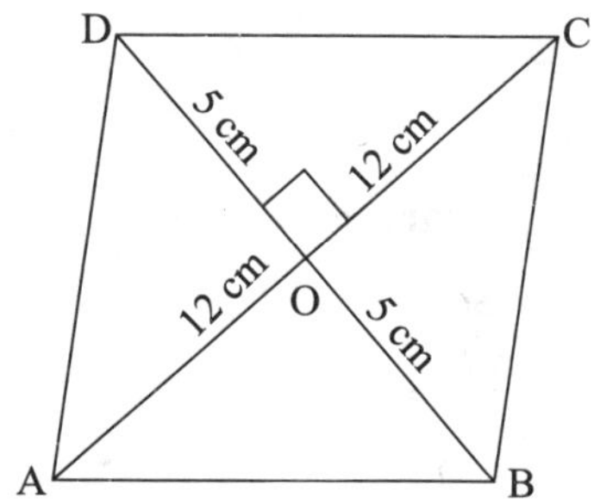

Sol : $A = \frac{1}{2} \times d_1 \times d_2$

$$120 = \frac{1}{2} \times 24 \times d_2$$

$$d_2 = 10 \text{ cm.}$$

In fig $d_1 = AC = 24$ cm

$d_2 = BD = 10$ cm

Now applying pythagoras by in ΔCOD

$$\angle DOC = 90°$$

$\therefore$ $DO^2 + OC^2 = DC^2$

$$5^2 + 12^2 = DC^2$$

$$25 + 144 = DC^2$$

$$DC^2 = 169$$

$$DC = 13 \text{ cm}$$

$$P = 4a$$

$= 4 \times 13 =$ **52 cm.**

Example 10. Each side of equilateral Δ is 20 cm. Find its area.

Sol : Area of equilateral $\Delta = \frac{\sqrt{3}}{4} \times (\text{side})^2$

$$= \frac{\sqrt{3}}{4} \times 20 \times 20 = 100\sqrt{3} \text{ cm}^2$$

Example 11. Three sides of a triangle are 8 cm, 7 cm, and 9 cm. Find its area.

Sol : $\mathbf{A = \sqrt{S(S-a)(S-b)(S-c)}}$

$$S = \frac{a+b+c}{2}$$

$$= \frac{8+7+9}{2} = 12$$

$$A = \sqrt{12(12-8)(12-7)(12-9)}$$

$$= \sqrt{12 \times 4 \times 5 \times 3}$$

$$= \sqrt{720} = \mathbf{26.83\ cm^2}$$

Example 12. Two opposite parallel sides of trapezium are 6cm and 8cm and its area is 49 cm. Find the distance between two parallel sides.

Sol :

$$\text{Area of Trapezium} = \frac{1}{2} \times (a+b) \times h$$

$$49 = \frac{1}{2} \times (6+8) \times h$$

$$49 = 7h$$

$$h = \mathbf{7cm.}$$

Example 13. Area of a IIgm is 144 cm^2. Its base is 9 cm. Find its height.

Sol :

$$b \times h = 144\ \text{cm}^2$$

$$9 \times h = 144$$

$$h = \frac{144}{9} = \mathbf{16\ cm}$$

Example14. Area of a triangle is 142 cm^2. Its base is 42 cm. Find its height.

Sol :

$$A = \frac{1}{2} \times b \times h$$

$$142 = \frac{1}{2} \times 42 \times h$$

$$h = \frac{142}{21} = \mathbf{6\frac{16}{21}\ cm}$$

Type II

Before doing **questions** on this type **remember** the following :

1. Whenever we have to **plough** or **cultivate** a field its **area** is required and **per unit cost** is given in terms of **per unit area** e.g., **Rs. 2 per m^2.**
2. When we are have to **fence** any figure we are concerned with its **perimeter** and **per unit cost** is in terms of **per unit length** e.g., **Rs. 2.50 per meter.**
3. Price of **carpets, flooring** etc. are marked in terms of **per unit length** while their **width remain fixed** and we take **carpet** such that it **covers** our **floor area** and payment for **required length** is **paid**. This can be clearly understood by following **example** :-

Suppose we have to buy a carpet for our floor whose length is 20 m. and breadth 10 m as shown in the figure. The carpet avaliable in the market has fixed width of 2 m

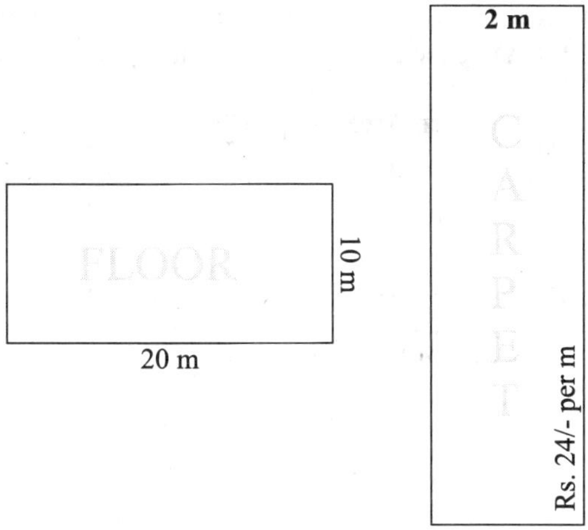

also shown in fig. It is quite clear that we have to buy the length of carpet which will cover our floor area. Let x meter of length serve this purpose.

$$\text{Area of carpet} = \text{Area of floor}$$
$$x \times 2 = 20 \times 10$$
$$x = \frac{20 \times 10}{2} = 100 \text{ m.}$$

Cost per m is Rs. 24 as shown in figure.

$\therefore$ Cost of carpeting the room $= 24 \times 100 =$ **2400 Rs.**

Example 15. A room is 30 m long and 18 m wide. If a carpet whose width is 2 m and which costs Rs. 35 per m is spread on the floor what is the total cost?

Sol : Area of carpet should be = Area of floor

Let length of carpet reqd. **x meter**

$$\therefore \quad 2x = 30 \times 18$$
$$x = \frac{30 \times 18}{2}$$
$$= 270 \text{ m.}$$

$\therefore$ Cost of carpet $= 270 \times 35 =$ **9450 Rs.**

Example 16. If it costs Rs. 320 to carpet a square room of 8 m broad what it would cost at the rate per square meter to carpet a room 6 m long and 5 m broad.

Sol : Since room is a **square**. $\therefore$ its length and breadth is equal, i.e., 8 m.

$$\therefore \quad \text{Area of room} = 8 \times 8 = 64 \text{ m}^2.$$
$$64 \text{ m}^2 \text{ costs} = 320 \text{ Rs.}$$
$$1 \text{ m}^2 \text{ costs} = \frac{320}{64} \text{ Rs.}$$
$$(6 \times 5) = 30 \text{ m}^2 \text{ costs} = \frac{320}{64} \times 30 = \textbf{150 Rs.}$$

Example 17. The cost of cultivating a field in the shape of parallelogram at 25 paise per metre square is Rs. 320. If the base of the field is 64 m. Find its height.

Sol : if it costs $\frac{25}{100}$ Rs. then Area $= 1\text{ m}^2$

if it costs 1 Rs. then Area $= 1 \times \frac{100}{25}$

if it costs 320 Rs. then Area $= 1 \times \frac{100}{25} \times 320 = 1280\text{ m}^2$

Area of IIgm $= b \times h$

$1280 = 64h$

$h = \frac{1280}{64} = \mathbf{20\ m}$

Type III

Example 18. The inner circumference of a circular field is 704 m. A road 7 m wide is constructed on the out side. Find the area of the road.

Sol : Let radius of inner circle is r_1

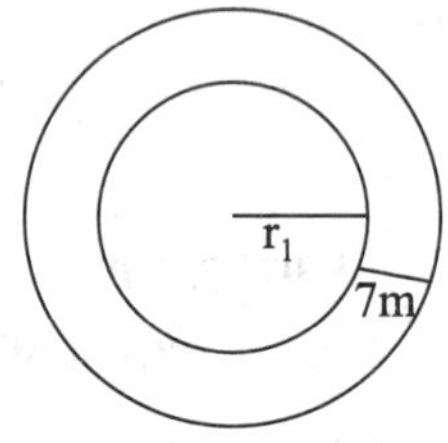

fig. 1

$$\therefore \quad 2\pi r_1 = 704\text{ m}$$

$$r_1 = \frac{704}{2\pi} = \frac{704 \times 7}{2 \times 22} = \mathbf{112\ m}$$

Let radius of outer circle be r_2

Then it is clear from the figure 2 that

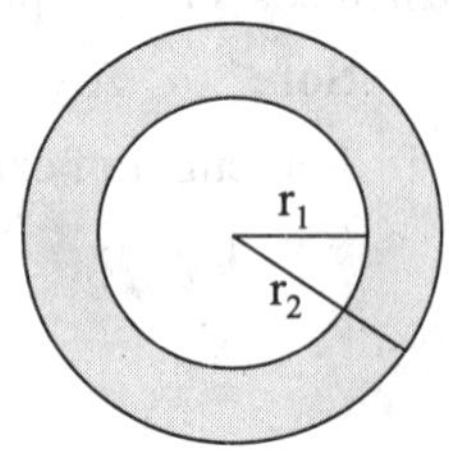

fig. 2

$$r_2 = r_1 + 7 = 112 + 7 = 119\text{ m}$$

Now shaded portion in the figure 2 is path

area of path = area of outer circle - area of inner circle

$$= \pi r_2^2 - \pi r_1^2$$

$$= \pi(r_2^2 - r_1^2)$$

$$= \pi(r_2 + r_1)(r_2 - r_1)$$

$$= \frac{22}{7}(119 + 112)(119 - 112)$$

$$= \frac{22}{7} \times 231 \times 7 = \mathbf{5082\ m^2}$$

Example 19. A path of uniform width 5 m. sorrounds a rectangular field 45 m × 35 m. Find area of path.

Sol :

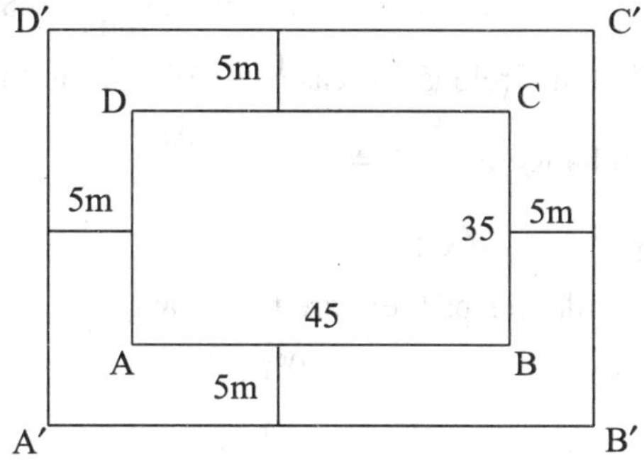

Area of path = Area of Rectangle A′ B′ C′ D′ – Area of rectangle ABCD

From figure now side A′B′ = 45 + 5 + 5 = 55 m

Similarly side C′D′ = 35 + 5 + 5 = 45 m

Area of path = 55 × 45 – 45 × 35

= 45[55 – 35]

= 45 × 20 = **900 m²**

Type IV

Example 20. If length of a square is increased by 40% and breadth increased by 20%. Find the percentage change in area.

Sol : Let side of square = 100 m

∴ Area of square $\mathbf{A_1}$ = 100 × 100 = **10000 m²**

Length of new square = 140 m

New breadth = 120 m

New area $\mathbf{A_2}$ = 140 × 120 = **16800 m²**

∴ Change in area = 16800 – 10000 = 6800 m²

$$\% \text{ change} = \frac{6800}{10000} \times 100 = \mathbf{68\%}$$

Shortcut : As discussed in the topic of **percentage** wherever there are two % changes

$$\text{Then net \% change} = A + B + \frac{AB}{100}$$

We will apply this shortcut here in this type of question.

First keep in mind that change in area involves two lengths and this shortcut is applicable in finding the percentage change in area of any two dimensional figure.

As discussed in the topic of **percentage** if there is an **increase or gain** in percentage then it is taken as **positive** if there is a **decrease or loss** in percentage then it is taken as **negative**.

In above example A = **+ 40**

B = **+ 20**

$$\text{Net \% change} = A + B + \frac{AB}{100}$$

$$= 40 + 20 + \frac{40 \times 20}{100} = \mathbf{68\%}$$

Example 22. If radius of circle is increased by 10%. Find the % charge in area.

Sol :

$$\% \text{ change in area} = A + B + \frac{AB}{100}$$

$$\text{Area} = \pi r^2$$

r comes twice in formula means there are two changes in r, i.e., of 10% and 10% resp.

$$\therefore \quad A = 10\%$$

$$B = 10\%$$

$$\text{Net \% change} = A + B + \frac{AB}{100}$$

$$= 10 + 10 + \frac{10 \times 10}{100} = \mathbf{21\%}$$

So it is hoped that now you have understood the concept.

Type V (Misc)

Example 23. The area of longest circle that can be drawn in a square of side 14 cm.

Sol : As shown in adjacent figure the diameter of circle will be equal to the side of square.

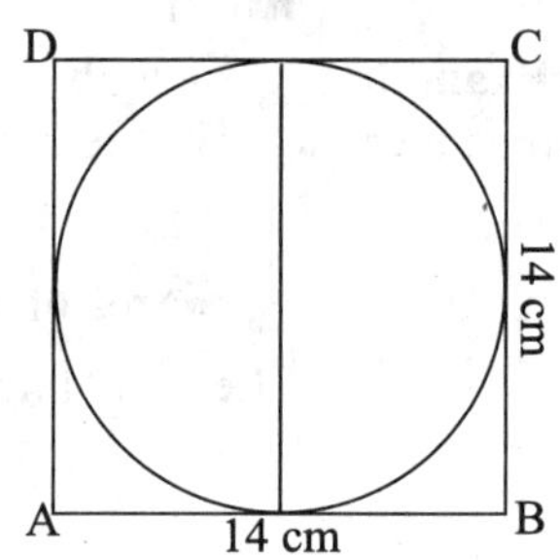

$$2r = 14 \text{ cm.}$$

$$\therefore \quad r = 7 \text{ cm.}$$

$$\text{Area} = \pi r^2$$

$$= \pi \times 7^2$$

$$= \frac{22}{7} \times 7^2$$

$$= \mathbf{154\ cm^2}$$

Example 24. A wire is in the form of a circle of radius 56 cm. What is the area of square into which it is bent.

Sol : The circumference of circle should be equal to the perimeter of square means

$$2\pi r = 4a$$

$$2 \times \frac{22}{7} \times 56 = 4a$$

$$352 = 4a$$

$$a = \frac{352}{4}$$

$$= 88 \text{ cm}$$

$$\text{Area of Square} = a^2$$

$$= 88^2 = \mathbf{7744\ cm^2}.$$

Example 25. A room is 5.44 m long and 3.74 m broad. This room is to be paved with square tiles. Find the edge of longest square tile which will exactly fit and also find the number of tiles.

Sol : Length of longest tile will the highest common factor of length and breadth.

$$l = 5\text{m } 44 \text{ cm} = 544 \text{ cm.}$$

$$b = 3 \text{ m } 74 \text{ cm} = 374 \text{ cm.}$$

HCF of 544 m and 374 cm is found as shown below :

```
374) 544 (1
     374
     170) 374 (2
          340
           34 ) 170 (5
                170
                 x
```

HCF of 374 cm & 544 cm is 34 cm

∴ Length of the longest tile that will exactly fit in the floor is **34 cm.**

Now

$$\text{no. of tiles} = \frac{\text{Area of floor}}{\text{Area of one tile}}$$

$$= \frac{374 \times 544}{34 \times 34} = \mathbf{176}$$

Example 26. Find the radius of a circular field whose area is equal to the sum of area of three circles having radius equal to 8 metre, 9 metre and 12 metres.

Sol :

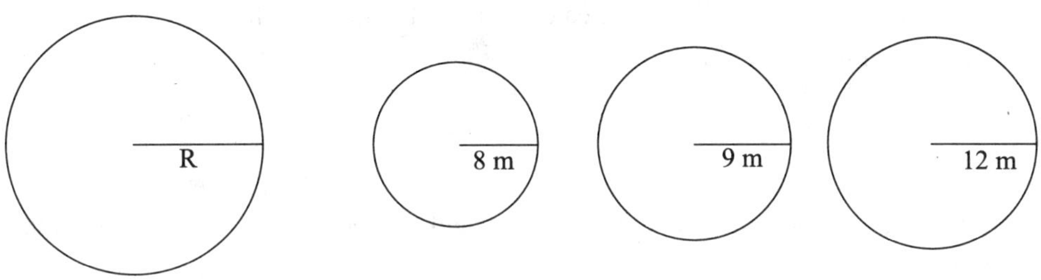

Let R be the required radius then

$$\pi R^2 = \pi 8^2 + \pi 9^2 + \pi 12^2$$

$$R^2 = 8^2 + 9^2 + 12^2$$

$$= 64 + 81 + 144 = 289 \text{ m}$$

$$R = \mathbf{17\ m}$$

Example 27. Four circles of radius r cm. are placed in such a way on the plane paper that each touches the other two. Find the area of space left in between four circles.

Sol :

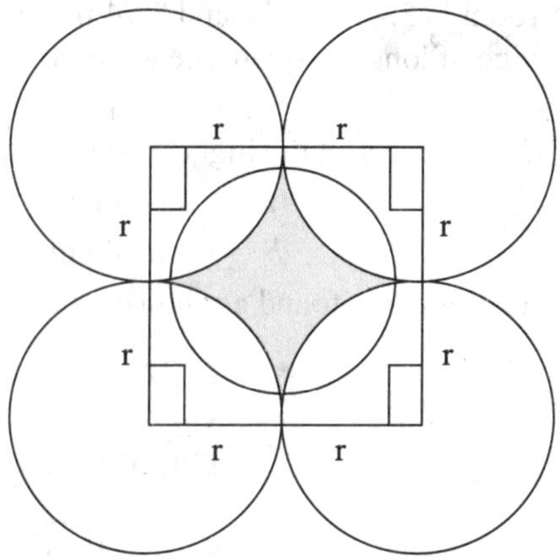

Required area = **Area of square of side 2r – Area of 4 sectors.**

$$= (2r)^2 - 4 \times \frac{\pi r^2}{4}$$

$$= 4r^2 - \pi r^2$$

$$= r^2 (4 - \pi)$$

$$= r^2(4 - 3.14) = .86\, r^2$$

Area of space left in between = $.86r^2$

This result can be **remembered** as a **general formula** and can be used directly to do this type of questions.

Example 28. Four circles of radius 2 cm. are placed in such a way on plain paper that each touches the other two. Find the area of space left in between four circles.

Sol : From last example

$$\text{Required area} = .86\, r^2$$

$$= .86 \times 2^2$$

$$= .86 \times 4 = \mathbf{3.24\ cm^2}$$

Example 29. The perimeter of rectangular field is 1200 m. Find the greatest possible area it may contain.

Sol : Remember : For a given perimeter square has the largest area.

We convert this perimeter into a square field.

$\therefore$ perimeter of square = 1200 m

$$4a = 1200 \text{ m}$$

$$a = \frac{1200}{4} = 300 \text{ m}$$

$$\text{Area of square} = a^2$$

$$= 300^2 = \mathbf{90000\ m^2}$$

EXERCISE

Q.1. The area of a rectangular field 2800 m^2. Its length is 70 m. Find its breadth

(a) 30 m (b) 40 m (c) 36 m (d) none

Q.2. The diagonal of a rectangle is 29 m. its breadth is 20 m. Find its length.

(a) 24 m (b) 21 m (c) 21.2 m (d) none

Q.3. The diagonal of the square is 47cm. Find the area of square.

(a) 1050.5 cm^2 (b) 1104.5 cm^2 (c) 1120 cm^2 (d) none

Q.4. The circumference of a circle is 264 cm. **Find its area.**

(a) 5544 cm^2 (b) 4418 cm^2 (c) 2100 cm^2 (d) none

Q.5. Area of a rhombus is 1000 cm^2. If one of the diagonals is 50cm. Find its another diagonal.

(a) 20cm (b) 40cm. (c) 70cm (d) none

Q.6. The perimeter of a rhombus is 68cm. If one of the diagonal is 30cm. FInd its area.

(a) 68 cm^2 (b) 272 cm^2 (c) 136cm^2 (d) none

Q.7. Two parallel sides of a trapezium are 11 cm and 17cm resp. Its height is 4 cm. Find its area.

(a) 56 cm^2 (b) 15 cm^2 (c) 40 cm^2 (d) none

Q.8. Area of a circle 2464 cm^2. Find its circumference.

(a) 174 cm (b) 176 cm (c) 352 cm (d) none

Q.9. The area of a circular field is 138600 m^2. Find the cost of fencing it at a rate of 20 paise per metre

(a) 132 Rs. (b) 210 Rs. (c) 264 Rs. (d) none

Q.10. The circumference of a circular field is 440 m. A circular path of 10 metre width runs around the out side of the field. Find the cost of cultivating the path at the rate of 70 paise per square metre.

(a) 2200 Rs. (b) 3300 Rs. (c) 4400 Rs. (d) none

Q.11. The cost of cultivating a field in the shape of a parallelogram at 2Rs. per metre square is Rs. 850. Its height is 17m. Find its base.

(a) 18.5 m (b) 28.5 m (c) 25 m (d) none

Q.12. What will be the breadth of a room 2500 m long if it costs Rs. 480 to cover it with tiles each having an area of 50 m^2 at 60 paise each.

(a) 1600 m (b) 160 m (c) 1.6 m (d) none

Q.13. The cost of carpeting a room 30 m long is Rs. 80. Had the breadth been 3 metre less the cost could have been Rs. 20. Find the area of floor.

(a) 40 m^2 (b) 240 m^2 (c) 120 m^2 (d) none

Q.14. Find the cost of fencing a semicircular field of radius 42 metre at the rate of Rs. 2 per metre.

(a) 216 Rs. (b) 432 Rs. (c) 108 Rs. (d) none

Q.15. The rectangular field, the sides of which are in the ratio 6 : 2 costs Rs. 360 for cultivating it at 30 paise per metre^2. Find the length of the field.

(a) 60 m (b) 40 m (c) 120 m (d) none

Q.16. A rectangular plot of grass is 80 metres long and 60m broad. From the centre of each side a path of 3 metre wide goes across the centre of the opposite side. Find the area of these paths.

(a) 420 m^2 (b) 411 m^2 (c) 402 m^2 (d) none

Q.17. A rectangular path 100 metres long and 80 metre wide has with in a travel path 8 metres running around it. Find the area of path.

(a) 2624m^2 (b) 2644m^2 (c) 2602m^2 (d) none

Q.18. Length of a rectangle is increased by 25% and breadth by 20%. Find the % charge in area.

(a) 50% (b) 45% (c) 47.5% (d) none

Q.19. The length of a rectangle is increased by 25% and breadth decreased by 20% . Find the % charge in area.

(a) 5% (b) no change (c) 2.5% (d) none

Q.20. Four circles of each of radius 1cm are placed on plane paper such that each touches the other two. Find the area of space left in betwen these circles.

(a) 4 cm^2 (b) 1.72 cm^2 (c) .86cm^2 (d) none

Q.21. The radii of two circles are in the ratio 4 : 5. The area of 1st field is what % less than the area of second.

(a) 12% (b) 64% (c) 36 % (d) none

Q.22. A room whose length is 10.98 m and breadth is 7.74 m is to be paved square tiles. Find the length of longest tile that will exactily fit.

(a) 18 cm (b) 9 cm (c) 36 cm (d) none

Q.23. The area of a rectangular field is 2304 m^2. The ratio of length and breadth is 16 : 9. Find its perimeter.

(a) 100 m (b) 150 m (c) 175 m (d) none

Q.24. The length of a room is 3 times the breadth. The cost of carpeting the room at ₹7.50 per square metre is ₹562.50. Find the length of the room.

(a) 5 m (b) 7.5 m (c) 10 m (d) none

Q.25. How long it will take for a boy to run around a square field of an area of 1 hectare at a rate of 4 km/h.

(a) 10 minute (b) 8 minute (c) 7 minute (d) none

Q.26. The area of the largest triangle that can be inscribed in a semi circle of radius r cm is ....

(a) r^2 (b) $2r^2$ (c) $r/2$ (d) none

SOLUTIONS

1. (b)

$$lb = 2800 \text{ m}^2$$

$$b = \frac{2800}{70} = \mathbf{40\ m}$$

2. (b)

$$\sqrt{l^2 + b^2} = 29 \text{ m}$$

$$l^2 + b^2 = 841 \text{ m}^2$$
$$l^2 + 400 = 841 \text{ m}^2$$
$$l^2 = 441 \text{ m}^2$$
$$l = \mathbf{21 \text{ m}}$$

3. (b)

$$D = 47 \text{ cm}$$
$$\sqrt{2a} = 47 \text{ cm}$$
$$a = \frac{47}{\sqrt{2}} \text{ cm}$$
$$\text{Area} = a^2 = \left(\frac{47}{\sqrt{2}}\right)^2$$
$$= \frac{47 \times 47}{2} = \mathbf{1104.5 \text{ cm}^2}$$

4. (a)

$$2\pi r = 264 \text{ cm}$$
$$r = \frac{264 \times 7}{2 \times 22} = 42 \text{ cm}$$
$$\text{Area} = \pi r^2$$
$$= \frac{22}{7} \times 42 \times 42 = \mathbf{5544 \text{ cm}^2}$$

5. (b)

$$\frac{1}{2} \times d_1 \times d_2 = 1000$$
$$\frac{1}{2} \times 50 \times d_2 = 1000$$
$$d_2 = \mathbf{40 \text{ cm.}}$$

6. (d)

$$4a = 68 \text{ cm}$$
$$a = 17 \text{ cm}$$

In ΔDOC

$$DC^2 = DO^2 + OC^2$$
$$289 = DO^2 + 225$$
$$64 = DO^2$$
$$8 \text{ cm} = DO$$
$$\therefore \quad x = 8 \text{ cm}$$
$$BD = 2x = 16 \text{ cm}$$
$$\therefore \quad \text{Area} = \frac{1}{2} \times 30 \times 16$$
$$= 240 \text{ cm}^2$$

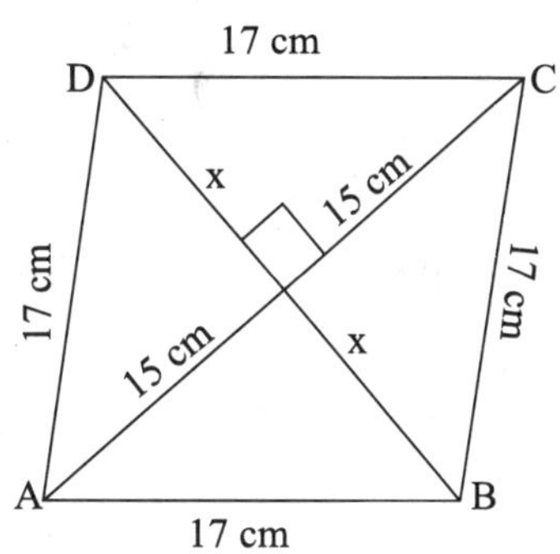

7. (a)

$$\text{Area of Trapezium} = \frac{1}{2}(a+b)h$$
$$= \frac{1}{2}(11+17)4 = 56 \text{ cm}^2$$

8. (b)

$$\pi r^2 = 2464 \text{ cm}^2$$

$$r^2 = \frac{2464}{22} \times 7$$

$$= 784 \text{ cm}^2$$

$$r = 28 \text{ cm}$$

$$C = 2\pi r$$

$$= 2 \times \frac{22}{7} \times 28 = \mathbf{176\ cm}$$

9. (c)

$$\pi r^2 = 138600 \text{ m}^2$$

$$r^2 = \frac{138600}{\pi} = 138600 \times \frac{7}{22}$$

$$r^2 = 44100 \text{ m}^2$$

$$r = \mathbf{210\ m}$$

$$\text{Circumference} = 2 \times \frac{22}{7} \times 210 = 1320 \text{ m}$$

$$\text{Cost of fencing} = 1320 \times \frac{1}{5} = \mathbf{264\ Rs.}$$

10. (b) Let r be the radius of circular field.

$\therefore$

$$2\pi r = 440 \text{ m}$$

$$r = \frac{440 \times 7}{2 \times 22} = 70 \text{ m}$$

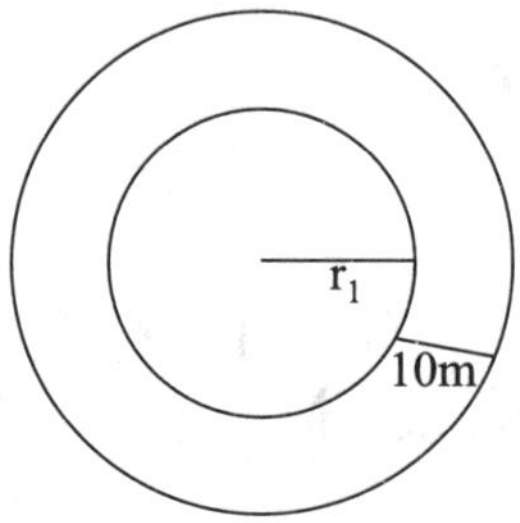

Area of path = Area of outer circle – Area of inner circle

$$= \pi\, 80^2 - \pi\, 70^2$$

$$= \pi(80^2 - 70^2)$$

$$= \pi(80 + 70)\,(80 - 70)$$

$$= \frac{22}{7} \times 150 \times 10 \text{ m}^2$$

$$\text{Cost of path} = \frac{22}{7} \times 150 \times 10 \times \frac{70}{100} = \mathbf{3300\ Rs.}$$

11. (c)

$$\text{Area of IIgm} = \frac{\text{Total cost}}{\text{Per unit cost}} = \frac{850}{2} = \mathbf{425\ m^2}$$

$$b \times h = 425 \text{ m}^2$$

$$17 \times h = 425$$

$$h = \frac{425}{17} = \mathbf{25\ m}$$

12. (d)

$$\text{No. of tiles} = \frac{480}{60/100} = 800$$

$$\text{Area of room} = 800 \times 50$$

$$= 40000 \text{ m}^2$$

$$lb = 40000$$
$$2500 \times b = 40000 \text{ m}^2$$
$$b = \frac{40000}{2500} = \mathbf{16\ m}$$

13. (c) Let breadth of room $= x$ m

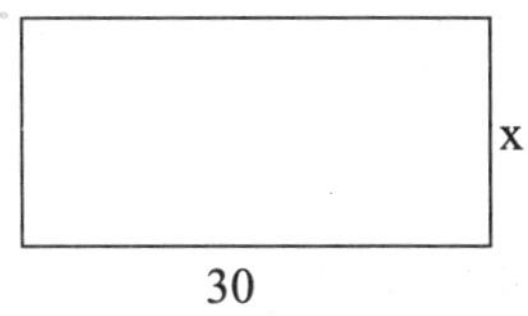

$$\text{Area} = 30x \text{ m}^2$$
$$\text{Area when breadth is } x - 3 \text{ is} = 30(x-3) \text{ m}^2$$
$$30x \text{ m}^2 \text{ costs} = 80 \text{ Rs.}$$
$$1 \text{ m}^2 \text{ costs} = \frac{40}{30x}$$
$$30(x-3) \text{ costs} = \frac{80}{30x} \times 30(x-3)$$
$$= \frac{80x - 240}{x} \text{ Rs.}$$

According to statement

$$\frac{80x-240}{x} = 20 \text{ Rs.}$$
$$80x - 240 = 20x$$
$$60x = 240$$
$$x = 4 \text{ m}$$
$$\therefore \quad \text{Area of floor} = 30 \times 4 = \mathbf{120\ m^2}$$

14. (b) Circumference of semicircle $= \pi r + 2r$

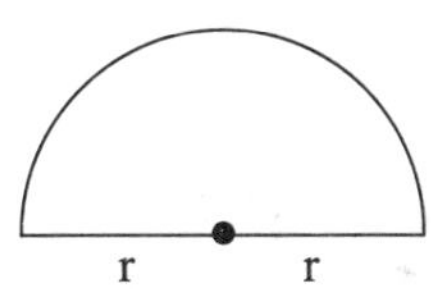

$$= \frac{22}{7} \times 42 + 2 \times 42$$
$$= 132 + 84 = 216 \text{ m}$$
$$\text{Cost of fencing} = 216 \times 2 = \mathbf{432\ Rs.}$$

15. (a)

16. (b) Area of Path = Area of EFGH

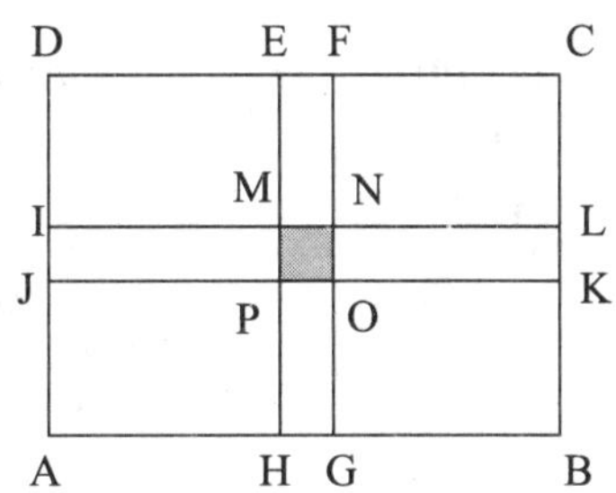

+ area of IJKL – Area of MNOP

$= 3 \times 80 + 3 \times 60 - 3 \times 3$

$= 240 + 180 - 9$

$= 420 - 9 =$ **411 m^2**

17. (a) Area of path $= 100 \times 80 - 84 \times 64$

$= 8000 - 5376$

$=$ **2624 m^2**

18. (a) Net % change $= A + B + \frac{AB}{100}$

$$= 25 + 20 + \frac{25 \times 20}{100}$$

$$= 45 + 5 = \mathbf{50\%}$$

19. (b)

$$\text{Net \% change} = A + B + \frac{AB}{100}$$

$$= 25 - 20 + \frac{25 \times -20}{100}$$

$$= 5 - 5 = 0\% = \textbf{no change}$$

20. (c)

$$\text{Required Area} = .86 \times r^2$$

$$= .86 \times 1^2 = \mathbf{.86\ cm^2}$$

21. (c)

$$A_1 = \pi(4x)^2$$

$$= \boldsymbol{\pi 16\, x^2}$$

$$A_2 = \pi(5x)^2$$

$$= \boldsymbol{\pi 25\, x^2}$$

$$\text{\% decrease} = \frac{\pi 9x^2}{\pi 25x^2} \times 100 = \mathbf{36\%}$$

22. (a) HCF of 1098 cm. and 774cm is 18cm.

23. (d) Let

$$l = 16x$$

$$b = 9x$$

$$\therefore \quad 144x^2 = 2304$$

$$x^2 = \frac{2304}{144} = 16$$

$$x = 4 \text{ m}$$

$$l = 64 \text{ m}$$

$$b = 36 \text{ m}$$

$$P = 2(l + b)$$

$$= 2(64 + 36)$$

$$= 2 \times 100 = \mathbf{200\ m}$$

24. (d)

$$\text{Area of room} = \frac{562.50}{7.5} = 75 \text{ m}^2$$

Let

$$b = x$$

$$l = 3x$$

$$\therefore \quad 3x^2 = 75$$

$$x^2 = 25$$

$$x = 5$$

$$\therefore \quad l = \mathbf{15\ m}$$

25. (d)

$$1 \text{ Hectare} = 10000 \text{ m}^2$$

$$\therefore \quad a^2 = 10000 \text{ m}^2$$

$$a = 100 \text{ m}$$

$$\text{Perimeter} = 4 \times 100 = 400 \text{ m}$$

$$\text{time} = \frac{\text{distance}}{\text{speed}}$$

$$= \frac{400}{4 \times 1000} h = \frac{1}{10} h$$

$$= \frac{1}{10} \times 60 = \textbf{6 minutes}$$

26. (a)

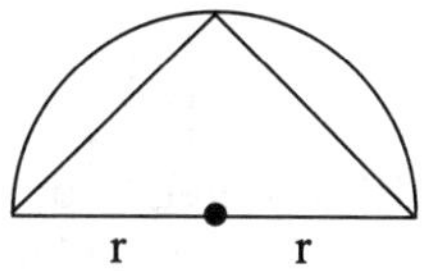

$$\text{Area} = \frac{1}{2} \times b \times h$$

$$= \frac{1}{2} \times 2r \times r$$

$$= \boldsymbol{r^2}$$

Day 18

Mensuration (3D)

Today we will do the topic of **mensuration (3-D)**. 3 Dimensional means figures having **length, breadth and height** or figures which occupy **space**. The 3-D figures we will discuss are **Cube, Cuboid, Cylinder, Cone** and **Sphere**. As discussed in last chapter, in mensuration you should know and understand the basic **formulae**. So we will first discuss the basic **formulae** in 3-D.

CUBE

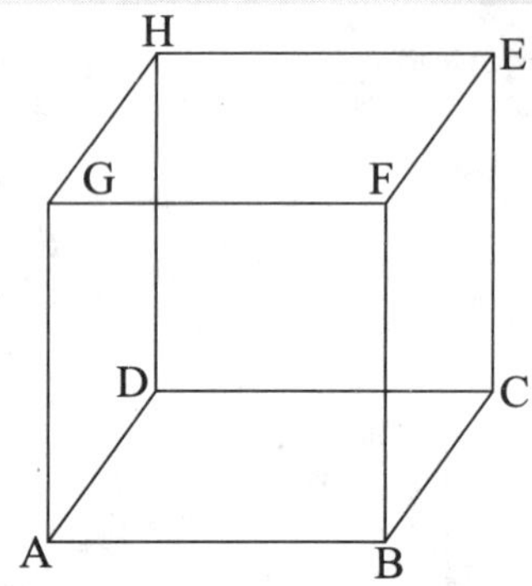

Adjacent fig is a **cube**. It is the three dimensional expansion of a square. In cube like square all sides are equal. First we will find the total surface area of a cube. Now what is total surface area? Total surface area means **area of all the surfaces**. A cube has **six surfaces** and **12 edges**. Let each edge of each surface is $\boldsymbol{a}$. Each surface is a square and area of one surface is $\boldsymbol{a^2}$

Therefore area of 6 surfaces is $6a^2$

So total **surface area of cube is $\boldsymbol{6a^2}$**

T.S.A. of cube = $\boldsymbol{6a^2}$

Now next comes the volume? **What is volume? Volume is the capacity of 3-D figure**, or **how much a 3-D figure can contain**. It is the product of length, breadth, and height and in case of cube all are the equal.

$$\text{Volume of cube} = a \times a \times a = a^3$$

$\boldsymbol{V = a^3}$

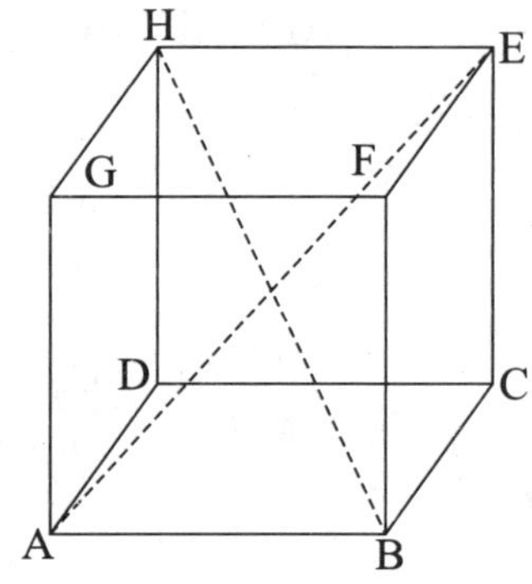

Now comes the **diagonal**.

Diagonal is the longest possible straight distance between two points of the cube or Diagonal is the length of longest rod that can be kept in a cube. In adjacent figure AE and BH are diagonals.

$$\text{Diagonal of cube} = \sqrt{a^2 + a^2 + a^2} = \sqrt{3a^2} = \sqrt{3}\,a$$

$$\textbf{diagonal of cube} = \sqrt{3}\,a$$

CUBOID

Cuboid is the 3 dimensional expansion of a rectangle.

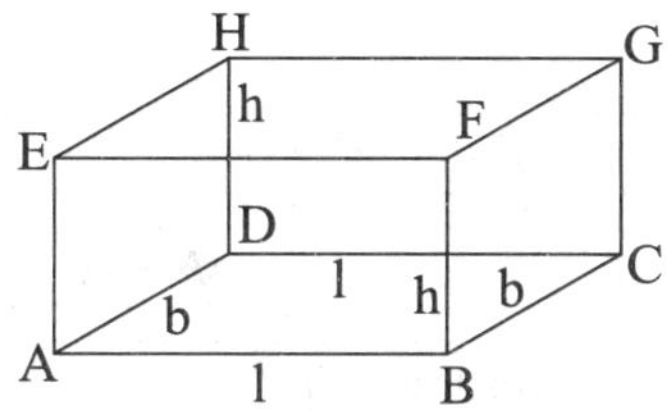

$$\text{Total surface area of a cuboid} = 2lb + 2bh + 2lh$$
$$= \mathbf{2(lb + bh + lh)}$$
$$\text{Area of four walls} = 2bh + 2lh$$
$$= \mathbf{2h(l + b)}$$
$$\text{Volume of cuboid} = \mathbf{lbh}$$
$$\text{Diagonal of cuboid} = \sqrt{l^2 + b^2 + h^2}$$

Therefore for cuboid

$$\textbf{T.S.A} = \mathbf{2(lb + bh + lh)}$$

$$\textbf{Area of four walls} = \mathbf{2h(l + b)}$$

$$\mathbf{V} = \mathbf{lbh}$$

$$\mathbf{D} = \sqrt{\mathbf{l^2 + b^2 + h^2}}$$

CYLINDER

A cylinder is made by folding a rectangular paper or sheet as shown below :

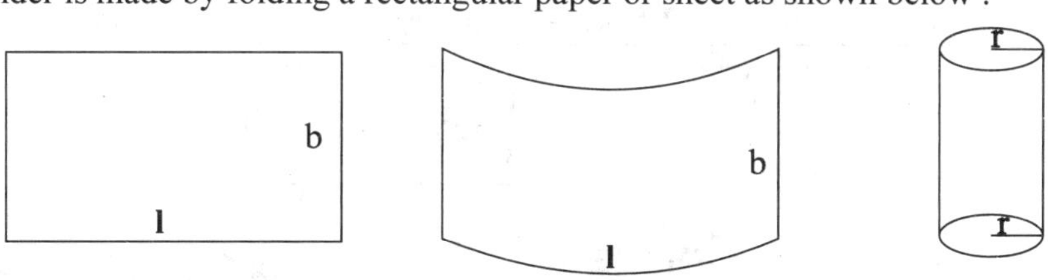

Rectangular Sheet .... By Folding .... becomes .... cylinder

As shown in above figure, on folding **length of rectangle** has become **circumfrence of circle** and **breadth of rectangle** has become **height of cylinder**.

$$\text{Area of rectangle} = \text{Curved Surface area of cylinder}$$
$$l \times b = 2\pi r \times h$$
$$\textbf{Curved surface area} = \mathbf{2\pi rh}$$

2. Total surface area : In total surface area, **area of two openings** (i.e., two circles) is also added in the **area of curved surface**.

$$\textbf{Total surface area} = 2\pi rh + 2\pi r^2$$
$$= 2\pi r\,(h + r)$$

3. Volume : Volume of cylimder $= \pi r^2 h$

So remember

(i) **C.S.A of cylinder = $2\pi rh$**

(ii) **T.S.A of cylinder = $2\pi r\,[h + r]$**

(iii) **Volume of cylinder = $\pi r^2 h$**

CONE

If a right angled triangle is revolved around the **side forming the right angle**, then solid so formed is called **Right circular cone**. In **right angle triangle, ABO, side AB** is revolved around the **side AO**, then **AB** will form **curved surface** area of the **cone**. **Side BO** will be the right **circular base. A** is called the **vertex** of the **cone. OA** is the **verticle height** (***h***) and **AB** is the **slant height (*l*)**.

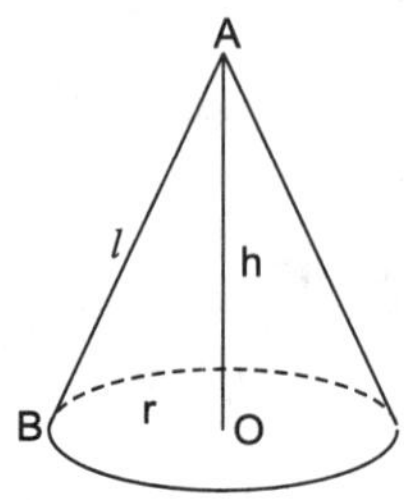

$$AB^2 = AO^2 + BO^2$$
$$l^2 = r^2 + h^2$$

Curved surface area of cone $= \pi rl$

Total surface area of cone $= \pi rl + \pi r^2$

$= \pi r\,(l + r)$

Volume of cone $= \frac{1}{3}\pi r^2 h$

So remember

1. **C.S.A. of cone = πrl**
2. **T.S.A. = $\pi r\,(l + r)$**
3. **V of cone = $\frac{1}{3}\pi r^2 h$**

SPHERE

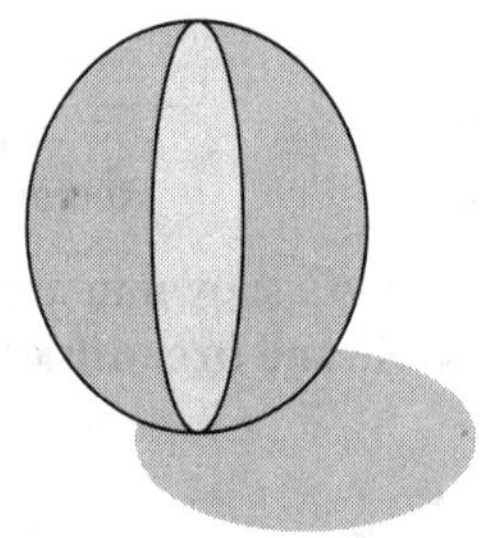

Sphere is the 3 D expansion of circle.

Example of sphere is a ball as shown in the adjacent figure. Let R be radius of sphere. There are two formulae in sphere to be remembered.

1. **Total surface area of Sphere = $4\pi R^2$**

2. **Volume of sphere = $\frac{4}{3}\pi R^3$**

HEMISPHERE

Hemisphere means half sphere as shown in adjacent figure. Volume of hemisphere is half the volume of sphere.

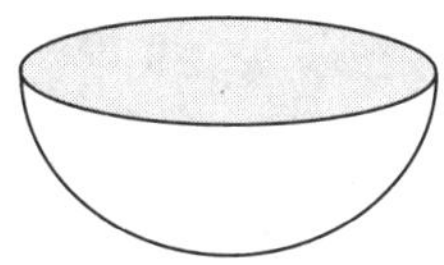

$$\text{Volume of hemisphere} = \frac{1}{2}\times\frac{4}{3}\pi R^3$$
$$= \frac{2}{3}\pi R^3$$

But in surface area one additional shaded surface is exposed.

$$\text{T.S.A. of hemisphere} = 2\pi r^2 + \pi r^2$$
$$= \mathbf{3\pi r^2}$$

Remember

1. **Volume of sphere = $\frac{4}{3}\pi R^3$**

2. **T.S.A. of hemisphere = $3\pi r^2$**

Now as in the previous chapter you must learn the formula by heart in this chapter also. So take a break and revise the formula.

:

:

:

Break for formulae.

:

:

:

It is hoped that you have gone through all the formulae. Now let us have a simple test.

EXERCISE

Fill in the blanks :

1. Volume of cuboid = ...
2. Volume of hemisphere = ...
3. T.S.A. of cube = ...
4. T.S.A. of cone = ...

5. Volume of cone = ...
6. Volume of Cylinder = ...
7. Volume of Sphere = ...
8. C.S.A of Cylinder = ...
9. C.S.A of Cone = ...
10. Diagonal of Cuboid = ...
11. Diagonal of Cube = ...

SOLUTIONS

1. lbh 2. $\frac{2}{3}\pi R^3$ 3. $6a^2$ 4. $\pi r(l + r)$

5. $\frac{1}{3}\pi R^2 h$ 6. $\pi r^2 h$ 7. $\frac{4}{3}\pi R^3$ 8. $2\pi rh$

9. πrl 10. $\sqrt{l^2 + b^2 + h^2}$ 11. $\sqrt{3}\, a$

If all the answers in the above exercise are correct then you should proceed to the next part of the chapter otherwise once again revise the formulae.

Types of Question

Type I : (Simple Question, based on direct formulae)
Type II : Where 3D box is to be made
Type III : When 3D solid is melted to form an other 3D solid
Type IV : % change in area or volume with the change in dimensions
Type V : Misc. Question

Type I

Example 1. The edge of a cube is 16 cm. Find its total surface area, volume and diagonal.

Sol :

$$\text{Total Surface Area} = 6a^2 = 6 \times 16^2 = 6 \times 256 = \mathbf{1536\ cm^2}$$

$$\text{Volume} = a^3 = 16^3 = \mathbf{4096\ cm^3}$$

$$\text{Diagonal} = \sqrt{3}\, a = \mathbf{16\sqrt{3}\ cm}$$

Example 2. The length, breadth and height of a cuboid are 3 cm, 4 cm and 5 cm. resp. Find its total surface area, volume and diagonal.

Sol :

$$\text{T.S.A.} = (3 \times 4 + 4 \times 5 + 3 \times 5)$$
$$= 2(12 + 20 + 15)$$

$$= 2(47) = \mathbf{94\ cm^2}$$

$$\text{Volume} = 3 \times 4 \times 5 = \mathbf{60\ cm^3}$$

$$\text{Diagonal} = \sqrt{3^2 + 4^2 + 5^2}$$

$$= \sqrt{9 + 16 + 25}$$

$$= \sqrt{50} = \mathbf{5\sqrt{2}\ cm.}$$

Example 3. The radius of base of a cylinder is 21 cm. and its height is 4 cm. Find its curved surface area and volume.

Sol :

$$\text{CSA of cylinder} = 2\pi rh$$

$$= 2 \times \frac{22}{7} \times 21 \times 4 = \mathbf{528\ cm^2}$$

$$\text{Volume of cylinder} = \pi r^2 h$$

$$= \frac{22}{7} \times 21 \times 21 \times 4 = \mathbf{5544\ cm^3}$$

Example 5. The radius of base of right circular cone is 6 cm. its height is 8 cm. Find its curved surface area and volume.

Sol :

$$l^2 = r^2 + h^2$$

$$= 6^2 + 8^2$$

$$= 36 + 64$$

$$l^2 = 100$$

$$l = 10\ \text{cm.}$$

$$\text{C.S.A} = \pi rl$$

$$= \frac{22}{7} \times 6 \times 10 = \mathbf{188.5\ cm}$$

$$\text{Volume of cone} = \frac{1}{3}\pi r^2 h$$

$$= \frac{1}{3} \times \frac{22}{7} \times 6 \times 6 \times 8 = \mathbf{301.71\ cm^3}$$

Example 6. The radius of a sphere is 28 cm. Find its total surface area and volume.

Sol :

$$\text{T.S.A.} = 4\pi r^2$$

$$= 4 \times \frac{22}{7} \times 28 \times 28 = \mathbf{9856\ cm^2}$$

$$\text{Volume} = \frac{4}{3}\pi r^3$$

$$= \frac{4}{3} \times \frac{22}{7} \times 28 \times 28 \times 28 = \mathbf{91989.3\ cm^3}$$

Example 7. The curved surface area of a cylindrical pillar is 66 m^2 and its volume is 231 m^3. Find its radius.

Sol : $2\pi rh = 66$ ...**(1)**

$\pi r^2 h = 231$ ...**(2)**

Divide **(1)** by **(2)**

$$\frac{2\pi rh}{\pi r^2 h} = \frac{66}{231}$$

$$\frac{2}{r} = \frac{66}{231}$$

$$r = \frac{2 \times 231}{66} = \mathbf{7\ m}$$

Example 8. Two cones have their height in the ratio 1:2 and the ratio of the radius of their bases is 2:1. Find the ratio of their volumes.

Sol : 1st cone 2nd cone

Let $h_1 = x$ $h_2 = 2x$

$r_1 = 2y$ $r_2 = y$

$V_1 = \pi r_1^2 h_1$ $V_2 = \pi r_2^2 h_2$

$$\frac{V_1}{V_2} = \frac{\frac{1}{3}\pi r_1^2 h_1}{\frac{1}{3}\pi r_2^2 h_2}$$

$$= \frac{r_1^2 h_1}{r_2^2 h_2}$$

$$= \frac{4y^2 \times x}{y^2 \times 2x}$$

$$= \frac{2}{1} = \mathbf{2 : 1}$$

Example 9. Find the length of the longest rod that can be placed in a room whose length is 17 m. breadth 15 m and height is 8 m.

Sol : Length of the longest rod that can be kept in a room is equal to the length of its diagonal.

$$D = \sqrt{l^2 + b^2 + h^2}$$

$$= \sqrt{17^2 + 15^2 + 8^2}$$

$$= \sqrt{289 + 225 + 64}$$

$$= \sqrt{578}$$

$$= \sqrt{289 \times 2} = \mathbf{17\sqrt{2}\ m}$$

Example 10. The breadth of a room is twice its height and half its length and volume of the room is 729 cubic metre. How high is the roof of the room?

Sol :

$$\text{Breadth} = 2\text{ height} = \frac{\text{length}}{2} = \text{x}$$

$\therefore$ $$\text{breadth} = \boldsymbol{x}$$

$$2\text{ height} = x$$

$$\text{height} = \frac{\boldsymbol{x}}{\mathbf{2}}$$

$$\frac{\text{length}}{2} = x$$

$$\text{length} = \mathbf{2x}$$

$$\text{volume} = 729\text{ m}^3 \quad \textbf{... (Given)}$$

Or $$\text{lbh} = 729\text{ m}^3$$

$$2x \times \frac{x}{2} \times x = 729\text{ m}^3$$

Or $$x^3 = 9^3$$

$$x = \mathbf{9\ m}$$

$\therefore$ $$\text{height} = \frac{9}{2} = \mathbf{4.5\ m}$$

Example 11. The surface area of a sphere is S. What is its volume?

Sol :

$$4\pi R^2 = S$$

$$R^2 = \frac{S}{4\pi}$$

$$R = \sqrt{\frac{S}{4\pi}}$$

$$\text{volume} = \frac{4}{3}\pi R^3$$

$$= \frac{4}{3}\pi\left(\sqrt{\frac{S}{4\pi}}\right)^3$$

$$= \frac{4}{3}\pi\frac{S}{4\pi}\sqrt{\frac{S}{4\pi}} = \frac{\mathbf{S}}{\mathbf{3}}\sqrt{\frac{\mathbf{S}}{\mathbf{4\pi}}}$$

Example 12. The area of two spheres is in the ratio 4:9, then the ratio of their volume is ....

Sol :

$$\frac{4\pi R_1^2}{4\pi R_2^2} = \frac{4}{9}$$

$$\frac{R_1^2}{R_2^2} = \frac{4}{9}$$

$$\frac{R_1}{R_2} = \frac{2}{3}$$

$$\frac{V_1}{V_2} = \frac{\frac{4}{3}\pi R_1^3}{\frac{4}{3}\pi R_2^3}$$

$$= \frac{R_1^3}{R_2^3}$$

$$= \left(\frac{R_1}{R_2}\right)^3$$

$$= \left(\frac{2}{3}\right)^3 = \mathbf{\frac{8}{27}}$$

Example 13. Height of a cylinder is 14 cm. and its curved surface is 594 square cm if $\pi = 22/7$, then radius of its base is ....

Sol :
$$2\pi rh = 594 \text{ cm}^2$$
$$2 \times \frac{22}{7} \times r \times 14 = 594 \text{ cm}^2$$
$$r = \frac{594 \times 7}{2 \times 22 \times 14} = \mathbf{6.75\ cm}$$

Example 14. The surface area of a cube is 384 cm^2. then its volume is ....

Sol :
$$6a^2 = 384$$
$$a^2 = \frac{384}{6} = 64 \text{ cm}^2$$
$$a = 8 \text{ cm}$$
$$V = a^3 = 8^3 = \mathbf{512\ cm^3}$$

Example 15. The perimeter of one face of a cube is 40 cm. The volume of cube is ....

Sol :
$$4a = 40 \text{ cm}$$
$$a = 10 \text{ cm}$$
$$V = a^3 = 10^3 = \mathbf{1000\ cm^3}.$$

Example 16. The diameter of the base of a circular cylinder whose volume is 648 π cm^3 and where height is 8 cm, is ....

Sol :
$$\pi r^2 h = 648\,\pi$$
$$r^2 h = 648$$
$$r^2 = \frac{648}{8} = 81 \text{ cm}^2$$
$$r^2 = 81 \text{ cm}^2$$
$$r = \mathbf{9\ cm}$$
$$\therefore \quad \text{Diameter} = 18 \text{ cm}$$

Example 17. The radius and height of right circular cone are in the ratio 5 :12. If its volume is 314 cubic m. Find the diameter of its base.

Sol :

$$\frac{1}{3}\pi r^2 h = 314 \text{ m}^3$$

$$\frac{1}{3}\pi (5x)^2 \times 12x = 314 \text{ m}^3$$

$$\frac{1}{3}\pi\, 25x^2 \times 12x = 314 \text{ m}^3$$

$$\frac{1}{3} \times 3.14 \times 300x^3 = 314 \text{ m}^3$$

$$x^3 = \frac{314 \times 3}{3.14 \times 300} = 1$$

$$x^3 = 1$$

$$x = \mathbf{1\ m}$$

$$\text{diameter} = 2 \times 5x = 10x = \mathbf{10\ m}$$

Type II

Let us clear this concept. Suppose you are asked to make a cubical box and are asked to bring the required matellic sheet. The first question you will ask the dimensions of the cube. Suppose the edge of the cube to be made is 20 cm. **To make a box you will require its surface area** which comes out to be $6a^2$

$$= 6 \times 20^2 = 6 \times 400 = \mathbf{2400 cm^2}$$

So you require a sheet whose total area is **2400 cm²**. The width of the sheet available in the market is fixed. So you buy the length of sheet which will cover your required area i.e., 2400 cm^2.

Let the width of sheet be 10 cm. So you require $\frac{2400}{10}$ = **240 cm** length of sheet.

Now solve the following example:

Example 18. How much length of iron sheet 11 cm wide is required for making an cube whose edge is 22 cm.

Sol : As discussed above, for making a cube we need its total surface area.

$$\text{T.S.A.} = 6\,a^2$$

$$6a^2 = 6 \times 22 \times 22 = 2904 \text{ cm.}$$

$$\text{length of sheet required} = \frac{2904}{11} = \mathbf{264\ cm}.$$

Example 19. How much length of iron sheet 168 cm. wide is required to make an open cylinder whose base radius is 21 cm. and height is 42 cm.

Sol : To make open cylinder we require its curved surface area

$$\text{Length of sheet reqd.} = \frac{\text{CSA}}{\text{width of sheet}}$$

$$= \frac{2\pi rh}{168}$$

$$= \frac{2 \times 22 \times 21 \times 42}{7 \times 168} = \mathbf{33\ cm.}$$

Example 20. A room is 15 m long, 4 m broad and 5 m high. Find the cost of painting its four walls at 50 p per metre^2

Sol : Area of four wall $= 2h(l + b)$

$= 2 \times 5(15 + 4)$

$= 10 \times 19 = 190 \text{ m}^2$

cost of washing $= 190 \times \frac{50}{100} = \mathbf{95\ Rs.}$

Example 21. The height of a room is 5 m and length is twice the breadth, If 120 m paper of width 1 m. are required for papering its walls. Find the area of floor.

Sol : Let $b = x$

Then $l = 2x$

And $h = 5$ m

Area of four walls = Area of paper required for papering its four walls.

$= 120 \times 1 = 120 \text{ m}^2$

$\therefore$ $2h(l + b) = 120$

$2 \times 5(2x + x) = 120$

$30x = 120$

$x = 4$

$\therefore$ length $= 2x = 2 \times 4 = \mathbf{8\ m}$

breadth $= x = \mathbf{4\ m}$

Area of floor $= lb = 8 \times 4 = \mathbf{32\ m^2}$

Type III

When one solid is melted to convert into another solid then their volumes are made equal and before starting these types of questions you must revise the cubes upto 15 given below:

1^3	=	**1**	2^3	=	**8**	3^3	=	**27**
4^3	=	**64**	5^3	=	**125**	6^3	=	**216**
7^3	=	**343**	8^3	=	**512**	9^3	=	**729**
10^3	=	**1000**	11^3	=	**1331**	12^3	=	**1728**
13^3	=	**2197**	14^3	=	**2744**	15^3	=	**3375**

Now solve the following examples

Example 22. A cuboid whose edges are 9 cm, 8 cm and 3 cm resp. This solid is melted to form another cube. Find the edge of the cube.

Sol :

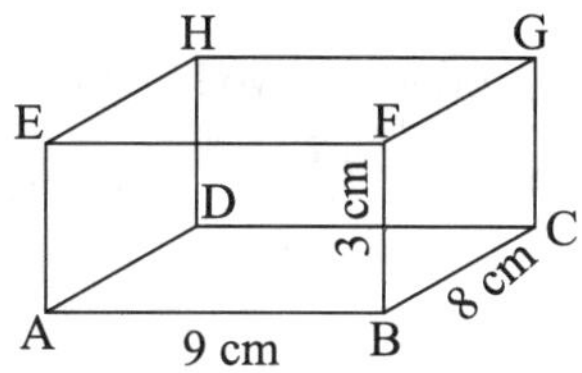

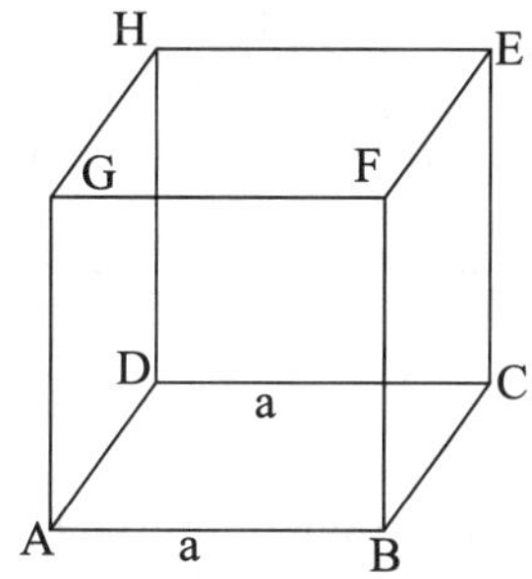

Let a be the edge of the cube

Then
$$a^3 = lbh$$
$$= 9 \times 8 \times 3 = 216 = 6^3$$
$$a^3 = 6^3$$
$$a = \mathbf{6\ cm}$$

Example 23. Three solid spherical balls of 3 cm, 4 cm and 5 cm are melted into a spherical ball. Its radius in cm is ....

Sol : Let R be the required radius

$$\therefore \quad \frac{4}{3}\pi R^3 = \frac{4}{3}\pi 3^3 + \frac{4}{3}\pi 4^3 + \frac{4}{3}\pi 5^3$$
$$= \frac{4}{3}\pi (3^3 + 4^3 + 5^3)$$
$$R^3 = 3^3 + 4^3 + 5^3$$
$$R^3 = 27 + 64 + 125$$
$$= 216 = 6^3$$
$$R = \mathbf{6\ cm}$$

Example 24. How many cubes of edge 20 cm can be cut out of a cube of edge 100 cm.

Sol : Let ***n*** be the number of smaller cubes of edge 20 cm

$$\therefore \quad n \times 20^3 = 100^3$$
$$n = \frac{100^3}{20^3} = \mathbf{125}$$

Example 25. How many sphere of radius 4 cm can be made by melting a sphere of radius 8 cm.

Sol : Let n be the no. of spheres of radius 4 cm

$$\therefore \quad n \times \frac{4}{3}\pi 4^3 = \frac{4}{3}\pi 8^3$$
$$n = \frac{8^3}{4^3} = \mathbf{8}$$

Type IV

Before starting this topic let us understand these type of question.

As discussed in the last chapter, area is the product of two lengths. So % change in area is used. If the radius of a sphere is increased by 10% then % change in surface area $A + B + \frac{AB}{100}$ will be $10 + 10 + \frac{10 \times 10}{100} = 21\%$.

Now volume is product of three lengths. If **we have to find the % change in volume of sphere** when radius is increased by 10% then we will proceed as follows:

$$V = \frac{4}{3}\pi R^3 = \frac{4}{3} R \times R \times R$$

Here radius is involved three times. In other words we have to find the net % change when there are three % changes of 10%, 10% and 10% resp. are there.

The method is same $A + B + \frac{AB}{100}$ i.e., we will first find the net% change by taking first two changes and then the net % change by taking this result and last change i.e., So net % change in volume is found as follows.

1st step

$$10 + 10 + \frac{10 \times 10}{100} = \mathbf{21\%}$$

Now we will find net % change of 21% and 10%

which is $= 21 + 10 + \frac{21 \times 10}{100}$

$= 31 + 2.1 = \mathbf{33.1\%}$

Example 26. The radius of cylinder is increased by 10% and height by 20%. Find the % change in volume.

Sol : Volume of cylinder $= \pi r^2 h = \pi \times r \times r \times h$

Here r is involved two times and h once. So we have to find the net % change of 10%, 10% and 20%

1st step

Net % change of 10% and 10% $= 10 + 10 + \frac{10 \times 10}{100} = 21\%$

2nd step

Net % change of 21% and 20% $= 21 + 10 + \frac{21 \times 10}{100}$

$= 41 + 4.2 = \mathbf{45.2\%}$

(**Note :** We will find the same result **if we solve in any order** i.e., **20%, 10%, 10%** or **10%, 20%, 10% etc.**)

Example 27. The edge of a cube is increased by 20% . Find the % change in volume.

Sol : As volume of cube is a^3. So it is clear that we have to find the **net% change** of **three changes 20%, 20%** and **20%**.

1st step

$$\text{Net \% change of 20\% and 20\%} = 20 + 20 + \frac{20 \times 20}{100}$$

$$= 40 + 4 = \mathbf{44\%}$$

2nd step

$$\text{Net change of 44\% and 20\%} = 44 + 20 + \frac{44 \times 20}{100}$$

$$= 64 + 8.8 = \mathbf{72.8\%}$$

Type V

Example 1. A wooden box is made of a board 1 cm thick. It is 120 cm long 80 cm wide and 60 cm high. Find the capacity of box and quantity of wood used

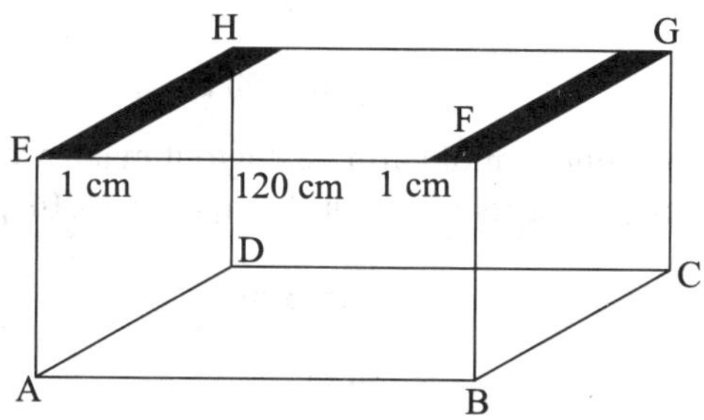

Sol :

External length = 120 cm

Internal length = 120 – 2 = 118 cm

(1 cm is subtracted form both sides as shown in fig and similar is the case of breadth and height)

External Breadth = 80 cm

Internal breadth = 80 – 2 = 78 cm

External height = 60 cm

Internal height = 60 – 2 = 58 cm

capacity of the box = Internal volumes

= **internal length × internal breadth × internal height**

= $118 \times 78 \times 58 = \mathbf{533832\ cm^3}$

Quantity of wood used = External volume - Internal volume

= $120 \times 80 \times 60 - 118 \times 78 \times 58 = \mathbf{42168\ cm^3}$

Example 29. A field is 500 m long and 30 m broad. A tank 30 m long 20 m broad and 12 deep is in the field and earth taken out of it is spread equally over the field. How much is the level of field raised.?

Sol :

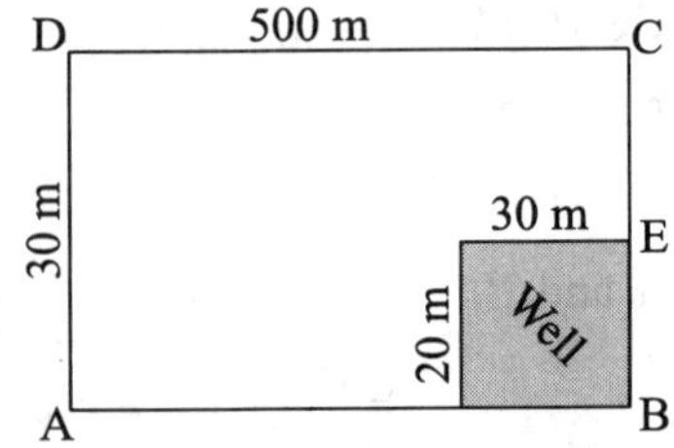

$$\text{Area of field} = 500 \times 30 = 15000 \text{ m}^2$$

$$\text{Area of tank} = 30 \times 20 = 600 \text{ m}^2$$

$$\text{Area of remaining field} = 15000 - 600 = 14400 \text{ m}^2$$

$$\text{Volume of earth taken out} = 30 \times 20 \times 12 = 7200 \text{ m}^3$$

Let h be level of field raised

$$14400\,h = 7200$$

$$h = \frac{7200}{14400}$$

$$= \frac{1}{2} \text{ m} = \mathbf{50 \text{ cm}.}$$

Example 30. How many cubic m of earth be dug out make a well 14 m deep and 6 m in diameter. What will it cost to plaster its inner curved surface at 50P per m^2.

Sol :

$$r = \frac{6}{2} = 3 \text{ m}$$

$$h = 14 \text{ m}$$

$$\text{Vol} = \pi r^2 h$$

$$= \frac{22}{7} \times 3 \times 3 \times 14 = \mathbf{396 \text{ m}^3}$$

$$\text{Inner curved surface} = 2\pi rh$$

$$= 2 \times \frac{22}{7} \times 3 \times 14 = 264 \text{ m}^2$$

$$\text{Cost of plastering} = 264 \times \frac{50}{100} = \mathbf{132 \text{ Rs.}}$$

EXERCISE

Q.1. The total surface area of a cube is 96 cm^2. Find its volume.

(a) 16 cm^3 (b) 32 cm^3 (c) 64 cm^3 (d) none

Q.2. The diagonal of a cube is 15 $\sqrt{3}$ cm. Find its total surface area.

(a) 1350 cm^2 (b) 675 cm^2 (c) 650 cm^2 (d) none

Q.3. The curved surface area of cylinder is 440 sq. cm. If the height of this cylinder is 10 cm then diameter of cylinder is ....

(a) 14 cm (b) 7cm (c) 9 cm (d) none

Q.4. The height of a right circular cone is 21 cm If the diameter of base is $2r$ cm then its volumes is (in sq. cm)

(a) $14\pi r^2$ (b) $7\pi r^2$ (c) $2\pi r^2$ (d) none

Q.5. From a solid cylinder whose height is 8 cm and base radius 6 cm a conical cavity height 8 cm and base radius 6 cm is hauled out. Find the volume of remaining solid.

(a) 200.2 cm^3 (b) 100 cm^3 (c) 603.4 cm^3 (d) none

Q.6. Two cones have their heights in the ratio 1 : 3 and ratio of their bases is 3:1. Find the ratio of their volumes.

(a) 2 : 1 (b) 3 : 1 (c) 3 : 2 (d) none

Q.7. In one cm of rainfall, the volume of water that falls on 1 hectare of ground is

(a) 10 m^3 (b) 50 m^3 (c) 1000 m^3 (d) none

Q.8. Volumes of two cubes are in the ratio 64 : 125. What will be the ratio of their surface areas?

(a) 16 : 25 (b) 9 : 25 (c) 4 : 5 (d) none

Q.9. The area of three faces of a cuboid are in the ratio 2 : 3 : 4 if its volume is 9000 cm^3. What is its height?

(a) 20 cm (b) 15 cm (c) 30 cm (d) none

Q.10. Given the volumes of matellic spherical surface is 38808 cm^3. Find its surface area.

(a) 5444 cm^2 (b) 5544 cm^2 (c) 5244 cm^2 (d) none

Q.11. The radius of a conical vessel is 6 cm and the vertical height is 8 cm. What is the length of cloth used to cover its surface if width of cloth is 5 cm.

(a) 34.25 cm (b) 36.2 cm (c) 37.7 cm (d) none

Q.12. A circus tent is conical in shape having base radius 70 m and slant height 40 m. The length of canvas 10 m wide required to make the tent ....

(a) 840 m (b) 880 m (c) 800 m (d) none

Q.13. Find the length of the metal sheet of width 2 m required to make a closed hollow cone of base radius 21 m and verticle height 20 m

(a) 1650 m (b) 165 m (c) 16.5 m (d) none

Q.14. Two cubes each of edge 6 cm. are joined end to end. Find the surface area of resulting cuboid.

(a) 180 cm^2 (b) 360 cm^2 (c) 365 cm^2 (d) none

Q.15. Three cubes of edge 3 cm, 4 cm, and 5 cm. are melted to form a single cube. Find the surface area of new cube.

(a) 216 cm^3 (b) 64 cm^3 (c) 125 cm^3 (d) none

Q.16. Three cubes of metal whose edges are in the ratio 3 : 4 : 5 are melted to form a single cube whose diagonal is $6\sqrt{3}$ cm. Find the edge of greatest of three cubes.

(a) 10 cm (b) 5 cm (c) 20 cm (d) none

Q.17. How many balls of radius 1 cm. can be made from a copper sphere whose radius is 4 cm.

(a) 64 (b) 512 (c) 68 (d) none

Q.18. 2200 cm^3 brass is to be drawn into cylinderical wire .25 cm in diameter. Find the length of wire.

(a) 448 m (b) 640 m (c) 480 m (d) none

Q.19. A cylnderical bucket is 28 cm in diameter and 72 cm high is full of water. The water is emptied into rectangular vessal 16 cm long and 28 cm wide. Find the height of water in the tank.

(a) 99 cm (b) 49.5 cm (c) 52.5 cm (d) none

Q.20. Each edge of cube is decreased by 10%. Find the % change in volume.

(a) 27.1% (b) 28.9% (c) 30% (d) none

Q.21. The radius of a cone is increased by 20% and height is decreased by 10%. Find the % change in volumes.

(a) 29.6% (b) 28.6% (c) 10% (d) none

Q.22. The radius of a sphere is increased by 30%. Find the % change in its volume.

(a) 60% (b) 119.7% (c) 90% (d) none

Q.23. The length, breadth and height of a cuboid are increased by 10%, 20% and 30% resp. Find the % change in its volume.

(a) 70.6% (b) 71.4% (c) 71.6% (d) none

Q.24. How many cubic m of earth be dug out to make a well 14 m deep and 4 metres in diameter.

(a) 88 m^3 (b) 154 m^3 (c) 616 m^3 (d) none

Q.25. A closed wooden box 22 cm. long 12 cm high and 17cm broad is made of board 1 cm thick. Find the capacity of box.

(a) 3000 cm^3 (b) 30000 cm^3 (c) 300 cm^3 (d) none

Q.26. A field is 160 m long and 130 m broad. A tank 80 m long 10 m broad and 4 m deep is dug in the field and earthtaken out of it is spread evenly over the field. How much is the level of field raised?

(a) 32 cm (b) 80 cm (c) 6 cm (d) none

Q.27. A rectangle piece of paper is 44 cm long and 10 cm wide. A cylinder is formed by rolling the paper along its length. Find the volume of cylinder.

(a) 385 cm^3 (b) 175 cm^3 (c) 350 $cm.^3$ (d) none

SOLUTIONS

1. (c)

$$6a^2 = 96$$
$$a^2 = 16$$
$$a = 4$$
$$v = a^3 = \mathbf{64\ cm^3}$$

2. (a)

$$\sqrt{3}\,a = 15\sqrt{3}\text{ cm}$$
$$a = 15$$
$$\text{T.S.A} = 6a^2$$
$$= 6 \times 15^2 = 6 \times 225 = \mathbf{1350\ cm^2}$$

3. (a)

$$2\pi rh = 440$$
$$r = \frac{440}{2\pi h} = \frac{440 \times 7}{2 \times 22 \times 10} = 7\text{ cm.}$$
$$D = 2 \times 7 = \mathbf{14\ cm}$$

4. (b)

$$h = 21 \text{ cm} \qquad D = 2r$$

$$\text{Radius} = \frac{D}{2} = \frac{2r}{2} = r$$

$$\text{volume} = \frac{1}{3}\pi r^2 h = \frac{1}{3}\pi r^2 \times 21 = 7\pi r^2 = \mathbf{7\pi r^2}$$

5. (c)

$$\text{Required volume} = \pi r^2 h - \frac{1}{3}\pi r^2 h$$

$$= \frac{2}{3}\pi r^2 h = \frac{2}{3} \times \frac{22}{7} \times 6 \times 6 \times 8 = \mathbf{603.4\ cm^3}$$

6. (b)

	1st case	**2nd case**
Let	$h_1 = x$	$h_2 = 3x$
	$r_1 = 3y$	$r_2 = y$
	$v_1 = \frac{1}{3}\pi r_1^2 h_1$	$v_2 = \frac{1}{3}\pi r_2^2 h_2$

$$\frac{v_1}{v_2} = \frac{\frac{1}{3}\pi r_1^2 h_1}{\frac{1}{3}\pi r_2^2 h_2}$$

$$= \frac{r_1^2 h_1}{r_2^2 h_2} = \frac{9y^2 x}{y^2 3x} = \frac{9}{3} = \mathbf{3:1}$$

7. (d)

$$1 \text{ Hectare} = 10000 \text{ m}^2$$

$$1 \text{ cm} = \frac{1}{100} \text{ m}$$

$$v = \text{area} \times \text{height}$$

$$= 10000 \times \frac{1}{100} = \mathbf{100\ m^3}$$

8. (a)

9. (c)

$$lb = 2x \qquad \textbf{...(1)}$$

$$bh = 3x \qquad \textbf{...(2)}$$

$$lh = 4x \qquad \textbf{...(3)}$$

Multiply **(1), (2) and (3)**

$$(lbh)^2 = 24x^3$$

$$(9000)^2 = 24x^3$$

$$x^3 = \frac{9000 \times 9000}{24}$$

$$= 9000 \times 375$$

$$= \underline{5^3 \times 2^3 \times 3^2 \;\; 5^3 \times 3}$$

$$x^3 = 5^3 \times 2^3 \times 3^3 \times 5^3$$

$$x = 5 \times 2 \times 3 \times 5 = 150 \text{ cm.}$$

$$lbh = 9000 \text{ cm}^3$$

$$lb = 150 \times 2 = 300 \text{ cm}^2$$

$$h = \frac{lbh}{lb} = \frac{9000}{300} = \mathbf{30\ cm}$$

10. (b)

$$\frac{4}{3}\pi r^3 = 38808$$

$$r^3 = \frac{38808 \times 3 \times 7}{4 \times 22} = 441 \times 21 = 21^3$$

$$r = 21 \text{ cm}$$

$$\text{TSA} = 4\pi r^2$$

$$= 4 \times \frac{22}{7} \times 21 \times 21 = \mathbf{5544\ cm^2}$$

11. (c)

$$l = \sqrt{6^2 + 8^2} = \sqrt{100} = 10 \text{ cm}$$

$$\text{length of cloth used} = \frac{\text{CSA}}{\text{Width of cloth}}$$

$$= \frac{22 \times 6 \times 10}{7 \times 5} = \frac{264}{7} = \mathbf{37.7cm}$$

12. (b)

$$\text{Length of cloth reqd.} = \frac{\text{CSA}}{\text{Width of cloth}}$$

$$= \frac{22 \times 70 \times 40}{7 \times 10} = \mathbf{880\ m}$$

13. (a) In this we require the T.S.A. which is = p$r(l + r)$

and $l^2 = h^2 + r^2 = 20^2 + 21^2$

$$= 841 \text{ m}^2 = 29 \text{ m}$$

$$\text{metal sheet required in this case} = \frac{\text{CSA}}{\text{Width of cloth}} = \frac{22 \times 21 \times 50}{7 \times 2} = \mathbf{1650\ m}$$

14. (b)

$$l = 12 \text{ cm}$$

$$b = 6 \text{ cm}$$

$$h = 6 \text{ cm}$$

$$\text{T.S.A} = 2(lb + bh + lh)$$

$$= 2(12 \times 6 + 6 \times 6 + 12 \times 6)$$

$$= 2(72 + 36 + 72) = \mathbf{360\ \ cm^2}$$

15. (a)

$$a^3 = 3^3 + 4^3 + 5^3$$

$$= 27 + 64 + 128 = 216 \text{ cm}^3$$

$$a^3 = 6^3$$

$$a = 6 \text{ cm}$$

$$\text{T.S.A} = 6a^2 = \mathbf{216\ cm^2}$$

16. (b)

$$\sqrt{3}\, a = 6\sqrt{3} \text{ cm}$$

$$a = 6 \text{ cm}$$

$$(3x)^3 + (4x)^3 + (5x)^3 = 6^3$$
$$216x^3 = 216$$
$$x^3 = 1$$
$$x = 1 \text{ cm}$$
$$\text{length of greatest cube} = 1 \times 5 = \mathbf{5\ cm.}$$

17. (a)
$$n \times \frac{4}{3}\pi r^3 = \frac{4}{3}\pi 4^3$$
$$n = \frac{4^3}{1^3} = \mathbf{64}$$

18. (a)
$$\pi\left(\frac{.25}{2}\right)^2 h = 2200$$
$$h = 44800 \text{ cm} = \mathbf{448\ m}$$

19. (a)

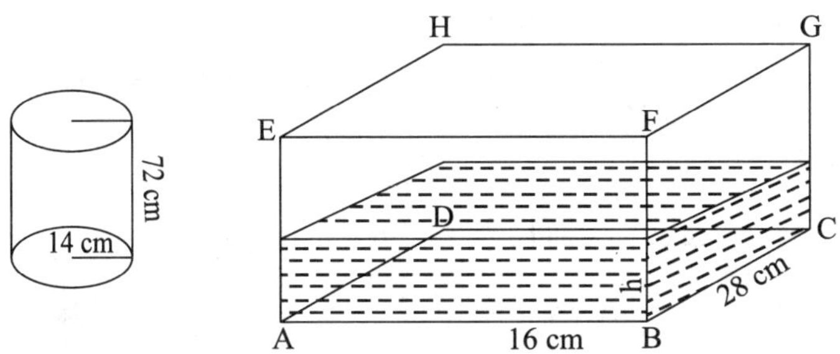

Let h be the level of water in tank as shown in figure.

volume of water in cylinder = volume of water in vessel

$$\pi r^2 h = lbh$$
$$\frac{22}{7} \times 14 \times 14 \times 72 = 16 \times 28 \times h$$
$$h = \frac{22 \times 14 \times 14 \times 72}{7 \times 16 \times 28} = \mathbf{99\ cm}$$

20. (a)
$$V = a^3$$

a is decreased by 10% means for volume there are three changes of –10%, –10% and –10%.

1st Step

$$\text{Net change of} -10\% \text{ and} -10\% = -10 - 10 + \frac{10 \times -10}{100}$$
$$= -20 + 1 = \mathbf{-19\%}$$

2nd Step

$$\text{Net \% change of} -19\% \text{ and } 10\% = -19 - 10 + \frac{-19 \times -10}{100}$$
$$= -29 + 1.9 = -27.1\%$$

$$\text{So change in volume} = \mathbf{27.1\%}$$

21. (a)

$$V = \frac{1}{3}\pi r^2 h$$

r comes twice and h once in the formula and % increase in r is 20% and % decrease in h is 10%. So there are three % changes of + 20% + 20% and – 10% resp.

1st Step

$$\text{Net \% change of + 20\% and + 20\%} = 20 + 20 + \frac{20 \times 20}{100}$$
$$= 40 + 4 = \mathbf{44\%}$$

2nd Step

Net % change of 44% and –10%

$$= 44 - 10 + \frac{44 \times -10}{100}$$
$$= 34 - 4.4 = \mathbf{29.6\%}$$

22. (b)

$$\text{Volume of sphere} = \frac{4}{3}\pi r^3$$

where r is increased by 30%. So to find % change in volume we have to find net % change of 30%, 30% and 30%.

1st Step

$$\text{Net \% change of 30\% and 30\%} = 30 + 30 + \frac{30 \times 30}{100}$$
$$= 60 + 9 = \mathbf{69\%}$$

2nd Step

$$\text{Net \% change of 69\% and 30\%} = 69 + 30 + \frac{69 \times 30}{100}$$
$$= 69 + 30 + 20.7 = \mathbf{119.7\%}$$

23. (c) We have to find net % change of 10%, 20% and 30% resp.

1st Step

$$\text{Net \% change of 10\% and 20\%} = 10 + 20 + \frac{10 \times 20}{100} = \mathbf{32\%}$$

2nd Step

$$\text{Net \% change of 32\% and 30\%} = 32 + 30 + \frac{32 \times 30}{100}$$
$$= 62 + 9.6 = \mathbf{71.6\%}$$

24. (d)

$$\text{Earth dugout} = \text{volume of well}$$
$$= \pi r^2 h = \frac{22}{7} \times 2 \times 2 \times 14 = \mathbf{176\ m^3}$$

25. (a)

$$\text{Internal length} = 22 - 2 = 20 \text{ cm}$$
$$\text{Internal breadth} = 17 - 2 = 15 \text{ cm}$$
$$\text{Internal height} = 12 - 2 = 10 \text{ cm}$$
$$\text{capacity} = 20 \times 15 \times 10 =$$

3000 cm^3

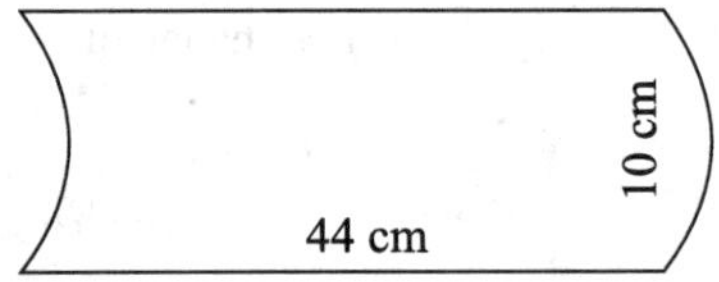

26. (d)

$$\text{Area of field} = 160 \times 130 = 20800 \text{ m}^2.$$

$$\text{Area of Tank} = 80 \times 10 = 800 \text{ m}^2$$

$$\text{Area of remaining field} = 20800 - 800 = 20000 \text{ m}^2$$

$$\text{Volume of earth taken out} = 80 \times 10 \times 4 = 3200 \text{ m}^3$$

Net h be the level of field raised then

$$20000\,h = 3200 \text{ m}^3$$

$$h = \frac{3200}{20000}\text{ m} = \frac{16}{100}\text{ m} = \mathbf{16\ cm}$$

27. (c) By rolling along length breadth becomes circumference of base of cylinder and length becomes height of cylinder.

$$\therefore \quad 2\pi r = 10 \text{ cm}$$

$$r =$$

$$h = 44 \text{ cm}$$

$$\text{volume} = \pi r^2 h$$

$$= \pi \times \frac{10}{2\pi} \times \frac{10}{2\pi} \times 44$$

$$= \frac{10 \times 10 \times 44 \times 7}{4 \times 22} = \mathbf{350\ cm^3}$$

Day 19

Geometry–1

Today we will do the topic of Geometry. **Geo** means **Earth** and **Metry** means to **measure**. To measure the earth this study was introduced thousands years back. This is of utmost important for Architect, Engineers etc. So the understanding of this topic is must for a person to be successful in practical life.

Geometry is a very vast topic but we have summarize this topic from the competitive exams point of view.

The topics we will do in Geometry are **Angles**, **Lines**, **Parallel Lines**, **Triangle**, **quadrilateral**, **Polygon**, **Circle** & **Locus**. We had divided these topics in two days i.e., **day 19** and **day 20**. On day 19 we will do angles, lines, parallel lines, triangle & quadrilateral and on day 20 we will do polygon, circle & locus.

It is very difficult to do the questions of Geometry without knowing the Basic Concepts, Terminology, Axioms, Theorems & Prepositions. So First we will do these basics (with proofs whereever necessary) & then we will do the examples & questions.

Now the topic we will do today are angles, lines, parallel lines, triangle & quadrilateral. First Topic is Angle.

ANGLE

Angle : When two non parallel lines intersect Angle is formed. There are six types of angles on the basis of their measurement which are as follows :

1. Acute Angle	**2.** Obtuse Angle	**3.** Right Angle
4. Straight Angle	**5.** Reflex Angle	**6.** Complete Angle

We will explain each of them one by one.

1. **Acute Angle:** The Angle which lies between 0° & 90° is called Acute Angle. We write it as $0° < \theta < 90°$

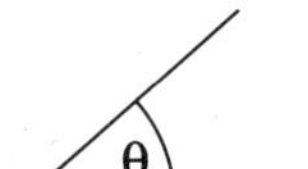

Acute angle

2. **Obtuse Angle:** The Angle which lies between 90° & 180° is known as Obtuse Angle. We write it as $90° < \theta < 180°$. Which means θ or the Angle lies between $90° < \theta < 180°$.

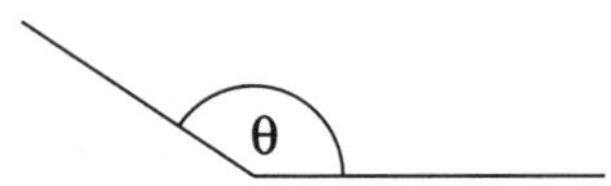

Obtuse angle

3. **Right Angle :** The Angle whose value is 90° is called Right Angle.

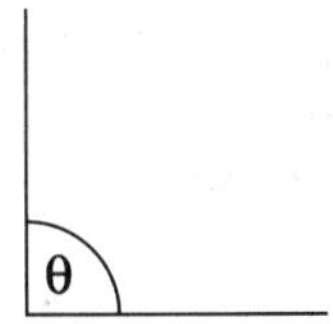

Right angle

4. **Straight Angle :** The Angle whose value is 180° is called Straight Angle.

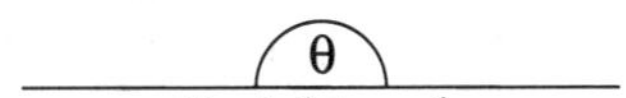

Straight angle

5. **Reflex Angle :** The Angle which lies between 180° & 360° is known as Reflex Angle. We write it as $180° < \theta < 360°$. Which means θ or the Angle lies between 180° & 360°.

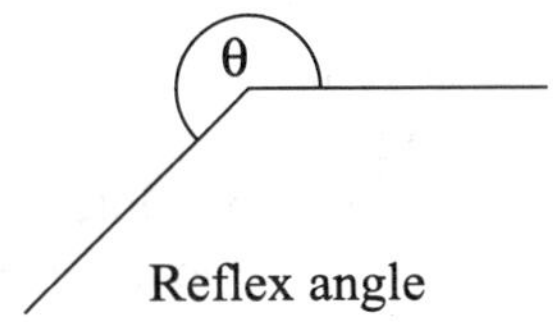

Reflex angle

6. **Complete Angle :** The Angle whose value is 360° is called Complete Angle.

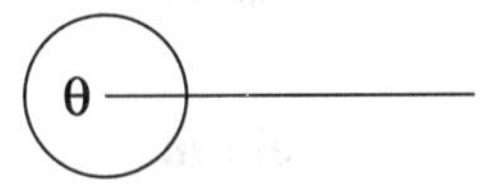

Complete angle

These were the angles on the basis of their measurement. Now we will discuss

1. Complementary angle
2. Supplementary angle

1. **Complementary Angles:** Two angles sum of whose measure is 90° are called complementary angles. Each angle is complement of the other angle e.g., complement of θ is $90 - \theta$. If it is asked what is the complement angle of 30° you will straightway say 60° because $30° + 60° = 90°$.

2. **Supplementary Angle:** Two angles are said to be supplementary if sum of the measure of two is 180°. Each angle is known as the suppliment of other. Now if it is asked what is the suppliment angle of 120° you will straightway say it is 60° because $120° + 60° = 180°$.

3. **Adjacent Angle:** In figure ∠1 and ∠2 are adjacent. ∠2 and ∠3 are adjacent. So the angles which are adjacent or next to each other as shown in fig. are adjacent angles. Technically two angle are adjacent if they have

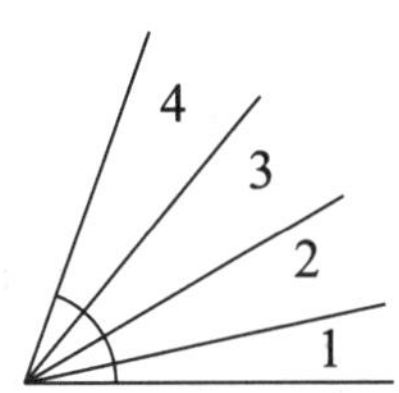

   (i) Same vertex (vertex; in fig. O is vertex)

   (ii) They have a common arm.

   (iii) Other arms are on either side of common arm. In this OA, OB and OC have common initial point O or they have a common vertex O. They have common arm OC and the distinct arms OA and OB lie on the different side of line OC.

   ∴ ∠1 and ∠2 are adjacent.

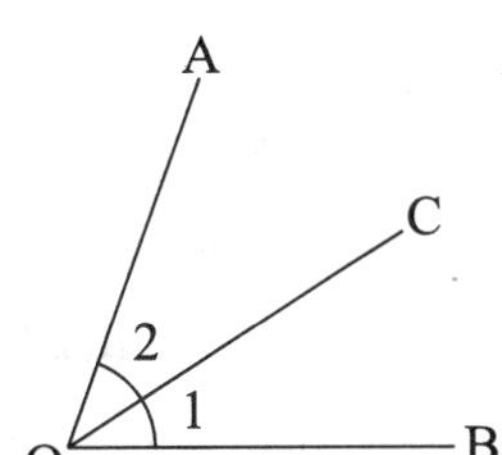

Linear pair of angles. Two adjacent angles are set to form a linear pair of angles if there sum is 180°. In figure ∠1 & ∠2 are linear pair of angles.

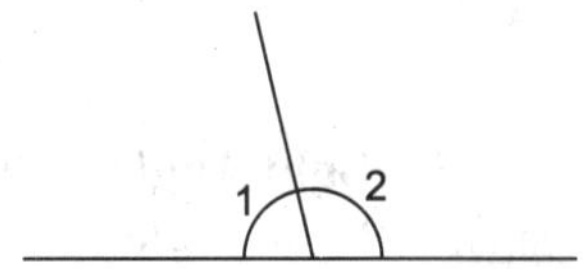

LINES

First we will discuss a few facts about intersecting & parallel lines. Now first about intersecting lines:

1. Two different lines cannot have more than one point in common.
2. Two intersecting lines cannot be parallel to same line.
3. If two lines intersect than vertically opposite angles are equal. In figure ∠1 and ∠2 and ∠3 and ∠4 are vertically opposite angle and are equal.

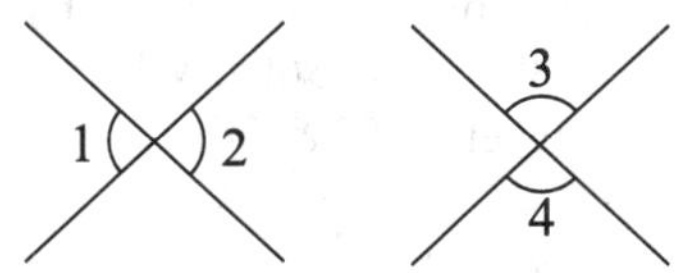

1. **Parallel Lines:** Two lines are parallel if they lie in a plane and do not intersect.

   Two lines which are both parallel to same line are parallel to each other.

   In figure *l* & *m* are parallel to *n*. Therefore they are parallel each other.

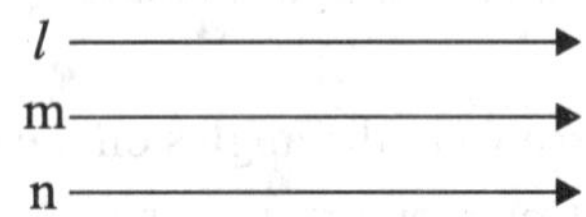

2. **Transversal:** A line which cuts two or more given lines (whether parallel or non-parallel) is called trans-versal. In figure AB & CD are two lines. So it is clear that EF is the transversal which cuts AB & CD in G & H respectively.

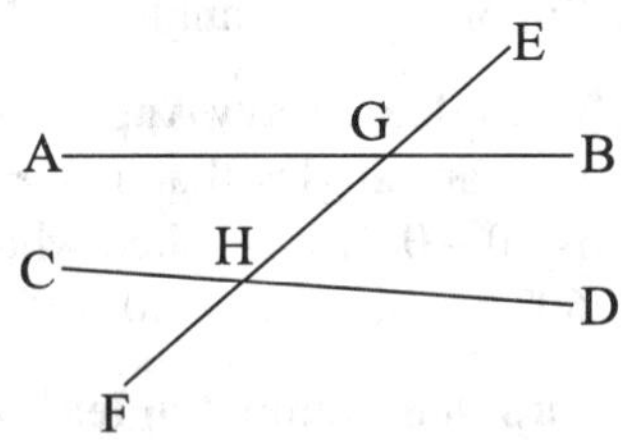

Following pair of angles are formed when a transversal intersect two lines.

1. Alternate Angles
2. Corresponding Angles
3. Consecutive Interior Angles

The angles & their properties are explained below :

AB & CD are two lines transversal LM intersects these lines in P & Q res. Four angles (∠1, ∠2, ∠3 & ∠4) are formed at P and four angles (∠5, ∠6, ∠7 & ∠8) are formed at Q. Some of these pair form alternate angle, some corresponding & some consecutive interior angles. So

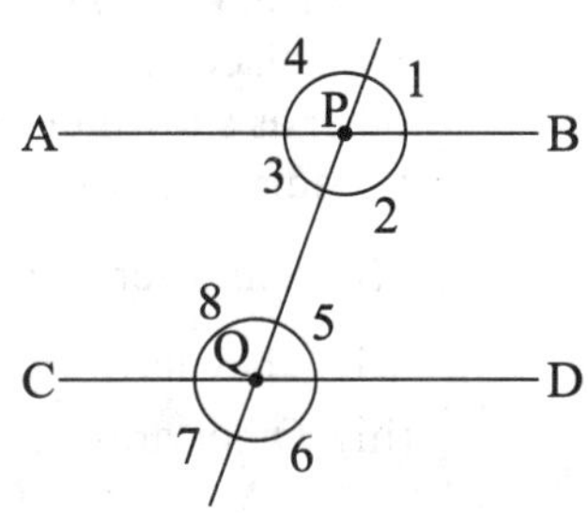

1. **Alternate Angle:** Following pairs of angles are alternate angle.

   (i) ∠2 & ∠8

   (ii) ∠3 & ∠5

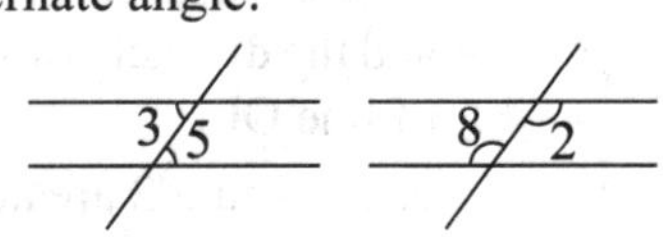

2. **Corresponding Angle :** Following pairs of angles are corresponding angle.

   (i) $\angle 1$ & $\angle 5$

   (ii) $\angle 4$ & $\angle 8$

   (iii) $\angle 3$ & $\angle 7$

   (iv) $\angle 2$ & $\angle 6$

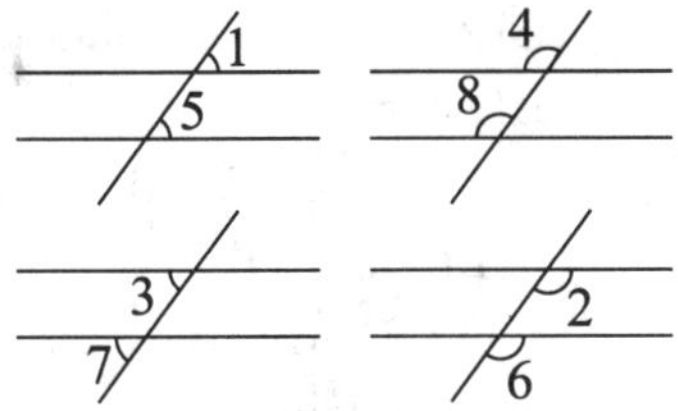

3. **Consecutive Interior Angle :** Following pairs are pairs of consecutive interior angles.

   (i) $\angle 2$ & $\angle 5$

   (ii) $\angle 3$ & $\angle 8$

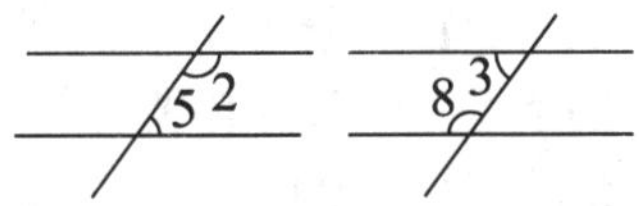

So it is hoped that you can now identify the different pair of angles.

Now **if two lines are parallel then**:

1. Each pair of Alternate angles are equal.
2. Each pair of Corresponding angles are equal.
3. Each pair of Consecutive interior angle are supplementary i.e., sum of the consecutive interior angles are 180°.

Explanation:

In figure if AB & CD are parallel

then $\angle 3 + \angle 8 = 180°$

and $\angle 2 + \angle 5 = 180°$

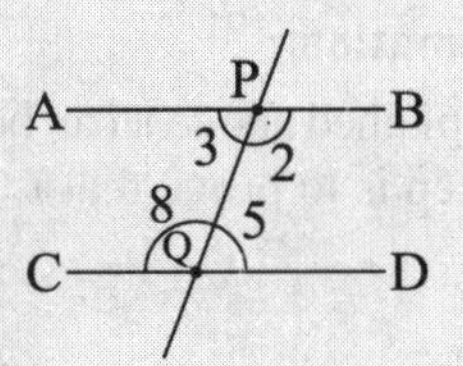

Converse of these is also true which means:

1. If a transversal cut two lines in such a way that each pair of alternate angles are equal then two lines are parallel.
2. If a transversal cuts two lines in such a way that sum of each pair of consecutive interior angle is 180° then two lines are parallel.

Further:

1. If two parallel lines are cut by a transversal then bisector of interior angles encloses a **rectangle**.

Explanation:

In figure AB & CD are two parallel lines cut by a transversals at L&M Bisectors of interior angle makes a quadrilateral LPMO.

We have to prove that this quadrilateral is a rectangle. For this first we will prove that it is a parallelogram and then if one of the angle of parallelogram is 90° then it will become a rectangle.

(Because then all the angles will be of 90°)

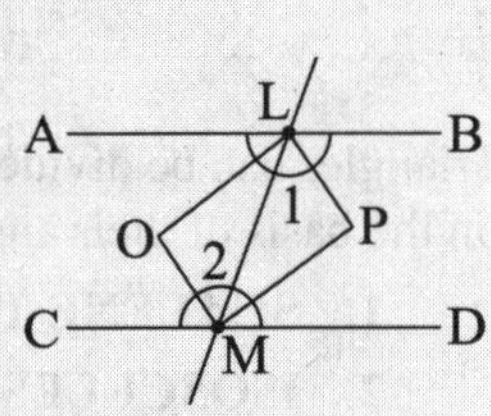

Ist Step:

$\angle BLM = \angle LMC$ (alternate angle)

Therefore $\frac{1}{2}\angle BLM = \frac{1}{2}\angle LMC$

$\angle 1 = \angle 2$ But these are alternate angles for LP&MO

Therefore LP ll MO

Similarily LO ll MP

Therefore LOMP is a parallelogram

2nd Step:

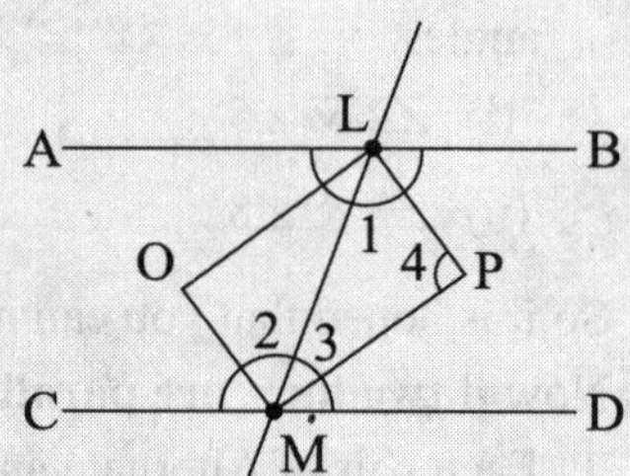

$\angle BLM + \angle LMD = 180°$

or $\frac{1}{2}(BLM + LMD) = \frac{1}{2} \times 180°$

$\angle 1 + \angle 3 = 90°$

But $\angle 1 + \angle 3 + \angle 4 = 180°$

Therefore $\angle 4 = 90°$

Hence LOMP is a rectangle.

2. If two parallel lines are cut by a perpendicular then bisectors of interior angles encloses a **square**.

Explanation :

As explained above it can be proved that LOMP is a rectangle last step is to prove that LO, OM, MP & PL are equal.

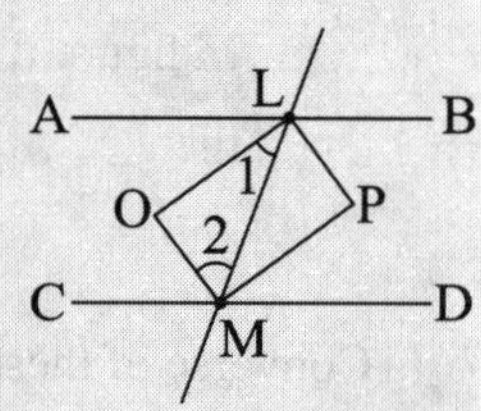

then $\angle 1 = 45°$(half of 90°)

$\angle 2 = 45°$(half of 90°)

$\therefore \angle 1 = \angle 2$

$\therefore$ LO = OM

but LO = PM & OM = LP

$\therefore$ All sides are equal. Hence LOMP is a square.

TRIANGLE

Now we discuss geometrical figures which are found by the intersection of more than two lines. The simplest of these is **TRIANGLE**.

Tri means three so a geometrical figure in a plane having three angles or three sides is called **TRIANGLE**.

Types of Triangle

Triangles can be divided in the following six types. Three on the basis on their sides and three on the basis of their angles. The types are :

1. **SCALENE TRIANGLE**
2. **ISOSCELES TRIANGLE**

3. **EQUILATERAL TRIANGLE**
4. **ACUTE TRIANGLE**
5. **RIGHT TRIANGLE**
6. **OBTUSE TRIANGLE**

Now we will explain each of the above one by one.

1. **SCALENE TRIANGLE:** A triangle no two of whose sides are equal is called **scalene triangle.**

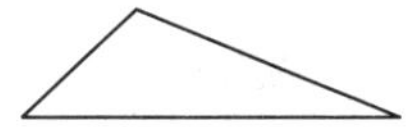

Scalean triangle

2. **ISOSCELES TRIANGLE:** A triangle two of whose sides are equal is called **isosceles triangle.**

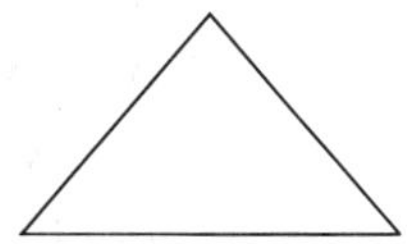

Isosceles triangle

3. **EQUILATERAL TRIANGLE:** A triangle all of whose sides are equal is called **equilateral triangle**.

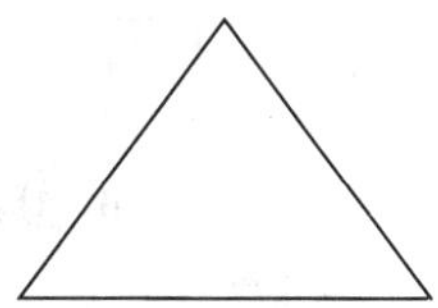

Equilateral triangle

4. **ACUTE TRIANGLE:** A triangle each of whose angle is acute i.e., less than 90° is called **acute triangle**.

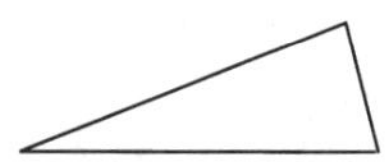

Acute triangle

5. **RIGHT TRIANGLE:** A triangle with one angle a right angle i.e., 90° is called **right triangle**.

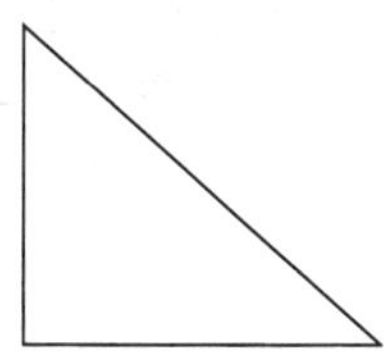

Right triangle

6. **OBTUSE TRIANGLE:** A triangle with one angle obtuse i.e., greater than 90° is called **obtuse triangle**.

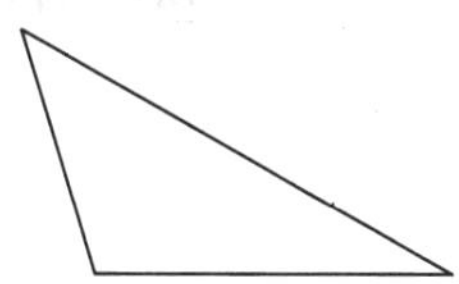

Obtuse triangle

Now First of all we will do a few basic theorems (with proofs wherever necessary).

1. **Sum of Angles of a triangle is 180°.**

Proof : Given: a Triangle ABC

To prove : $\angle 1 + \angle 2 + \angle 3 = 180°$

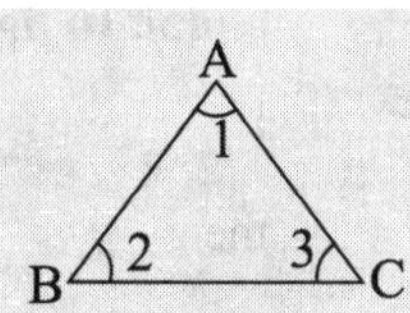

Construction : Draw a line l parallel to BC

Proof :

$\angle 2 = \angle 5$ (alternate angles)

$\angle 4 = \angle 3$ (alternate angles)

but $\angle 1 + \angle 4 + \angle 5 = 180°$

$\therefore$ $\angle 1 + \angle 2 + \angle 3 = 180°$

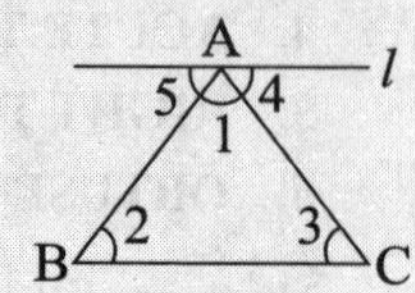

2. If side of a triangle is produced then the exterior angle so formed is equal to sum of two interior opposite angles.

Given : A Triangle ABC in which BC is produce to D.

To Prove

$\angle 4 = \angle 1 + \angle 2$

To Prove $\angle 1 + \angle 2 + \angle 3 = 180°$ ...(i)

(sum of Angles of a triangle)

$\angle 3 + \angle 4 = 180°$ ...(ii)

(Angles of a straight line)

From (i) and (ii)

$\angle 1 + \angle 2 + \angle 3 = \angle 3 + 4$

$\therefore$ $\angle 1 + \angle 2 = \angle 4$

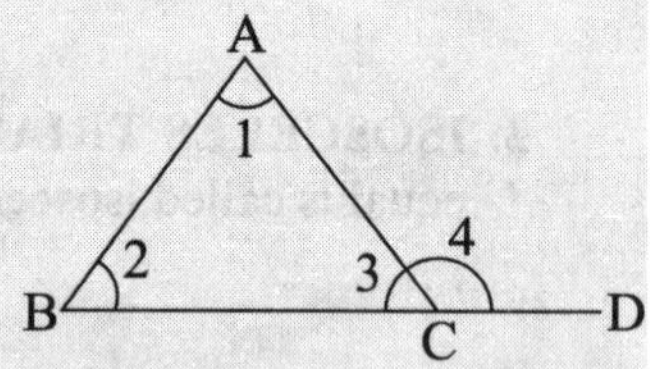

Further an exterior angle of a Triangle is Greater than the either of two interior opposite angles. In figure $\angle 4 > \angle 1$ & $\angle 2$.

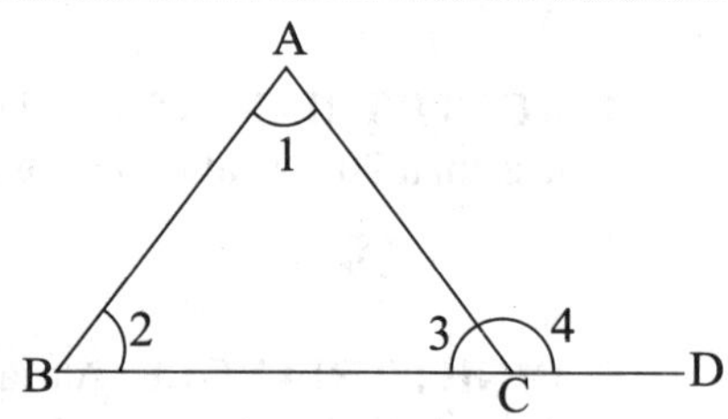

3. Sum of any two sides of a triangle is greater than the third side. In figure:

AB + BC > AC

AB + AC > BC

AC + BC > AB

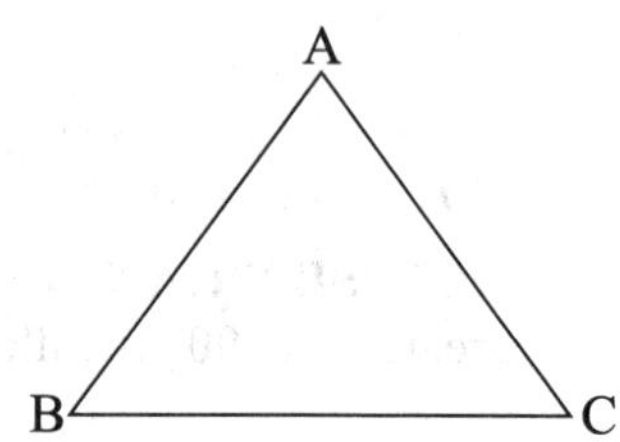

4. Angles Opposite to equal sides of a triangle are equal. In figure:

If AB = AC

then $\angle 1 = \angle 2$

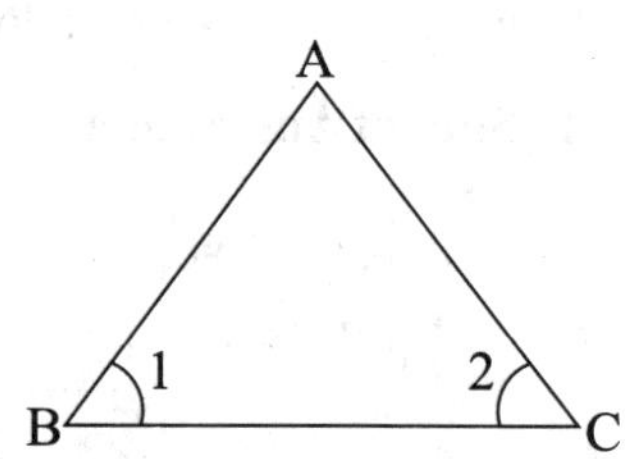

5. Sides opposite to equal angles of a triangle are equal. In figure:

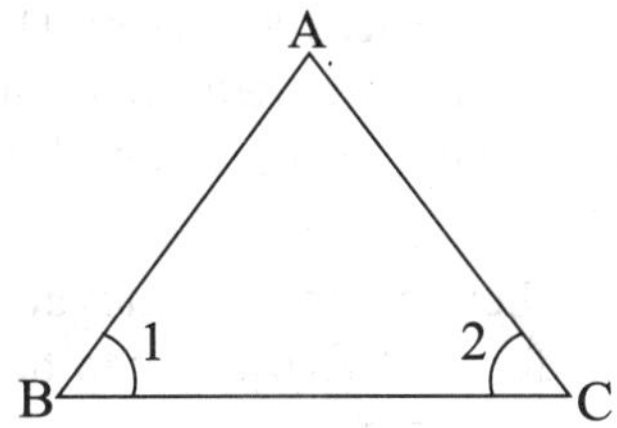

If $\angle 1 = \angle 2$

then $AC = AB$

6. If two sides of a triangle are unequal then longer side has greater angle opposite to it.

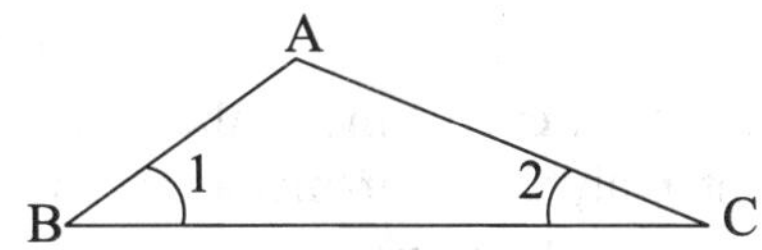

In triangle ABC :

If $AC > AB$

$\therefore$ $\angle 1$ which is opposite to AC is greater than $\angle 2$ which is opposite to A.

7. Of all the line segments that can be drawn to a given line from a point (not lying on it) the perpendicular line segment is shortest.

Given a line *l*.

P is a point from which many line segments are drawn of which PN is perpendicular.

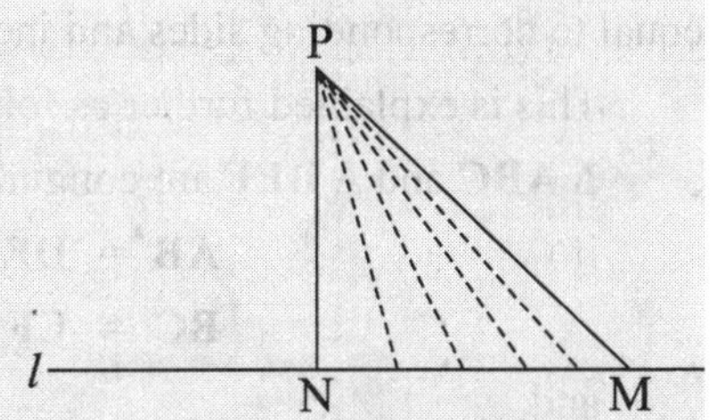

To Prove PN is shortest.

Proof : Take a line segment PM in ΔPMN

$\angle N$ is 90°

$\therefore$ $\angle M$ is acute.

$\therefore$ $\angle N$ is $> \angle M$

$\therefore$ PM which is opposite to $\angle N$ is greater than PN which is opposite to angle M or $PN < PM$

Similarly PN is shorter than all other line segments that can be drawn from P to line L.

So PN is shortest.

Now further this topic is divided in two parts. In first part we will discuss **(1) Congurency** and **similarity** and theorem & facts based on these. In second part we will discuss concept of **(2) Median, circumcenter, orthocenter, incenter** and **Ecenter** etc. and facts & theorems based on these.

PART I
(Congurency and Similarity)

First we will do **Congurency.**

CONGURENCY : Congurency simply means two geometrical figures having same shape and size. Two Triangles are congurant if each triangle can hide the other if they are placed on each other one by one.

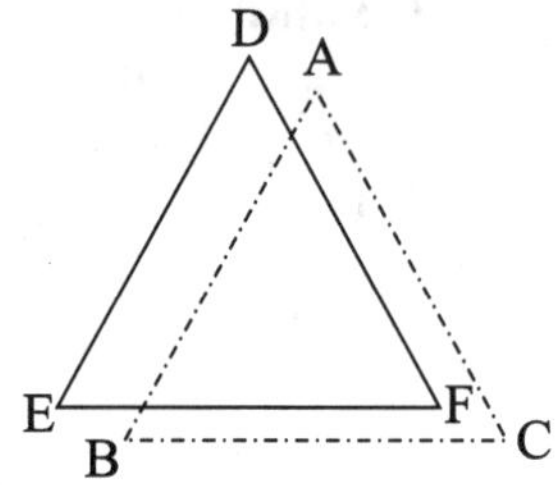

Let vertices A falls on D, B falls on E and C falls on F then Δ ABC will completly Super impose Δ ABC so automatically AB will be equal to BE, BC will be equal to EF and AC will be equal to DF.

Therefore three angles and three corresponding sides are to be equal for the two triangle to be congurant.

If Δ ABC is congurant to Δ DEF then it is written as

Δ ABC ≅ Δ DEF

Now every time all the six parameter are not measured to prove two triangle to be congurant. Only three parameters are measured with some condition which are given below:

1. **SAS Congurency**
2. **SSS Congurency**
3. **ASA Congurency**

We will explain each of them one by one.

1. SAS Congurency : Two triangles are congurent if two sides and included angle are equal to corresponding sides and included angle of other triangle.

This is explained further as follows

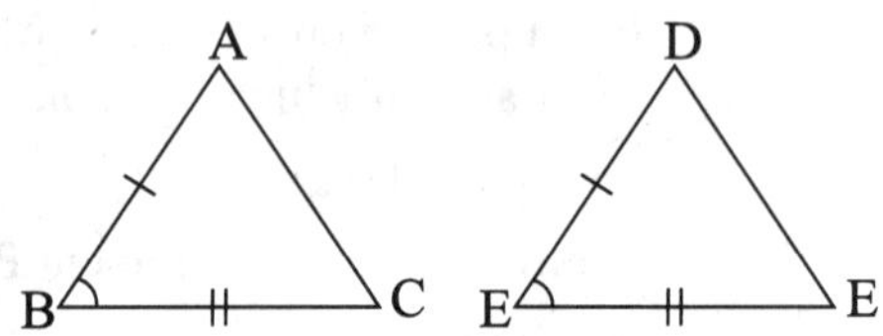

Δ ABC and **Δ DEF** are congurent

if **AB = DE**

BC = EF

and

their included angles **∠B = ∠E**

then we say **Δ ABC ≅ Δ DEF**

then the remaining part will automatically be equal i.e., **AC** will be equal to **DF**

and **∠A** will be equal to **∠D**

and **∠C** will be equal to **∠F**

We write this as **c.p.c.t.**

i.e., **corresponding parts of congurent triangles.**

2. SSS Congurency : Two triangles are congurent if three sides of one triangle are equal to three corresponding sides of the other triangle.

Explanation :

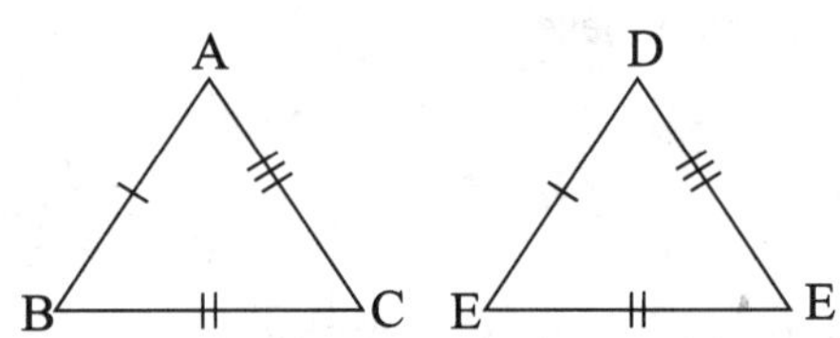

In Δ ABC & Δ DEF

Given AB = DE

BC = EF

and AC = DF

∴ Δ ABC ≅ Δ DEF

∴ ∠A = ∠D

∠B = ∠E

∠C = ∠F **(c.p.c.t.)**

3. ASA Congurency : Two triangles are congurent if two angles and included side of one triangle is equal to the corresponding two angles and included side of the other.

Explanation :

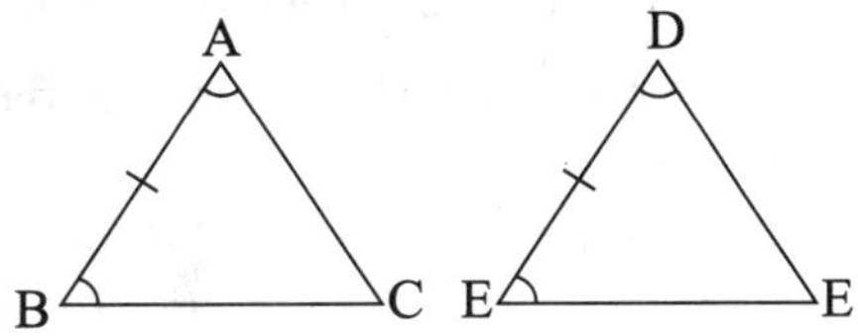

In Δ ABC & Δ DEF

Given $\angle A = \angle D$

$\angle B = \angle E$

& included side AB = DE

∴ Δ ABC ≅ Δ DEF

∴ BC = EF

AC = DF

$\angle C = \angle F$ **(c.p.c.t.)**

So remember only three types of congurency are there.

1. SAS Congurency i.e., side – angle – side

2. SSS Congurency i.e., side – side – side

3. ASA Congurency i.e., angle – side – angle

Next we will discuss **Similarity:**

SIMILARITY : It means same shape but does not mean same size.

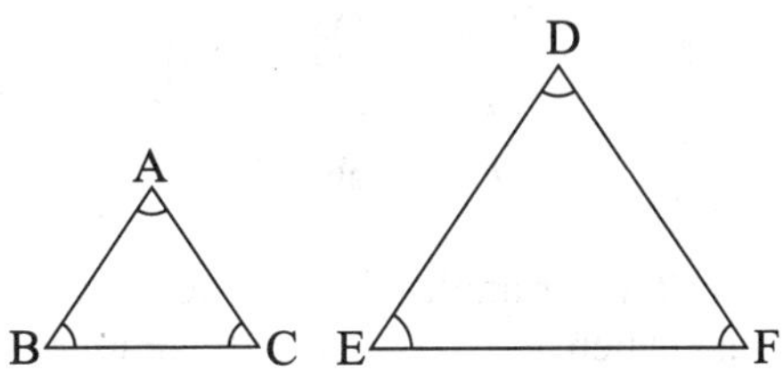

In similiar triangle ABC & DEF

1. Corresponding sides are proportional means

$$\frac{AB}{DE} = \frac{BC}{EF} = \frac{AC}{DF}$$

2. Corresponding angles are equal i.e.,

$$\angle A = \angle D$$
$$\angle B = \angle E$$
and $$\angle C = \angle F$$

3. If two triangles are similiar then **Ratio of Area of two triangles is equal to the ratio of squares of corresponding sides.** In other words in similiar Δ ABC & Δ DEF

$$\frac{\text{Area of Triangle ABC}}{\text{Area of Triangle DEF}} = \frac{AB^2}{DE^2} = \frac{BC^2}{EF^2} = \frac{AC^2}{DF^2}$$

Further like congurency there are conditions of similarity which are as follows :

1. AAA Similarity 2. SSS Similarity 3. SAS Similarity

We will explain each of them one by one.

1. AAA Similarity : If in two triangles all corresponding angles are equal then triangle are similar.

Explanation : Triangle ABC & Triangle DEF

If $\angle A = \angle D$

$\angle B = \angle E$

$\angle C = \angle F$

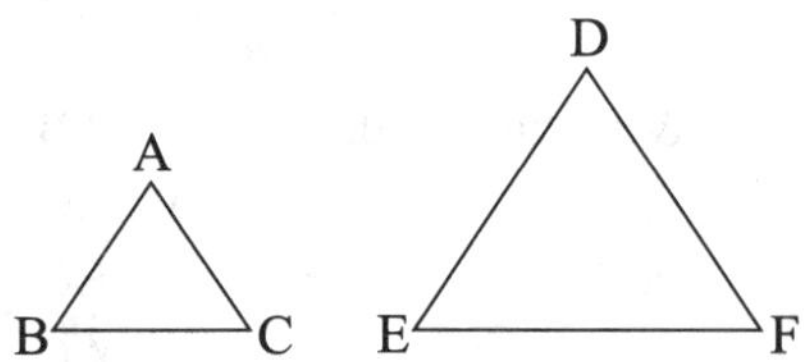

Then Triangles are **similar**.

∴ Their corresponding sides are proportional.

2. SSS Similarity : If Corresponding sides of Two Triangles are proportional then Triangles are similar.

Explanation : In Triangle ABC & Triangle DEF

if $$\frac{AB}{DE} = \frac{BC}{EF} = \frac{AC}{DF}$$

then triangles are similar.

Therefore their corresponding angles will be equal

i.e., $\angle \mathbf{A} = \angle \mathbf{D}$

$\angle \mathbf{B} = \angle \mathbf{E}$

$\angle \mathbf{C} = \angle \mathbf{F}$

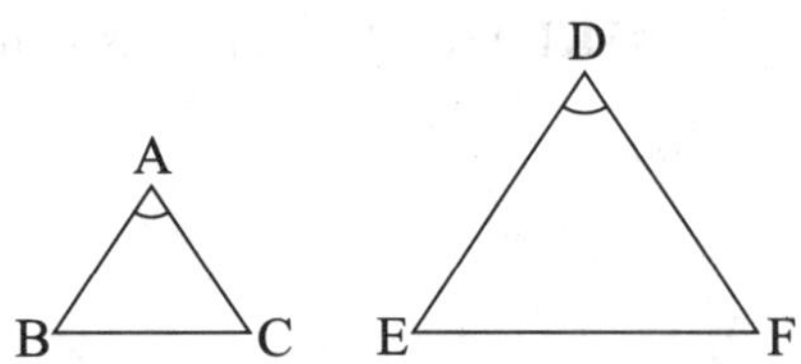

3. SAS Similarity : If in Two Triangles a pair of corresponding sides are proportional and their including angle is equal. Then Two Triangles are similar.

Explanation: In Triangle ABC & Triangle DEF

if $$\frac{AB}{DE} = \frac{AC}{DF}$$

and their included angles $\angle A = \angle D$.

Then Triangles are similar.

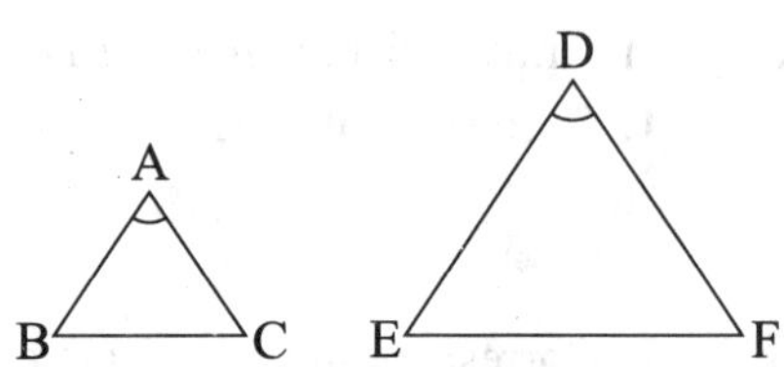

Therefore their remaining correspon-ding sides are proportional and remaining correspondence angles are equal.

In addition of these there are few basic theorams which we will do under this topic.

1. **Thales Theoram :** It states that **in a triangle if a line is drawn parallel to one side, to intersect the other two sides in distinct points then it divides the two sides in same ratio.**

   Explanation : In Triangle ABC DE || BC

$$\therefore \qquad \frac{\mathbf{AD}}{\mathbf{DB}} = \frac{\mathbf{AE}}{\mathbf{EC}}$$

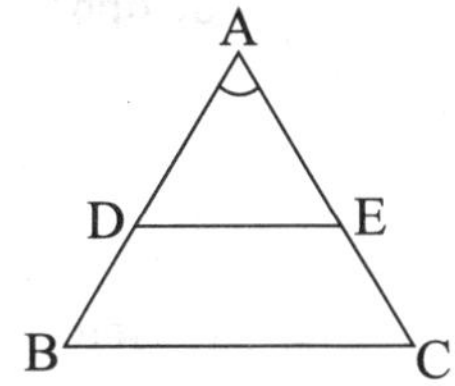

The converse of this theorem is also true. In a triangle ABC a line is drawn such that

$$\frac{\mathbf{AD}}{\mathbf{DB}} = \frac{\mathbf{AE}}{\mathbf{EC}}$$

then DE is parallel to DC.

2. **Bisector Theorem :** It states that **the Bisector of vertical angle divide the base in the ratio of line making the angle.**

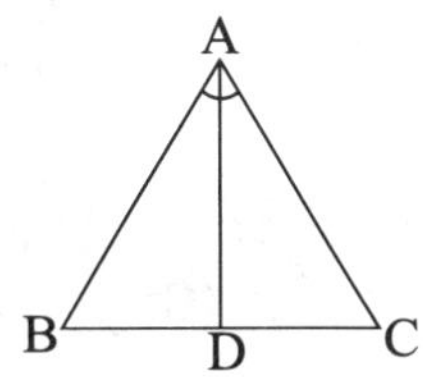

Explanation : In Triangle ABC

AB is the Bisector of angle A then

$$\therefore \qquad \frac{\mathbf{AB}}{\mathbf{AC}} = \frac{\mathbf{BD}}{\mathbf{DC}}$$

3. **Pythagoras Theorem :** It states that **in a right angle triangle square of the hypotenuse is equal to the sum of squares of other two sides.**

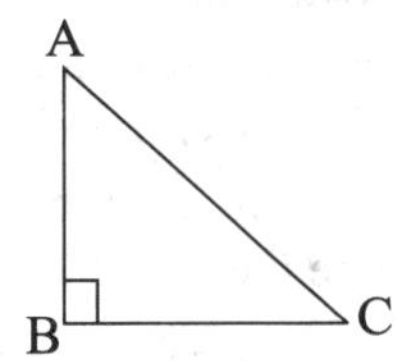

Explanation : Given Triangle ABC

Where $\angle B = 90°$

Side opposite to $\angle B$ that is AC is hypotenuse

then $\mathbf{AC^2 = AB^2 + BC^2}$

Converse of the above theorem is also true which states that if **in a triangle ABC $\mathbf{AC^2 = AB^2 + BC^2}$** then angle opposite to AC is 90°.

Few more facts related to right angle triangle.

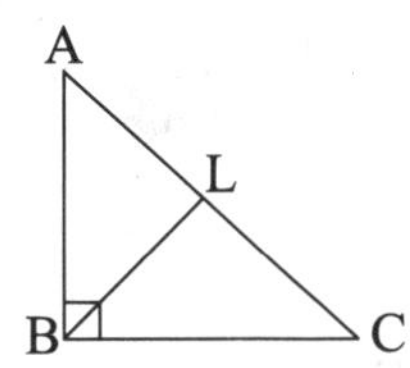

(i) If angle B is 90° and BL is perpendicular to AC then

$$\mathbf{BL^2 = AL \times LC}$$

(ii) In right angle triangle ABC if E is the Mid point of hypotenuse AC then

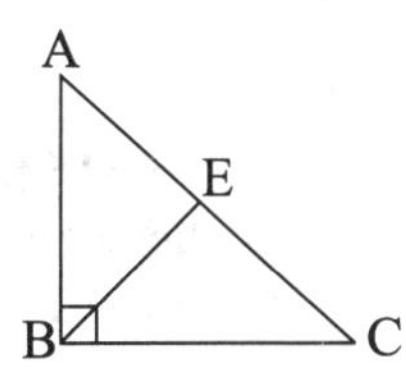

$$\mathbf{AE = EC = BE}$$

i.e., midpoint of hypotenuse is equi distance from vertices of triangle.

(iii) In a right angle triangle if one of the angle is 30° then side opposite to it is half the hypotenuse. In figure

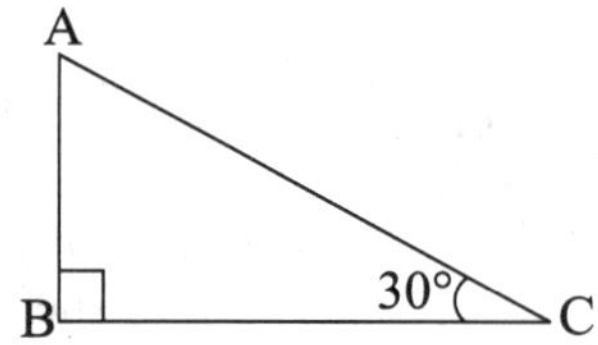

$$\angle C = 30°$$

$$\text{then } \mathbf{AB} = \frac{1}{2}\mathbf{AC}$$

4. **Midpoint Theorem :** It states that **line joining the mid-points of two sides of a triangle is parallel to the third side & half of it.**

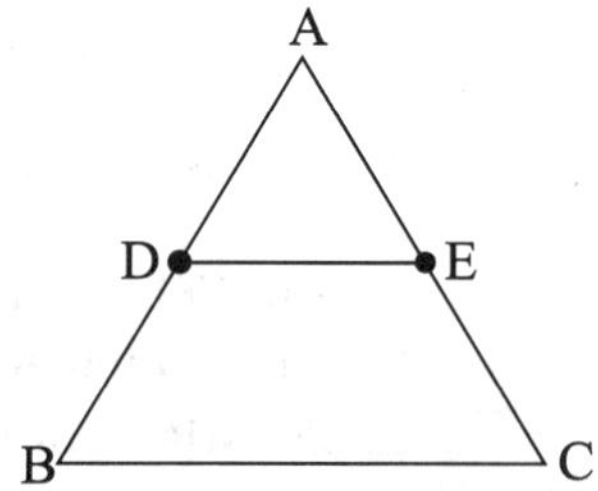

Explanation : Given Triangle ABC, D & E are midpoins of sides AB & AC respectively.

$\therefore$ $\quad$ DE || BC and $\mathbf{DE} = \frac{1}{2}\mathbf{BC}$

PART II
(Circumcenter, Incenter, Orthocenter, Median Ecenter etc.)

First we will discuss **Circumcenter.**

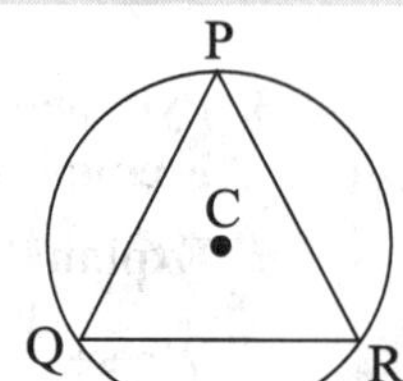

CIRCUMCENTER : Circumcenter means center of circumcircle. Now what is circum circle? Circumcircle means Outercircle. So centre of outercircle of a triangle is called circumcenter as shown in figure. C is the circumcenter of triangle PQR.

To draw a outer circle first we will draw perpendicular bisector of line QR.

Q. How to draw perpendicular bisector of a line?

Ans. Following steps are involved to draw perpendicular bisector of line QR.

First Step : From Point Q draw an arc after opening the compass more than half of QR.

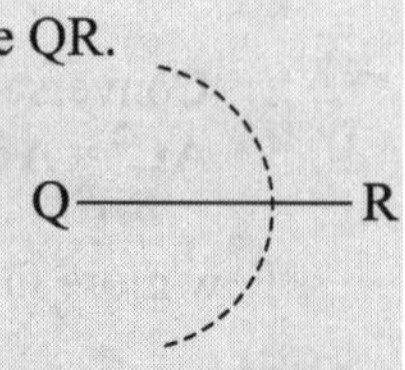

Second Step : Put the compass with same opening at R & draw another arc. Which will cut the previous arc at A & A′.

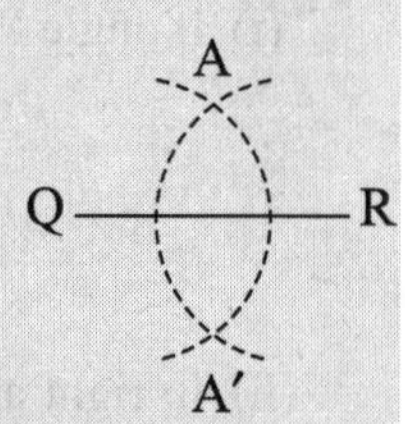

Third Step : Join A & A′. AA′ will be perpendicular bisector of QR. Which means **QO = OR and ∠1 = ∠2 = 90°**

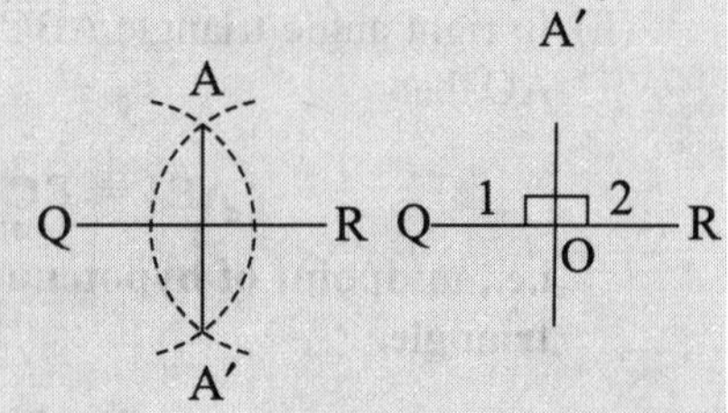

Now similarily draw perpen-dicular bisectors of BC and AC they are concurrent. (Concurrent means they meet in the same point. In this case at C)

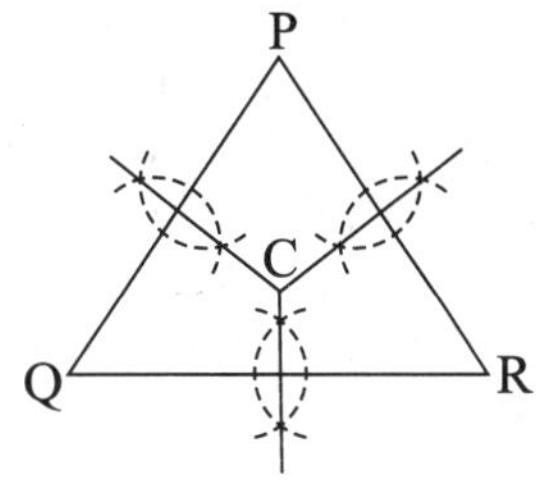

From C, CP, CQ & CR will be at the same distance. So from C if we draw a circle taking CP, CQ or CR as a radius it will touch the vertices P, Q, R of the circle so **C is the circumcenter of ΔPQR** and also CP, CQ and CR are called circum radius. Their is an important formula for finding circum radius which is given below.

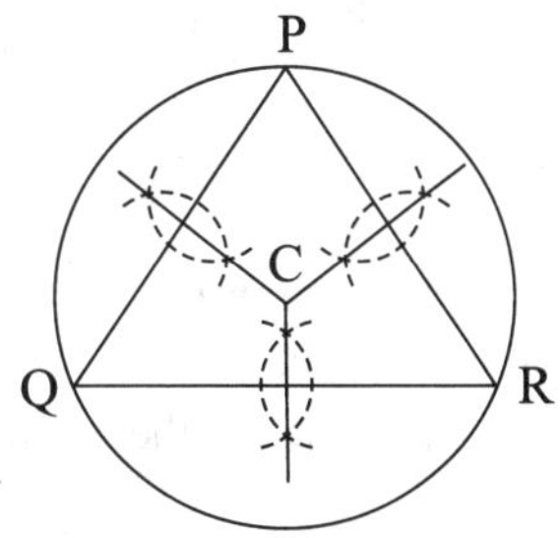

CIRCUM RADIUS R = $\frac{abc}{4\Delta}$ where a, b and c are the three sides of the triangle and Δ is the area of the triangle.

So remember : CIRCUMCENTER is the point of intersection of **perpendicular bisector** of three sides of a triangle.

Next comes the incenter

INCENTER : Incenter means centre of incircle. What is incircle?

In Circle of a triangle is the circle which is inside the triangle and just touches the three sides of a triangle ABC as shown in figure.

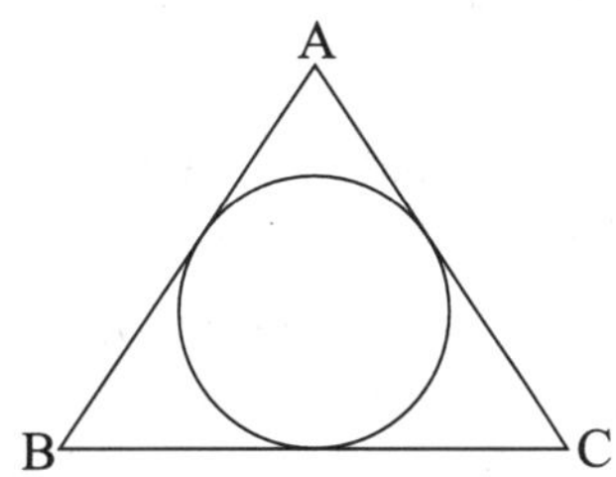

Now how to draw In Circle of ABC First of all we will draw angle bisector of $\angle$B.

Q. How to draw angle bisector of angle B?

Ans. Following steps are involved to draw angle bisector of angle B.

First Step : Open the compass and draw an arc with cuts AB in D and BC in D′.

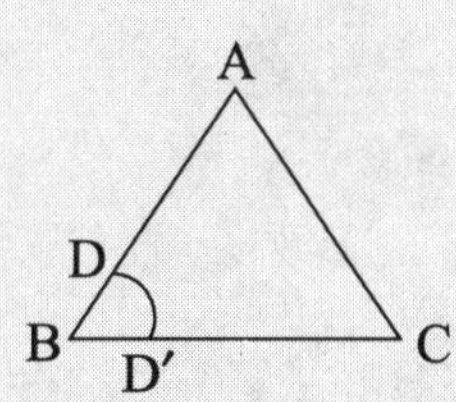

Second Step : From D draw an arc with some opening of compass. Draw an arc from D′ with same opening which cuts the earlier arc at P. So BP is angle bisector of angle B. Means divide angle B into two equal parts.

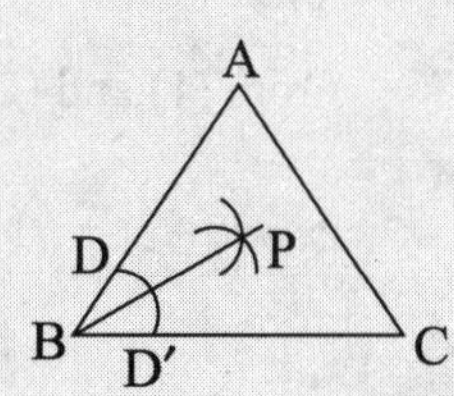

Similarly draw angle bisector of $\angle C$ & $\angle A$. These angle bisectors are also concurrent means they meet at same point. Say I from I as center if we take the distance from AB, BC and AC, they are same (they are IL, IM and IN res.), they will be equal and the circle drawn will touch the sides at LM and N res. So this is how the incircle is drawn and also IL, IM and IN are called **inradius (*r*)** and there is an important formula for finding the inradius which is given below

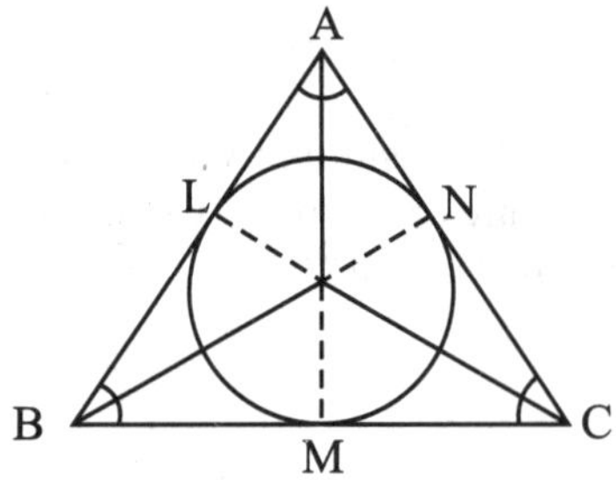

INRADIUS (r) = $\frac{\mathbf{A}}{\mathbf{S}}$ where A is the area of triangle and S is its semiperimeter

$$\left(S = \frac{\text{sum lengths of 3 sides of } \Delta}{2}\right)$$

So remember : INCENTER is the point of intersection of **angle bisectors** of three angles of a triangle.

Also if I is the angle bisector of triangle ABC then

$$\angle BIC = 90 + \frac{1}{2}\angle A$$

$$\angle AIC = 90 + \frac{1}{2}\angle B$$

and $$\angle AIB = 90 + \frac{1}{2}\angle C$$

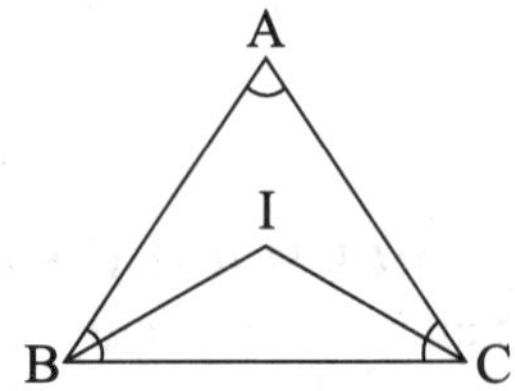

[V. Imp.]

Proof : in ΔABC $\angle A + \angle B + \angle C = 180°$

$$\frac{1}{2}(\angle A + \angle B + \angle C) = \frac{1}{2} \times 180° = 90°$$

$$\frac{1}{2}\angle A + \angle B + \frac{1}{2}\angle C = 90°$$

$$\frac{1}{2}\angle B + \frac{1}{2}\angle C = 90° - \frac{1}{2}\angle A$$

$$\angle 1 + \angle 2 = 90° - \frac{1}{2}\angle A \quad \text{... (1)}$$

But $$\angle 1 + \angle 2 + \angle BIC = 180°$$

$\therefore$ $$\angle 1 + \angle 2 + \angle BIC = 180°$$

$\therefore$ $$\angle 1 + \angle 2 = 180° - \angle BIC \quad \text{... (2)}$$

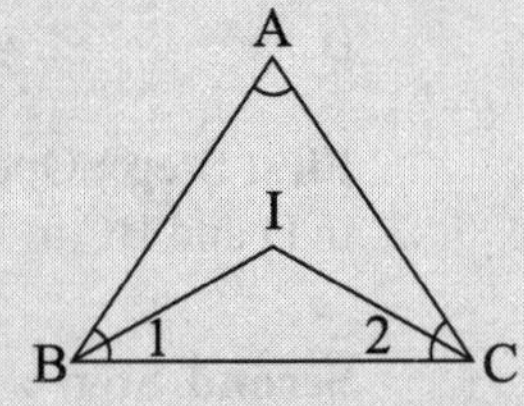

From (1) and (2)

$$180° - \angle BIC = 90° - \frac{1}{2}\angle A$$

$$180° - 90° + \frac{1}{2}\angle A = \angle BIC$$

or $$\angle BIC = 90° + \frac{1}{2}\angle A$$

Similarily we can prove that

$$\angle AIC = 90° + \frac{1}{2}\angle B$$

and $$\angle AIB = 90° + \frac{1}{2}\angle C$$

ORTHOCENTER : Point of intersection of altitudes (perpendiculars) of a triangle is called **orthocenter**.If we draw altitude of Δ **ABC** i.e., from A **AL**, from B **BM** and from C **CN**. They are concurrent. Means they meet at same point say H. So H is the orthocenter of triangle ABC

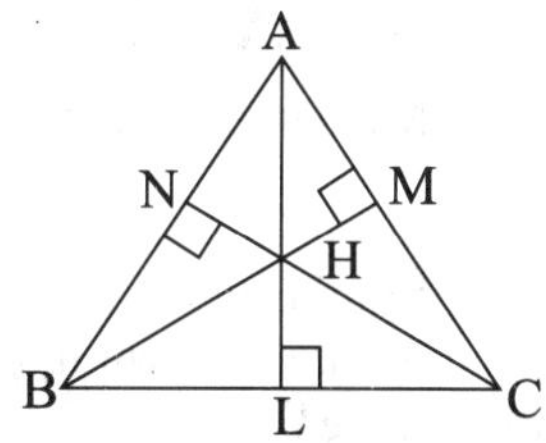

Q. How to draw an altitude from a point to a line?

Ans. Following steps are involved

First Step : Draw an arc on BC from A after opening some compass draw an arc on DC Which will cut BC in D & D′.

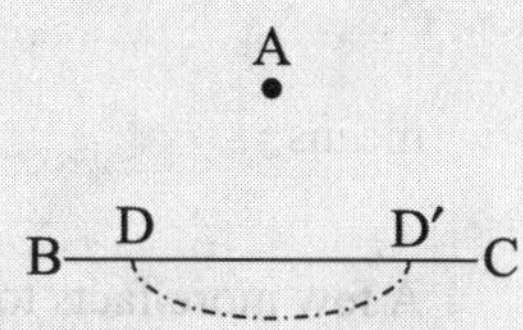

Second Step : From D draw an arc after opening of compass and with same opening with compass draw another arc. Both will cut at O′ and A are joint making altitude on BC from A.

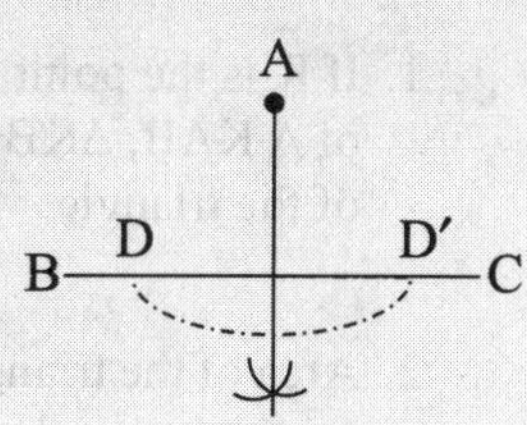

A few more facts about orthocenter

1. Ortho center of an acute triangle lies in side the triangle as shown in figure.

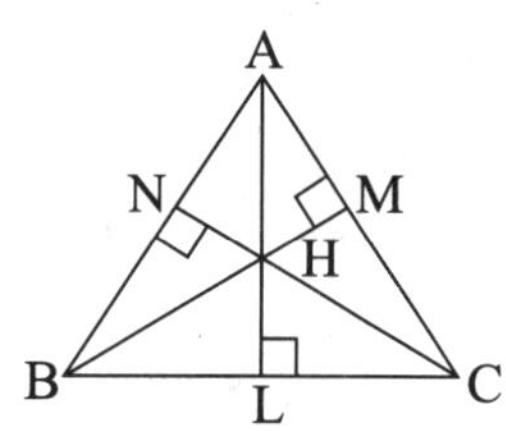

2. Orthocenter of a obtuse angle triangle lies out side the triangle as shown in figure.

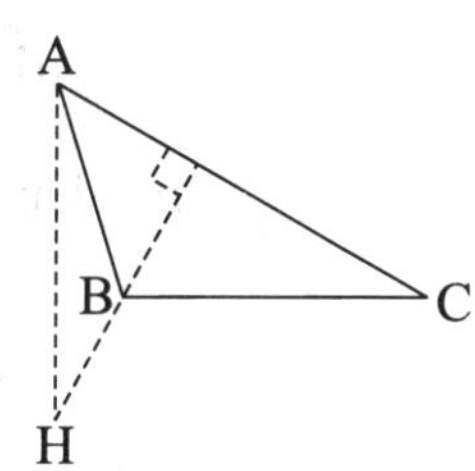

3. Orthocenter of a right angle triangle lies on the vertex making the right angle as shown in figure.

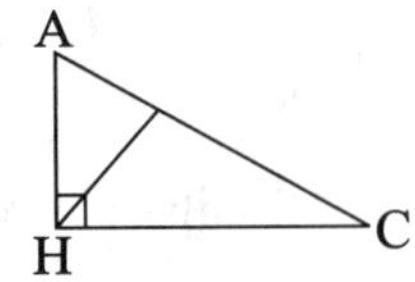

CENTROID : It is a **point of intersection of medians of a triangle.** Now what is **Median**?

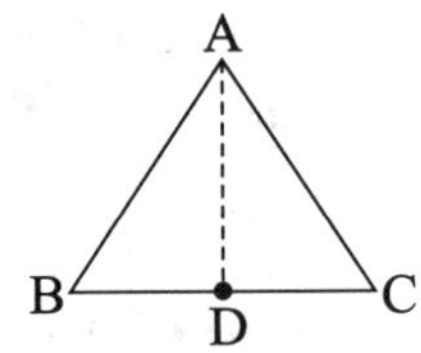

To understand it take the mid point of BC, it is D. Join AD. So AD is the Median.

Median divides the triangle in to two equals halves.

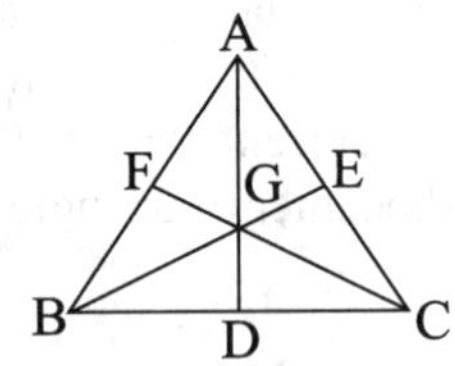

Similarily BE and CF are also median. Medians of a triangle are concurrent as shown in figure and meet in point G. G is called **centroid** of triangle.

Centroid divides the median in the ratio **2 : 1** and bigger parts is towards the vertex.

means $$\frac{\mathbf{AG}}{\mathbf{GD}} = \frac{\mathbf{2}}{\mathbf{1}}\ \frac{\mathbf{BG}}{\mathbf{GE}} = \frac{\mathbf{2}}{\mathbf{1}} \text{ and } \frac{\mathbf{CG}}{\mathbf{GF}} = \frac{\mathbf{2}}{\mathbf{1}}$$

A few more facts to be noted

1. If K is the point that is taken inside the triangle such that area of Δ KAB, ΔKBC and Δ KAC are equal then K is the centroid of the triangle.

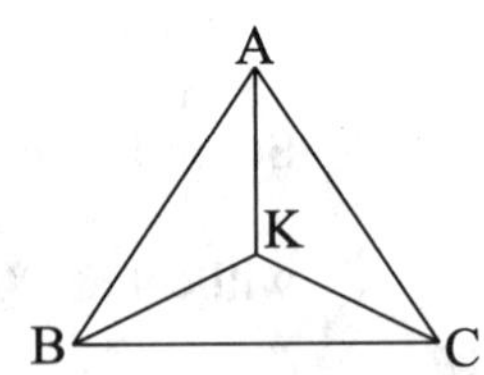

2. Area of the triangle formed by the medians is $\frac{3}{4}$ of the area of original triangle

Explanation : It means that if we make a triangle with three sides as AD, BE and CF means whose length are equal to AD, BE and CF respectively then area of triangle so formed is equal to $\frac{\mathbf{3}}{\mathbf{4}}$ or 75% of the area of triangle ABC. Suppose area of triangle ABC is 12 Sq. cm then area of the triangle formed by the medians is $\frac{\mathbf{3}}{\mathbf{4}}\times\mathbf{12} = \mathbf{9}$ **sq. cm**

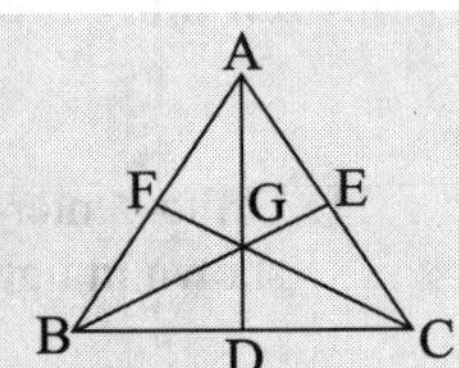

E-CENTER: The circle which touches the sides of a triangle when produced externally is called **escribed** circle (as shown in figure) and its centre is called E Centre.

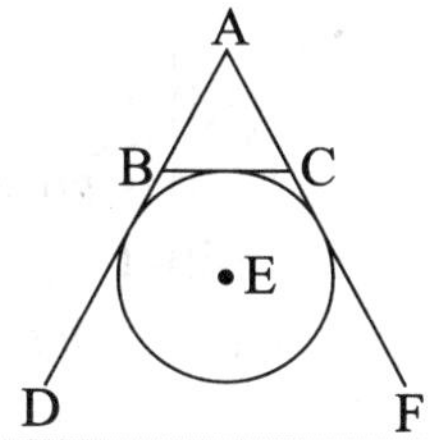

To Draw E Circle : Draw the angle bisectors of $\angle CBD$ and $\angle BCF$ they meet at E. From E take perpendicular distanceof ABD, BC and ACF. They are equal and shown as EL, EM and EN res. So when we draw the E Circle taking E as centre it will touch the side of triangle externally at L, M and N res.

And for E Circle $\angle \mathbf{BEC = 90° - \frac{1}{2}\angle A}$

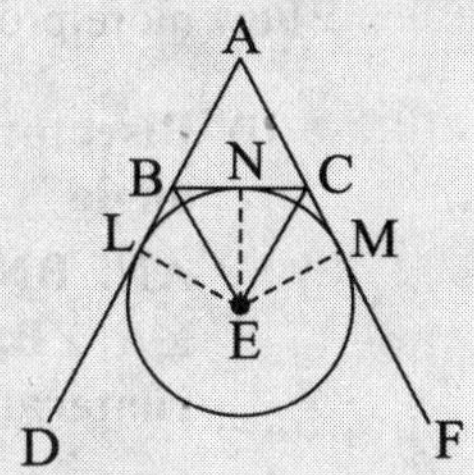

Last Comes the Quadrilateral

QUADRILATERALS : Quadrilaterals is a figure formed by four straight lines. We have already discussed about square rectangle ||gm rhombus etc. in the topic of **mensuration**. We will once again revise the basic properties of these one by one.

1. **Square :** In square all sides are equal and each angle is of 90° means

$$\mathbf{AB = BC = CD = DA}$$

and $$\mathbf{\angle A = \angle B = \angle C = \angle D = 90°}$$

diagonals are equal means AC = BD and they bisect each other at 90°

means $$AO = OC \text{ and } BO = OD$$

and $$\mathbf{\angle 1 = \angle 2 = \angle 3 = \angle 4 = 90°}$$

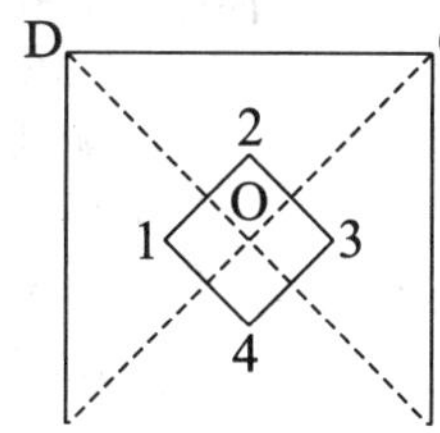

2. **In Rectangle:** Opposite sides are equal and all angles are 90° and diagonal bisect are each other (but not at 90°) that is **AB = CD and AD = BC**

also $$\mathbf{\angle A = \angle B = \angle C = \angle D = 90°}$$

$$AO = OC \ \& \ BO = OD$$

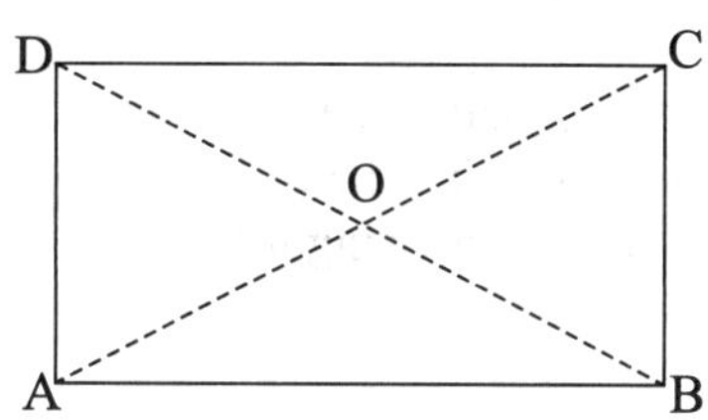

3. **Rhombus:** All sides are equal and diagonal bisect each other at 90°.

That is $$\mathbf{AB = BC = CD = DA}$$

and $$\mathbf{AO = OC, \quad BO = OD}$$

also $$\mathbf{\angle 1 = \angle 2 = \angle 3 = \angle 4 = 90°}$$

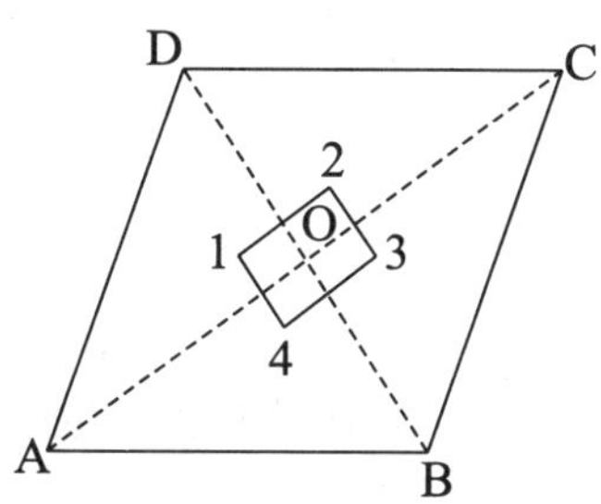

4. In ||gm:

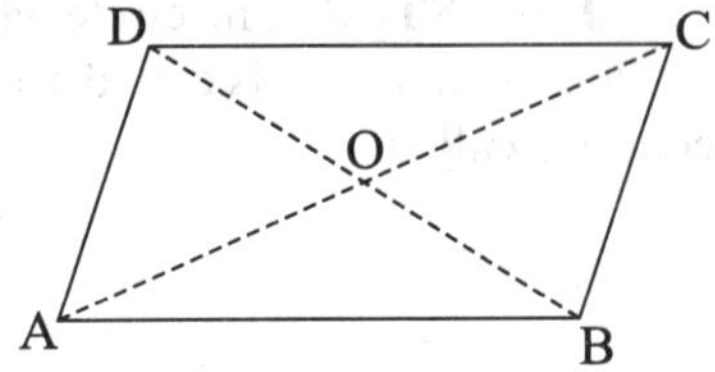

(i) Opposite side are equal.
(ii) Opposite angles are equal.
(iii) Diagonals Bisect each other.
(iv) Diagonals need not to be equal.

Few more properties regarding ||gm

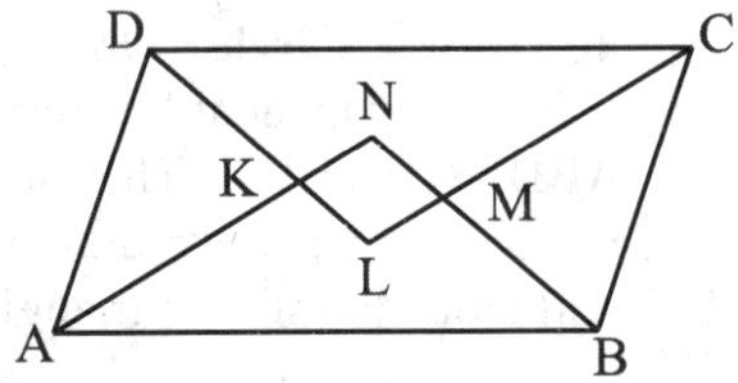

(i) Bisector of angle of ||gm encloses rectangle
AN, BN, CL and DL are Bisectors of ∠A, ∠B, ∠C and ∠D res. Forming a quadrilaterals KLMN. KLMN is a rectangle.

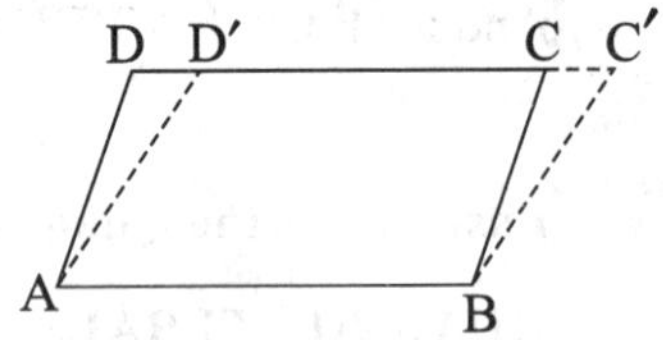

(ii) Two ||gm between same base and between same parallel lines are equal in area. ABCD and ABC′ D′ are two ||gm on same base AB and between same parallel lines.

∴ Area of IIgm ABCD = Area of ||gm ABC′ D′

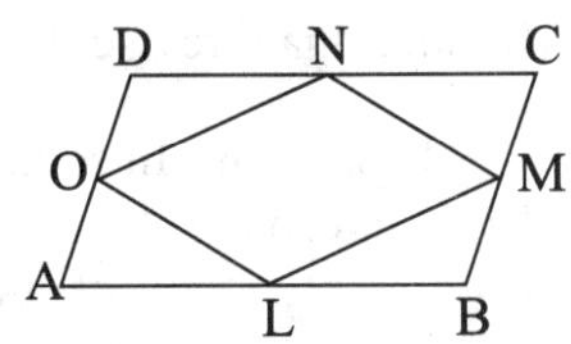

(iii) Quadrilateral formed by joining the mid points of pair of adjacent sides of a rectangle is a rhombus. L, M, N and O are mid points of sides AB, BC, CD and DA res

∴ LMNO is a rhombus.

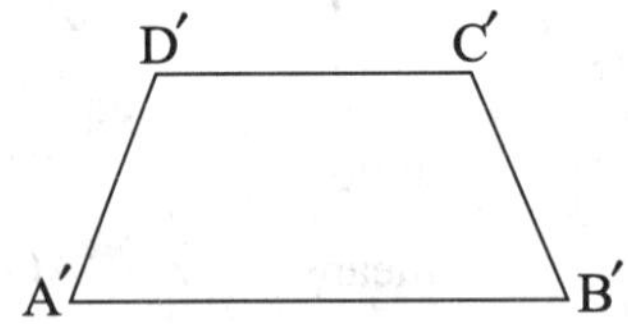

5. Trapezium : In Trapezium one pair of opposite sides is parallel and other is non parallel. In figure A′B′C′D′ is a trapezium.

As said earlier you should have all the above basics on your fingre tips. So it is advise that before starting the questions you should revise the basic once again and should devote 15-20 minutes in revising the basics.

REVISE THE BASICS

It is hoped that you have gone through the basics. Now let us have a simple test to check your revision. Now solve the following exercise test.

EXERCISE

Fill in the blanks :

1. The angle which lies between 0° and 90° is called .... angle.
2. Two angles sum of whose measure is 90° are called .... angles.
3. Angle of ....° is called complete angle.
4. The angle which lies between 90° and 180° is called .... angle.
5. Two angles sum of whose measures is 180° are called .... angle.
6. If two lines intersect then vertically opposite angles are ....
7. Two lines which are parallel to same line are .... to each other.
8. If two parallel lines are cut by a transversal then bisector of interior angles encloses a ....
9. If two parallel lines are cut by a perpendicular transversal then bisectors of interior angles encloses a ....
10. A triangle two of whose sides are equal is called .... triangle.
11. A triangle no two of whose sides are equal is called .... triangle.
12. A triangle whose one angle is more then 90° is called .... triangle.
13. Sum of the three angles of a triangle is ....
14. If the side of a triangle is produced then exterior angle so form is equal to sum of two .... angles.
15. Of all the lines that can be drawn to a given line from a point, the .... is shortest.
16. Three conditions of congurency are (i) .... (ii) .... (iii) ....
17. In two similar triangles corresponding sides are ....
18. In two similar triangles corresponding angles are ....
19. The point of intersection of perpendicular bisector of the sides of a triangle is called ....
20. The point of intersection of three angle bisectors of a triangle is called ....
21. Point of intersection of altitudes of a triangle is called ....
22. Point of intersection of medians of a triangle is called ....
23. Centroid divides the medians in the ratio ....
24. The area of triangle formed by the medians is .... % of the area of original triangle.
25. **Orthocenter of .... triangle is outside the triangle.**
26. **Orthocenter of .... triangle is inside the triangle.**
27. **Orthocenter of right angle triangle is at ....**
28. **Center of escribed circle is known as ....**
29. In ||gm opposite angles are ....
30. Bisector of interior angles of a ||gm makes a ....
31. In .... diagonals bisects each other at 90°.

32. The diagonals of a rectangle are ....

33. The diagonals of a square are ....

34. The diagonals of a rhombus are ....

35. All sides of a rhombus are ....

SOLUTIONS

1. Acute	**2.** Complimentry	**3.** 360°	**4.** Obtuse
5. Supplementary	**6.** Equal	**7.** Parallel	**8.** Rectangle
9. Square	**10.** Isosceles	**11.** Scalene	**12.** Obtuse
13. 180°	**14.** Interior Opposite	**15.** Perpendicular	
16. (i) SAS; (ii) SSS; (iii) ASA		**17.** Proportional	**18.** Equal
19. Circumcentre	**20.** In Centre	**21.** Orthocenter	**22.** Centroid
23. 2 : 1	**24.** 75%	**25.** Obtuse	**26.** Acute
27. vertex making right angle		**28.** Ecenter	**29.** Equal
30. Rectangle	**31.** Square & Rhombus	**32.** Equal	**33.** Equal
34. Unequal	**35.** Equal		

If your answers in the above exercise are correct then you should do the next part of the chapter otherwise you should revise the basics again.

Now we will do the types of question.

Types of Questions

Here We have divide the question in two types:

1. **Simple :** Direct application of facts and basics.
2. **Typical :** Typical application of facts and basics.

We will now do few examples of above two types which are followed by the exercise.

Type I

Example 1. An angle is five times its supplement. Find the angle.

Sol :

$$\begin{aligned} \text{Let one angle} &= x \\ \text{another angle} &= 5x \\ x + 5x &= 180^\circ \\ 6x &= 180^\circ \\ x &= 30^\circ \\ \therefore \quad \text{required angle} &= 5x = 5 \times 30^\circ = \mathbf{150^\circ} \end{aligned}$$

Example 2. Find the supplement of following angles.

(i) 120° (ii) 108° (iii) 67° 37′ (iv) 154° 20′

Sol : (i)

$$\begin{aligned} \text{Supplement of } \theta &= 180^\circ - \theta \\ \therefore \quad \text{Supplement of } 120^\circ &= 180^\circ - 120^\circ = \mathbf{60^\circ} \end{aligned}$$

(ii) Supplement of 108° $= 120° - 108° = \mathbf{72°}$

(iii) Supplement of 67°37′ $= 180° - 67°37' = \mathbf{112°23'}$

(iv) Supplement of 154°20′ $= 180° - 154°20' = \mathbf{25°40'}$

Example 3. AB and CD are two lines intersecting at O such that $\angle AOC + \angle BOC + \angle BOD = 280°$. Find $\angle AOD$

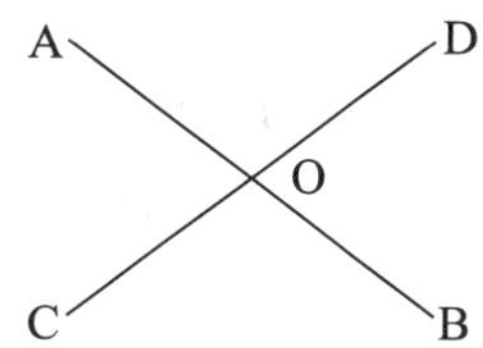

Sol : We know that

$$\angle BOD + \angle AOC + \angle BOC + \angle AOD = 360°$$
$$280° + \angle AOD = 360°$$
$$\angle AOD = 360° - 280° = \mathbf{80°}$$

Example 4. Find x in the following figure

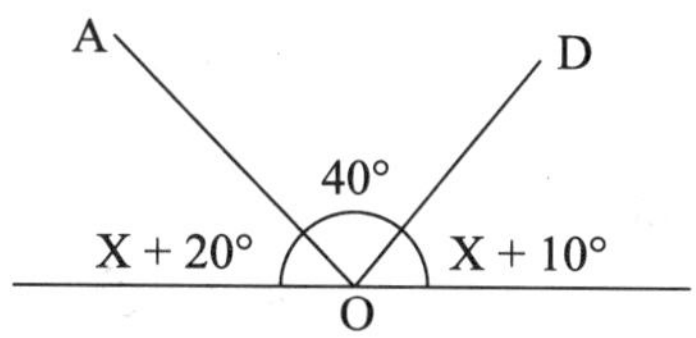

Sol :
$$x + 20° + 40° + x + 10° = 180°$$
$$2x + 70° = 180°$$
$$2x = 110°$$
$$x = \mathbf{55°}$$

Example 5. AB and CD are two parallel lines. Transversal LM cut these in E and F respectively. Given $\angle 1 = 110°$. Find $\angle 2, \angle 3, \angle 4, \angle 5, \angle 6, \angle 7, \angle 8$

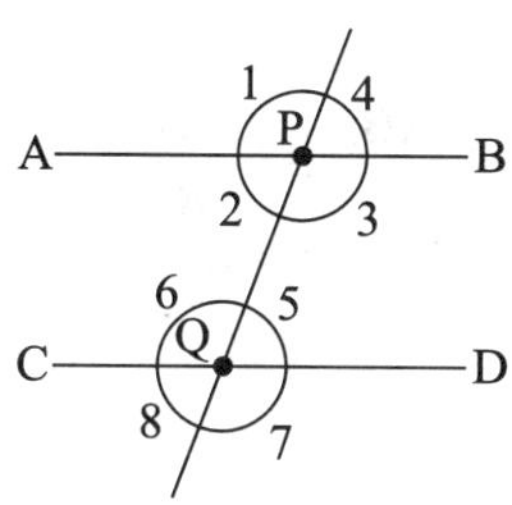

Sol : $\angle 1 = 110°$

$\therefore$ $\angle 3 = \mathbf{110°}$ (V.O.A.)

$\angle 6 = \mathbf{110°}$ (Corresponding angle)

Now $\angle 6 = 110°$

$\therefore$ $\angle 7 = \mathbf{110°}$ (V.O.A.)

Now $\angle 1 + \angle 4 = 180°$

$110° + \angle 4 = 180°$

$\therefore$ $\angle 4 = 180° - 110° = \mathbf{70°}$

Now	$\angle 4 = 70°$	
$\therefore$	$\angle 2 = \mathbf{70°}$	(V.O.A.)
and	$\angle 5 = \mathbf{70°}$	(Corresponding angle)
Now	$\angle 5 = 70°$	
$\therefore$	$\angle 8 = \mathbf{70°}$	(V.O.A.)

Now we have found that

$$\angle 1 = \angle 3 = \angle 6 = \angle 7 = 110°$$

and $$\angle 2 = \angle 4 = \angle 5 = \angle 8 = 70°$$

Example 6. In ΔABC

Given $\angle B = 90°$, $AB = 20$ cm

and $BC = 21$ cm.

Find $AC = ?$

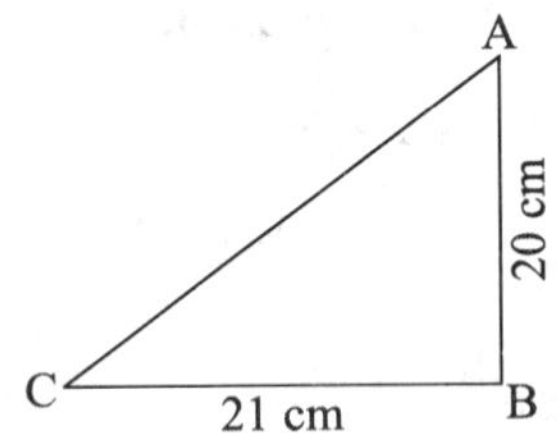

Sol : Pythagoras Theorem

$$AC^2 = AB^2 + BC^2$$
$$AC^2 = 20^2 + 21^2$$
$$AC^2 = 400 + 441$$
$$AC^2 = 841$$
$$AC^2 = 29^2$$
$$AC = \mathbf{29\ cm}$$

Example 7. In figure AD is bisector of $\angle A$. Find DC.

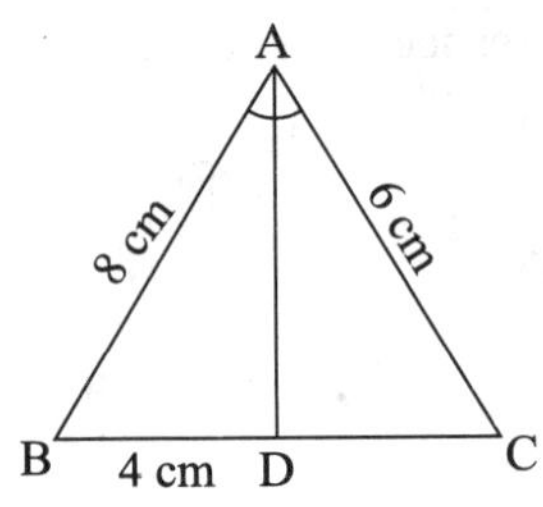

Sol : AD is Bisector of $\angle A$

$$\frac{AB}{AC} = \frac{BD}{DC}$$
$$\frac{8}{6} = \frac{4}{DC}$$
$$DC = \frac{4 \times 6}{8} = \mathbf{3\ cm}$$

Example 8. In figure XY || BC Find YC

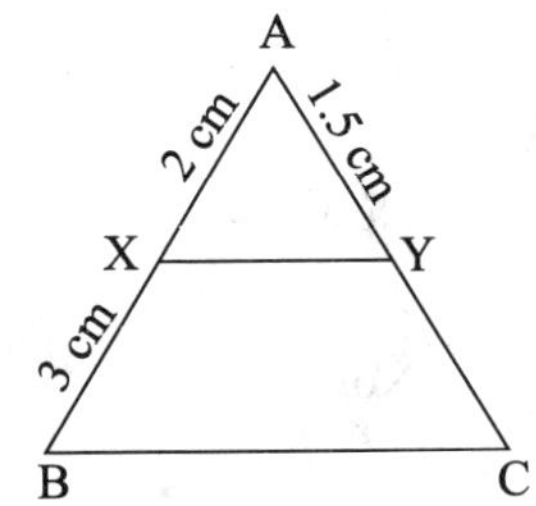

Sol : By Thales Theorem

$$\frac{AX}{XB} = \frac{AY}{YC}$$
$$\frac{2}{3} = \frac{1.5}{x}$$
$$x = \frac{1.5 \times 3}{2} = \mathbf{2.25\ cm}$$

Example 9. In figure G is Centroid of the triangle Find GD

Sol : We know that Centroid divides the medians in the ratio 2 : 1

$$\frac{AG}{GD} = \frac{2}{1}$$

$$\frac{3.5}{GD} = \frac{2}{1}$$

$$GD = \frac{3.5}{2} = \mathbf{1.75\ cm}$$

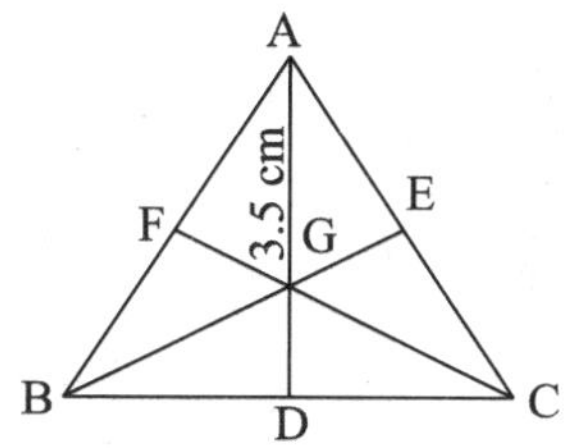

Example 10. In given figure ∠B = 90° and BM is the median. Find BM.

Sol :

$$AC^2 = AB^2 + BC^2$$

$$AC^2 = 8^2 + 15^2$$

$$AC^2 = 64 + 225$$

$$AC^2 = 289$$

$$AC = 17\ cm$$

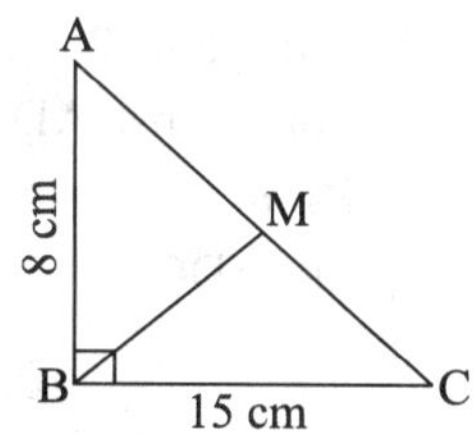

Also mid point of the hyp. of right angle triangle is equi distance from the vertices.

∴ $AM = BM = MC$

or $BM = \frac{17}{2} = \mathbf{8.5\ cm}$

Type II

Example 11 : In Equilateral triangle if AD ⊥ BC then prove that

$$3\,AB^2 = 4\,AD^2.$$

Sol : In ΔABD and ΔADC

$\angle B = \angle C$ (Given)

$\angle 3 = \angle 4$ (each 90°)

∴ $\angle 1 = \angle 2$

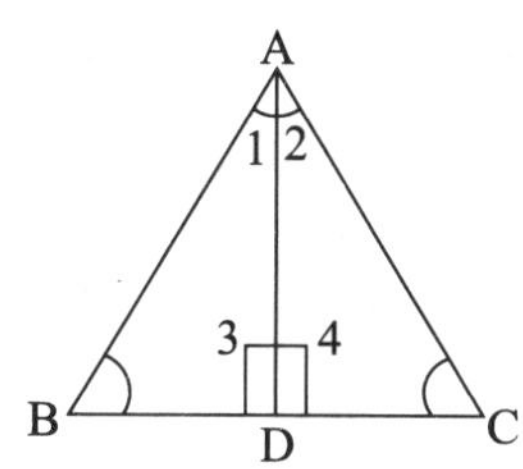

Now again in ΔABD and ΔADC

$\angle 1 = \angle 2$

$AD = AD$ (Common)

$\angle 3 = \angle 4$ (each 90°)

∴ By ASA Triangles are Congurent.

∴ $BD = DC = \frac{BC}{2} = \frac{AB}{2}$

Now in ΔABD

$$AB^2 = AD^2 + BD^2$$

$$AB^2 = AD^2 + \left(\frac{AB}{2}\right)^2$$

$$AB^2 = AD^2 + \frac{AB^2}{4}$$

$$AB^2 - \frac{AB^2}{4} = AD^2$$

$$\frac{3AB^2}{4} = AD^2$$

$$3AB^2 = 4AD^2$$

Example 12. Δ ABC is right angle at A and AD $\perp$ BC Prove that AD^2 = BD. DC

Sol : Const. AD $\perp$ BC

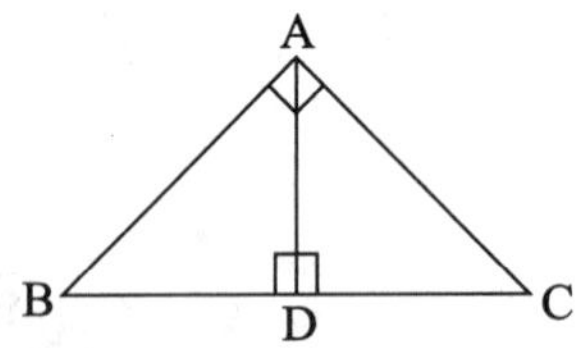

Proof: $AB^2 = AD^2 + BD^2$ ... (i)

In ΔADC

$AC^2 = AD^2 + DC^2$ ... (ii)

Add (i) and (ii)

$$AB^2 + AC^2 = 2AD^2 + BD^2 + DC^2$$

$$BC^2 = 2AD^2 + BD^2 + DC^2$$

$$(BD + DC)^2 = 2AD^2 + BD^2 + DC^2$$

$$BD^2 + DC^2 + 2BD.DC = 2AD^2 + BD^2 + DC^2$$

$$2BD.DC = 2AD^2$$

or $\mathbf{AD^2 = BD.\ DC}$

Example 13. A Triangle is right angle at A. The Perpendicular from A to BC is p. Prove that $\frac{1}{p^2} = \frac{1}{b^2} + \frac{1}{c^2}$

when a = BC

b = CA

and c = AB

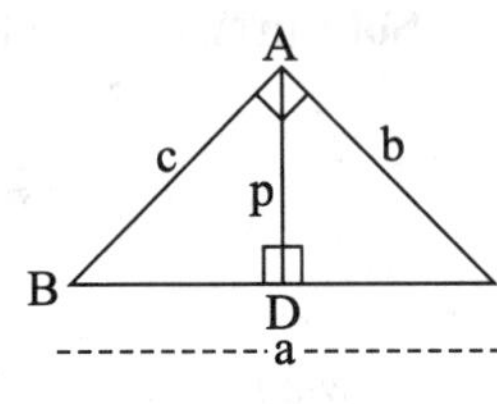

Sol : Area of $\Delta ABC = \frac{1}{2} \times BC \times AD$

$= \frac{1}{2} ap$ ... (i)

also area of $\Delta ABC = \frac{1}{2} \times AB \times AC$

$= \frac{1}{2} cb$ ... (ii)

From (i) and (ii)

$$ap = bc$$

$$\text{or} \quad p = \frac{bc}{a}$$

$$a = \frac{bc}{p}$$

$$BC^2 = AB^2 + AC^2$$

$$a^2 = c^2 + b^2$$

$$\left(\frac{bc}{p}\right)^2 = c^2 + b^2$$

$$\frac{b^2c^2}{p^2} = c^2 + b^2$$

Divide both sides by b^2c^2 we get

$$\mathbf{\frac{1}{p^2} = \frac{1}{b^2} + \frac{1}{c^2}}$$

Example 14. ABCD is a rhombus where AC and BD are its diagonal. Prove that $4AB^2 = AC^2 + BD^2$

Sol : In rhombus ABCD $\angle 1 = \angle 2 = \angle 3 = \angle 4$ (Diagonal of rhombus bisect each other at 90°)

In right triangle AOB $AO^2 + OB^2 = AB^2$

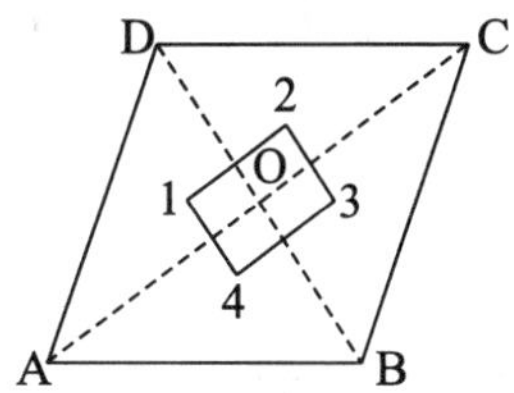

$$\left(\frac{AC}{2}\right)^2 + \left(\frac{BD}{2}\right)^2 = AB^2$$

$$\mathbf{AC^2 + BD^2 = 4\,AB^2}$$

Example 15. D is any point on BC, the base of Isosceles Triangle ABC. Prove that $AB^2 - AD^2 = BD.\ DC$

Sol : Const. Draw AR ⊥ BC

Proof: $\Delta ABR \cong \Delta ACR$ .... (SAS)

Congurency same as done in Example 11

Therefore $BR = RC$

$AB^2 = AR^2 + BR^2$ ... (i) (By Pythagoras)

also $AD^2 = AR^2 + RD^2$ ... (ii) (By Pythagoras)

Subtract (ii) from (i)

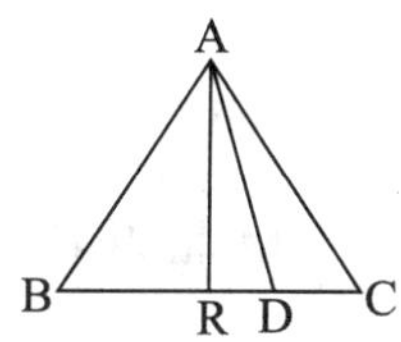

$$AB^2 - AD^2 = BR^2 - RD^2$$

$$AB^2 - AD^2 = (BR + RD)\,(BR - RD)$$

$$= BD\,(BR - RD)$$

$$= BD\,(RC - RD) \quad [\because \quad \mathbf{BR = RC}]$$
$$= \mathbf{BD.\,DC}$$

Example 16. In ΔABC E is the mid point of median AD. BE is joined and Produced to meet AC in F. Prove that

$$\frac{AF}{AC} = \frac{1}{3}$$

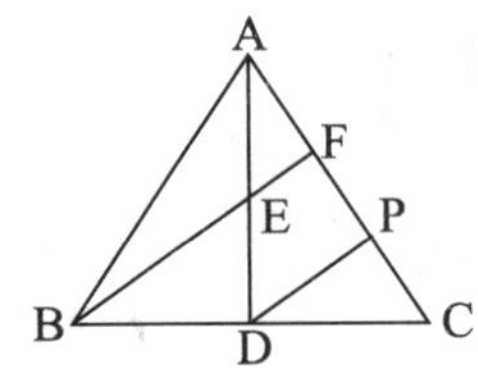

Sol : Const. Draw DP || BF

Proof: In ΔADP EF || DP

$$\therefore \quad \frac{AE}{ED} = \frac{AF}{FP}$$

also $$AE = ED$$

$$\therefore \quad \frac{AF}{FP} = 1$$

$$AF = FP$$

Now In ΔBFC DP || BF

$$\therefore \quad \frac{BD}{DC} = \frac{FP}{PC}$$

But $$BD = DC \qquad \text{(given)}$$

$$\therefore \quad \frac{FP}{PC} = 1$$

or $$FP = PC$$

$$\therefore \quad AF = FP = PC = \frac{1}{3}\mathbf{AC}$$

or $$\frac{\mathbf{AF}}{\mathbf{AC}} = \frac{\mathbf{1}}{\mathbf{3}}$$

Example 17. O is any point in ΔABC OD, OE and OF are bisectors of ∠AOB, ∠BOC and ∠COA. Prove that

$$AD.BE.CF = DB.EC.FA$$

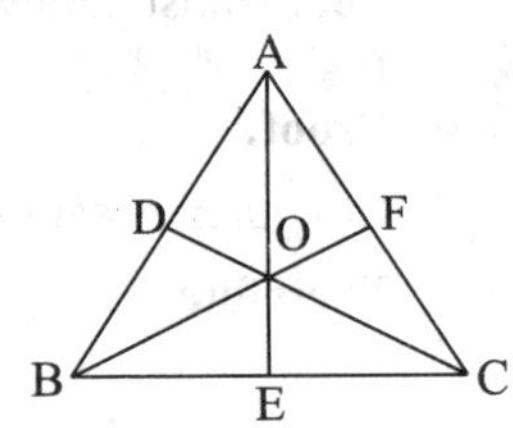

Sol : Because OD is the bisector of ∠AOB

$$\therefore \quad \frac{AO}{OB} = \frac{AD}{DB} \qquad \text{... (i)}$$

Similarly $$\frac{OB}{OC} = \frac{AE}{EC} \qquad \text{... (ii)}$$

and $$\frac{OC}{OA} = \frac{CF}{FA} \qquad \text{... (iii)}$$

Multiply (i), (ii) and (iii)

$$\frac{AO}{OB}\times\frac{OB}{OC}\times\frac{OC}{OA} = \frac{AD}{DB}\times\frac{BE}{EC}\times\frac{FC}{FA}$$

$$1 = \frac{AD}{DB}\times\frac{BE}{EC}\times\frac{FC}{FA}$$

$$\mathbf{AD.BE.CF = DB.EC.FA}$$

Example 18. H is orthocenter of ΔABC. X, Y and Z are mid points of AH, BH and CH then point H is .... of ΔABC.

Sol : In ΔBHC Y and Z are mid points of BH and CH

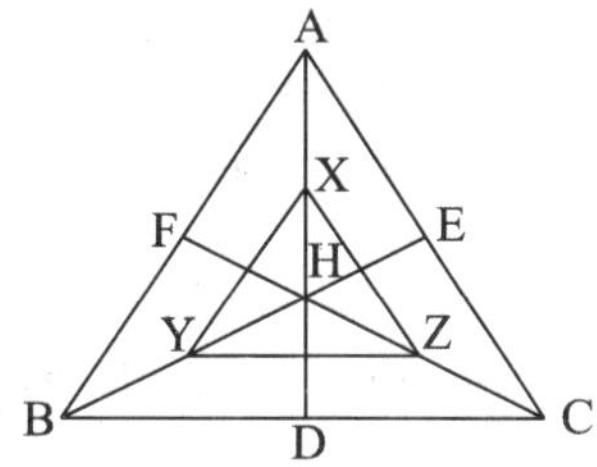

∴ YZ || BC

because AD ⊥ BC and YZ || BC

∴ AD ⊥ YZ

Similarily BE ⊥ XZ and CF ⊥ XY

Now in ΔXYZ Perpendicular on sides meet in H

∴ **H is orthocenter of ΔXYZ**

Now we will do the Exercise

EXERCISE

1. Which angle is two third of its supplement

(a) 72° (b) 108° (c) 35° (d) None

2. Which angle is four fifth of its supplement

(a) 88° (b) 80° (c) 84° (d) None

3. OE and OD are Bisectors of ∠AOC and ∠BOC res. then ∠DOE is equal to

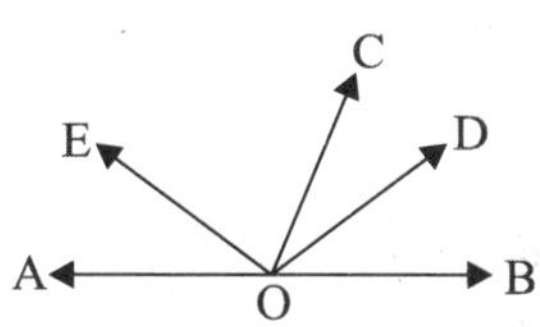

(a) 90° (b) 60° (c) 30° (d) 45°

4. In given figure ∠BAF = 40° ∠CFD = 105° then ∠FCD is

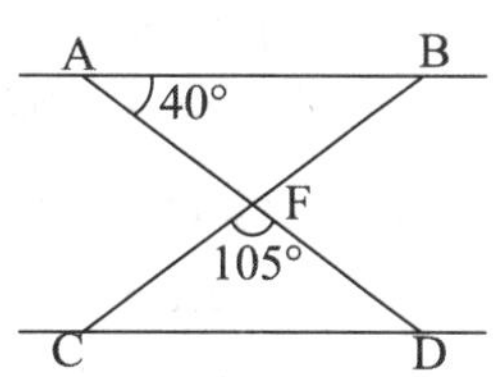

(a) 40°
(b) 25°
(c) 45°
(d) None

5. In given figure ∠3 = 110° and ∠2 = 60° then ∠1 is

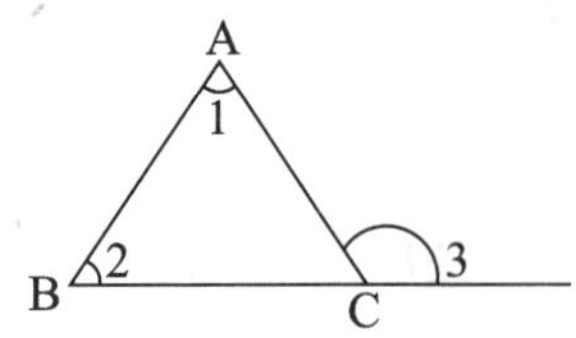

(a) 70°
(b) 50°
(c) 40°
(d) None

6. Ratio of corresponding sides of two similar triangles is 1 : 3. The ratio of their corresponding heights is

(a) 1 : 3 (b) 9 : 1 (c) 1 : 9 (d) None

7. If the angles of a triangle are in the ratio 3 : 4 : 5 then the smaller angle is

(a) 75° (b) 60° (c) 45° (d) None

8. The internal bisector of ∠B and ∠C meet in I. Then ∠BIC is

(a) $90° - \frac{1}{2}\angle A$ (b) $90° + \frac{1}{2}\angle A$ (c) 180° – A (d) None

9. The ratio of side of an equilateral triangle to its height is

(a) 1 : 1 (b) 1 : 2 (c) $2 : \sqrt{3}$ (d) None

10. The internal bisector of ∠B and ∠C meet in I. If ∠A is 70° Then ∠BIC is

(a) 125° (b) 135° (c) 65° (d) None

11. If D, E and F are mid points of sides BC, CA and AB res. & area of triangle ABC is 16 cm^2, then area of ΔDEF is

(a) 4 cm^2 (b) 8 cm^2 (c) 3 cm^2 (d) None

12. The bisector of any two adjacent sides of a ||gm intersect at

(a) 30° (b) 60° (c) 90° (d) None

13. ABC is equilateral triangle and AD is perpendicular on BC then AD^2 is equal to

(a) $7BD^2$ (b) $4BD^2$ (c) $3BD^2$ (d) None

14. ABC is a triangle and AD is its median. Then $AB^2 + BC^2$ is equal to

(a) $AD^2 + BD^2$ (b) $3(AD^2 + BD^2)$ (c) $2(AD^2 + BD^2)$ (d) None

15. In a parallelogram ABCD, X is the mid-point of AD and Y is the mid-point of BC. XY divides BD in the ratio

(a) 1 : 3 (b) 1 : 2 (c) 2 : 1 (d) None

16. Two isosceles triangles have equal vertical angles and their area is in the ratio 9 : 16, their heights are in the ratio

(a) 16 : 9 (b) 9 : 16 (c) 3 : 4 (d) None

17. The length of the two sides of a triangle are l and m res., the third side of the triangle is

(a) $l - m$ (b) $l + m$
(c) more than $l - m$ and less then $l + m$ (d) None

18. If the ratio of the base of isosceles triangle to corresponding altitude is $2 : \sqrt{3}$ the triangle is

(a) acute (b) equi angular (c) right angle
(d) obtused angle

19. ABC is an equi angular triangle in which BC is produced to K such that CK = $\frac{1}{2}$ BC then

(a) $AK^2 = 5\ CK^2$ (b) $AK^2 = 7\ CK^2$ (c) $AK^2 = 6\ CK^2$ (d) none

20. In an equi angular triangle ABC the side BC is trisected at D then

(a) $AD^2 = 5\ AB^2$ (b) $9\ AD^2 = 5\ AB^2$ (c) $9\ AD^2 = 7\ AB^2$ (d) none

21. In a triangle ABC a line XY is drawn parallel to BC meeting AB in X and AC in Y. The area of the triangle AXY is half the area of triangle ABC. X divides AB in the ratio

(a) $1 : \sqrt{2}$ (b) $\sqrt{2} : \sqrt{2} - 1$ (c) $1 : \sqrt{2} - 1$ (d) none

22. In triangle ABC, ∠B = 90°, AB = 20 cm, BC = 21 cm. What is the length of median BM?

(a) 14.5 cm (b) 15 cm (c) 15.5 cm (d) none

23. In triangle ABC, ∠A = 90°, D is the mid-point of AC then $BC^2 - BD^2$ is equal to

(a) AD^2 (b) $2AD^2$ (c) $3AD^2$ (d) none

24. In equilateral triangle ABC, D is mid-point of BC and E is mid-point of DC then $\frac{AE^2}{CE^2}$ is equal to

(a) 7 (b) 9 (c) 13 (d) none

25. In a right angle triangle square of the hypotenuse is equal to twice the product of other two sides. One of the following angle in the case is

(a) 60° (b) 15° (c) 45° (d) none

SOLUTIONS

1. (a)

$$\text{Let one angle} = x$$

$$\therefore \quad \text{Another angle} = \frac{2}{3}x$$

According to statement

$$x + \frac{2}{3}x = 180°$$

$$\frac{5}{3}x = 180°$$

$$x = 180° \times \frac{3}{5} = 180°$$

$$\therefore \quad \text{Reqd. Angle} = 108 \times \frac{2}{3} = \mathbf{72°}$$

2. (b) same as above

3. (a)

$$\angle AOC + \angle BOC = 180°$$

$$\frac{1}{2}(\angle AOC + \angle BOC) = \frac{1}{2} \times 180°$$

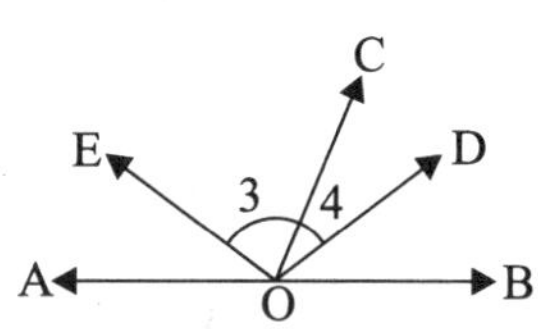

or $\frac{1}{2}\angle AOC + \angle BOC = 90°$

or $\angle 3 + \angle 4 = 90°$

or $\angle DOE = \mathbf{90°}$

4. (d) $\angle BAF = \angle FDC$

$\therefore$ $\angle FDC = 40°$

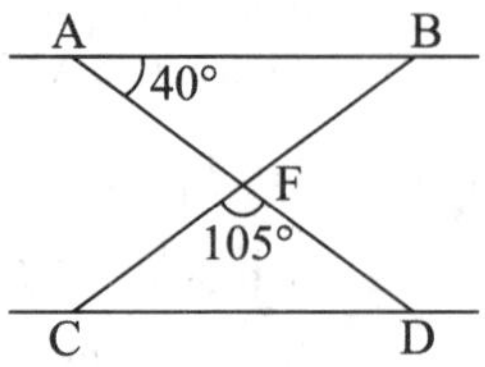

In triangle FCD

$$\angle FCD + \angle CFD + \angle FDC = 180°$$

$$\angle FCD + 40° + 105° = 180°$$

$$\angle FCD = 180 - 40° - 105° = \mathbf{35°}$$

5. (b)

$$\angle 3 = \angle 1 + \angle 2$$

$$110° = \angle 1 + 60°$$

$$\angle 1 = 110° - 60° = \mathbf{50°}$$

6. (a) $\because$ in similar triangles ratio of corresponding sides is equal to ratio of corresponding heights.

7. (c)

$$3x + 4x + 5x = 180°$$

$$12x = 180°$$

$$x = 15°$$

$\therefore$ Smallest angle $= 15° \times 3 = \mathbf{45°}$

8. (b) From Basics

9. (c) Equilateral ΔABC in which AD ⊥ BC

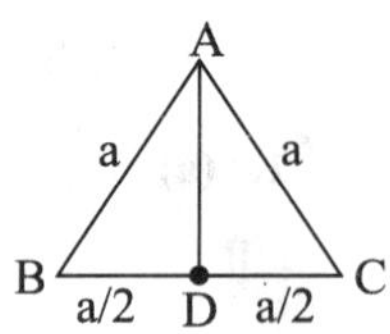

Let each side $= a$

and in equilator Δ ⊥ ar bisects the base.

$\therefore$ $BD = DC = a/2$

Now in right angle ΔABD

$$AB^2 = AD^2 + BD^2$$

$$a^2 = AD^2 + \left(\frac{a}{2}\right)^2$$

$$AD^2 = a^2 - \frac{a^2}{4} = \frac{3a^2}{4}$$

or $AD = \frac{\sqrt{3}}{2}a$ and side $AB = a$

$\therefore$ $\frac{AB}{AD} = \frac{a}{\frac{\sqrt{3}}{2}a} = \frac{2}{\sqrt{3}} = \mathbf{2:\sqrt{3}}$

10. (a) From Basics $\angle BIC = 90° + \frac{1}{2}\angle A$

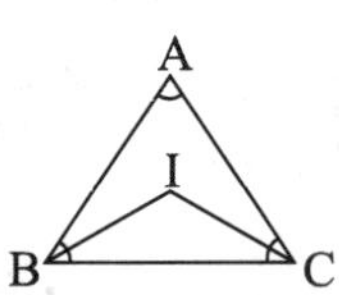

$$\angle BIC = 90° + \frac{1}{2} \times 70°$$

$$\angle BIC = 90° + 35°$$

$$\angle BIC = \mathbf{125°}$$

11. (a) Area of Triangle DEF is 1/4th of area of ΔABC

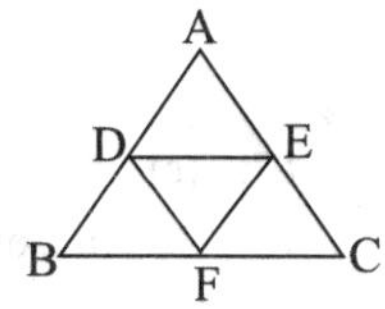

$\therefore$ Area of $\Delta DEF = \frac{1}{4} \times 16 = \mathbf{4\ cm^2}$

12. (c) $\angle ABD + \angle BAC = 180°$

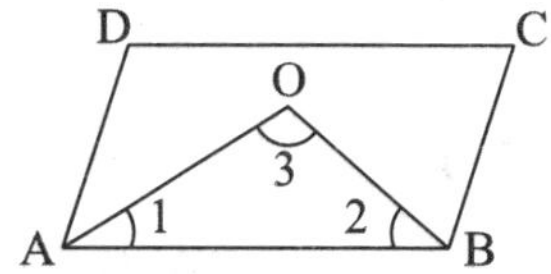

$$\frac{1}{2}\angle ABD + \frac{1}{2}\angle BAC = 90°$$

$$\angle 1 + \angle 2 = 90°$$

But $\angle 1 + \angle 2 + \angle 3 = 180°$

or $\angle 3 = 180° - 90° = 90°$

$\therefore$ **Bisector of two angles of a parallelogram meet at 90°**

13. (c)

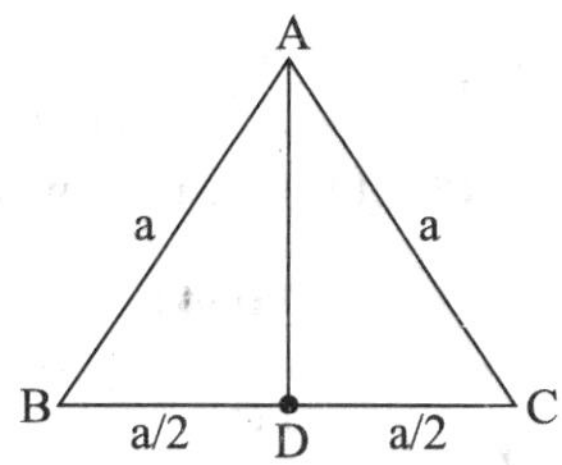

$$AD^2 = AB^2 - BD^2$$

$$AD^2 = a^2 - \frac{a^2}{4} = \frac{3a^2}{4}$$

$$BD^2 = \frac{a^2}{4}$$

$\therefore$ $\frac{AD^2}{BD^2} = 3$

$\therefore$ $\mathbf{AD^2 = 3BD^2}$

14. (c) ΔABC in which AD is the median **Const.** Draw AE ⊥ BC

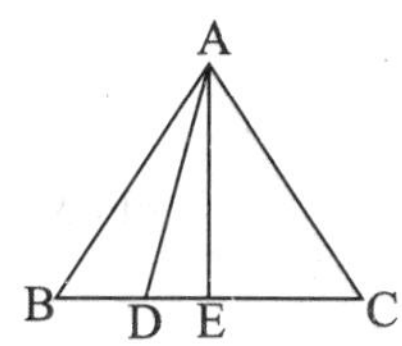

Proof: In ∠ADE $AD^2 = AE^2 + DE^2$ ... (i)

In ∠ABE $AB^2 = AE^2 + BE^2$

$AB^2 = AE^2 + (BD + DE)^2$

$AB^2 = AE^2 + BD^2 + DE^2 + 2BD.\ DE$ ... (ii)

From (i) $AD^2 = AE^2 + DE^2$

$\therefore$ $AB^2 = AD^2 + BD^2 + 2BD.\ DE$ ... (iii)

Similarly in ΔADC $AC^2 = AD^2 + DC^2 - 2DC.\ DE$

or $AC^2 = AD^2 + DC^2 - 2BD.\ DE$ ($\because$ BD = DC) ... (iv)

Add (iii) and (iv) $AB^2 + AC^2 = 2AD^2 + 2DC^2$

$= \mathbf{2\,(AD^2 + DC^2)}$

15. (d) None (1 : 1)

16. (c) They have equal vertical angle $\therefore$ both are similar.

$$\frac{\text{Area of } \Delta ABC}{\text{Area of } \Delta DEF} = \frac{AL^2}{DM^2}$$ (Where AL and DM are heights)

or $\quad \dfrac{9}{16} = \dfrac{AL^2}{DM^2}$

$$\frac{AL}{DM} = \frac{3}{4}$$

17. (c)

18. (b) In isosceles ΔABC $AL \perp BC$

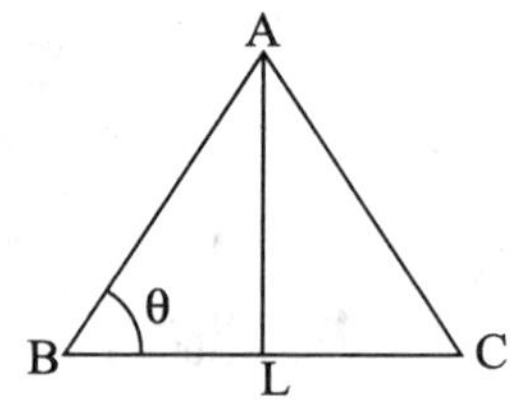

and $\quad BC : AL = 2 : \sqrt{3}$

In isosceles Δ $\perp$ ar bisect the base

$\therefore \quad BL : AL = 1 : \sqrt{3}$

In ΔABL $\quad \dfrac{BL}{AL} = \cot\theta$

$$\frac{1}{\sqrt{3}} = \cot\theta$$

$\therefore \quad \theta = 60°$ means $\angle B = 60°$ and $\angle C = 60°$

$\therefore$ **Δ is equi-angular**

19. (b) **Const.** Draw $AD \perp BC$

Proof : $\quad CK = \dfrac{1}{2} BC$ means $BD = DC = CK$

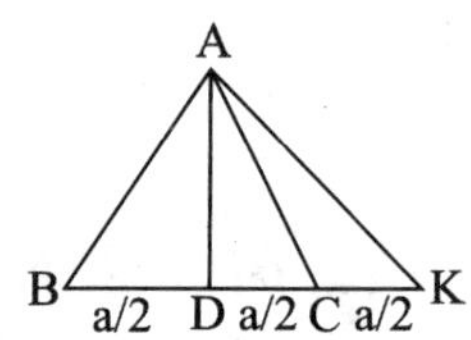

Let $\quad AB = a$

$\therefore \quad BD = DC = CK = \dfrac{a}{2}$

Now $\quad AD^2 = AB^2 - BD^2 = a^2 - \dfrac{a^2}{4} = \dfrac{3a^2}{4}$

In ΔADK $\quad AK^2 = AD^2 + DK^2$

$$AK^2 = AD^2 + (DC + CK)^2$$

$$AK^2 = AD^2 + DC^2 + CK^2 + 2DC.\,CK$$

$$AK^2 = AD^2 + DC^2 + CK^2 + 2CK.\,CK \quad (\because \; DC = CK)$$

$$= AD^2 + DC^2 + 3CK^2$$

$$= \frac{3a^2}{4} + CK^2 + 3CK^2$$

$$= 3\left(\frac{a}{2}\right)^2 + 4\,CK^2$$

$$= 3\,CK^2 + 4\,CK^2 = \mathbf{7\,CK^2}$$

20. (c) **Const.** Draw $AL \perp BC$

Proof : $\quad$ Let $AB = a$

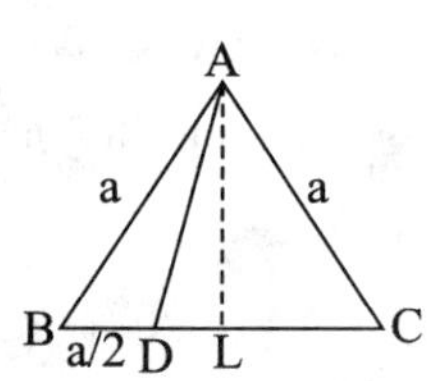

$$BD = \frac{a}{3}\; CL = BL\; \frac{a}{2} =$$

$$\therefore \quad DL = BL - BD = \frac{a}{2} - \frac{a}{3} = \frac{\boldsymbol{a}}{\mathbf{6}}$$

In ΔABL
$$AL^2 = AB^2 - BL^2$$
$$= a^2 - \frac{a^2}{4} = \frac{3a^2}{4}$$

In ΔADL
$$AD2 = AL^2 + DL^2$$
$$= \frac{3a^2}{4} + \frac{a^2}{36} = \frac{27a^2 + a^2}{36} = \frac{28a^2}{36} = \frac{7a^2}{9}$$
$$9AD^2 = 7a^2$$
$$\mathbf{9AD^2 = 7AB^2}$$

21. (c)
$$\frac{\text{Area of } \Delta ABC}{\text{Area of } \Delta AXY} = \frac{AB^2}{AX^2}$$

or
$$\frac{AB^2}{AX^2} = \frac{2}{1}$$
$$\frac{AB}{AX} = \frac{\sqrt{2}}{1}$$
$$\frac{AB}{AX} - 1 = \frac{\sqrt{2}}{1} - 1$$
$$\frac{AB - AX}{AX} = \sqrt{2} - 1$$
$$\frac{BX}{AX} = \sqrt{2} - 1$$
$$\mathbf{\frac{AX}{BX} = \frac{1}{\sqrt{2} - 1}}$$

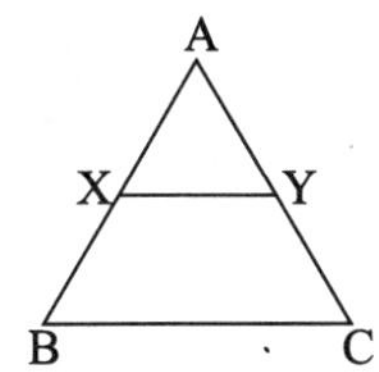

22. (a) In right angle triangle median is half the hypotenuse and hypotenuse $\sqrt{20^2 + 21^2}$ = 29.

$$\therefore \quad \text{Length of Median} = \frac{29}{2} = \mathbf{14.5\ cm}$$

23. (c)
$$BC^2 = AB^2 + AC^2$$
$$= AB^2 + (2AD)^2$$
$$= AB^2 + 4AD^2$$
$$= \underline{AB^2 + AD^2} + 3AD^2$$
$$= BD^2 + 3\,AD^2$$
$$\mathbf{BC^2 - BD^2 = 3AD^2}$$

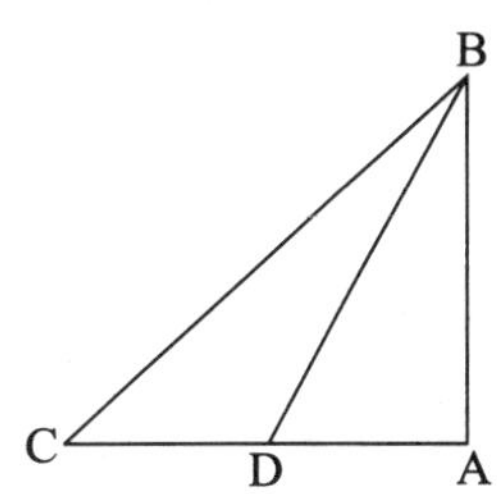

24. (c)
$$AD^2 = a^2 - \frac{a^2}{4} = \frac{3a^2}{4}$$

$$DE^2 = \left(\frac{a}{4}\right)^2 = \frac{a^2}{16}$$

$$AE^2 = AD^2 + DE^2$$

$$= \frac{3a^2}{4} + \frac{a^2}{16} = \frac{13a^2}{16}$$

$$= 13\left(\frac{a}{4}\right)^2 = 13\ CE^2$$

$$\mathbf{\frac{AE^2}{CE^2} = 13}$$

25. (c) $AB^2 + AC^2 = BC^2$

also $BC^2 = 2AB.\ AC$

$\therefore$ $AB^2 + AC^2 = 2AB.\ AC$

$\therefore$ $AB^2 + AC^2 - 2AB.\ AC = 0$

or $(AB - AC)^2 = 0$

or $AB - AC = 0$

$\therefore$ $AB = AC$

$\therefore$ Each of two angles is **45°**

Day 20

Geometry–2

Today we will do **Polygon, Circle & Locus**. First we will discuss **polygon**.

POLYGON

Polygon : A closed geometrical figure made by three or more line segments is called Polygon. Smallest of these is called Triangle and next one is called quadrilateral. These two we have already discussed on Day 19.

So today we will do the Polygon of more than Four Sides named as follows.

5 sided polygon which is called **Pentagon**.
6 sided polygon which is called **Hexagon**.
7 sided polygon which is called **Heptagon**.
8 sided polygon which is called **Octagon**.
9 sided polygon which is called **Nonagon**.
10 sided polygon which is called **Decagon**.

Polygon are of two types

1. **Concave**
2. **Convex**

We will discuss each of them one by one.

1. Concave Polygon :

A Polygon in which atleast one angle is more than 180° is called Concave Polygon as shown in figure.

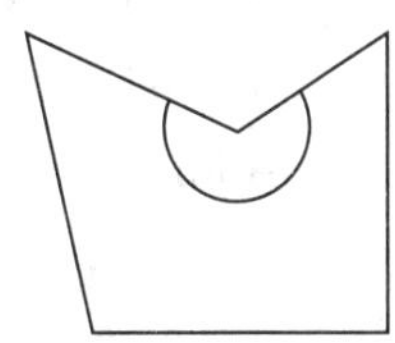

Further in concave polygon a straight line can cut the polygon in more than two points as shown in figure.

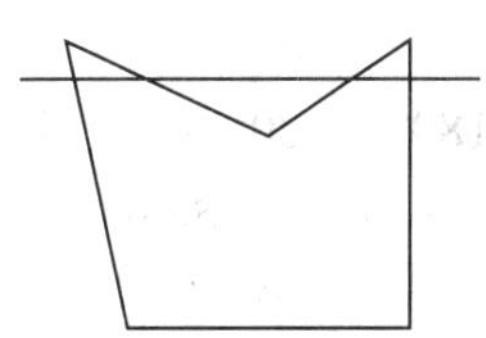

2. Convex Polygon :

A Polygon in which none of its angle is more than 180° is called Convex Polygon as shown in figure.

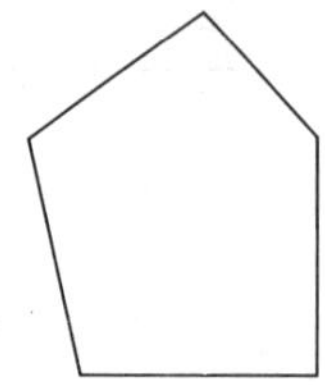

Further in convex polygon a straight line can cut the polygon at the most two points as shown in figure.

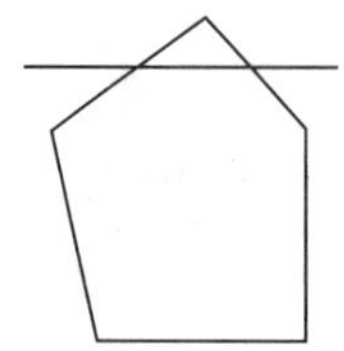

Further we will dealing with **Convex Polygon** only and as such simply a **Polygon** means a **Convex Polygon**.

Regular Polygon : A polygon is called regular polygon if all of its sides and all of the angles are equal as shown in figure.

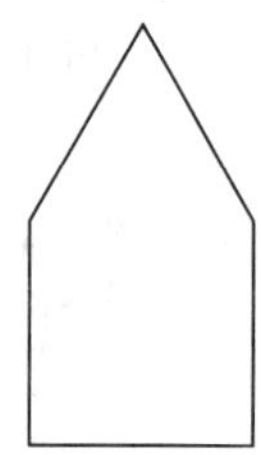

Interior Angle of Regular Polygon : Now we will find the interior angle of small n sided regular polygon.

In figure a pentagon ABCDE is shown. Join AC and AD as shown in figure.

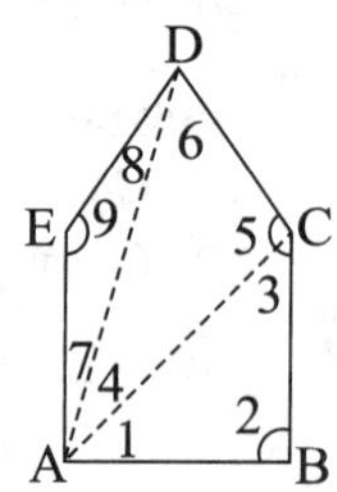

In Δ ABC $\quad \angle 1 + \angle 2 + \angle 3 = 180°$ ...(i)

In Δ ACD $\quad \angle 4 + \angle 5 + \angle 6 = 180°$ ... (ii)

In Δ ADE $\quad \angle 7 + \angle 8 + \angle 9 = 180°$ ...(iii)

Adding (i), (ii) and (iii) we get

$$\angle 1 + \angle 2 + \angle 3 + \angle 4 + \angle 5 + \angle 6 + \angle 7 + \angle 8 + \angle 9 = 180° + 180° + 180°$$

It can be written as

$$\underline{\angle 1 + \angle 4 + \angle 7} + \angle 2 + \underline{\angle 3 + \angle 5} + \underline{\angle 6 + \angle 8} + \angle 9 = 3 \times 180°$$

or $\quad \angle A + \angle B + \angle C + \angle D + \angle E = 3 \times 180°$

So we have seen that sum of the angles of five sided polygon *i.e.* of Pantagon is 180° + 180° + 180° = 540° i.e., sum of the angles of three triangles.

So in 5 sided polygon 5 – 2 = 3 triangles are formed and sum of its angles is equal to sum of the angles of **3 triangles**.

So we have seen that

in **5** sided polygon	*i.e.* **Pantagon**	**5 – 2** = **3** △'s are formed.
Similarly		
in **6** sided polygon	*i.e.* **Hexagon**	**6 – 2** = **4** Δ's are formed.
in **7** sided polygon	*i.e.* **Heptagon**	**7 – 2** = **5** Δ's are formed.
in **8** sided polygon	*i.e.* **Octagon**	**8 – 2** = **6** Δ's are formed.
in **9** sided polygon	*i.e.* **Nonagon**	**9 – 2** = **7** Δ's are formed.
in ***n*** sided polygon		***n* – 2** Δ's are formed.

So sum of angles of ***n*** sided polygon is **$(n-2)\times 180°$**

So if the polygon is regular then each angle will be equal and is $\dfrac{(n-2)\times 180°}{n}$

it can be written as $\dfrac{(n-2)\times 2\times 90}{n} = \dfrac{(2n-4)\times 90°}{n}$.

∴ **Each angle of n sided regular polygon is $\dfrac{(2n-4)\times 90°}{n}$.**

This is very important formula. It can be used to find each angle of the polygon when its sides are given or can be used to find the number of sides when each angle of regular polygon is given. This is further explained in the following examples.

Example 1. Find the each angle of a regular hexagon.

Sol : Each angle of regular polygon $= \dfrac{(2n-4)\times 90°}{n}$.

Since hexagon has 6 sides

∴ $n = 6$

Putting this value in above formula we get

$$\frac{(2\times 6-4)\times 90}{6}. = \frac{8\times 90°}{6} = \mathbf{120°}$$

Example 2. If each angle of a regular polygon is 108° find the number of sides of a polygon.

Sol : Each angle of regular polygon $= \dfrac{(2n-4)\times 90°}{n}$.

$$108° = \frac{(2n-4)\times 90°}{n}.$$

$$108\,n = 180n - 360$$

$$72\,n = 360$$

$$n = \mathbf{5}$$

Therefore it is **Pantagon**.

Further we will discuss a few more points regarding polygon.

1. Sum of exterior angle of a polygon (of any number of sides) is 360° and for regular polygon each exterior angle is $\frac{360°}{n}$.

   ∴ Each exterior angle is of a regular polygon $\frac{360°}{n}$.

2. If the alternate sides of a polygon are produced they form a star shaped figure.
   Adj. figure shows the pantagon and a hexagon where alternate sides are produced to form in star shape figure.

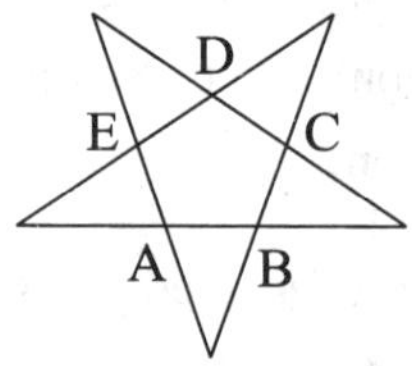

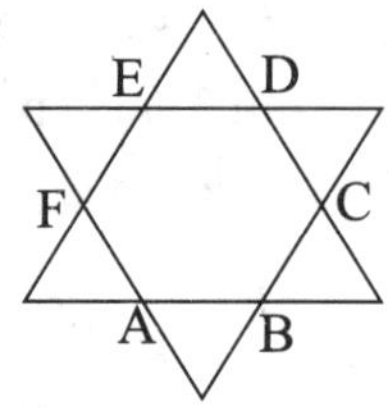

3. First we will find the sum of angles of vertices of star shape pentagon then we will find its general formula.
   In adj. figure a pentagon is shown in which alternate sides are produced to form a star shape figure.
   In figure

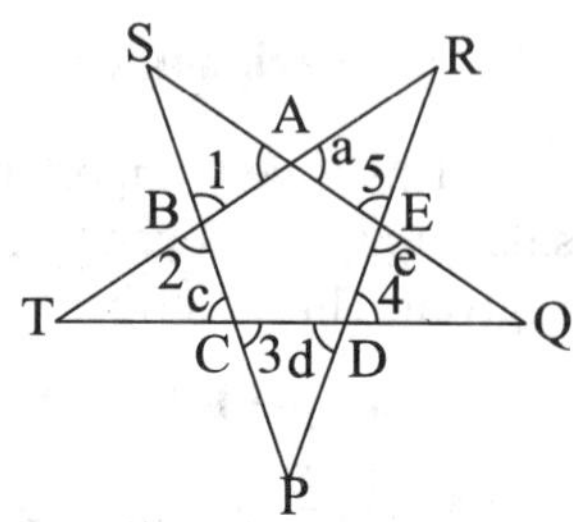

$$\angle 3 = \angle S + \angle Q \quad ...(i)$$

(Exterior ∠ of CSQ)

$$\angle d = \angle R + \angle T \quad ...(ii)$$

(Exterior ∠ of TDR)

$$\angle P = \angle P \quad ...(iii)$$

Adding (i), (ii), (i) we get

$$\angle 3 + \angle d + \angle P = \angle P + \angle Q + \angle R + \angle S + \angle T$$

or

$$180° = \angle P + \angle Q + \angle R + \angle S + \angle T$$

∴ Sum of vertices of a star shaped pengagon is 180°.

The general formula for **sum of angles of vertices of n sided star shaped regular polygon. (when its alternate sides are produced) is $(2n - 8) \times 90°$.**

So remember

1. **Each angle of a regular n sided polygon** $= \frac{(2n-4)\times 90°}{n}$.
2. **Sum of angles of vertices of n sided star shaped regular polygon. (when its alternate sides are produced) is $(2n - 8) \times 90°$.**

CIRCLE

You have seen many figures in daily life which are circular like wheel, plates etc. also you have done in the chapter of mensuration where two basic formulae of circle

1. **Circumference of Circle = $2\pi r$**
2. Area of Circle $= \pi r^2$

were discussed. Now here we will do some more properties of circle.

Circle : If a point moves in a plane such that its distance from a fix point remain constant than the path traced by it is called **circle** as shown in figure. The fixed point is called **center** and the distance from it is called **radius**. In figure O is center, $OA = r$ is called its radius.

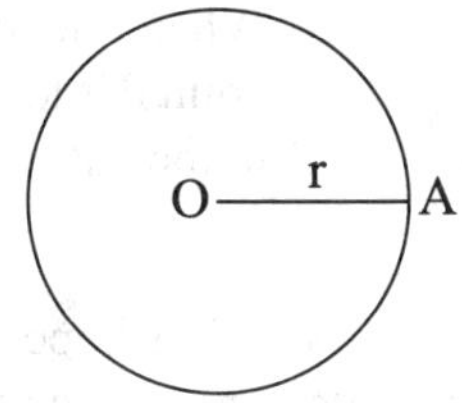

Now we will discuss some basic terms which are used in this topic like arc, chord, minor segment, major segment, tangent etc.

1. **Circumference :**

   Boundry of Circle is called **circumference.**

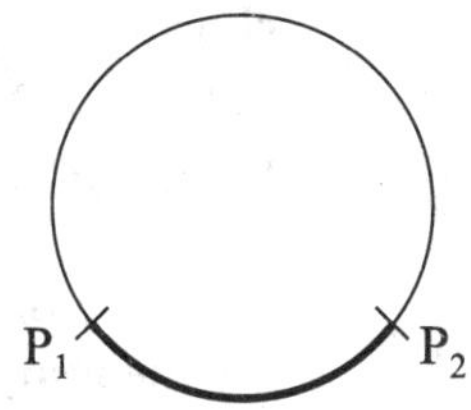

2. **Arc :**

   Any part of the circumfrance is called **arc** in figure P_1 P_2 is the **arc**.

3. **Chord :**

   A line joining the two points of an arc is called **chord.**

   Remember chord is not a part of circle. A chord passing through the centre of a circle is called diameter. Diameter is the longest chord in any circle.

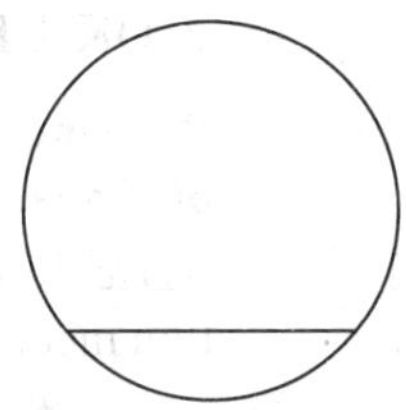

4. **Segment :**

   In figure PQ is a chord. It divides a circle into two parts as shown in figure. Each part is called **segment**. The bigger part is called **major segment** and the smaller one is called **minor segment as shown in Fig.**

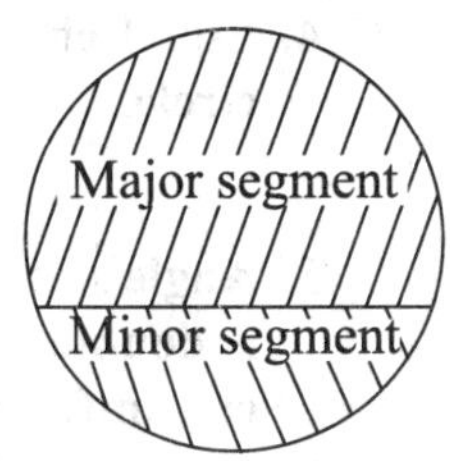

Now we will define secant and tangent.

Before this consider a line and a circle. Following three cases arises

1. Line does not intersect the circle at all.

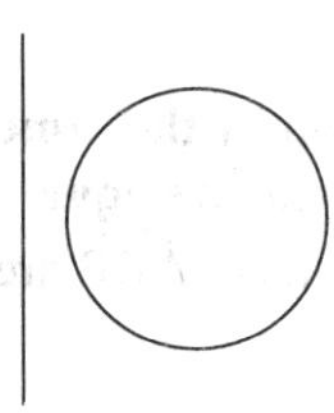

2. Line intersect the circle in two different points P and Q. This line is called **secant.**

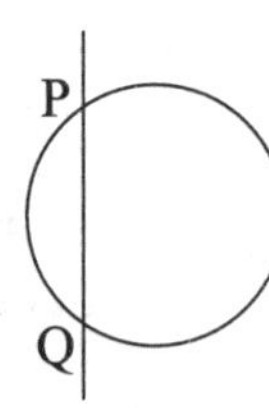

3. When line touches the circle at a point say P as shown in figure. The point P is known as the **point of contact** and the line is called **tangent** to the circle.

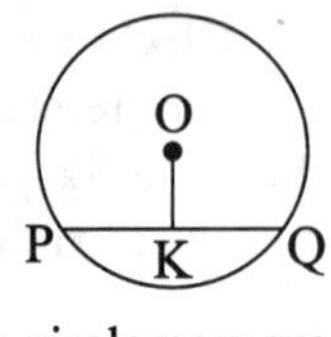

Now it is hoped that you have understood the basic terminology related to circle now we will do some basic properties of circle.

1. **Perpendicular from the centre of a circle to a chord bisects the chord.**
   In figure PQ is a chord and O is the centre of circle.
   From O, OK ⊥ PQ. Hence **PK = KQ.**

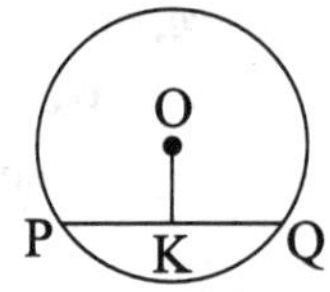

2. Converse of the above is also true which means that **the line joining the mid-point of a chord to the center is perpendicular to the chord.**
   PQ is the chord of which K is the mid-point of PQ. OK is joined
   ∴ OK ⊥ PQ.

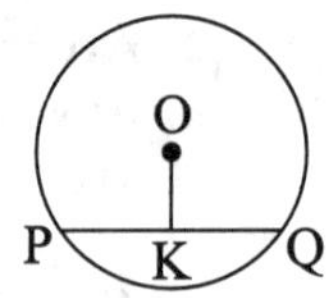

3. **Perpendicular bisectors of two chords intersect at the centre of circle.**
   AB and BC are two chords of the circle. Their perpendicular bisectors meet at O.
   ∴ O is the centre of the circle.
4. **Equal chords of circle are equi distance from the centre of circle.**

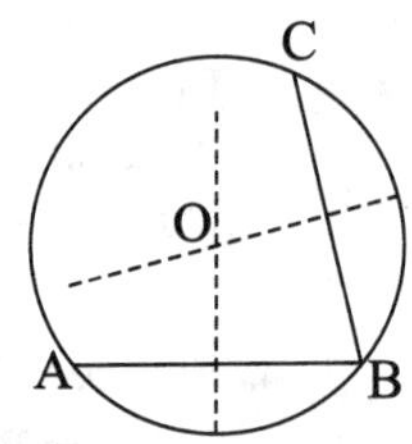

5. **Angle subtanded by the arc or chord of a circle at the center is double the angle subtanded by the arc at any point on its circumference.**
   In figure ∠ **AOB = 2 ∠ ACB.**

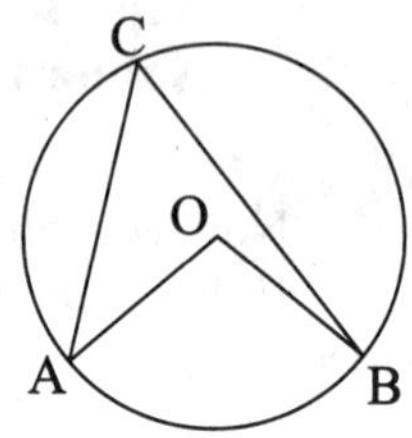

6. **Angles substanded by the same arc or chord of a circle on the same side of a circle are equal.**
   In figure ∠ ACB and ∠ ADB are equal.

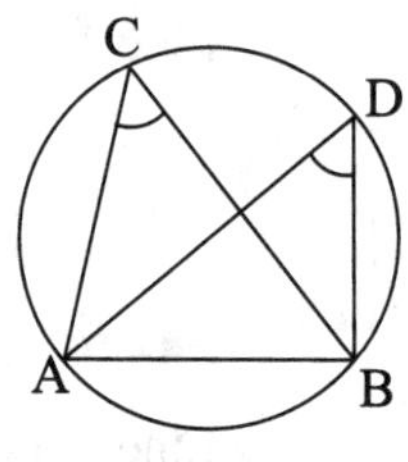

7. **An angle in the semi-circle is a right angle or 90°.**
   In figure ∠ ACB = 90°

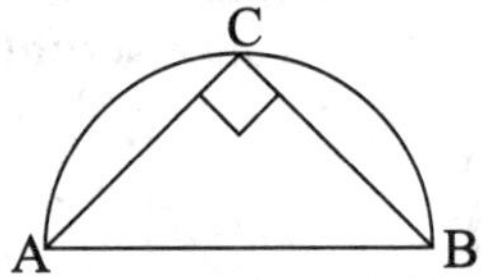

8. **Any angle subtanded by in minor arc in the alternate segment is acute and any angle subtanded by the major arc in the alternate segment is obtuse.**

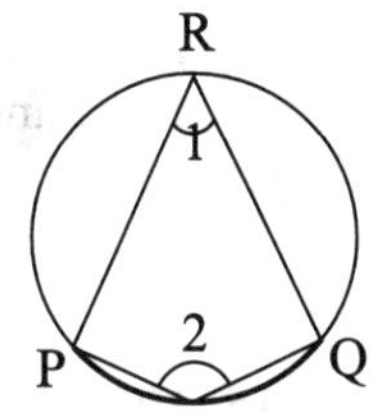

In figure PQ is minor means smaller arc.

Therefore ∠ 1 which is made in the alternate segment (means another segment) is less than 90° i.e., acute. QP is the major arc. ∴ ∠ 2 is obtuse.

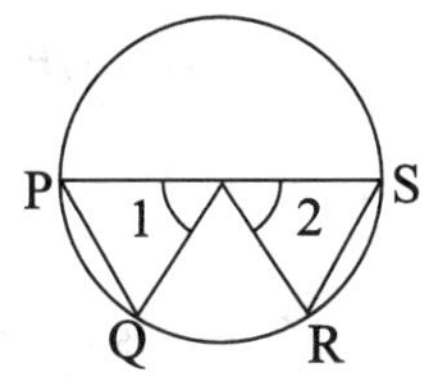

9. **Equals chords of a circle subtandes equal angles at the center.**

PQ and SR are equal chords of the circle with center O.

∴ ∠ 1 = ∠ 2

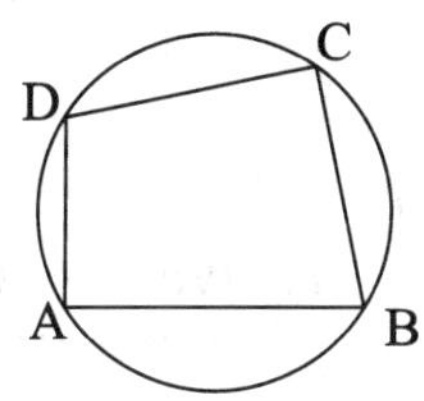

10. **A quadrilateral is called cyclic if its all the four vertices lies on circle.**

In figure ABCD is a cyclic quadrilateral.

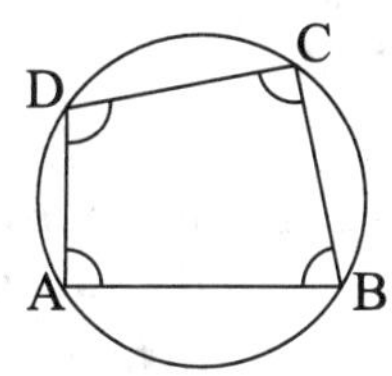

11. **The sum of pairs of opposite angles of a cyclic quardi-lateral is supplemenry.**

In figure

∠A + ∠C = ∠B + ∠D = 180°

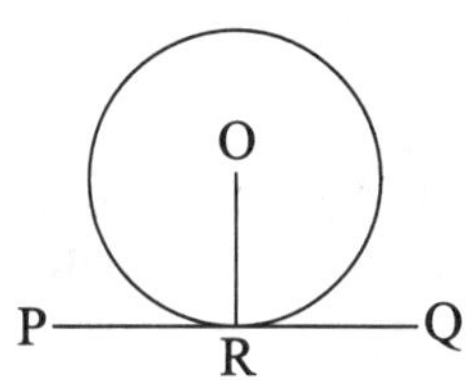

12. **A tangent to a circle is perpendicular to the radius through the point of contact.**

In figure PQ is the tangent and R is the point of contact.

∴ **OR ⊥ PQ**

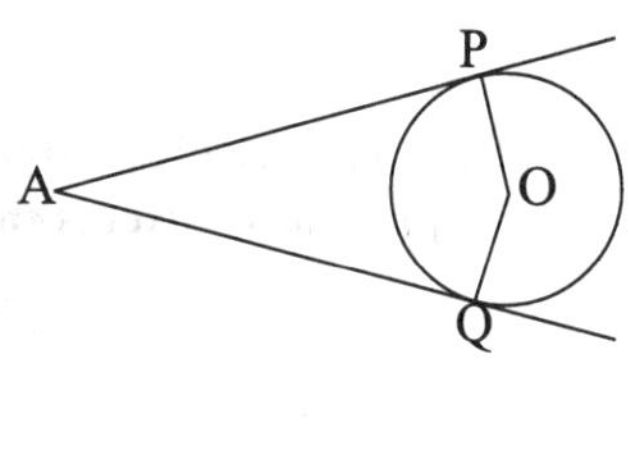

13. **Length of the tangents drawn from an external point to a circle are equal.**

In figure AP and AQ are two tangents to a circle from external point A.

∴ **AP = AQ**

14. **Angles made by a chord of a circle with the tangent and that made by chord in the alternate segment are equal.**

In figure ∠ 1 is made between tangent and chord and ∠ 2 is the angle made by a chord in the alternate segments.

∴ ∠ 1 = ∠ 2

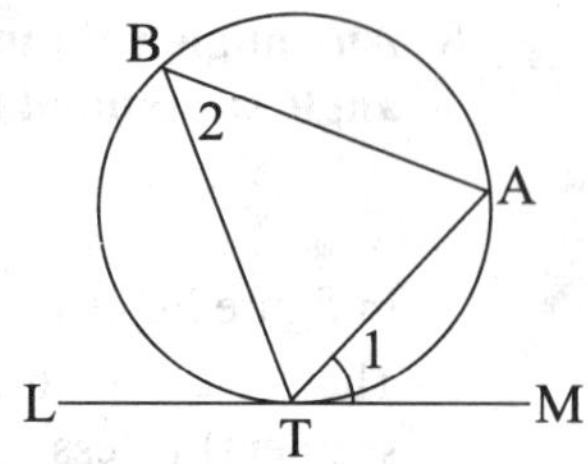

15. **If two chords AB and CD of a circle intersect internally at O then**

AO × OB = CO × OD

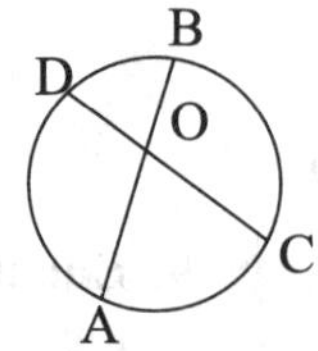

16. **OT is a tangent and OAB is a secant then**

$\mathbf{OT^2 = OA \times OB}$

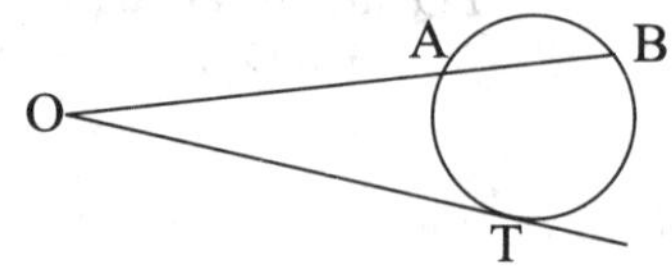

17. **If the distance between the centers of two circles is equal to the difference of their radii then circles are touching internally.**

In figure AB = OA – OB and circles are touching internally.

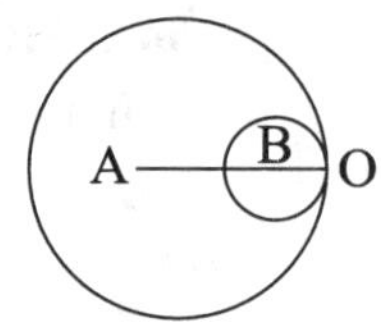

18. **If the distance between the center of the circles is equal to the sum of their radii then circles are touching externally.**

In figure AB = OA + OB and circles are touching externally.

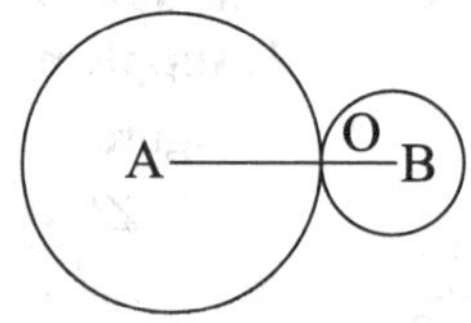

19. **The Incenter of triangle ABC touches the side AB, BC and CA in points P, Q and R res. then**

AP + BQ + CR = PB + QC + RA = ½ perimeter of Δ ABC

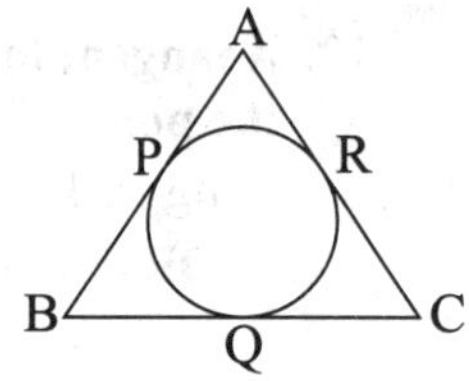

20. **If two circles intersects then their common chords when produced bisects the common tangent.**

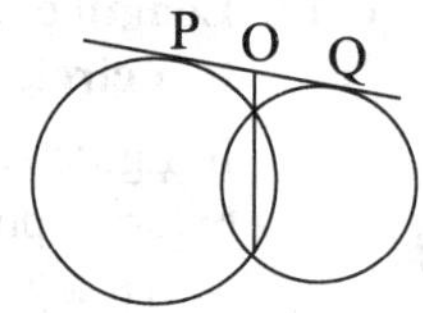

21. **Four common tangents can be drawn to the two non-intersecting circles as shown in figure.**

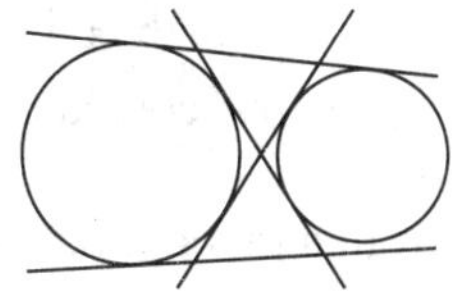

22. **Three common tangents can be drawn to the two externally touching circles as shown in figure.**

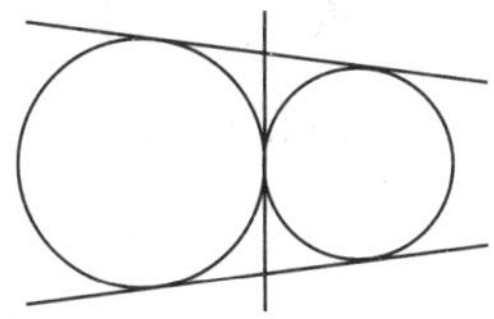

23. **Two common tangents can be drawn to the two intersecting circles as shown in figure.**

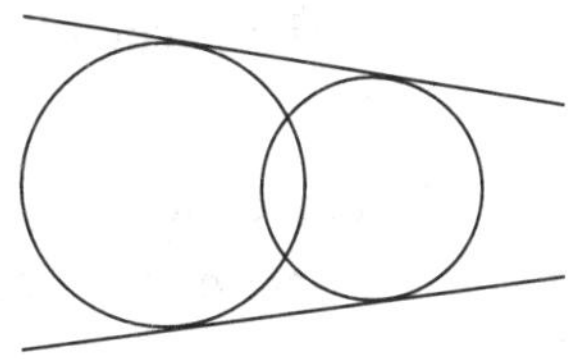

24. **Only one common tangent can be drawn to the two internally touching circles as shown in figure.**

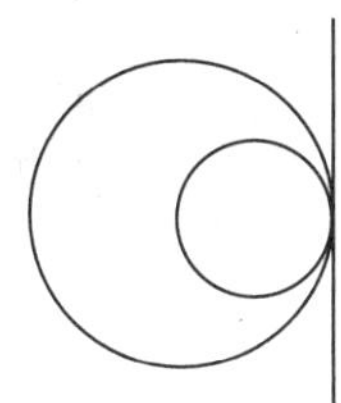

These were the basics about circle.

Now we will discuss the next topic which is **locus**.

LOCUS

Locus : Locus is the path alongwhich a point moves under certain geometrical conditions.

Let us understand it with a daily life simple example.

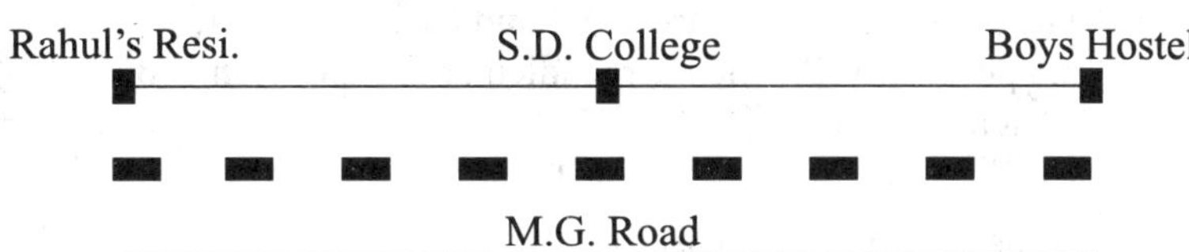

Above **M.G. Road** is shown on which **Rahul's House**, His **College** and **Boys Hostel** are located. Rahul starts from his house at 9 am attends college and after attending college leaves his friend on his bike to hostel and following the same path returned to his house i.e., the M.G. Road at 4 pm. This is his daily routine come what may.

Now if we ask about the movements of Rahul between 9 am to 4 pm then instead of explaining the whole above story we can reply in a single word that is "M.G. Road". So Locus of Rahul i.e., path followed or traced by Rahul in the prescribed time is **M.G. Road**.

Similarly in **Geometry** when a point moves according to some given **geometrical condition** then the path traced by moving point is called **Locus**.

The path can be curved, circular or straight line.

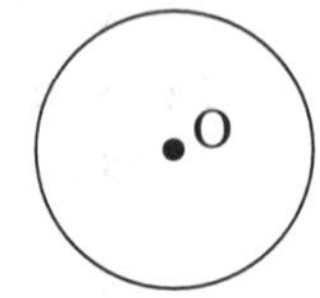

Now consider a circle. In this any point on the circumference is equi distance from the centre O. So if a point moves in such a way that it's distance from a given fixed point is always constant then Locus of point is **circle**.

Now we will do a few basic theorem of Locus.

1. If we are to find the Locus of point which is equi distance from given two fixed points. It will be shown as dotted line in the figure. This dotted line is the perpendicular bisector of AB. **So Locus of point equi distance from two given points is the perpendicular bisector of the line joining these two points.**

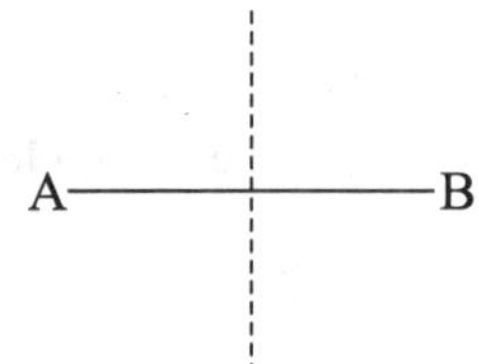

2. **The Locus of a point equi distance from two intersecting line is the pair of lines bisecting the angles formed by the given lines.**

Explanation :

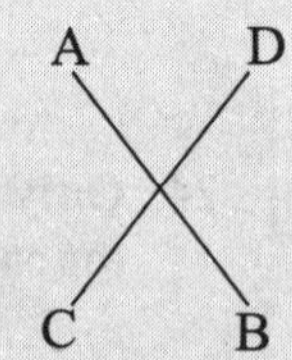

Given two intersecting lines AB and CD intersecting at O as shown in figure 1. These line form four angles i.e.,

$\angle AOC$ $\angle BOC$ $\angle BOD$ and $\angle AOD$

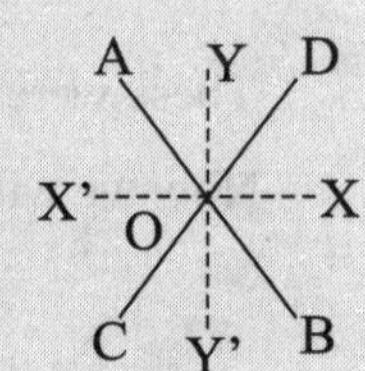

Now X′OX and Y′OY are two pair of lines bisecting the four angles.

Any point on these lines will be equal distance from AB and CD which is further explained.

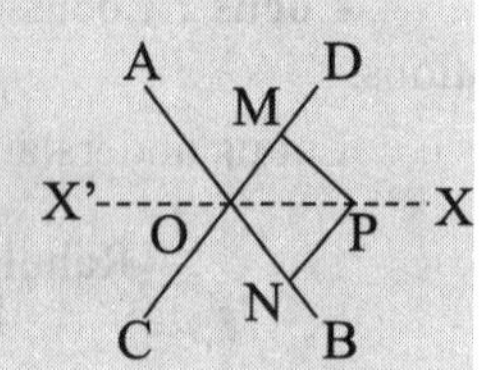

In angle BOD, OX is the bisector we take any point P on it and from P if you draw PM and PN perpendicular than PM = PN or P will be equal distance from AB and CD. Similary if we take any point on Y′OY then perpendicular distance from AB and CD will be same

Now we will discuss few more cases of Locus.

1. **The Locus of points equi distance from a pair of parallel line is a line parallel to two given lines.**
2. **AB is a fixed line Locus of** point **P** such that $AB^2 = AP^2 + BP^2$ is a **circle with AB as diameter.**

Explanation :

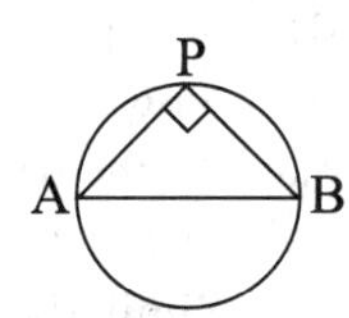

If AB is the diameter then $\angle APB$ will always be equal to 90° (angle in the semi circle) now by pythagoras

$AB^2 = AP^2 + BP^2$

3. **The Locus of centers of all the circles passing through two fix point A and B is the perpendicular bisector of AB.**
4. **The Locus of mid-points of parallel chords of a circle is the diameter of the circle perpendicular to parallel chord.**
5. **The Locus of vertices of isosceles triangles having a common base is the perpendicular bisector of the base.**

Now it is hoped that you have gone through all the basics of Polygon Circle and Locus. As stated earlier you should have all the basics on your finger tips so before going to the question you should revise the basics once again and should devote 15 – 20 minutes for this purpose and then should proceed.

REVISE THE BASICS

It is hope that you have gone through all the basics. Let us have a simple test to check your revision.

Now solve the following exercise.

EXERCISE

Fill in the blanks

1. Six sided Polygon is called _______.
2. Ten sided Polygon is called _______.
3. Eight sided Polygon is called _______.
4. The Polygon whose none of its angle is more than 180° is called _______.
5. A straight line can cut the convex polygon at the most _______ points.
6. A polygon whose all sides are equal is called _______.
7. Formula for each angle of a n sided regular polygon is _______.
8. Formula for sum of the angles of vertices of a star shape n sided regular polygon (when its alternate sides are produced) is _______.
9. Each exterior angle of a regular polygon is _______.
10. Boundry of a circle is called _______.
11. The perpendicular from the centre of circle to a chord _______ the chord.
12. Perpendicular bisectors of two chords of a circle meet at _______ of the circle.
13. Equal chords of a circle are _______ from centre.
14. Equal chord of a circle subtand _______ angle at the centre.
15. The sum of pair of opposite angles of a cyclic quardilator is _______ .
16. Length of tangents from the external points to a circle are _______ .

17. The maximum number of tangents that can be drawn to the internally touching circles are ______.

18. The maximum number of tangents that can be drawn to the externally touching circles are _______.

19. If a point moves according to some given geometrical condition then the path traced by moving point is called _______.

20. The Locus of point equi distance from two given points is the _______ of line segment joining the two given points.

21. AB is a fixed line Locus of point P such that $AB^2 = AP^2 + BP^2$ is a circle with AB as _______ .

SOLUTIONS

1. Hexagon	**2.** Decagon	**3.** Octagon	**4.** Convex
5. Two	**6.** Regular	**7.** $\frac{(2n-4)\times 90}{n}$	**8.** $(2n-8)\times 90$
9. $\frac{360}{n}$	**10.** Circumference	**11.** Bisect	**12.** Centre
13. Equi-Distance	**14.** Equal	**15.** 180°	**16.** Equal
17. 4	**18.** 3	**19.** Locus	
20. Perpendicular Bisector	**21.** Diameter.		

If your answers to this exercise are correct then you should do the next part of the chapter otherwise you should revise the basics once again.

Now comes the types of question.

Types of Questions

In this chapter we have divided the questions on the basis of their topics i.e., in three types.

Type I. Questions related to **Polygon**.

Type II. Questions related to **Circle**.

Type III. Questions related to **Locus**.

We will first do a few examples of above three types followed by the exercise.

TYPE I
(Questions related to Polygon)

Example 1. Find the each interior angle of a regular Decagon.

Sol : We know that

$$\text{Each angle of regular polygon} = \frac{(2n-4)\times 90}{n}$$

Since decagon has 10 sides

$$\therefore \quad n = 10$$

Putting this value in above formula we get $\frac{(2\times 10-4)\times 90}{6} = \frac{16\times 90}{10}$ = **144°.**

Example 2. If each angle of a regular polygon is 120° find the number of sides of a polygon.

Sol : Each angle of regular polygon = $\frac{(2n-4)\times 90}{n}$

$$120° = \frac{(2n-4)\times 90}{n}$$

$$120\,n = 180\,n - 360$$

$$60\,n = 360$$

$$n = \mathbf{6}$$

Therefore it is **Hexagon**.

Example 3. If each angle of a regular polygon is 156° find the number of sides of a polygon.

Sol : Each angle of regular polygon = $\frac{(2\text{n}-4)\times 90}{\text{n}}$

$$156° = \frac{(2\text{n}-4)\times 90}{\text{n}}$$

$$156\,n = 180\,n - 360$$

$$24\,n = 360$$

$$n = \mathbf{15}.$$

Example 4. If each exterior angle of a regular polygon is 36° find the number of sides of a polygon.

Sol : Each exterior angle of regular polygon = $\frac{360°}{\text{n}}$

$$36° = \frac{360°}{\text{n}}$$

$$n = \frac{360°}{36}$$

$$n = \mathbf{10}$$

Therefore it is **Decagon**.

Example 5. The alternate sides of a Pentagon are produced to meet so as to form a star shaped figure. The sum of the angles of the vertices of the star shaped figure is equal to .... ?

Sol : We know that

Sum of angles of vertices of star shaped figure

$$= (2n - 8) \times 90°$$
$$= (2 \times 5 - 8) \times 90°$$
$$= 2 \times 90° = \mathbf{180°}$$

TYPE II
(Questions related to Circle)

Example 6. The length of a chord of a circle is 8 cm and it is at a distance of 3 cm from the centre. Find the radius of the circle.

Sol : The diagram of the question is shown in figure.

$$OL \perp AD$$
$$\therefore \quad AL = LB = 4 \text{ cm}$$

In right angle Δ OAL

$$OA^2 = OL^2 + AL^2$$
$$= 3^2 + 4^2$$
$$= 9 + 16$$
$$= 25$$
$$\therefore \quad OA = \mathbf{5\ cm}$$

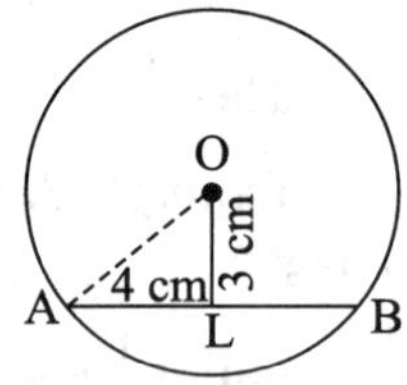

Example 7. In adj. figure

$$OT = 12 \text{ cm}$$
$$OA = 8 \text{ cm}$$

Find AB

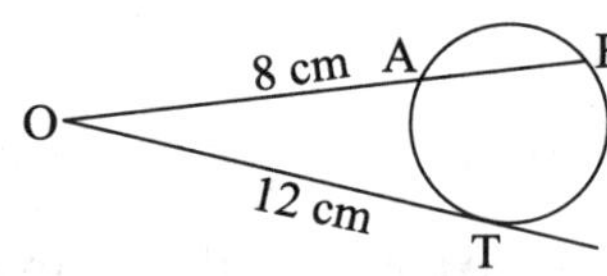

Sol :

$$OT^2 = OA \times OB$$
$$12^2 = 8 \times OB$$
$$OB = \frac{144}{8}$$
$$= 18 \text{ cm}$$
$$AB = OB - OA$$
$$= 18 - 8 = \mathbf{10\ cm}$$

Example 8. If one angle of a cyclic trapezium is triple the other then the smaller angle is ....

Sol : Let smallar angle $= x$

$\therefore$ Greater angle $= 3x$

$\therefore$ $x + 3x = 180°$

$4x = 180°$

$\therefore$ $x = \mathbf{45°}$

Example 9. Find $\angle$ OBC in the figure.

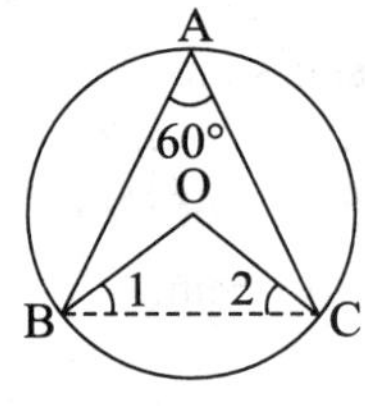

Sol : We know that $\angle BOC = 2 \angle BAC$

$$\angle BOC = 2 \times 60° = 120°$$

Now in $\angle OBC$ $\quad OB = OC$ (Radius)

$\therefore \quad \angle 1 = \angle 2$

But $\quad \angle 1 + \angle 2 + \angle BOC = 180°$

$\therefore \quad \angle 1 + \angle 2 = 180° - 120° = 60°$

$\therefore \quad \angle 1 + \angle 1 = 60°$

$\angle 1 = 60°$

$\angle 1 = \frac{60}{2} = 30°$

$\therefore \quad \angle 1$ or $\angle$ **BOC** = **30°**

Example 10. Find x in the figure.

Sol : We know that $\quad OA \times OB = OC \times OD$

$2 \times x = 3 \times 4$

$\therefore \quad x = \frac{3 \times 4}{2}$ = **6 cm**

TYPE III
(Questions related to Locus)

The questions related to Locus are only theoretical. The examples are as follows :

Example 11. A point moves such that its distance from two given points A and B are equal. The Locus of the point P is

(a) A line passing through A and B.

(b) A straight line (which is right bisector of AB).

(c) A circle with center at A.

(d) none of the above.

Sol : Obviously the answer is B. As done in the basics the Locus of the point which is equi-distance from two give points is the right bi-sector of the line segment joining these two points.

Example 12. The Locus of point equi-distance from the two sides making an angle is

(a) Circle touching the sides.

(b) Perpendicular bisector of the line.

(c) Angle bisector

(d) None.

Sol : Obviously the answer is C i.e., the angle bisector of these two lines.

Now we will do the exercise.

EXERCISE

Q.1. The no. of sides of a regular polygon whose each interior angle is 140°.

(a) 10 (b) 8 (c) 9 (d) none

Q.2. The number of sides of a regular polygon whose each interior angle is 170° are

(a) 9 (b) 36 (c) 30 (d) none

Q.3. Each interior angle of a regular polygon of 30 sides is

(a) 168° (b) 178° (c) 188° (d) none

Q.4. Each interior angle of a regular polygon of 18 sides is

(a) 168° (b) 160° (c) 140° (d) none

Q.5. In a polygon each angle is three times the exterior angle. Find the number of sides .

(a) 8 (b) 6 (c) 5 (d) none

Q.6. The no. of sides of a regular polygon is double that of another and magnitude of its angle is also double that of the other. Find the number of sides of a larger sided polygon.

(a) 5 (b) 6 (c) 8 (d) 10

Q.7. The number of sides of a nonagon are produced to meet so as to form a star shaped polygon. The each of the angle of the vertices of a star is equal to

(a) 108° (b) 110° (c) 115° (d) none

Q.8. The alternate sides of a pentagon are produced to meet so as to make a star shaped figure. The sum of the of vertices of star is equal to

(a) 180° (b) 360°

(c) 540° (d) none

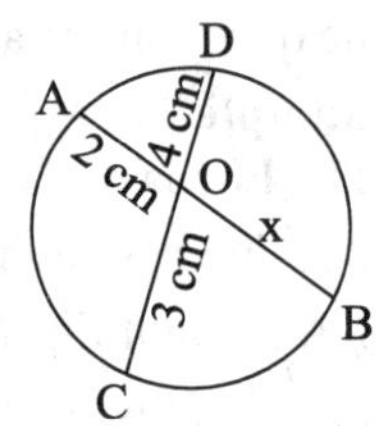

Q.9. Find PD in the adjcent figure

(a) 10 cm (b) 18 cm

(c) 14 cm (d) none

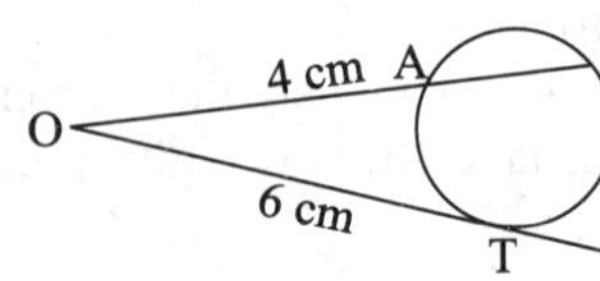

Q.10. Find AB in fig.

(a) 9 cm (b) 5cm

(c) 3 cm (d) none

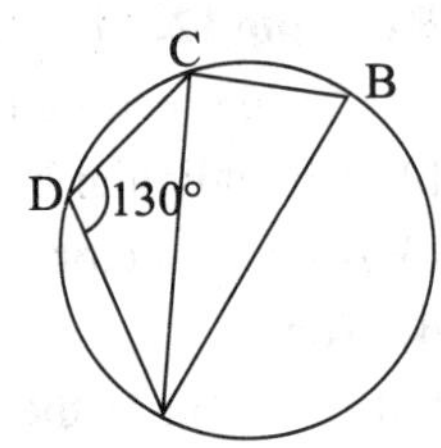

Q.11. In fig AB is diameter and ∠ADC = 130°. Then ∠ BAC is

(a) 40° (b) 50°

(c) 60° (d) none

Q.12. Two opposite angle of a Cyclic quadrilateral are in ratio 1 : 2 the smaller angle is

(a) 60° (b) 70° (c) 80° (d) none

Q.13. In a circle with centre O, AB and CD are two diameter perpendicular to each other. The length, of chord AC is

(a) AB (b) $\sqrt{2}$ AB (c) $\frac{AB}{\sqrt{2}}$ (d) none

Q.14. In a circle of radius is 5 cm a chord is at a distance of 4 cm from the centre of circle. The length of chord is

(a) 10 cm (b) 3 cm (c) 6 cm (d) none

Q.15. The radius of a circle is $\sqrt{2}$ cm. The circle is divided into two segments by a chord of length 2 cm. The angle in smallest segment is

(a) 90° (b) 45° (c) 135° (d) none

Q.16. Two circles with radii r_1 and r_2 are externally touching and $r_1 > r_2$ PT is their common tangent then PT^2 is

(a) $r_1 + r_2$ (b) $r_1 - r_2$ (c) $4\, r_1\, r_2$ (d) none

Q.17. The locus of point equidistant from fixed point is

(a) A circle passing through fixed point

(b) a straight line passing through fixed point

(c) a circle with fixed point as its centre

(d) none

Q.18. What is the locus of centre of circles of equal radius passing the fixed point

(a) A parallel line passing through fixed point

(b) a straight line passing through fixed point

(c) a circle with fixed point as its center

(d) none

Q.19. A wheel of radius 20 cm roll on a plane st. road , the locus of centre of wheel

(a) A circle of radius 10 cm

(b) A st. line parallel to the path of wheels and at distance of 20 cm

(c) A straight line passing to the path of wheel at a 40 cm from it

(d) none of these

SOLUTIONS

1. (c)

$$\left(\frac{(2n-4)}{n}\right) 90° = 140$$

$$(2n-4)\, 90° = 140n$$

$$180n - 360 = 140n$$

$$40m = 360$$

$$x = \frac{360}{40} = \mathbf{9}$$

2. (b)

$$\theta = 170$$

$$180n - 360 = 170n$$

$$10n = 360$$

$$n = \mathbf{36}$$

3. (a)

$$\text{Each angle} = \frac{(2n-4)\times 90}{4}$$

$$= \frac{(2 \times 30 - 4) \times 90}{30}$$

$$= \frac{56 \times 90}{30} = 56 \times 3 = \mathbf{168°}$$

4. (b) Each angle $= \frac{(2n - 4) \times 90}{4}$

$$= \frac{(2 \times 18 - 4) \times 90}{18} = \frac{32 \times 90}{18} = \mathbf{160°}$$

5. (a) Let Exterior angle $= x$

$\therefore$ interior angle $= 3x$

$\therefore$ $x + 3x = 180$

$4x = 180$

$x = \frac{180}{4} = 45°$

$\therefore$ angle $= 45 \times 3 = 135°$

$\therefore$ $135\text{ x} = \frac{(2n - 4) \times 90}{4}$

$135n = 180\,n - 360$

$180\,n - 135\,n = 360$

$45\,n = 360$

$n = \frac{360}{45} = \mathbf{8}$

6. (b) Let $n_1 = n$ $\therefore$ $n_2 = 2n$

$\theta_1 = \theta$ $\therefore$ $\theta_2 = 2\theta$

$\theta_1 = \frac{(2n_1 - 4)}{n_1} \times 90°;$ $\theta_2 = \frac{2n_2 - 4}{n_2} \times 90°$

Divided (i) by (ii) we get

$$\frac{\theta_1}{\theta_2} = \frac{2n_1 - 4}{2n_2 - 4} \times \frac{n_2}{n_1}$$

$$\frac{\theta}{2\theta} = \frac{2n - 4}{4n - 4} \times \frac{2n}{n}$$

$$\frac{1}{2} = \frac{4n - 8}{4n - 4}$$

Close multiply

$4n - 4 = 8x - 16$

$4n = 12$

$n = 3$

$\therefore$ Polygon with larger sides $= 2n = 2 \times 3 = 6$

7. (d) $\frac{(2 \times 9 - 8) \times 90}{9} = \frac{10 \times 90}{9} = 100°$

8. (a) $(2n - 8) \times 90 = (2 \times 5 - 8) \times 90 = \mathbf{180^\circ}$

9. (c) $PD = \dfrac{16 \times 7}{8} = 14$

10. (b) $OT^2 = OA \times OB$

We get $OB = 9$ cm

$AB = OB - OA = 9 - 4 = \mathbf{5\ cm}$

11. (a) $\angle ADC = 130^\circ$

$\therefore$ $\angle ABC = 50^\circ$

$\angle ACB = 90^\circ$ ($\angle$ in semi-circle)

$\angle BAC = 180^\circ - 90^\circ - 50^\circ = \mathbf{40^\circ}$

12. (a) $x + 2x = 180^\circ$

$3x = 180^\circ$

$x = 60^\circ$

13. (c) $AC^2 = OC^2 + OA^2$

$= x^2 + x^2 = 2x^2$

$AC = \sqrt{2}\,x = \dfrac{AB}{\sqrt{2}}$

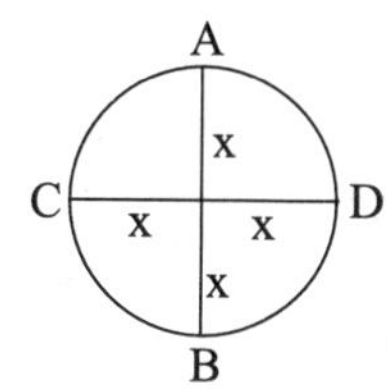

14. (c) $OA^2 = OC^2 + AC^2$

$5^2 = 4^2 + AC^2$

$AC = 3$ cm

$\therefore$ $AB = 2 \times 3 = 6$ cm

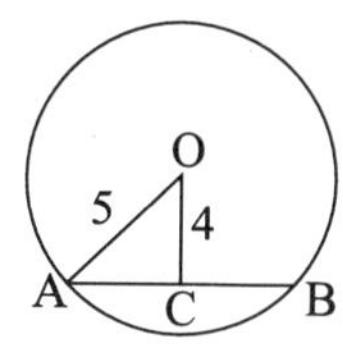

15. (a) $OA = \sqrt{2}$

$AC = 1$

$\therefore$ $OC = 1$ (by Pythagoras)

$AC = OC$

$\angle 1 = \angle 3 = \angle 45^\circ$

but $\angle 1 = \angle 2$

$\angle 2 = 45^\circ$

hence $AOC = \angle 1 + \angle 2 = \mathbf{90^\circ}$

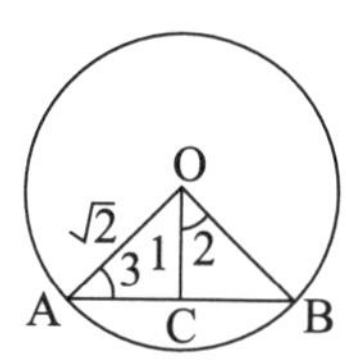

16. (c) $PC = BT$

as shown in fig. $AC = r_1 - r_2$

$AB = r_1 + r_2$

by Pythagoras

$BC^2 = AB^2 - AC^2$

$= (r_1 + r_2)^2 - (r_1 - r_2)^2$

$= \mathbf{4r_1r_2}$

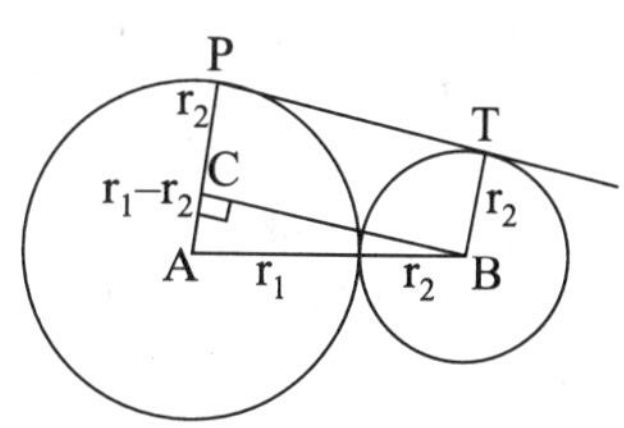

17. (c) **18. (c)** **19. (b)**

Section III

Test Yourself

1. ABCD is a quadrilateral in which measures of angle D and angle C are 60 degree, and 100 degree respectively. If the internal bisectors of angle A and angle B meet at P, then measure of angle APB is **(CLAT 2018)**

   (a) 80 (b) 90 (c) 100 (d) 110

2. The parallel sides of a trapezium shaped field are 25 m and 10 m and non parallel sides are 14 m and 13 m. What is the area (in m^2) of the field? **(CLAT 2018)**

   (a) 196 m^2 (b) 176 m^2 (c) 126 m^2 (d) 296 m^2

3. PQR is an equilateral triangle whose side is 10 cm. What is the value (in cm.) of the in radius of Triangle PQR? **(SSC CGL Tier-II 2018)**

   (a) $\frac{5}{\sqrt{3}}$ (b) $10\sqrt{3}$ (c) $\frac{10}{\sqrt{3}}$ (d) $5\sqrt{2}$

4. A hall is 5 m long, 4 m wide and 3 in high. What is the cost of white washing its walls and roof at a rate Rs. 25/m^2 and laying of marble floor at a rate Rs. 175/m^2? **(NIFT UG 2017)**

   (a) Rs. 4675 (b) Rs. 7000 (c) Rs. 5350 (d) Rs. 8350

5. What is the area (in cm.) of the circumcircle of a triangle whose sides are 6 cm., 8 cm. and 10 cm. respectively? **(SSC CGL Tier-II 2018)**

   (a) $\frac{275}{7}$ (b) $\frac{550}{7}$ (c) $\frac{2200}{7}$ (d) $\frac{1100}{7}$

6. What can be the maximum number of common tangents which can be drawn to two non-intersecting circles? **(SSC CGL Tier-II 2018)**

   (a) 2 (b) 4 (c) 3 (d) 6

7. Instead of walking along two adjacent sides of a rectangular field, a boy took a short cut along the diagonal and saved a distance equal to half the longer side. Then the ratio of the shorter side to the longer side is ________ **(NIFT UG 2017)**

   (a) 1/2 (b) 2/3 (c) 1/4 (d) 3/4

8. PQR is a right angle triangle in which PQ = QR. If the hypotenuse of the triangle is 20 cm., then what is the area (in $cm.^2$) of the triangle PQR? **(SSC CGL Tier-II 2018)**

   (a) $100\sqrt{2}$ (b) 100 (c) $50\sqrt{2}$ (d) 50**38.**

9. There is circular garden of radius 21 meters. A path of width 3.5 meters is constructed just outside the garden. What is the area (in $meter^2$) of the path? **(SSC CGL Tier-II 2018)**

   (a) 500. 5 (b) 507.56 (c) 152.12 (d) 256.07

10. The height of a cone is 24 cm and the area of the base is 154 cm. What is the curved surface area (in $cm.^2$) of the cone? **(SSC CGL Tier-II 2018)**

    (a) 484 (b) 550 (d) 525 (c) 515

11. A solid hemisphere has radius 14 cm. It is melted to form a cylinder such that the ratio of its curved surface area and total surface area is 2:3. What is the radius (in cm.) of its base?

(SSC CGL Tier-II 2018)

(a) $\frac{10}{\sqrt[3]{3}}$ (b) $\frac{14}{\sqrt[3]{3}}$ (c) $\frac{7}{\sqrt[3]{3}}$ (d) $\frac{21}{\sqrt[3]{3}}$

12. The length of a rectangular field is 75% more than the breadth of the same field. If the total cost of fencing the field at the rate of 5 per meter was ₹220, what is the area of the field? (in m^2)

(SBI Clerk 2016)

(a) 112 (b) 72 (c) 114 (d) 96
(e) 128

13. The radius of the circular field is equal to the side of a square field. If the difference between the area of the circular field and area of the square field is 105 m^2, what is the perimeter of the circular field? (in meters) **(LIC AAO March 2016)**

(a) 132 (b) 80 (c) 44 (d) 176
(e) 112

14. In measuring the sides of a rectangle, one side is taken 5% in excess, and the other 4% in deficit. Find the error per cent in the area calculated from these measurement **(NCHMCT 2016)**

(a) 1% (b) 0.9% (c) 0.8% (d) 0.85%

15. If the area of a circle is 9π sq.cm then its circumference is **(RRB Non. Tec March 2016)**

(a) 9 cm (b) 6π cm (c) 3π cm (d) 6 cm

16. The numerical value of the perimenter of a rectangular field is less than that of its area by 128. If the breadth of the same field is halved, then the numerical value of the perimeter of the field becomes 40 less than its area. What is the area of the rectangle?

(NABARD Asst. Manager 2016)

(a) 192 (b) 220 (c) 176 (d) 208

17. A rectangular garden of length 12 m is surrounded by a 2 m wide path. If the area of the garden is 84 m^2 and the cost of gravelling is ₹ 8 per m^2, what is the total cost of gravelling the path? (In ₹)? **(SBI Clerk 2016)**

(a) 780 (b) 742 (c) 724 (d) 775
(e) 736

18. The length of a rectangle is increased to 3/2 times of its length and breadth is reduced to 1/3rd of its breadth. What is the change in area? **(MAT May 2016)**

(a) 1/4 (b) 1/3 (c) 1/2 (d) No change

19. The ratio of the volume of a cube to that of the sphere which can fit inside the cube is_______. **(MAT May 2016)**

(a) 21 : 11 (b) 3 : 4 (c) 1 : 2 (d) 4 : 3

20. The length of a rectangle is 1.5 times its breadth and the side of a square is equal to the length of the rectangle, If the area of the square is 147 m^2 more than the area of the rectangle, what is the area of the rectangle? **(IBPS Clerical 2015)**

(a) 280 m^2 (b) 292 m^2 (c) 326 m^2 (d) 312 m^2
(e) 294 m^2

21. If the volume and curved surface area of a cylinder are 616 m^3 and 352 m^3 respectively, what is the total surface area of the cylinder? (in m^2)

(a) 429 (b) 419 (c) 435 (d) 421
(e) 417

22. A courtyard is 25 m long and 16 m board is to be paved with bricks of dimensions 20 cm by 10 cm. What is the total number of bricks required? **(IBPS RRB Office Asst. Exam 2015)**
(a) 16000 (b) 18000 (c) 20000 (d) 22000
(e) 25000

23. The diameter of a circle is equal to the perimeter of a square whose area is 3136 cm'. What is the circumference of the circle? **(IBPS RRB Office Asst. Exam 2015)**
(a) 352 cm (d) 704 cm (c) 39424 cm (d) 1024 cm
(e) Other than those given as options

24. The base of a triangle is 15 cm and height is 12 cm. The height of another triangle of double the area having base 20 cm is **(IBPS RRB Office Asst. Exam 2015)**
(a) 22 cm (b) 20 cm (c) 18 (d) 10 cm
(e) 19 cm

25. A slice from a circular pizza of diameter 14 inches is cut in such a way, that the slice forms a sector of angle measuring 45°. Find the measure of area of the cut slice of the pizza. (in inches square)? (Use $\pi = \frac{22}{7}$) **(RBI Asst. 2015)**
(a) 17.5 (b) 18.25 (c) 19.25 (d) 19.75
(e) 17.75

26. A rectangular plot 55 m long and 45 m broad, has two concrete crossroads (of equal width) running in the middle of it one parallel to the length and the other parallel to the breadth. The rest of the plot is used as a lawn. If the area of the lawn is 1911 m^2, what is the width of each of the crossroads? (in m) **(SBI PO Prelim Exam 2015)**
(a) 5 (b) 5.5 (c) 6 (d) 4
(e) 4.5

27. A square field has its area equal to 324 m^2. The perimeter of the field is: **(CLAT 2014)**
(a) 36 m (b) 72 m (c) 18 m (d) 6561 m

28. A closed metal box measure 30 cm × 20 cm × 10 cm. Thickness of the metal is 1 cm. The volume of the metal required to make the box is **(CLAT 2014)**
(a) 1041cm^3 (b) 6000 cm^3 (c) 4536 cm^3 (d) 1968 cm^3

29. The difference between the the area of a square field and the area of circular field is 168 sq. m. Side of the square field is equal to the diameter of the circular field. What is the cost of fencing the square field at the rate of ₹ 20 per meter? (in ₹) **(IBPS Bank of Maharashtra 2016)**
(a) 2480 (b) 2420 (c) 2520 (d) 2240
(e) None of these

30. The areas of three consecutive faces of a cuboid are 12 cm^2, 20 cm^2 and 15 cm^2, then the volume (in cm^3), then the volume (in cm^3) of the cuboid is **(NICL AO 2013)**
(a) 3600 (b) 100 (c) 80 (d) 60

31. The wheel of a motor car makes 1000 revolutions in moving 440 m. The diameter (in metre) of the wheel is **(NICL AO 2013)**
(a) 0.44 (b) 0.14 (c) 0.24 (d) 0.34

32. Diagonals of a rhombus are 1 meter and 1.5 meter in lengths. The area of the rhombus is **(CLAT 2013)**

(a) 0.75 m^2 (b) 2.25 m^2 (c) 1.5 m^2 (d) 0.375 m^2

33. An angle in a semi circle is **(CLAT 2013)**

(a) π (b) $\pi/4$ (c) $\pi/2$ (d) 2π

34. In an equilateral triangle ABC, whose length of each side is 3 cm, D is a point on BC such that BD = $^1/_2$ CD. What is the length of AD? **(Xat 2010)**

(a) $\sqrt{5}$ cm (b) $\sqrt{6}$ cm (c) $\sqrt{7}$ cm (d) $\sqrt{8}$ cm

(e) None of the above

35. A tank internally measuring 150 cm × 120 cm × 100 cm has 1281600 cm^3 water in it. Porous bricks are placed in the water until the tank is full up to its brim. Each brick absorbs one tenth of its volume water. How many bricks, of 20 cm × 6 cm × 4 cm, can be put in the tank without spilling over the water? **(Xat 2010)**

(a) 1100 (b) 1200 (c) 1150 (d) 1250

(e) None of the above

DIRECTIONS for questions (36-40): Study the figure and information given below carefully and answer the questions that follow. **(ATMA 2010)**

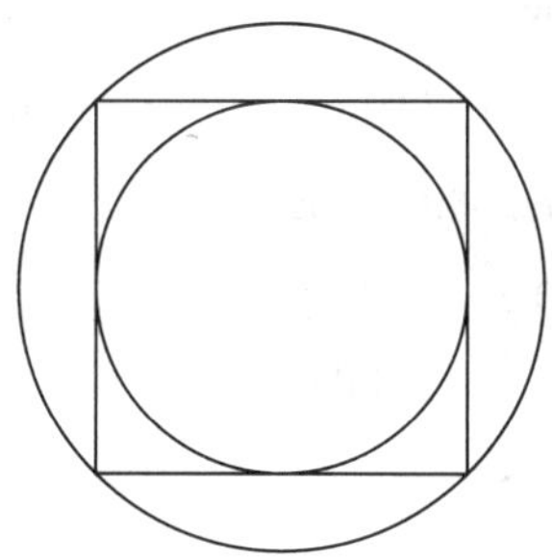

The figure above shows two concentric circles. The larger circle has diameter equal to 28 cms. The square is inscribed in the larger circle whereas the smaller circle is inscribed within the square.

36. What is the area of the larger circle? **(ATMA 2010)**

(a) 412 sq. cms (b) 541 sq. cms (c) 616 sq. cms (d) 654 sq. cms

37. What is the **approximate** measure of the side of the square? **(ATMA 2010)**

(a) 19.8 cms (b) 15.2 cms (c) 12.6 cms (d) 21.7 cms

38. What is the **approximate** circumference of the smaller circle? **(ATMA 2010)**

(a) 49 cms (b) 54 cms (c) 68 cms (d) 62 cms

39. What is the **approximate** radius of the smaller circle? **(ATMA 2010)**

(a) 6.2 cms (b) 9.9 cms (c) 11.4 cms (d) 12.6 cms

40. What is the **approximate** perimeter of the square? **(ATMA 2010)**

(a) 82.2 cms (b) 56.4 cms (c) 64.8 cms (d) 79.2 cms

41. If the perimeter of a rectangular plot is 78 meters and its area is 350 meters, what is the measure of each of its longer side? **(ATMA 2010)**

(a) 14 meters (b) 18 meters (c) 25 meters

(d) Cannot be determined

42. If the sum of the interior angles of a regular polygon measures up to 1800 degrees, how many sides does the polygon have? **(ATMA 2010)**

(a) 18 (b) 8 (c) 12 (d) 10

DIRECTIONS ***for questions (10 to 14):*** Study the figure and information given below carefully and answer the questions that follow –

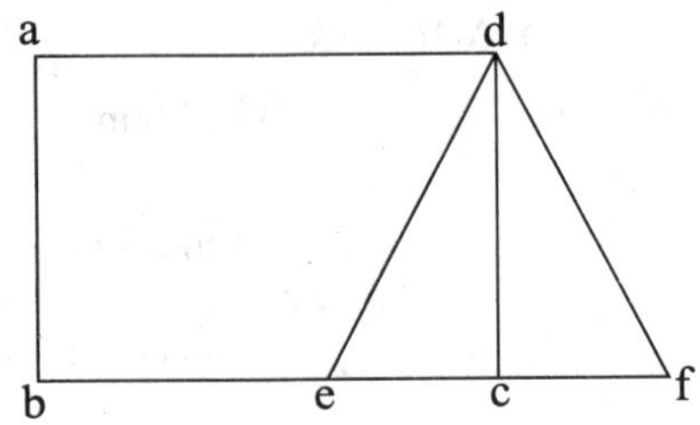

The figure above shows a square abed and an equilateral triangle def. The segment dc is a perpendicular line drawn from the point d. The length of the segment ec is 5 cms.

43. What is the area of the square? **(ATMA 2010)**

(a) 75 sq. cms (b) 100 sq. cms (c) 50 sq. cms (d) 125 sq. cms

44. What is the area of the triangle def ? **(ATMA 2010)**

(a) $25\sqrt{3}$ sq. cms (b) 50 sq. cms (c) $15\sqrt{3}$ sq. cms (d) 30 sq. cms

45. What is the measure of the side of the square ? **(ATMA 2010)**

(a) $3\sqrt{3}$ cms (b) 10 cms (c) $5\sqrt{3}$ cms (d) 20 cms

46. What is the measure of the diagonal of the square ? **(ATMA 2010)**

(a) 12 cms (b) $6\sqrt{5}$ cms (c) 10 cms (d) $5\sqrt{6}$ cms

47. What is the perimeter of the triangle dec? **(ATMA 2010)**

(a) $5\sqrt{3}(3+\sqrt{3})$ cms (b) $5(3+\sqrt{3})$ cms

(c) $5(1+\sqrt{3})$ cms (d) $5(3+\sqrt{3})$ cms

48. If the mean of numbers 11, x, 58, 56 and 33 is 44, then what is the mean of 155,263,148,222 and x? **(ATMA 2010)**

(a) 166 (b) 184 (c) 142 (d) 170

49. What would be the circumference of a circle with area equal to 1122.66 sq. cms? **(ATMA 2010)**

(a) 106.4 cms (b) 148.2 cms (c) 118.8 cms (d) 124.6 cms

50. The radius of a circle is so increased that its circumference increased by 5%. The area of the circle then increases by **(SNAP 2009)**

(a) 12.5% (b) 10.25% (c) 10.5% (d) 11.25%

SOLUTIONS

1. (1)

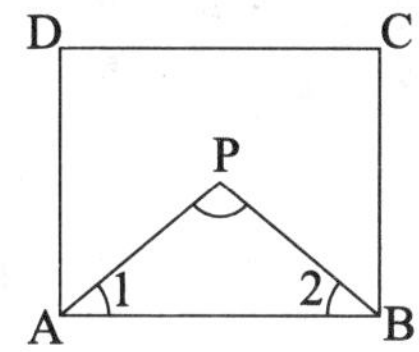

$$\angle A + \angle B + \angle C + \angle D = 360°$$
$$\angle A + \angle B + 100° + 60° = 360°$$
$$\angle A + \angle B = 200°$$
$$\frac{1}{2}\angle A + \frac{1}{2}\angle B = 100°$$

In Δ ABP

$$\angle 1 + \angle 2 + \angle APB = 180°$$
$$100° + \angle APB = 180°$$
$$\Rightarrow \quad \angle APB = 80°$$

2. (1)

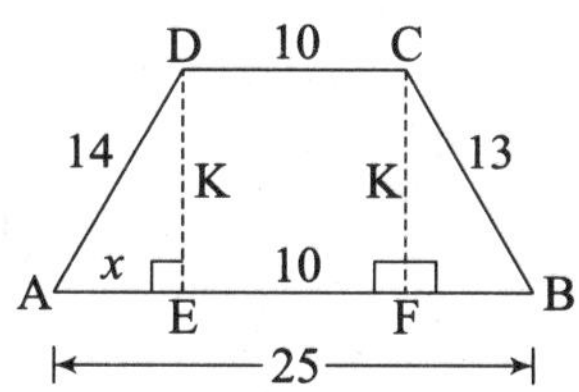

$\Rightarrow \quad \angle APB = 80°$

Let $\quad AE = x$

$\therefore \quad FB = 25 - 10 - x = 15 - x$

In ΔADE

$$14^2 = k^2 + x^2$$
$$\Rightarrow \quad k^2 = 196 - x^2 \qquad ...(1)$$

In ΔBCF

$$k^2 + (15 - x)^2 = 13^2$$
$$\Rightarrow \quad k^2 = 169 - (15 - x)^2 \qquad ...(2)$$

from (1) and (2)

$$196 - x^2 = 169 - (15 - x)^2$$
$$196 - x^2 = 169 - 225 - x^2 + 30x$$
$$30x = 196 - 169 + 225$$
$$= 252$$
$$x = \frac{252}{30} = 8.4$$

$\therefore \quad k^2 = 196 - (8.4)^2$

$= 196 - 70.56$

$= 125.44$

$\Rightarrow \quad k = 11.2$

$\therefore$ Area of trapezium

$= \frac{1}{2}(25 + 10) \times 11$

$= 35 \times 5.6$

$= 196 \text{ m}^2$

3. (1) $\text{Inradius} = \frac{\text{Area of } \Delta}{\text{semi perimeter}}$

$$= \frac{\frac{\sqrt{3}}{4} \times 10^2}{\frac{10+10+10}{2}}$$

$$= \frac{\frac{\sqrt{3}}{4} \times 100}{15}$$

$$= \frac{25\sqrt{3}}{15}$$

$$= \frac{5}{3}\sqrt{3} = \frac{5}{\sqrt{3}}$$

4. (3) Area of four walls + Area of roof

$= 2h(\ell + b) + \ell b$

$= 2 \times 3(5 + 4) + 5 \times 4$

$= 54 + 20 = 74$ sqm

Area of floor $= 5 \times 4 = 20$ sqm

Total cost $= 74 \times 25 + 20 \times 175$

$= 1850 + 3500 = 5350$ Rs.

5. (2) 6, 8, 10 will be

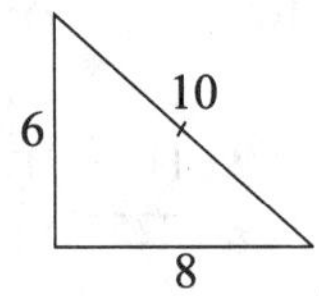

In right angles Δ circumradius is half of hypotenuse

i.e., $\frac{10}{2} = 5$

Required Area $= \pi r^2$

$$= \frac{22}{7} \times 25 = \frac{550}{7}$$

6. (2) 4 common to get's can be drawn as shown below

7. (4) By option (4)

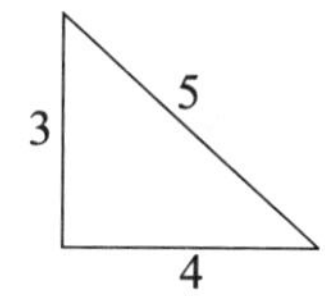

$$4 + 3 - 5 = 2$$

$$\text{which is } \frac{4}{2} = 2 \text{ half of longer side}$$

8. (2)

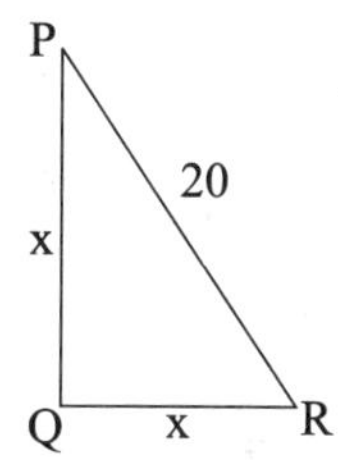

$$x^2 + x^2 = 20^2$$

$$2x^2 = 400$$

$$x^2 = 200$$

$$x = 10\sqrt{2}$$

$$\therefore \quad \text{Area of} = \frac{1}{2} \times 10\sqrt{2} \times 10\sqrt{2}$$

$$= 100 \text{ m}^2$$

9. (1)

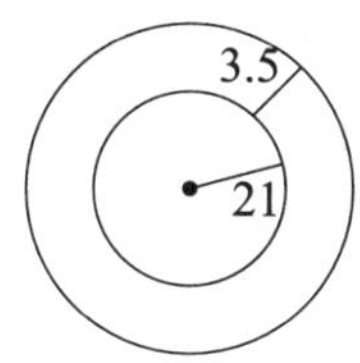

$$\text{Area} = \pi\, 24.5^2 - \pi\, 21^2$$

$$= x\,(24.5 + 21)\,(24.5 - 21)$$

$$= \frac{22}{7} \times 45.5 \times 3.5$$

$$= 500.5 \text{ Sqm.}$$

10. (2)

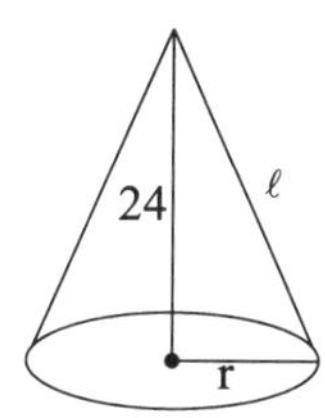

$$\pi r^2 = 154$$

$$\frac{22}{7} \times r^2 = 154$$

$$r^2 = 49$$

$$\therefore \quad r^2 + 24^2 = \ell^2$$

$$49 + 576 = \ell^2$$

$$\Rightarrow \quad \ell = 25$$

$$\therefore \quad \text{CSA} = \pi r l = \frac{22}{7} \times 7 \times 25 = 550 \text{ cm}^2$$

11. (2) According to question

For cylinder

$$\frac{2\pi rh}{2\pi r(h + r)} = \frac{2}{3}$$

$$\Rightarrow \quad \frac{h}{h+r} = \frac{2}{3}$$

$$3h = 2h + 2r$$

$$h = 2r$$

$\therefore$ Volume of hemi sphere = Volume of cylinder

$$\frac{2}{3}\pi(14)^3 = \pi r^2 h$$

$$\frac{2}{3} \times 14^3 = r^2 \times 2r$$

$$\pi r^3 = \frac{14^3}{3}$$

$$\therefore \quad r = \frac{14}{3\sqrt{3}}$$

12. (a) Let $b = 4x$

$$\therefore \quad l = 7x$$

$$\text{Perimeter} = \frac{220}{5} = 44 \text{ m}$$

$$\therefore \quad 2(7x + 4x) = 44$$

$$\Rightarrow \quad x = 2$$

$\therefore$ length = 14 m; breadth = 8 m

$\therefore$ Area = $14 \times 8 = 112 \text{ m}^2$

13. (c)

$$\pi r^2 - a^2 = 105$$

$$\pi r^2 - r^2 = 105 \quad (\because a = r)$$

$$r^2(\pi - 1) = 105$$

$$r^2 \times \frac{15}{7} = 105$$

$$r^2 = 49$$

$$\Rightarrow \quad r = 7$$

$$\therefore \quad \text{Circumfrence} = 2 \times \frac{22}{7} \times 7 = 44\text{m}$$

14. (c)

$$\%\text{ Error} = 5 - 4 + \frac{5 \times -4}{100}$$

$$= 1 - .2 = .8$$

15. (b)

$$\pi r^2 = 9\pi$$

$$r^2 = 9$$

$$r = 3$$

$\therefore$ Circumfrence = $2\pi \times 3 = 6\pi$

16. (a)

$$lb - 2(l + b) = 128$$

$$\frac{lb}{2} - 2\left(l + \frac{b}{2}\right) = 40$$

Solving we get

$$l = 24$$
$$b = 8$$
$$\therefore \text{Area} = 24 \times 8 = 192$$

17. (e)

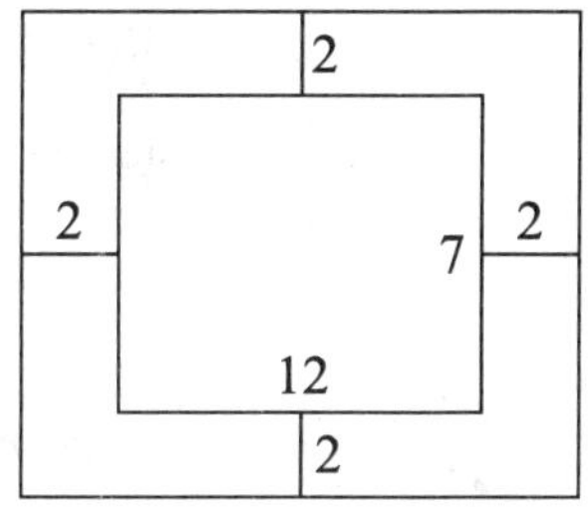

$$\text{Breadth of garden} = \frac{84}{12} = 7 \text{ m}$$
$$\text{Area of path} = 16 \times 11 - 12 \times 7$$
$$= 176 - 84 = 92 \text{ m}^2$$
$$\therefore \text{Cost} = 92 \times 8 = ₹\, 736$$

18. (c)

$$\text{Let length} = x$$
$$\text{breadth} = y$$
$$\therefore \text{Area} = xy$$
$$\text{New length} = \frac{3}{2}x$$
$$\text{breadth} = \frac{y}{3}$$
$$\therefore \text{New Area} = \frac{3}{2}x \times \frac{y}{3} = \frac{xy}{2}$$
$$\therefore \text{Change in Area} = xy - \frac{xy}{2}$$
$$= \frac{1}{2}xy$$
$$= \frac{1}{2} \text{ of Area.}$$

19. (a)

$$\text{Let edge of cube} = 2a$$
$$\therefore \text{Diameter of sphere} = 2a$$
$$\text{Radius of sphere} = a$$
$$\therefore \text{Required Ratio} = \frac{(2a)^3}{\frac{4}{3}\pi a^3}$$
$$= \frac{8a^3}{\frac{4}{3} \times \frac{22}{7} \times a^3}$$
$$= \frac{8 \times 3 \times 7}{4 \times 22} = \frac{21}{11}$$
$$= 21 : 11$$

20. (e) Let breadth of rectangle = x

Length = $1.5x$

Area of rectangle = $1.5x^2$

Now side of square = $1.5x$

Area of square = $(1.5x)^2 = 2.25\,x^2$

According to question

$$2.25\,x^2 - 1.5x^2 = 147\text{ m}^2$$
$$.75x^2 = 147$$
$$x = 14$$

Area of rectangle = $1.5x^2 = 1.5 \times 196 = 294\text{ m}^2$

21. (a) $$\pi r^2 h = 616 \;\&\; 2\pi rh = 352$$

$$\frac{\pi r^2 h}{2\pi rh} = \frac{616}{352}$$

Solving this we get $r = 7/2$

$$\text{T.S A} = \text{C.S A} + (\text{Area of base}) \times 2 = 352 + 2 \times \frac{22}{7} \times \frac{7}{2} \times \frac{7}{2} = 429$$

22. (c) Total area to be paved = $(2500 \times 1600)\text{ cm}^2$

Area of one brick = 200 cm^2

Bricks required = 4000000/200 = 20000

23. (b) Side of square = $\sqrt{3136} = 56$ cm

Perimeter of square = Diameter of circle = $56 \times 4 = 224$;

Radius of circle = 112;

Circumference of circle = $2 \times \pi \times 112 = 2 \times 22/7 \times 112 = 704$ cm

24. (c) Area of the first triangle = $1/2 \times 15 \times 12 = 90$;

Area of second triangle is twice i.e. 180.

So, $$180 = 1/2 \times 20 \times h$$
$$h = 180/10;$$
$$h = 18$$

25. (c) Radius of the circle is 7 inch.

$$\text{Area of the sector} = \pi \times 7^2 \times \frac{45}{360} = \frac{22}{7} \times 49 \times \frac{1}{8} = 19.25 \text{ sq. inch.}$$

26. (c) The area of the whole lawn comprises of the area of the crossroads and the area of the lawn. The area of the plot = $55 \times 45 = 2475$, area of lawn = 1911 and area of crossroads taking the width of each road as 'x' will be = $55x + 45x - x^2$. So $100x - x^2 + 1911 = 2475$

Solving this we get $x = 6$

27. (b) The side of the field = $\sqrt{324} = 18$ m

Perimeter = $4 \times 18 = 72$m

28. (d) The volume of metal = outer volume – inner volume = $30 \times 20 \times 10 - (30 - 2) \times (20 - 2) \times (10 - 2) = 1968\text{ cm}^3$

29. (d) Let radius of circular field = r

Side of square = $a = 2r$

According to question,

$$a^2 - \pi r^2 = 168$$
$$(2r)^2 - \pi r^2 = 168$$

$$4r^2 - \pi r^2 = 168$$
$$r^2(4 - \pi) = 168$$
$$r^2(4 - 22/7) = 168$$
$$r^2 \times \frac{6}{7} = 168$$
$$r^2 = \frac{168 \times 7}{6} = 196$$
$$\Rightarrow \quad r = 14 \text{ m}$$

So side of square $= 2 \times 14 = 28$ m

Hence cost of fencing square field $= 4 \times 28 \times 20 =$ ₹ 2240

30. (d) Let l, b and h be the length, breadth and height respectively

Therefore, $lb = 12$, $bh = 20$ and $lh = 15$

Multiply all the above equations, we get $(lbh)^2 = 12 \times 20 \times 15$

Thus, volume $= lbh = \sqrt{12 \times 20 \times 15}$

31. (b) Let d be the diameter

Therefore, $2 \times 22/7 \times d/2 \times 1000 = 440$

$$d = 0.14\text{m}$$

32. (a) Required area $= \frac{1}{2} \times d_1 \times d_2$

$$= \frac{1}{2} \times 1 \times 1.5 = .75\text{m}^2$$

33. (a) Angle in a semicircle $= 180° = \pi$

34. (c) Using cosine formula

$$\text{Cos B} = \frac{3^2 + 1^2 - x^2}{2 \times 3 \times 1}$$

$$\frac{1}{2} = \frac{10 - x^2}{6} \Rightarrow x = \sqrt{7}$$

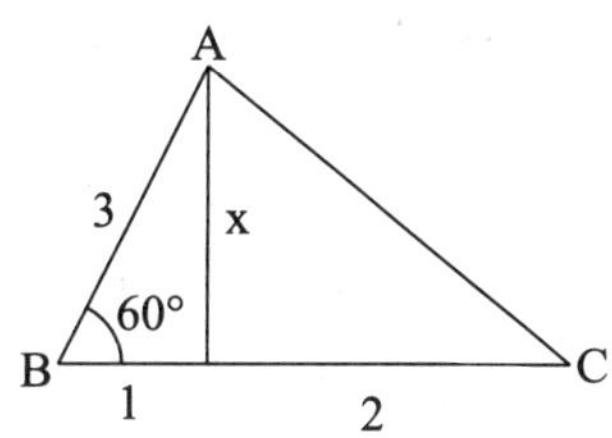

35. (b) Volume of Tank $= 150 \text{ cm} \times 120 \text{ cm} \times 100 \text{ cm} = 1800{,}000 \text{ cm}^3$

Qty of water $= 1281600 \text{ cm}^3$, Left over $= 18{,}000{,}00 - 1281600 = 518400 \text{ cm}^3$

Volume of Bricks $= 20 \text{ cm} \times 6 \text{ cm} \times 4 \text{ cm} = 180 \text{ cm}^3$

Absorbed $= 48 \text{ cm}^3$, Volume Replaced $= 480 - 48 = 432 \text{ cm}^3$

No. of Bricks $= \frac{518400}{432} = 1200$

36-40 Diameter of larger circle = Diagonal of the square = 28 cms. So radius of the larger circle R = 14 cms.

Let the side of the square = a cms.

Diagonal of square $= \sqrt{2}$ a = 28 cms

$a = 14\sqrt{2}$ cms. The diameter of the smaller circle = side of the square. So radius of the smaller circle = $7\sqrt{2}$ cms.

36. (c) Area of the larger circle = $\pi R^2 = 616$ cms.
Option (c).

37. (a) Side = $14\sqrt{2}$ cms = 19.8 cms. Option (a).

38. (d) Circumference of the smaller circle = $2\pi r = 62$ cms. Option (d).

39. (b) Radius of smaller circle = $7\sqrt{2}$ cms.= 9.9 cms. Option (b).

40. (d) Perimeter of the square = $4a = 4 \times 14\sqrt{2}$ cms = 79.2 cms. Option (d).

41. Here $2(L + B) = 78$ & $LB = 350$. On solving we get L = 25 m. Hence option (c).

42. Sum of interior angles of a triangle = $(n - 2) \times 180°$ where n is the no of sides.
So, $(n - 2) \times 180° = 1800°$ n = 12. Hence option (b).

43-47

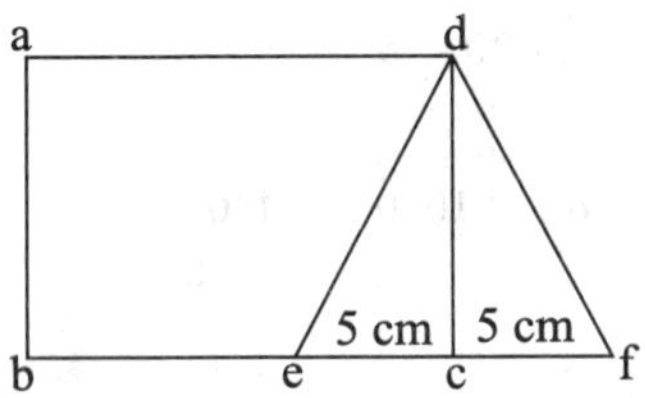

$$dc = \text{height of equilateral triangle} = \frac{\sqrt{3}}{2}a$$

$$= \frac{\sqrt{3}}{2} \times 10 = 5\sqrt{3} \text{ m}$$

$$dc = 5\sqrt{3} \text{ cm} = \text{side of the square.}$$

In the triangle edc. By Pythagoras theorem.

$$(de)^2 = (dc)^2 + (ec)^2$$

$$de = 10 \text{ cm.}$$

43. (a) Area of square = $(5\sqrt{3})^2 = 75$ sq. cm.
Hence option (a).

44. (a) Area of triangle def = $\frac{1}{2} \times \text{base} \times \text{height}$

$$= \frac{1}{2} \times 10 \times 5\sqrt{3} = 25\sqrt{3} \text{ cms}^2\text{. Hence option (a).}$$

45. (c) Side of the square = $5\sqrt{3}$ cm hence option (c)

46. (d) Side of the square = $5\sqrt{3}$

Diagonal = $\sqrt{2} \times \text{side} = 5\sqrt{6}$ cm.

Hence option (d)

47. (b) Perimeter of the triangle dec = de + ec + dc

$$= 5\sqrt{3} + 5 + 10 \text{ cm}$$

$$= 15 + 5\sqrt{3} \text{ cm}$$

Hence option (b)

48. (d) average $= \dfrac{11 + x + 58 + 56 + 33}{5} = 44$

Solving, x = 62

So mean of x, 155, 263, 148, 222

$$= \frac{62 + 155 + 263 + 148 + 222}{5} = 170$$

Hence option (d)

49. (c) Area = $\Pi r^2 = 1122.66$

$$r = 18.9 \text{ apprx}$$

$$\text{now circumference} = 2\Pi r = 2 \times \frac{22}{7} \times 19$$

$$= 118.89 \text{ cm}$$

Hence option (c)

50. (b) The circumference of the circle = 2π r. If the circumference increases by 5% then it means that the radius also increases by 5%. So % change in the area of the circle

$$= \left(r - r + \frac{r \times r}{100}\right)\% = \left(5 + 5 + \frac{5 \times 5}{100}\right)\% = 10.25\%$$

SECTION-IV

ALGEBRA

Day 21

Factorization, Remainder Theorem and Conditional Identities

Today we will do the first day of Algebra. As discussed we have divided the algebra into four days. So on day I that is today we will do **Factorization, Remainders Theorem and Conditional Identities.** First we will discuss **Factorization**.

FACTORIZATION

Factorization: In numerical problem if you are asked to make factors of 540 then you make it as follows :

2	540
2	270
3	135
3	45
3	15
	5

$$\mathbf{540 = 2 \times 2 \times 3 \times 3 \times 3 \times 5}$$

So factorization of numerical problem is done as shown above.

Now if you are asked to do the factorization of $x^2 + 8x + 15$ then you can't do as above. There are proper methods to factorize algebric expression. We have divided the factorization of algebric expression into following four types.

TYPES OF FACTORIZATION

Type I : Factorization by Grouping

Type II : Factorization by splitting middle term (of type $x^2 + px + q$)

Type III : Factorization by Algebric formulae

Type IV : Factorization of cyclic expression

TYPE I (Factorization By Grouping)

This is simplest of all the types and can be clearly understood by following examples.

Example 1. Factorize $ax + bx + ay + by$

Sol : $ax + bx + ay + by = x(a + b) + y(a + b)$

$= (a + b)(x + y)$

$= \mathbf{(a + b)(x + y)}$

Example 2. Factorize $1 - x^5 - x + x^4$

Sol : $1 - x^5 - x + x^4 = x^4 - x^5 + 1 - x$

$= x^4(1 - x) + 1(1 - x)$

$= \mathbf{(1 - x)(x^4 + 1)}$

Example 3. Factorize $x^4 + x^3 + 2x^2 + x + 1$

Sol : $x^4 + x^3 + 2x^2 + x + 1 = x^4 + x^3 + x^2 + x^2 + x + 1$

$= x^2(x^2 + x + 1) + 1(x^2 + x + 1)$

$= \mathbf{(x^2 + x + 1)(x^2 + 1)}$

Example 4. Factorize $xy(p^2 + q^2) + pq(x^2 + y^2)$

Sol : $xy(p^2 + q^2) + pq(x^2 + y^2) = xyp^2 + xyq^2 + pqx^2 + pqy^2$

$= p^2xy + pqx^2 + q^2xy + pqy^2$

$= px(py + qx) + qy(qx + py)$

$= \mathbf{(px + qy)(py + qx)}$

TYPE II
(Factorization By Splitting Middle Term)

This can be clearly under stood by following examples :

Example 5. Factorize $x^2 + 8x + 15$

Sol : If we compare it with $\mathbf{x^2 + px + q}$ then we split $\mathbf{p}$ into two parts such that **product** of these two parts is equal to $\mathbf{q}$. This can be further explained by taking $\mathbf{x^2 + 8x + 15}$ into consideration.

Here we break **8** into two parts *i.e.*, **+ 5 and + 3** such that its **sum** is **5 + 3 = 8** and **product** is **5 × 3 = 15** and we solve it as follows.

$x^2 + 8x + 15 = x^2 + 5x + 3x + 15$

$= x(x + 5) + 3(x + 5)$

$= \mathbf{(x + 5)(x + 3)}$

Now we will do few more examples

Example 6. Factorize $x^2 + 6x + 8$

Sol : $x^2 + 6x + 8 = x^2 + 2x + 4x + 8$

$= x(x + 2) + 4(x + 2)$

$= \mathbf{(x + 2)(x + 4)}$

Example 7. Factorize $x^2 - 5x + 6$

Sol : $x^2 - 5x + 6 = x^2 - 3x - 2x + 6$

$= x(x - 3) - 2(x - 3)$

$= \mathbf{(x - 3)(x - 2)}$

Example 8. Factorize $6x^2 - x - 2$

Sol : Here we are to split the middle term x into two parts such that its sum is -1 and product equal to $6 \times -2 = -12$

We solve it as follows : $6x^2 - x - 2 = 6x^2 + 3x - 4x - 2$

$$= 3x(2x + 1) - 2(2x + 1)$$

$$= \mathbf{(2x + 1)(3x - 2)}$$

Example 9. Factorize $14x^2 + 19x - 3$

Sol :

$$14x^2 + 19x - 3 = 14x^2 + 21x - 2x - 3$$

$$= 7x(2x + 3) - 1(2x + 3)$$

$$= \mathbf{(2x + 3)(7x - 1)}$$

Example 10. Factorize $4\sqrt{3}x^2 + 5x - 2\sqrt{3}$

Sol :

$$4\sqrt{3}x^2 + 5x - 2\sqrt{3} = 4\sqrt{3}x^2 + 8x - 3x - 2\sqrt{3}$$

$$= 4x(\sqrt{3}x + 2) - \sqrt{3}(\sqrt{3}x + 2)$$

$$= \mathbf{(\sqrt{3}x + 2)(4x - \sqrt{3})}$$

TYPE III
(Factorization By Formulae)

Before starting this type let us once again revise the basic formulae.

$$(a + b)^2 = a^2 + b^2 + 2ab$$

$$(a - b)^2 = a^2 + b^2 - 2ab$$

$$a^2 - b^2 = (a + b)(a - b)$$

$$a^3 + b^3 = (a + b)(a^2 - ab + b^2)$$

$$a^3 - b^3 = (a - b)(a^2 + ab + b^2)$$

In following example we would take help of the given formulae.

Example 11. Factorize $x^2 + 4x + 4$

Sol : $x^2 + 4x + 4 = x^2 + 2.2x + 2^2$

By using formula $x^2 + 2ax + a^2 = (x + a)^2$

We get $x^2 + 2.2x + 2^2 = \mathbf{(x + 2)^2}$

Example 12. Factorize $4x^2 - 12x + 9$

Sol :

$$4x^2 - 12x + 9 = (2x)^2 - 2.2x.3 + 3^2$$

$$= \mathbf{(2x - 3)^2}$$

Example 13. Factorize $25a^2 - 16b^2$

Sol :

$$25a^2 - 16b^2 = (5a)^2 - (4b)^2$$

$$= \mathbf{(5a + 4b)(5a - 4b)} \quad [\because \ a^2 - b^2 = (a + b)(a - b)]$$

Example 14. Factorize $4a^2 - 9b^2$

Sol :

$$4a^2 - 9b^2 = (2a)^2 - (3b)^2$$

$$= \mathbf{(2a + 3b)(2a - 3b)}$$

Example 15. Factorize $x^8 - y^8$

Sol :

$$\begin{aligned} x^8 - y^8 &= (x^4)^2 - (y^4)^2 \\ &= (x^4 + y^4)(x^4 - y^4) \\ &= (x^4 + y^4)[(x^2)^2 - (y^2)^2] \\ &= (x^4 + y^4)[(x^2 + y^2)(x^2 - y^2)] \\ &= x^4 + y^4 [(x^2 + y^2)(x + y)(x - y)] \\ &= \mathbf{(x^4 + y^4)(x^2 + y^2)(x + y)(x - y)} \end{aligned}$$

Example 16. Factorize $a^3 - 8b^3$

Sol :

$$\begin{aligned} a^3 - 8b^3 &= (a)^3 - (2b)^3 \\ &= (a - 2b)(a^2 + 2ab + 4b^2) \\ &\quad [\because x^3 - y^3 = (x - y)(x^2 + xy + y^2)] \\ &= \mathbf{(a - 2b)(a^2 + 2ab + 4b^2)} \end{aligned}$$

Example 17. Factorize $x^3 + 64$

Sol :

$$\begin{aligned} x^3 + 64 &= (x)^3 + (4)^3 \\ &= \mathbf{(x + 4)(x^2 - 4x + 16)} \end{aligned}$$

Example 18. Factorize $x^6 - y^6$

Sol :

$$\begin{aligned} x^6 - y^6 &= (x^2)^3 - (y^2)^3 \\ &= (x^2 - y^2)(x^4 + x^2y^2 + y^4) \\ &= \mathbf{(x + y)(x - y)(x^4 + x^2y^2 + y^4)} \end{aligned}$$

Example 19. Factorize $x^4 + 8x$

Sol :

$$\begin{aligned} x^4 + 8x &= x(x^3 + 8) \\ &= x[(x)^3 + (2)^3] \\ &= x[(x + 2)(x^2 - 2x + 4)] \\ &= \mathbf{x(x + 2)(x^2 - 2x + 4)} \end{aligned}$$

TYPE IV
(Factorization of Cyclic Expression)

Before doing **factorization** of **cyclic expression** Ist we will understand what is **cyclic expression**?

Consider the expression

$\boldsymbol{a(b + c) + b(c + a) + c(a + b)}$

It contains 3 terms

(i) $a(b + c)$ (ii) $b(c + a)$ and (iii) $c(a + b)$

Now if in **Ist term** i.e., $\boldsymbol{a(b + c)}$ if we change **a by b, b by c** and **c by a** we get **b(c + a)** i.e., **second term**. In the same manner we also get **second term** changed to **third term** and **third term** changed to **Ist term**. This type of expression is **cyclic expression.**

So the expression in which **each of the term** can be obtained from **previous term** by changing **first letter** by **second**, the **second letter** by **third** and so on till the **last by the first** is called **cyclic expression**.

A few more examples of cyclic expression are :

(i) $a^3(b-c)+b^3(c-a)+c^3(a-b)$

(ii) $x^2(y-z)^3+y^2(z-x)^3+z^2(x-y)^3$

(iii) $a(b^2+c^2)+b(c^2+a^2)+c(a^2+b^2)$ etc

Now we will do the factorization of cyclic expression.

Factorization of Cyclic Expression

For this following steps are taken

(i) First we open the brackets.

(ii) We will write the terms of the expression according to ascending or desending powers of one of the letters.

(iii) By taking two or three letters together we make one factor common and take it out.

(iv) Now we do the above same procedure for next letter and do till the factorization is done.

The above process can be more clearly understood by following example.

Example 20. Factorize $x(y^2+z^2)+y\,(z^2+x^2)+z(x^2+y^2)+2xyz$

Sol : $= x(y^2+z^2)+y(z^2+x^2)+z(x^2+y^2)+2xyz$

Ist Step : We open the brackets

$$xy^2+xz^2+yz^2+x^2y+x^2z+y^2z+2xyz$$

IInd Step : We arrange in the descending power of x as shown below :

x^2y+x^2z	+	xy^2+xz^2+2xyz	+	y^2z+yz^2
Highest power of x		Next highest power of x		No power of x

IIIrd Step : We take out the common factor

$$= x^2(y+z)+x(y^2+z^2+2yz)+yz(y+z)$$

$$= x^2(y+z)+x(y+z)^2+yz(y+z)$$

Now we take out y + z as common

$$= (y+z)[x^2+x(y+z)+yz]$$

$$= (y+z)(x^2+xy+xz+yz)$$

IVth Step : Now we arrange in the descending power of y and proceed as follows :

$$= (y+z)[xy+yz+x^2+xz]$$

$$= (y+z)[y(x+z)+x(x+z)]$$

$$= (y+z)[(x+z)\,(y+x)]$$

$$= \mathbf{(y+z)(x+z)(x+y)}$$

$$= \mathbf{(x+y)(y+z)(z+x)}$$

Now it is hoped that you have clearly understood how to factorize a cyclic expression.

Now we will do few more examples of cyclic expression.

Example 21. Factorize $a^2(b-c)+b^2(c-a)+c^2(a-b)$

Sol : Ist we **remove bracket** and get

$$= a^2b - a^2c + b^2c - b^2a + c^2a - c^2b$$

Now we arrange in the **descending power of a**

$$= a^2b - a^2c + ac^2 - ab^2 + b^2c - c^2b$$
$$= a^2(b-c) - a(b^2-c^2) + bc(b-c)$$
$$= a^2(b-c) - a(b+c)(b-c) + bc(b-c)$$
$$= (b-c)[a^2 - a(b+c) + bc]$$
$$= (b-c)[a^2 - ab - ac + bc]$$

Now arranging second factor according to **descending power of b**, we get

$$= (b-c)[bc - ab + a^2 - ac]$$
$$= (b-c)[b(c-a) - a(c-a)]$$
$$= (b-c)(b-a)(c-a)$$
$$= \mathbf{-(a-b)(b-c)(c-a)}$$

Example 22. Factorize $(a+b+c)(ab+bc+ca) - abc$

Sol : $(a+b+c)(ab+bc+ca) - abc$

$$= a^2b + abc + a^2c + ab^2 + b^2c + abc + abc + bc^2 + ac^2 - abc$$
$$= a^2b + a^2c + ab^2 + b^2c + bc^2 + c^2a + 2abc$$

Arranging in the **descending power of a** we get

$$= a^2b + a^2c + ab^2 + ac^2 + 2abc + b^2c + bc^2$$
$$= a^2(b+c) + a(b^2+c^2+2bc) + bc(b+c)$$
$$= a^2(b+c) + a(b+c)^2 + bc(b+c)$$
$$= (b+c)[a^2 + a(b+c) + bc]$$
$$= (b+c)[a^2 + ab + ac + bc]$$

Now arranging with **descending power of *b*** we get

$$= (b+c)[ab + bc + a^2 + ac]$$
$$= (b+c)[b(a+c) + a(a+c)]$$
$$= (b+c)[(a+c)(b+a)]$$
$$= (b+c)(a+c)(a+b)$$
$$= \mathbf{(a+b)(b+c)(c+a)}$$

Example 23. Factorize $a^3(b-c)+b^3(c-a)+c^3(a-b)$

Sol : $a^3(b-c)+b^3(c-a)+c^3(a-b)$

$$= a^3b - a^3c + b^3c - b^3a + c^3a - c^3b$$

Arranging in the terms according to descending power of a we get

$$= a^3b - a^3c - ab^3 + ac^3 - c^3b + cb^3$$
$$= a^3(b-c) - a(b^3 - c^3) + bc(b^2 - c^2)$$
$$= a^3(b-c) - a(b-c)(b^2 + bc + c^2) + bc(b+c)(b-c)$$
$$= (b-c)[\,a^3 - a(b^2 + bc + c^2) + bc(b+c)]$$
$$= (b-c)[a^3 - ab^2 - abc - ac^2 + b^2c + bc^2]$$

Now arranging in the **descending power of b** we get

$$= (b-c)[b^2c - b^2a + bc^2 - abc + a^3 - ac^2]$$
$$= (b-c)[b^2(c-a) + bc(c-a) - a(c^2 - a^2)]$$
$$= (b-c)[b^2(c-a) + bc(c-a) - a(c+a)(c-a)]$$
$$= (b-c)[(c-a)(b^2 + bc - a(c+a)]$$
$$= (b-c)[(c-a)(b^2 + bc - ac - a^2)]$$

Now arranging in the **descending power of c** we get

$$= (b-c)(c-a)[bc - ac + b^2 - a^2]$$
$$= (b-c)(c-a)[(b-a)(c+b+a)]$$
$$= (b-c)(c-a)(b-a)(a+b+c)$$
$$= \mathbf{-(a-b)(b-c)(c-a)(a+b+c)}$$

REMAINDER AND FACTOR THEOREM

Now before starting this topic divide $x^2 - 5x + 2$ by $x - 2$.

We do it as follows :

$$
\begin{array}{r|l}
x-2 & x^2 - 5x + 2 \;(x - 3 \\
 & \underline{x^2 - 2x} \\
 & \quad -3x + 2 \\
 & \quad -3x + 6 \\
 & \quad \underline{\;+ \quad\; -} \\
 & \quad\quad\; \underline{-4}
\end{array}
$$

So when $x^2 - x + 2$ is divided by $x - 2$ the remainder is -4

Now there is a theorem **(Remainder Theorem)** by which we can find the **remainder** without actually performing the **process of division.** This theorem states that **when a polynomial p(x) with degree greater than one is divided by a binomial of the form (x – a) then remainder is obtained by putting x = a in p(x).**

So when $\mathbf{p(x) = x^2 - 5x + 2}$ was divided by $\boldsymbol{x - 2}$ then remainder can be obtained by putting $\boldsymbol{x = 2}$ in $p(x)$ i.e., in $\boldsymbol{x^2 - 5x + 2}$.

So remainder $= (2)^2 - 5x\,2 + 2$

$= 4 - 10 + 2$

$= 6 - 10$

$= \mathbf{-4}$

Which is same as obtained by actual process of division.

Now we will do few more examples based on above theorem.

Example 24. Find remainder when $x^3 + 3x^2 - 12x + 6$ is divided by $x - 2$

Sol : Let $p(x) = x^3 + 3^2x - 12x + 6$

Remainder is obtained by putting

$$x - 2 = 0 \quad \text{i.e., } x = 2 \text{ in } p(x).$$

$$\begin{aligned} p(2) &= 2^3 + 3.2^2 - 12.2 + 6 \\ &= 8 + 12 - 24 + 6 \\ &= 26 - 24 = \mathbf{2} \end{aligned}$$

$\therefore$ remainder is **2**

Example 25. Find remainder when $x^3 - 14x^2 + 36x - 60$ is divided by $x + 1$

Sol : Let $p(x) = x^3 - 14x^2 + 36x - 60$

Remainder is obtained by putting

$$x + 1 = 0 \quad \text{i.e., } x = -1 \text{ in } p(x).$$

$$\begin{aligned} \therefore \quad p(-1) &= (-1)^3 - 14(-1)^2 + 36(-1) + 60 \\ &= -1 - 14 - 36 + 60 \\ &= 60 - 51 = \mathbf{9} \end{aligned}$$

$\therefore$ remainder is **9**

Now we will do factor theorem

Factor Theorem

If there is **no remainder** obviously the given polynomial is **completly divisible** by the given divisor. It means **divisor** is a **factor** of given polynomial.

So we have the following theorem.

Let **$p(x)$ be a polynomial and a be the real number, then $(x - a)$ is a factor of $p(x)$ only if $p(a) = 0$.**

i.e., (i) If $(x - a)$ is a factor of $p(x)$ then $p(a) = 0$

(ii) If $p(a) = 0$ Then $(x - a)$ is a factor of $p(x)$

Now we will do few examples based on factor theorem

Example 26. Find which one of the following is a factor of $f(x) = x^2 - 5x + 6$

(a) $(x - 1)$ (b) $(x - 2)$ (c) $(x - 5)$ (d) $(x - 7)$

Sol : Let $x - 1$ is a factor then putting $x = 1$ must make $f(x)$ equal to zero.

i.e.,

$$\begin{aligned} f(1) &= 1^2 - 5 \times 1 + 6 \\ &= 1 - 5 + 6 \\ &= 7 - 5 = 2 \end{aligned}$$

Which is not equal to zero.

$\therefore$ $x - 1$ is not a factor of $x^2 - 5x + 6$

Now we will check with second option

Let $x - 2$ is a factor then putting $x = 2$ must make $f(x)$ equal to zero.

$$\therefore \quad \text{if}(2) = 2^2 - 5 \times 2 + 6$$
$$= 4 - 10 + 6$$
$$= 10 - 10 = 0$$

$\therefore$ **$x - 2$ is a factor of $f(x)$**

Example 27. Find the value of k for which $x^3 - 2kx^2 + kx - 1$ is a factor of $x - 2$

Sol : Let $p(x) = x^3 - 2kx^2 + kx - 1$

if $x - 2$ is a factor of $p(x)$ then $p(2) = 0$

i.e., $2^3 - 2 \times k \times 2^2 + 2k - 1 = 0$

i.e., $8 - 8k + 2k - 1 = 0$

i.e., $-6k + 7 = 0$

i.e., $-6k = -7$

i.e., $k = \frac{7}{6}$

Example 28. Find the value of a if $x - a$ is a factor of $x^5 - a^2x^3 + 2x + a + 2$

Sol : Let $p(x) = x^5 - a^2x^3 + 2x + a + 2$

if $x - a$ is factor of $p(x)$ then $p(a) = 0$

i.e., $(a)^5 - a^2a^3 + 2a + a + 2 = 0$

i.e., $a^5 - a^5 + 3a + 2 = 0$

i.e., $3a + 2 = 0$

i.e., $3a = -2$

i.e., $a = \frac{-2}{3}$

Example 29. If $(x - 1)$ and $(x - 2)$ are factors of $p(x) = lx^2 + mx + 6$ then find the value of l and m.

Sol : In this question we will make two equations

First by putting $\boldsymbol{p(1) = 0}$ and then by putting $\boldsymbol{p(2) = 0}$

As $x - 1$ is factor of $p(x) = lx^2 + mx + 6$

$\therefore \quad p(1) = 0$

$\therefore \quad l(1)^2 + m(1) + 6 = 0$

$\therefore \quad l + m = -6$ ... (i)

Further $p(2) = 0$

$\therefore \quad l(2)^2 + m(2) + 6 = 0$

$\therefore \quad 4l + 2m = -6$

$\therefore \quad 2l + m = -3$ ... (ii)

Now we solve (i) and (ii)

$$\begin{aligned} l + m &= -6 \\ 2l + m &= -3 \\ - \; - \quad & \quad + \\ \hline -l &= -3 \\ \hline l &= 3 \end{aligned}$$

$\therefore \quad 3 + m = -6$

$\therefore \quad m = -6 - 3$

$\therefore \quad m = -9$

$$\boldsymbol{l = 3}$$

$$\boldsymbol{m = -9}$$

So it is hoped that now you can solve the equation based on factor and remainder theorem.

Now we will do a few theorems on divisibility.

Theorem 1

The expression $\boldsymbol{a^n + b^n}$ is exactly divisible by $\boldsymbol{a + b}$, If ***n* is odd**, but **not if n is even**.

e.g., $a^3 + b^3$, $a^7 + b^7$, $a^9 + b^9$ are exactly divisible by $a + b$ where as $a^2 + b^2$, $a^4 + b^4$, $a^6 + b^6$ are not.

Theorem 2

The expression $\boldsymbol{a^n + b^n}$ is never divisible by $\boldsymbol{a - b}$, Whether ***n* is even or odd**.

e.g., $a^2 + b^2$, $a^3 + b^3$, $a^4 + b^4$ etc are never divisible by $a - b$.

Theorem 3

The expression $\boldsymbol{x^n - y^n}$ is always divisible by $\boldsymbol{x - y}$, If n is **+ ve integer even or odd**.

e.g., $x^2 - y^2$, $x^3 - y^3$, $x^4 - y^4$, $x^5 - y^5$ are exactly divisible by $x - y$.

Theorem 4

The expression $\boldsymbol{x^n - y^n}$ is exactly divisible by $\boldsymbol{x + y}$, When n is an **even + ve integer**, but not if n is odd.

e.g., $x^2 - y^2$, $x^4 - y^4$,$x^6 - y^6$ are divisible by $x + y$.

Whereas $x^3 - y^3$, $x^5 - y^5$ etc are never divisible by $x + y$.

Now solve the following examples.

Example 30. The expression $2^{6n} - 3^{3n}$ is divisible by what number. If n is any + ve integer.

Sol : $\quad 2^{6n} - 3^{3n} = (2^6)^n - (3^3)^n$

$\quad = 64^n - 27^n$

By Theorem 3 this expression is divisible by 64 – 27 *i.e.*, **37.**

Example 31. The expression $12^n - 1$ divisible by what number. If n is any + *ve* integer.

Sol : $\quad 12^n - 1 = 12^n - 1^n$

By Theorem 3 this expression is divisible by **12 – 1** i.e., **11.**

Example 32. The expression $17^n - 1$ is divisible by which number. If n is any + *ve* even integer.

Sol : $17^n - 1 = 17^n - 1^n$

Since n is + ve even integer

$\therefore$ $17^n - 1$ is divisible by $17 + 1 = 18$

Now will do the topic of Conditional Identities

CONDITIONAL – IDENTITIES

Conditional Identities are the two equality of two expression subject to given condition.

Now we will do a few example of Conditional Identities.

Example 33. If $a + b + c = 0$ then value of $(a + b)(b + c)(c + a)$ is

Sol : As $a + b + c = 0$

$$\therefore \quad a + b = -c$$
$$b + c = -a$$
$$a + c = -b$$
$$\therefore \quad (a + b)(b + c)(c + a) = (-a)(-b)(-c) = \boldsymbol{-abc}$$

Example 34. If $a + b + c = 0$ then value of $\dfrac{a^3 + b^3 + c^3}{abc}$ is ....

Sol : As $a + b + c = 0$

$$\therefore \quad a + b = -c$$

Cubing both sides

$$(a + b)^3 = (-c)^3$$
$$\therefore \quad a^3 + b^3 + 3ab(a + b) = -c^3$$
$$\therefore \quad a^3 + b^3 - 3abc = -c^3 \quad [\because a + b = -c]$$
$$\therefore \quad a^3 + b^3 + c^3 = 3\,abc$$
$$\therefore \quad \frac{a^3 + b^3 + c^3}{abc} = \mathbf{3}$$

Example 35. If $x + y + z = 0$ then value of $\dfrac{x^2 + y^2 + z^2}{xy + yz + zx}$ is ....

Sol : As $x + y + z = 0$

Squaring both sides

$$(x + y + z)^2 = 0$$
$$\therefore \quad x^2 + y^2 + z^2 + 2(xy + yz + zx) = 0$$
$$\therefore \quad x^2 + y^2 + z^2 = -2(xy + yz + zx)$$

$\therefore$ the value of $\dfrac{x^2 + y^2 + z^2}{xy + yz + zx} = \dfrac{-2(xy + yz + zx)}{xy + yz + zx} = \mathbf{-2}$

Example 36. If $a + b = 2c$ then value of $\frac{a}{a-c} + \frac{c}{b-c}$ is ....

Sol : As

$$a + b = 2c$$

$$\therefore \quad a + b - 2c = 0$$

$$\therefore \quad a + b - c - c = 0$$

$$(a - c) + (b - c) = 0$$

$$a - c = -(b - c)$$

$$\therefore \quad \frac{a}{a-c} + \frac{c}{b-c} = \frac{a}{a-c} + \frac{c}{-(a-c)}$$

$$= \frac{a}{a-c} - \frac{c}{(a-c)}$$

$$= \frac{a-c}{a-c} = 1$$

Example 37. If $a + b + 2c = 0$ then value of $a^3 + b^3 + 8c^3 =$ ....

Sol : If

$$x + y + z = 0$$

$$\therefore \quad x^3 + y^3 + z^3 = 3xyz$$

So

$$a + b + 2c = 0$$

$$\therefore \quad (a)^3 + (b)^3 + (2c)^3 = 3 \times a \times b \times 2c = \mathbf{6\,abc}$$

Example 38. If $x = a(b - c), y = b(c - a), z = c(a - b)$ then value of $\frac{abc}{xyz}\left[\left(\frac{x}{a}\right)^3 + \left(\frac{y}{b}\right)^3 + \left(\frac{z}{c}\right)^3\right]$ is ....

Sol : As $x = a(b - c), y = b(c - a), z = c(a - b)$.

$$\therefore \quad \frac{x}{a} = b - c$$

$$\therefore \quad \frac{y}{b} = c - a$$

$$\therefore \quad \frac{z}{c} = a - b$$

$$\therefore \quad \left(\frac{x}{a}\right) + \left(\frac{y}{b}\right) + \left(\frac{z}{c}\right) = b - c + c - a + a - b = 0$$

$$\therefore \quad \left(\frac{x}{z}\right)^3 + \left(\frac{y}{b}\right)^3 + \left(\frac{z}{b}\right)^3 = \frac{3xyz}{abc}$$

$$\therefore \quad \text{Value of } \frac{abc}{xyz}\left[\left(\frac{x}{a}\right)^3 + \left(\frac{y}{b}\right)^3 + \left(\frac{z}{c}\right)^3\right]$$

$$\frac{ahc}{xyz} \times \frac{3xyz}{abc} = \mathbf{3}$$

Example 39. If $a + b + c = 0$ then value of $\frac{(a+b)(a+b)(a+c)}{abc}$ is ....

Sol : As $a + b + c = 0$

$\therefore$ $a + b = -c$

$\therefore$ $b + c = -a$

$\therefore$ $c + a = -b$

$\therefore$ $$\frac{(a+b)(a+b)(a+c)}{abc} = \frac{-abc}{abc}$$

Now it is hoped that you have understood the concept of factorization, remainder and factor theorem and conditional identities and we have also done sufficient examples of above topic.

Now we will do the exercise.

EXERCISE

Q.1. The factors of $a^2 + ab + ac + bc$ are

(a) $(a - b)(a - c)$ (b) $(a + b)(a + c)$ (c) $(a - b)(a + c)$ (d) none

Q.2. The factors of $x^2y^2 + y^2 + x^2 + 1$ are

(a) $(x^2 + 1)(y^2 - 1)$ (b) $(x^2 - 1)(y^2 - 1)$ (c) $(x^2 + 1)(y^2 + 2)$ (d) none

Q.3. The factors of $x^2 + yz - xy - xz$ are

(a) $(x + y)(y + z)$ (b) $(x - y)(x - z)$ (c) $(x - z)(x + y)$ (d) none

Q.4. The factors of $x^2 - \left(a + \frac{1}{4}\right)x + 1$ are

(a) $(x - a)\left(x - \frac{1}{a}\right)$ (b) $(x + a)\left(x + \frac{1}{a}\right)$ (c) $\left(x + \frac{1}{a}\right)(x - a)$ (d) none

Q.5. The factors of $2x^2 - 2x - 24$ are

(a) $2(x + 2)(x - 3)$ (b) $(x + 2)(x - 3)$ (c) $2(x - 2)(x + 3)$ (d) none

Q.6. The factors of $6x^2 + x - 12$ are

(a) $(3x + 4)(2x - 3)$ (b) $(3x - 4)(2x + 3)$ (c) $(3x - 4)(2x - y)$ (d) none

Q.7. The factors of $6x^2 + x - 15$ are

(a) $(3x + 5)(2x - 3)$ (b) $(3x - 5)(2x + 3)$ (c) $(3x - 5)(2x - y)$ (d) none

Q.8. The factors of $y^2 - 10y + 24$ are

(a) $(y - 4)(y - 6)$ (b) $(y + 6)(y - 4)$ (c) $(y + 4)(y - 6)$ (d) none

Q.9. The factors of $\sqrt{3}\,y^2 + 11y + 6\sqrt{3}$ are

(a) $(y - 3\sqrt{3})(\sqrt{3}y + 2)$ (b) $(y - 3\sqrt{3})(\sqrt{3}y - 2)$ (c) $(y + 3\sqrt{3})(\sqrt{3}y + 2)$ (d) none

Q.10. The factors of $5x^2 - 16x - 21$ are

(a) $(x - 1)(5x + 21)$ (b) $(x + 1)(5x - 21)$ (c) $(x + 1)(5x + 21)$ (d) none

Q.11. The factors of $4x^2 + 4x - 3$ are

(a) $(2x + 1)(2x - 3)$ (b) $(2x - 1)(2x - 3)$ (c) $(2x - 1)(2x + 3)$ (d) none

Q.12. The factors of $14x^2 + 19x - 3$ are

(a) $(2x + 3)(7x - 1)$ (b) $(2x - 3)(7x + 1)$ (c) $(2x + 3)(7x + 1)$ (d) none

Q.13. The factors of $25x^2 - 10x + 1 - 36z^2$ are

(a) $(5x + 6z + 1)(5x + 6z - 1)$ (b) $(5x + 6z - 1)(5x - 6z - 1)$

(c) $(5x - 6z - 1)(5x - 6z + 1)$ (d) none

Q.14. The factors of $1 - 2ab - (a^2 + b^2)$ are

(a) $(1 - a - b)(1 + a - b)$ (b) $(1 - a - b)(1 - a + b)$

(c) $(1 + a + b)(1 - a - b)$ (d) none

Q.15. The factors of $a^2 + b^2 + 2(ab + bc + ca)$ are

(a) $(a - b)(a + b + 2c)$ (b) $(a + b)(a + b + 2c)$ (c) $(a + b)(a + b - 2c)$ (d) none

Q.16. The factors of $a^2 - 25b^2$ are

(a) $(a - 5b)(a + 3b)$ (b) $(a + 5b)(a - 3b)$ (c) $(a + 5b)(a - 5b)$ (d) none

Q 17. The factors of $a^3 + 125b^3$ are

(a) $(a + 5b)(a^2 + 25b^2 + b^2)$ (b) $(a + 5b)(a^2 - 5ab + 25b^2)$

(c) $(a^2 + 5b)(a^2 + 5ab + b^2)$ (d) none

Q.18. The factors of $x^6 - 64$ are

(a) $(x + 2)(x - 2)(x^2 - 2x + 4)(x^2 + 2x + 4)$ (b) $(x - 2)(x + 2)(x^2 + 2x - 4)(x^2 + 2x + 4)$

(c) $(x - 2)(x + 2)(x^2 + 2x + 4)(x^2 + 2x + 4)$ (d) none

Q.19. The factors of $x^6 + 26x^3 - 27$ are

(a) $(x - 1)(x^2 + x + 1)(x + 3)(x^2 - 3x + 9)$ (b) $(x + 1)(x^2 + x + 1)(x + 3)(x^2 - 3x + 9)$

(c) $(x + 1)(x^2 - x - 1)(x + 3)(x^2 - 3x - 9)$ (d) none

Q.20. The factors of $x^6 - 729$ are

(a) $(x + 3)(x^2 - 3x + 9)(x - 3)(x^2 + 3x + 9)$ (b) $(x + 3)(x^2 - 3x - 9)(x - 3)(x^2 + 3x - 9)$

(c) $(x^2 - 3)(x^4 - 9)$ (d) none

Q.21. The factors of bc $(b - c) + ca(c - a) + ab(a - b)$ are

(a) $(a - b)(b - c)(c - a)$ (b) $-(a - b)(b - c)(c - a)$

(c) $(a + b)(b + c)(c + a)$ (d) none

Q.22. The factors of $x(y^2 - z^2) + y(z^2 - x^2) + z(x^2 - y^2)$ are

(a) $(x - y)(y - z)(z - x)$ (b) $-(x - y)(y - z)(z - x)$

(c) $(x + y)(y + z)(z + x)$ (d) none

Q.23. The factors of $x^2(y - z) + y^2(z - x) + z^2(x - y)$ are

(a) $(x - y)(y - z)(z - x)$ (b) $-(x - y)(y - z)(z - x)$

(c) $(x + y)(y + z)(z + x)$ (d) none

Q.24. The factors of $x^3(y - z) + y^3(z - x) + z^3(x - y)$ are

(a) $(y - z)(z - x)(x - y)(x + y + z)$ (b) $(y + z)(z - x)(x - y)(x - y - z)$

(c) $-(y - z)(z - x)(x - y)(x + y + z)$ (d) none

Q.25. If $p(x) = 4x^3 - 3x^2 + 2x - 3$ is divided by $(x - 2)$. The remainder is

(a) 21 (b) 20 (c) 19 (d) none

Q.26. If $x^4 - 3x^2 + 3x - 5$ is divided by $(x - 1)$. The remainder is

(a) 3 (b) 4 (c) 5 (d) none

Q.27. The value of a if $(x - 1)$ is a factor of $x^3 - 3x^2 + a + 1$ is

(a) 0 (b) 1 (c) 2 (d) none

Q.28. The value of a if $x + a$ is a factor of $x^3 + ax^2 - 2x + a - 2$ is

(a) $\frac{2}{3}$ (b) $\frac{1}{3}$ (c) 2 (d) none

Q.29. The value of k if $(x - k)$ is a factor of $x^6 - kx^5 + x^4 - kx^3 + 3x - k + 2$ is

(a) 1 (b) 2 (c) -1 (d) none

Q.30. The value of k for which expression $x^2 - 5x + k$ is divisible by $(x - 3)$ is

(a) 4 (b) 5 (c) 6 (d) none

Q.31. For what value of k and m, $(x + 1)(x - 2)$ are factors of $x^3 + kx^2 + mx + 6$ is

(a) $k = 1$ $m = -4$ (b) $k = 4$ $m = -1$ (c) $k = -1$ $m = 4$ (d) none

Q.32. For what value of k and m, $(x - 1)$ and $(x - 2)$ are the factors of $x^3 + 2kx^2 + mx + 1$ is

(a) $-\frac{3}{4}, -\frac{3}{2}$ (b) $-3, -2$ (c) $-4, -2$ (d) none

Q.33. $a^n + b^n$ is divisible by $a + b$

(a) For all value of n (b) Only for even value of n

(c) Only for odd value of n (d) For no value of n

Q.34. $x^n + a^n$ is divisible by $x - a$

(a) For all value of n (b) For even value of n

(c) For odd value of n (d) For no value of n

Q.35. $a^n - b^n$ is divisible by $a - b$

(a) For odd value of n (b) For even value of n

(c) For all value of n (d) For no value of n

Q.36. $x^n - y^n$ is divisible by $x + y$ for

(a) All value of n (b) Only for even value of n

(c) Only for odd value of n (d) For no value of n

Q.37. If $ab + bc + ac = 0$ the value of $\frac{a^2 + b^2 + c^2}{(a + b + c)^2}$ is

(a) 1 (b) 0 (c) -1 (d) none

Q.38. If $(a + b + c) = 0$ the value of $\frac{a^2 + b^2 + c^2}{a^2 - bc}$ is

(a) 0 (b) 1 (c) 2 (d) none

Q.39. The value of $\frac{8abc}{a^3 + b^3 + c^3}$. If $a + b + c = 0$

(a) 3 (b) 8 (c) $\frac{8}{3}$ (d) none

Q.40. If $(x + y + z) = 0$ then value of $\dfrac{1}{x^2 + y^2 - z^2} + \dfrac{1}{y^2 + z^2 - x} + \dfrac{1}{z^2 + x^2 - y}$ is

(a) 0 (b) 1 (c) 2 (d) none

Q.41. if $\dfrac{a}{b+c} = x$, $\dfrac{a}{c+a} = y$, $\dfrac{c}{a+b} = 2$ then the value of $\dfrac{1}{1+x} + \dfrac{1}{1+y} + \dfrac{1}{1+z}$

(a) 0 (b) 1 (c) 2 (d) none

Q.42. If $x = a^2 - bc, y = b^2 - ca, z = c^2 - ab$ then value of $\dfrac{ax + by\,cz}{(a+b+c)(x+y+z)}$ is

(a) 3 (b) 2 (c) 1 (d) none

SOLUTIONS

1. (b)
$$\begin{aligned} a^2 + ab + ac + bc &= a(a+b) + c(a+b) \\ &= \mathbf{(a+b)(a+c)} \end{aligned}$$

2. (d)
$$\begin{aligned} x^2y^2 + y^2 + x^2 + 1 &= x^2y^2 + x^2 + y^2 + 1 \\ &= x^2(y^2+1) + 1(y^2+1) \\ &= \mathbf{(y^2+1)(x^2+1)} \end{aligned}$$

3. (b)
$$\begin{aligned} x^2 + yz - xy - xz &= x^2 - xz + yz - xy \\ &= x(x-z) - y(x-z) \\ &= (x-z)(x-y) = \mathbf{(x-y)(x-z)} \end{aligned}$$

4. (a)
$$\begin{aligned} x^2 - \left(a + \frac{1}{a}\right)x + 1 &= x^3 - ax - \frac{1}{a}x + 1 \\ &= x^2 - \frac{x}{a} - ax + 1 \\ &= x\left(x - \frac{1}{a}\right) - a\left(x - \frac{1}{a}\right) \\ &= \left(x - \frac{1}{a}\right)(x-a) = \mathbf{(x-a)\left(x - \frac{1}{a}\right)} \end{aligned}$$

5. (d)
$$\begin{aligned} 2x^2 - 2x - 24 &= 2(x^2 - x - 12) \\ &= 2(x^2 - 4x + 3x - 12) \\ &= 2[(x(x-4) + 3(x-4)] \\ &= \mathbf{2[(x-4)(x+3)]} \end{aligned}$$

6. (b)
$$\begin{aligned} 6x^2 + x - 12 &= 6x^2 + 9x - 8x - 12 \\ &= 3x(2x+3) - 4(2x+3) \\ &= (2x+3)(3x-4) \\ &= (3x-4)(2x+3) \end{aligned}$$

7. (a)
$$\begin{aligned} 6x^2 + x - 15 &= 6x^2 + 10x - 9x - 15 \\ &= 2x(3x+5) - 3(3x+5) \\ &= \mathbf{(3x+5)(2x-3)} \end{aligned}$$

8. (a)
$$\begin{aligned} y^2 - 10y + 24 &= y^2 - 6y - 4y + 24 \\ &= y(y-6) - 4(y-6) = (y-6)(y-4) \\ &= \mathbf{(y-4)(y-6)} \end{aligned}$$

9. (c)
$$\begin{aligned} y^2 + 11y + 6\sqrt{3} &= \sqrt{3}y^2 + 9y + 2y + 6\sqrt{3} \\ &= \sqrt{3}y(y - 3\sqrt{3}) + 2(y + 3\sqrt{3}) \\ &= \mathbf{(y + 3\sqrt{3})(\sqrt{3}y + 2)} \end{aligned}$$

10. (b)
$$\begin{aligned} 5x^2 - 16x - 21 &= 5x^2 - 21x + 5x - 21 \\ &= x(5x-21) + 1(5x-21) \\ &= (5x-21)(x+1) \\ &= \mathbf{(x+1)(5x-21)} \end{aligned}$$

11. (c)
$$\begin{aligned} 4x^2 + 4x - 3 &= 4x^2 + 6x - 2x - 3 \\ &= 2x(2x+3) - 1(2x+3) \\ &= \mathbf{(2x-1)(2x+3)} \end{aligned}$$

12. (a)
$$\begin{aligned} 14x^2 + 19x - 3 &= 14x^2 + 21x - 2x - 3 \\ &= 7x(2x+3) - 1(2x+3) \\ &= (7x-1)(2x+3) \\ &= \mathbf{(2x+3)(7x-1)} \end{aligned}$$

13. (b)
$$\begin{aligned} 25x^2 - 10x + 1 - 36z^2 &= (25x^2 - 10x + 1) - (36z^2) \\ &= [(5x)^2 - 2.5x + 12)] - (6z)^2 \\ &= (5x-1)^2 - (6z)^2 \\ &= (5x - 1 + 6z)(5x - 1 - 6z) \\ &= \mathbf{(5x + 6z - 1)(5x - 6z - 1)} \end{aligned}$$

14. (c)
$$\begin{aligned} 1 - 2ab - (a^2 + b^2) &= 1 - 2ab - a^2 - b^2 \\ &= 1 - (2ab + a^2 + b^2) \\ &= 1^2 - (a+b)^2 = \mathbf{(1 + a + b)(1 - a - b)} \end{aligned}$$

15. (b)
$$\begin{aligned} &= a^2 + b^2 + 2(ab + bc + ca) \\ &= a^2 + b^2 + 2ab + 2bc + 2ca \\ &= (a^2 + b^2 + 2ab) + (2bc + 2ca) \\ &= (a+b)^2 + 2c(a+b) \\ &= \mathbf{(a+b)(a+b+2c)} \end{aligned}$$

16. (c)
$$\begin{aligned} a^2 - 25b^2 &= (a)^2 - (5b)^2 \\ &= \mathbf{(a+5b)(a-5b)} \end{aligned}$$

17. (b)
$$\begin{aligned} a^3 + 125b^3 &= (a)^3 + (5b)^3 \\ &= \mathbf{(a+5b)(a^2 - 5ab + 25b^2)} \end{aligned}$$

18. (a)
$$\begin{aligned} x^6 - 64 &= (x^3)^2 - (2^3)^2 \\ &= (x^3 + 2^3)(x^3 - 2^3) \\ &= [(x)^3 + (2)^3)][(x)^3 - (2)^3)] \\ &= (x+2)(x^2 - 2x + 4)\ (x-2)(x^2 + 2x + 4) \\ &= \mathbf{(x+2)(x-2)(x^2 - 2x + 4)\ (x^2 + 2x + 4)} \end{aligned}$$

19. (a)
$$x^6 + 26x^3 - 27 = x^6 + 27x^3 - x^3 - 27$$

$= x^3(x^3 + 27) - 1(x^3 + 27) = (x^3 + 27)[(x)^3 - (1)^3]$

$= [(x + 3)(x^2 - 3x + 9)(x + 3)(x^2 - 3x + 9)]$

$= \mathbf{(x - 1)(x^2 + x + 1)(x + 3)(x^2 - 3x + 9)}$

20. (a) $x^6 - 729 = (x^3)^2 - (3^3)^2$

$= (x^3 + 3^3)(x^3 - 3^3)$

$= [(x + 3)^3)][(x)^3 - (3)^3]$

$= (x + 3)(x^2 - 3x + 9)(x - 3)(x^2 + 3x + 9)$

$= \mathbf{(x + 3)(x^2 - 3x + 9)(x - 3)(x^2 + 3x + 9)}$

21. (b) $= bc(b - c) + ca(c - a) + ab(a - b)$

$= b^2c - bc^2 + c^2a - a^2c + a^2b - ab^2$

arranging in descending power of a

$= a^2b - a^2c + ac^2 - ab^2 + b^2c - bc^2$

$= a^2(b - c) - a(b^2 - c^2) + bc(b - c)$

$= a^2(b - c) - a(b + c)(b - c) + bc(b - c)$

$= (b - c)[a^2 - a(b + c) + bc]$

$= (b - c)[a^2 - ab - ac + bc]$

arranging in descending power of b

$= (b - c)[bc - ab + a^2 - ac]$

$= (b - c)[b(c - a) - a(c - a)]$

$= (b - c)[(c - a)(b - a)]$

$= \mathbf{-(a - b)(b - c)(c - a)}$

22. (a) $= x(y^2 - z^2) + y(z^2 - x^2) + z(x^2 - y^2)$

$= xy^2 - xz^2 + yz^2 - yx^2 + zx^2 - zy^2$

arranging in descending power of x, we get

$= \underline{zx^2 - x^2y} + \underline{xy^2 - xz^2} + \underline{yz^2 - y^2z}$

$= x^2(z - y) - x(z^2 - y^2) + yz(z - y)$

$= (z - y)[x^2 - x(z + y) + yz)]$

$= (z - y)[y(z - x) - x(z - x)]$

$= \mathbf{(x - y)(y - z)(z - x)}$

23. (b) $x^2(y - z) + y^2(z - x) + z^2(x - y) = x^2y - x^2z + y^2z - y^2x + z^2x - z^2y$

arranging in descending order of x, we get

$= x^2y - x^2z - xy^2 + xz^2 + y^2z - yz^2$

$= x^2(y - z) - x(y^2 - z^2) + yz(y - z)$

$= (y - z)[x^2 - x(y + z) + yz)]$

$= (y - z)[x^2 - xy - xz + yz]$

$= (y - z)[x(x - y) - z(x - y)]$

$= (y - z)(x - y)(x - z)$

$= \mathbf{-(x - y)(y - z)(z - x)}$

24. (c) $x^3(y - z) + y^3(z - x) + z^3(x - y) = x^3y - x^3z + y^3z - y^3x + z^3x - z^3y$

$= x^3y - x^3z + xz^3 - xy^3 + y^3z - yz^3$

$$= x^3(y-z) - x(y^3 - z^3) + yz(y^2 - z^2)$$
$$= x^3(y-z) - x(y - z + z^2) + yz(y+z)(y-z)$$
$$= (y-z)[x^3 - x(y^2 + yz + z^2) + yz(y+z)]$$
$$= (y-z)[x^3 - xy^2 - xyz + xz^2 + y^2z + yz^2)]$$

now in arranging in descending power of y

$$= (y-z)[y^2z - xy^2 + yz^2 - xyz + x^3 - xz^2)]$$
$$= (y-z)[y^2(z-x) + yz(z-x) - x(z^2 - x^2)]$$
$$= (y-z)(z-x)[y^2 + yz - x(z+x)]$$
$$= (y-z)(z-x)[y^2 + yz - xz - x^2)]$$
$$= (y-z)(z-x)[yz - zx + y^2 - x^2]$$
$$= (y-z)(z-x)[z(y-x) + (y+x)(y-x)]$$
$$= (y-z)(z-x)[(y-x)(z+y+x)]$$
$$= (y-z)(z-x)(y-x)(x+y+z)$$
$$= \mathbf{-(x-y)(y-z)(z-x)(x+y+z)}$$

25. (a)

$$p(x) = 4x^3 - 3x^2 + 2x - 3$$
$$p(2) = 4(2)^3 - 3(2)^2 + 2(2) - 3$$
$$= 4 \times 8 - 3 \times 4 + 4 - 3$$
$$= 32 - 12 + 4 - 3 = 36 - 15 = \mathbf{21}$$

26. (d)

$$p(x) = x^4 - 3x^2 + 3x - 5$$
$$p(1) = (1)^4 - 3(1)^2 + 3(1) - 5$$
$$= 1 - 3 + 3 - 5 = \mathbf{-4}$$

27. (b) If $(x-1)$ is a factor of $p(x) = x^3 - 3x^2 + a + 1$

then $p(1) = 0$

$\therefore$ $p(1) = 1^3 - 3(1)^2 + a + 1 = 1 - 3 + a + 1$

$$= 2 - 3 + a$$

$p(1) = 0$ means

$\therefore$ $-1 + a = 0$

$$\boldsymbol{a} = \mathbf{1}$$

28. (a) If $x + a$ is a factor of $p(x) = x^3 + ax^2 - 2x + a - 2$

then $p(-a) = 0$

$$p(-a) = (-a)^3 + a(-a)^2 - 2(-a) + a - 2$$
$$= -a^3 + a^3 + 2a + a - 2 = 3a - 2$$
$$p(-a) = 0 \quad \therefore 3a - 2 = 0$$
$$a = \frac{2}{3}$$

29. (c) if $x - k$ is a factor of

$$p(x) = x^6 - kx^5 + x^4 - kx^3 + 3x - k + 2$$

then $p(k) = 0$

$$p(k) = k^6 - k(k)^5 + k^4 - k(k)^3 + 3x - k + 2$$
$$= k^6 - k^6 + k^4 - k^4 + 3k - k + 2 = 2k + 2$$
$$p(k) = 0 \quad \therefore 2k + 2 = 0$$
$$\boldsymbol{k} = \mathbf{-1}$$

30. (c) If $x - 3$ is a factor of

$$p(x) = x^2 - 5x + k$$

then $$p(3) = 0$$

$$p(3) = 3^2 - 5 \times 3 + k = 9 - 15 + k = -6 + k$$

as $$p(3) = 0 \quad \therefore -6 + k = 0$$

or $$\boldsymbol{k = 6}$$

31. (d) As $(x + 1)$ and $(x - 2)$ are factors of $p(x) = x^3 + kx^2 + mx + 6$

$\therefore p(-1)$ and $p(2)$ are equal to zero

$$\therefore \quad p(-1) = 0$$

i.e., $(-1)^3 + k(-1)^2 + m(-1) + 6 = 0$

$$-1 + k - m + 6 = 0$$

$$k - m = -5 \quad \text{...(i)}$$

$$p(2) = 0$$

$$\therefore \quad (+2)^3 + k(+2)^2 + m(2) + 6 = 0$$

$$8 + 4k + 2m + 6 = 0$$

$$4k + 2m = -14 \quad \text{...(ii)}$$

Solving **(i)** and **(ii)** we get

$$\boldsymbol{k = -4 \text{ and } m = 1}$$

32. (a) As $(x + 1)$ and $(x - 2)$ are factors of $p(x) = x^3 + 2kx^2 + mx + 1$

$$\therefore \quad p(-1) \text{ and } p(2) = 0$$

$$\therefore \quad p(-1) = 0$$

i.e., $(-1)^3 + 2k(-1)^2 + m(-1) + 1 = 0$

$$-1 + 2k - m + 1 = 0$$

$$2k - m = 0 \quad \text{...(i)}$$

$$p(2) = 0$$

$$(2)^3 + 2k(2)^2 + m(2) + 1 = 0$$

$$8 + 8k + 2m + 1 = 0$$

$$8k + 2m = -9 \quad \text{...(ii)}$$

Solving **(i)** and **(ii)**, we get

$$k = -\frac{3}{4}, m = -\frac{3}{2}$$

33. (c) By theorem 1.

34. (d) By theorem 2.

35. (c) By theorem 3.

36. (b) By theorem 4.

37. (a) $(a + b + c)^2 = a^2 + b^2 + c^2 + 2(ab + bc + ca)$

as $$ab + bc + ca = 0$$

$$\therefore \quad (a+b+c)^2 = a^2+b^2+c^2$$

$$\therefore \quad \frac{a^2+b^2+c^2}{(a+b+c)^2} = 1$$

38. (c)
$$a+b+c = 0$$
$$(a+b+c)^2 = 0$$

i.e., $a^2+b^2+c^2+2(ab+bc+ca)=0$

$$\begin{aligned} a^2+b^2+c^2 &= -2(ab+bc+ca) \\ &= -2[a(b+c)+bc] \\ &= -2[a(-a)+bc] \\ &= -2[-a^2+bc] = \mathbf{2[a^2-bc]} \end{aligned}$$

$\therefore$ given expression becomes $= -2$

39. (a) If $a+b+c=0$ then $a^3+b^3+c^3=3abc$

$$\therefore \quad \frac{8abc}{a^3+b^3+c^3} = \frac{8abc}{3abc} = \frac{8}{3}$$

40. (a)
$$\begin{aligned} x+y &= -z \\ x^2+y^2+2xy &= z^2 \\ x^2+y^2-z^2 &= -2xy \end{aligned}$$

Similarly $\quad y^2+z^2-x^2 = -2yz$ and so on.

$\therefore$ given expression become

$$\begin{aligned} &= -\frac{1}{2}\left[\frac{1}{xy}+\frac{1}{yz}+\frac{1}{zx}\right] \\ &= -\frac{1}{2}\left(\frac{x+y+z}{xyz}\right) \\ &= -\frac{1}{2}\left(\frac{z+x+y}{xyz}\right) \\ &= -\frac{1}{2}\times\frac{0}{xyz}=0 \end{aligned}$$

41. (c)
$$1+x = 1+\frac{a}{a+c} = \frac{a+b+c}{b+c}$$

$$\frac{1}{1+x} = \frac{b+c}{a+b+c}$$

Similarly, $\quad \dfrac{1}{1+y} = \dfrac{c+a}{a+b+c}$

or, $\quad \dfrac{1}{1+z} = \dfrac{a+b}{a+b+c}$

$$\therefore \quad \frac{1}{1+x}+\frac{1}{1+y}+\frac{1}{1+z} = \frac{a+c}{a+b+c}+\frac{c+a}{a+b+c}+\frac{a+b}{a+b+c}$$

$$= \frac{2(a+b+c)}{a+b+c} = \mathbf{2}$$

42. (c)

$$ax + by + cz = a(a^2 - bc) + b(b^2 - ca) + c(c^2 - ab)$$

$$= a^3 + b^3 + c^3 - 3abc$$

$$(a+b+c)(x+y+z) = (a+b+c)(a^2 - bc + b^2 - ca + c^2 - ab)$$

$$= (a+b+c)(a^2 + b^2 + c^2 - ab - bc - ca)$$

$$= a^3 + b^3 + c^3 - 3abc$$

$$\therefore \quad \frac{ax+by+cz}{(a+b+c)(x+y+z)} = \frac{a^3+b^3+c^3-3abc}{a^3+b^3+c^3-3abc} = \mathbf{1}$$

Day 22

Linear and Quadratic Equation

Today we will do **linear and quadratic equation.**

Before starting the topic we will define very basic few terms like polynomial, linear equation, quadratic equation etc. Let us start with first term i.e., polynomial.

Polynomial: Any expression of the form

$$p(x) = a_0 + a_1x + a_2x^2 \ldots + a_nx^n$$

Where $a_1, a_2 \ldots a_n$ are real numbers and n is non negative integer is called **polynomial.**

Let us explain the above statement with simple example. Consider the equation $5x^2 + 3x + 2$. It is a polynomial as it satisfies the conditions laid by the above definition.

Here a_1, a_2 and a_3 are 2, 3 and 5. As power of x are 1 and 2 i.e., powers are positive integers.

Let us explain it in more simple language that **any algebric expression is a polynomial if power of variables (i.e., x in this case) are positive whole number i.e., they should not be negative and should not be in fraction.**

For example

(i) $5x^2 - 3x^{\frac{1}{2}} + 1$ is not a polynomial because power of x is in fraction i.e., $\frac{1}{3}$.

(ii) $5x^{-2} + 3x + 5$ is not a polynomial because the power of x is negative i.e., –2.

(iii) $5x^2 + \sqrt{3}x + 2$ is a polynomial because the power of x are positive numbers ($\sqrt{3}$ is cofficient of x. It can be in **fraction irrational** etc. i.e., it can be any **real number**).

Now it is hoped that you can identify the polynomial. Now write which of the following are polynomial.

(i) $5x^2 - 3x + 6$

(ii) $\sqrt{5}x^3 + 2x^2 - 6x + 6$

(iii) $2x^{-5} + 3x^2 + 2$

(iv) $\sqrt{x+3}$

(v) $3x^2 + 5x^{-\frac{1}{2}} + 2$

Obviously (i) **and** (ii) are **polynomial** while rest (iii), (iv) **and** (v) are **not polynomial.**

Now comes the standard form of polynomial.

Standard Form: Consider the expression.

$$p(x) = 5x^6 + 9x^8 + 6x^7 + 3x^4 + 2x^3 + 6x + 5x^2 + 2 + 8x^5$$

It is sure that it is a polynomial because all the powers are positive integar. Now you may see that it is written in very hapzard manner. If we write it systematically i.e., either in **ascending power of x** or in **descending power of x** then it will be said to be written in **standard form** as shown below.

$$p(x) = 9x^8 + 6x^7 + 5x^6 + 8x^5 + 3x^4 + 2x^3 + 5x^2 + 6x + 2$$

(in descending order)

OR

$$p(x) = 2 + 6x + 2x^3 + 5x^2 + 3x^4 + 8x^5 + 5x^6 + 6x^7 + 9x^8$$

(in ascending order)

Above two forms shows the polynomial in **standard form.**

Now you can tell the highest power of x in the above expression. It is 8 (in $9x^8$). The highest power of a polynomial is called its **degree**. Here it is **8.**

So remember: **The highest power of variable in a polynomial is called its degree.**

Now define the degree of following polynomials

(i) $5x^6 + 3x^5 + 2x^4 + 6x + 7$

(ii) $9x^5 + 4x^4 + 2x^3 + x^2 + x + 1$

(iii) $2x^2 + 5x + 1$

(iv) $2x^2 - 3x + 6$

(v) $x - 4$

Ans. (i) 6 (ii) 5 (iii) 2 (iv) 2 (v) 1

Now it is hoped that you have understood what is polynomial, its standard form, its degree etc. Now it will be very easier to define the linear and quadratic equation.

Linear Polynomial: Polynomial of degree 1 is called linear polynomial.

Quadratic Polynomial: Polynomial of degree 2 is called quadratic polynomial.

If $p(x)$ is any polynomial of degree **1** then $p(x) = 0$ is called **Linear Equation.**

If $p(x)$ is any polynomial of degree **2** then $p(x) = 0$ is called **Quadratic Equation.**

Now we will cover the topic of Linear & Quadratic Equation. First comes the linear equation.

Linear Equation

$3x + 5 = 0$ is a **linear equation** because the **highest power of x** is 1 or its **degree** is 1. We will solve the questions with the help of linear equation of one variable in **arithmetic**. Now we will define the linear equation in two variables.

Linear Equation in Two Variables: Any equation of the form $ax + by = c$ is called linear equation in two variables **x and y.** eg. $3x + y = 4$ is linear equation in two variables x and y. We can draw the graph of above equation.

Steps of drawing the graph:

1. $3x + y = 4$ can be written as $y = 4 - 3x$
2. we put the various values of x and find the corresponding value of y

   (i) when $x = 0$ then $y = 4 - 3 \times 0 = 4$

(ii) when $x = 1$ then
$$\begin{aligned} y &= 4 - 3 \times 1 \\ &= 4 - 3 = \mathbf{1} \end{aligned}$$

(iii) when $x = 2$ then
$$\begin{aligned} y &= 4 - 3 \times 2 \\ &= 4 - 6 = \mathbf{-2} \end{aligned}$$

(iv) when $x = -1$ then
$$\begin{aligned} y &= 4 - 3 \times -1 \\ &= 4 + 3 = \mathbf{7} \end{aligned}$$

(v) when $x = -2$ then
$$\begin{aligned} y &= 4 - 3 \times -2 \\ &= 4 + 6 = \mathbf{10} \end{aligned}$$

Now we write the above cases in following tabular form.

x	0	1	2	–1	–2
y	4	1	–2	7	10

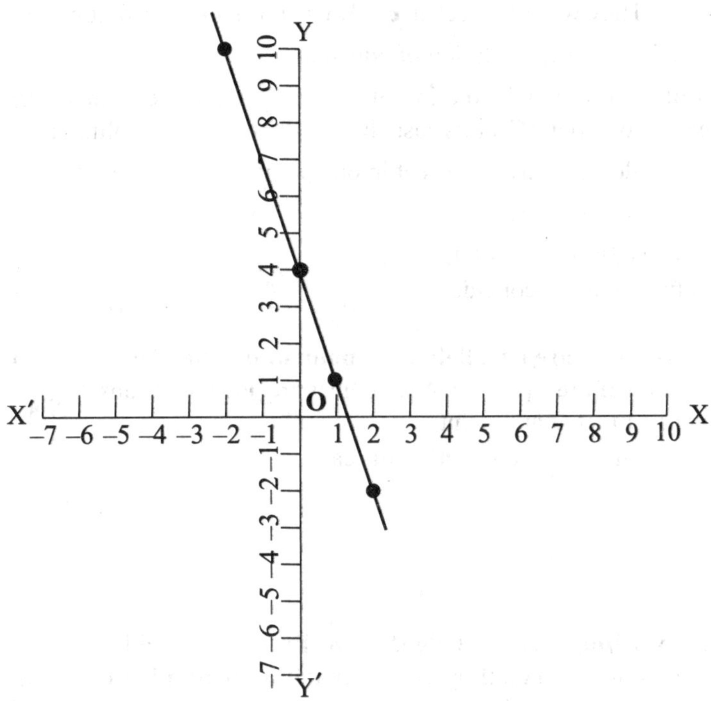

So graph of above equation is a straight line as shown in figure.

now consider following two linear equation.

$$3x + y = 4 \quad \text{...(i)}$$

$$5x - 2y = 3 \quad \text{...(ii)}$$

We can solve these by following method.

Multiple equation (i) by 2 and then adding with equation (ii), we get

$$\begin{aligned} 6x + 2y &= 8 \\ 5x - 2y &= 3 \\ \hline 11x &= 11 \end{aligned}$$

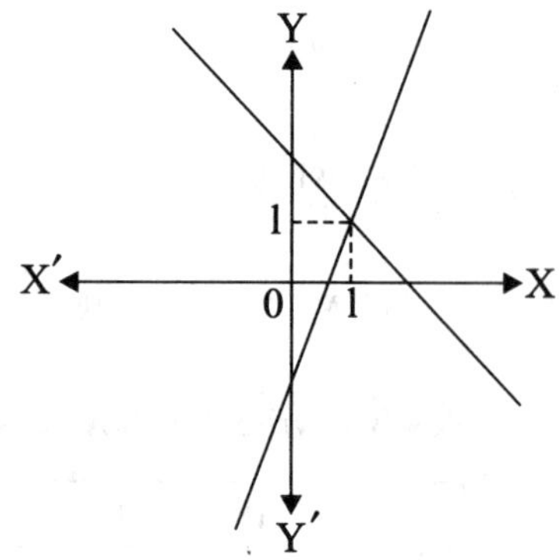

Now putting $x = 1$ in equation (i), we get

$$\begin{aligned} 3 \times 1 + y &= 4 \\ 3 + y &= 4 \\ y &= 4 - 3 \\ \mathbf{y} &= \mathbf{1} \end{aligned}$$

$\therefore$ $\boldsymbol{x = 1}$ and $\boldsymbol{y = 1}$ are the solution of above example.

Now we can solve the above example by graph

For that we draw the graph of **Ist equation** as discussed similarly we will draw the graph of **IInd equation**. They will intersect at $\boldsymbol{x = 1, y = 1}$ as shown in figure

$\therefore$ $x = 1$ and $y = 1$ is the solution of equation.

So point of intersection is solution of the equation. Since in this case two straight line intersect or meet at one point. In this case there will be a unique solution.

But two lines don't always intersect in one point.

There are two more cases:

(i) When two lines are parallel

(ii) When two lines are concident

(i) **When two lines are parallel:** Since **point of intersection** is the **solution of the equation**. As two lines are **parallel** means they **don't meet** at any point.

$\therefore$ There will be **no solution** in this case.

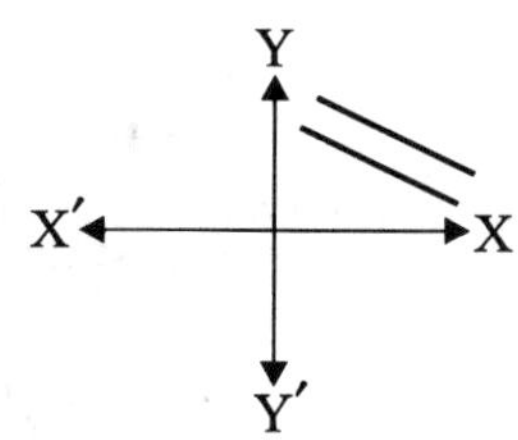

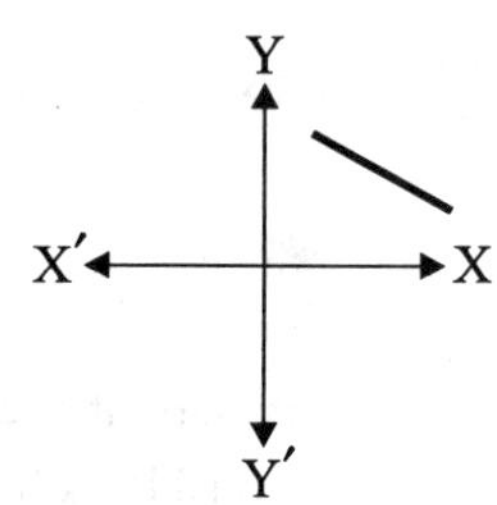

(ii) **When two lines are concident:** Two lines are concident means they are **intersecting** each other at **infinite number** of points.

$\therefore$ These have infinite many solutions.

There are few formulae in which we can just by **comparing the coefficient** can tell that whether two lines are **intersecting, parallel or concident**.

These are explained below

Explanation : Consider two equaltion

$$a_1x + b_1y = c_1 \quad ...(i)$$
$$a_2x + b_2y = c_2 \quad ...(ii)$$

Three cases arises:

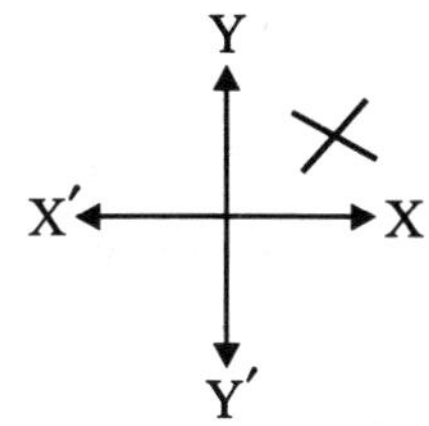

(i) if $\frac{a_1}{a_2} \neq \frac{b_1}{b_2}$

then the equation have **unique solution.**

∴ **graphs** will be **intersecting**

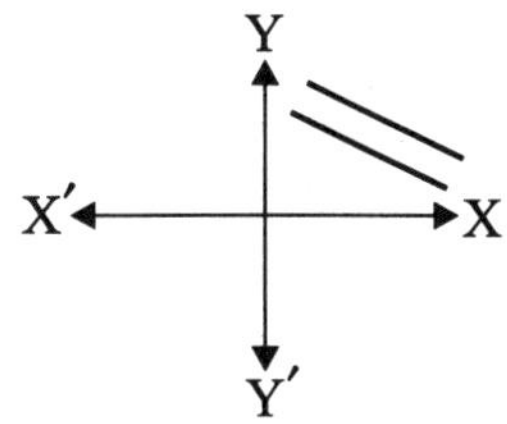

(ii) if $\frac{a_1}{a_2} = \frac{b_1}{b_2} \neq \frac{c_1}{c_2}$

then there is **no solution.** ∴ **lines** are **parallel**.

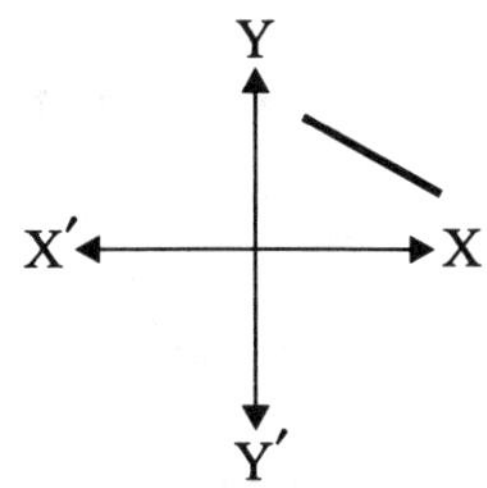

(iii) if $\frac{a_1}{a_2} = \frac{b_1}{b_2} = \frac{c_1}{c_2}$

there are **infinite many solution** and **lines are concident**.

So you have seen that how just by comparing the coefficients you can judge whether two linear are parallel, intersecting or concident.

Now comes the types of question.

Types of Questions

In linear equation we have divided the questions into following three types

Type I : Simple questions (where value of x and y are to be found)

Type II : Relating to comparision of coefficient

Type III : Relating to statement type question involing two variables.

We will do a few examples of each type of question one by one.

TYPE I
(Simple Question)

Example 1. Solve
$$3x - y = 23$$
$$4x + 3y = 48$$

Sol :
$$3x - y = 23 \quad ...(i)$$
$$4x + 3y = 48 \quad ...(ii)$$

In this we either make the coefficient of x **equal** or coefficient of y **equal**.

In this case if we **multiply** equation (i) by **3** then **coefficient of y** will become **equal**.

The question is solved as follows

$$3x - y = 23 \quad ...(i)$$
$$4x + 3y = 48 \quad ...(ii)$$

Multiply equation **(i)** by **3** and then **add** it to equation **(ii)**, we get

$$9x - 3y = 69$$
$$4x + 3y = 48$$
$$13x = 117$$
$$x = \frac{117}{13} = \mathbf{9}$$

Now put $x = 9$ in equation (i)

$$3 \times 9 - y = 23$$
$$27 - y = 23$$
$$y = 27 - 23 = 4$$

$\therefore$ **$x = 9$ and $y = 4$ is the solution**

Example 2. Solve
$$2x + 3y = 3$$
$$3x - 2y = 11$$

Sol :
$$2x + 3y = 3 \quad ...(i)$$
$$3x - 2y = 11 \quad ...(ii)$$

Multiply equation **(i)** by **2** and Multiply equation **(ii)** by **3** then **adding** we get

$$4x + 6y = 6$$
$$9x - 6y = 33$$
$$13x = 39$$
$$x = \frac{39}{13} = \mathbf{3}$$

Now put $x = 3$ in equation (i)

$$2 \times 3 + 3y = 3$$
$$3y = 3 - 6$$
$$3y = -3$$
$$y = \mathbf{-1}$$

$\therefore$ **$x = 3$ and $y = -1$ is the solution.**

Example 3. Solve
$$5x + 2y = 16$$
$$7x - 4y = 2$$

Sol :
$$5x + 2y = 16 \quad ...(i)$$
$$7x - 4y = 2 \quad ...(ii)$$

Multiply **(i)** by **2** and adding with **(ii)**, we get

$$\begin{aligned} 10x + 4y &= 32 \\ 7x - 4y &= 2 \\ \hline 17x &= 34 \\ \hline x &= \frac{34}{17} = \mathbf{2} \end{aligned}$$

Now put $x = 2$ in equation **(i)**, we get

$$\begin{aligned} 5 \times 2 + 2y &= 16 \\ 10 + 2y &= 16 \\ 2y &= 16 - 10 \\ 2y &= 6 \\ y &= 3 \end{aligned}$$

∴ $x = 2$ and $y = 3$ is the solution

TYPE II
(Related to Comparison of Coefficient)

Example 5. The equations $3x - y = 2$ and $4x + 7y = 1$ has

(a) infinite solution (b) unique solution (c) no solution (d) none

Sol : $$\frac{a_1}{a_2} = \frac{3}{4} \text{ and } \frac{b_1}{b_2} = \frac{-1}{7}$$

∴ $$\frac{a_1}{a_2} \neq \frac{b_1}{b_2}$$

so unique solution and **answer** is ***b***

Example 6. The equation $3x - 5y = 2$ and $6x - 10y = 4$ have

(a) no solution (b) infinite solution (c) unique solution (d) none

Sol : $$\frac{a_1}{a_2} = \frac{3}{6} = \frac{1}{2}$$

$$\frac{b_1}{b_2} = \frac{-5}{-10} = \frac{1}{2}$$

$$\frac{c_1}{c_2} = \frac{2}{4} = \frac{1}{2}$$

∴ $$\frac{a_1}{a_2} = \frac{b_1}{b_2} = \frac{c_1}{c_2}$$

So infinite many solution Hence **answer** is ***b***

Example 7. Find the value of k for which the system $kx + 2y = 3$ and $3x + y = 2$ have no solution

Sol : For no solution

$$\frac{a_1}{a_2} = \frac{b_1}{b_2} \neq \frac{c_1}{c_2}$$

i.e., $$\frac{k}{3} = \frac{2}{1}$$

i.e., $$\boldsymbol{k = 6}$$

Example 8. The graph of equation $3x - 4y = 2$ and $6x - 8y = 7$ is

(a) parallel (b) intersecting (c) concident (d) none

Sol : In this $$\frac{a_1}{a_2} = \frac{3}{6} = \frac{1}{2}$$

$$\frac{b_1}{b_2} = \frac{-4}{-8} = \frac{1}{2}$$

$$\frac{c_1}{c_2} = \frac{2}{7}$$

$\therefore$ $$\frac{a_1}{a_2} = \frac{b_1}{b_2} \neq \frac{c_1}{c_2} \text{ hence no solution}$$

In case of no solution **lines** are **parallel** Hence **answer** is ***a***

Example 9. Find the value of c for which equation $3x - 4y = 8$ and $6x - 8y = c$ have infinite solution.

Sol : For infinite solution

$$\frac{a_1}{a_2} = \frac{b_1}{b_2} = \frac{c_1}{c_2}$$

or $$\frac{1}{2} = \frac{8}{c}$$

$\therefore$ $$\boldsymbol{c = 16}$$

Example 10. Find the value of a if $2x + ay = 1$ and $3x - 5y = 7$ have unique solution.

Sol : System has unique solution

when $$\frac{a_1}{a_2} \neq \frac{b_1}{b_2}$$

$\therefore$ $$\frac{2}{3} \neq \frac{a}{-5} \quad \text{or } a \neq \frac{-10}{3}$$

Except $\frac{-10}{3}$ the system have unique solution for all the real numbers.

TYPE III
(Question based on Statement)

Example 11. when one is added to each of two numbers their ratio becomes 1:2 and when 5 is subtracted from each of them their ratio becomes 5:11. Find the numbers.

Sol : Let numbers are x and y

According to Ist condition

$$\therefore \quad \frac{x+1}{y+1} = \frac{1}{2}$$

cross multipling we get
$$2(x+1) = (y+1)$$
$$2x+2 = y+1$$
$$2x-y = -1 \quad \text{...(i)}$$

According to 2nd condition

$$\frac{x-5}{y-5} = \frac{5}{11}$$
$$11x-55 = 5y-25$$
$$11x-5y = 30 \quad \text{...(ii)}$$

Solving (i) and (ii) we get

$$x = \mathbf{35} \quad \text{and } y = \mathbf{71}$$

Example 12. There are two examination rooms A and B. If 10 students are sent from A to B then number in each room is same, while if 20 sent from B to A then number in A is double then that of number in B. Find the number of candidates in each room.

Sol : Let numbers of student in A $= x$

numbers of student in B $= y$

According to Ist condition

$$x-10 = y+10$$
$$x-y = 20 \quad \text{...(i)}$$

According to IInd condition

$$x+20 = 2(y-20)$$
$$x+20 = 2y-40$$
$$x-2y = -60 \quad \text{...(ii)}$$

Solving (i) and (ii) we get

$$\mathbf{x = 100 \quad and\ y = 80.}$$

$\therefore$ Number of student in **A** is **100** and in **B** is **80**

Example 13. The length of a room exceeds its breadth by 3 meter. If the length is increased by 3 m and breadth is decreased by 2 m the area would have remained the same. Find the length and breadth of the room.

Sol : Let length of room $= x$

breadth of room $= y$

According to Ist condition

$$x-y = 3 \quad \text{...(i)}$$

According to IInd condition

$$(x+3)(y-2) = xy$$
$$xy-2x+3y-6 = xy$$
$$-2x+3y = 6 \quad \text{...(ii)}$$

Solving **(i)** and **(ii)**, we get

$$x = \mathbf{15} \quad \text{and } y = \mathbf{12}$$

$\therefore$ Length of a room = **15 m**

and Breadth of room = **12 m**

Example 14. A said to B. "If you give me Rs 900, I shall be twice as rich as you are." B said "If you give me Rs. 100 I shall be three times as rich as you are". Find the amount of each.

Sol : Let money of A $= x$

money of B $= y$

So according to Ist condition

$$x + 900 = 2(y - 900)$$
$$x + 900 = 2y - 1800$$
$$x - 2y = -2700 \qquad \text{...(i)}$$

According to IInd condition

$$(y + 100) = 3(x - 100)$$
$$y + 100 = 3x - 300$$
$$3x - y = 400 \qquad \text{...(ii)}$$

Solving **(i)** and **(ii)**, we get

$$x = \mathbf{700}$$
$$y = \mathbf{1700}$$

$\therefore$ **A** has **Rs.700** and **B** has **Rs.1700**

We will do more question of above types in the exercise which will be followed after the topic of quadratic equation.

Now we will do quadratic equation.

Quadratic Equation

As discussed earlier if $p(x)$ is **polynomial** of **degree 2** then it is called **quadratic polynomial** and if $p(x) = 0$ then is called **quadratic equation.**

The general form of quadratic equation is

$$ax^2 + bx + c = 0$$

where ***a* and *b*** are coefficient of x^2 **and** x and c is **constant**. For example if we compare $x^2 - 5x + 6 = 0$ with $ax^2 + bx + c = 0$, we find that $a = 1, b = -5$ and $c = 6$.

Now we will discuss roots.

Roots: Consider the equation $x^2 - 5x + 6 = 0$

till now we solve it as follows

$$x^2 - 3x - 2x + 6 = 0$$
$$x(x - 3) - 2(x - 3) = 0$$
$$(x - 3)(x - 2) = 0$$

either $x - 3 = 0$ or $x - 2 = 0$

so we get $x = 3$ or $x = 2$

i.e., **two value of x** which we get after **solving the equation** are called **roots** of the equation.

Since x has a **degree 2**. $\therefore$ we will get **two roots**

In quadratic equation the general form of roots is $\boldsymbol{ax^2 + bx + c = 0}$ and $\mathbf{D} = \boldsymbol{b^2 - 4ac}$

Discriminant : In equation $\boldsymbol{ax^2 + bx + c = 0}$, the expression $\mathbf{D} = \boldsymbol{b^2 - 4ac}$ is called **discriminant. D** plays an important role in telling the **nature of roots.**

Following cases arises

(*i*) roots are **real and equal** if **D = 0**

(*ii*) roots are **real, unequal and rational** if **D > 0** and **D** is **perfect square.**

(*iii*) roots are **real, unequal and irrational** if **D > 0** but **D** is **not a perfect square.**

(*iv*) roots are **imaginary** if **D < 0**

The two roots α and β are given by

$$\alpha = \frac{-b+\sqrt{D}}{2a} = \frac{-\mathbf{b}+\sqrt{\mathbf{b}^2-\mathbf{4ac}}}{\mathbf{2a}}$$

$$\beta = \frac{-b+\sqrt{D}}{2a} = \frac{-\mathbf{b}-\sqrt{\mathbf{b}^2-\mathbf{4ac}}}{\mathbf{2a}}$$

Let us solve a simple example

Example 15. Solve $3x^2 - 5x + 2 = 0$

Sol : Ist we will check D

here $a = 3$, $b = -5$, $c = 2$

$$\therefore \quad D = b^2 - 4ac = 25 - 24 = 1$$

D > 0 and is a perfect square.

$\therefore$ roots are real, rational and unequal

$$\alpha = \frac{-b+\sqrt{D}}{2a}$$

$$= \frac{5+1}{6} = \frac{6}{6} = 1$$

$$\beta = \frac{5-1}{6} = \frac{4}{6} = \mathbf{\frac{2}{3}}$$

Now we will discuss sum and product of root

Sum and product of roots: we know that

$$\alpha = \frac{-b+\sqrt{D}}{2a}$$

$$\beta = \frac{-b-\sqrt{D}}{2a}$$

$$\alpha + \beta = \frac{-b+\sqrt{D}}{2a} + \frac{-b-\sqrt{D}}{2a}$$

$$= \frac{-b+\sqrt{D}-b-\sqrt{D}}{2a}$$

$$= \frac{-2b}{-2a} = \frac{-b}{a}$$

so **sum of roots** $= -\frac{\mathbf{b}}{\mathbf{a}}$

similarly $\alpha\beta = \frac{-b+\sqrt{D}}{2a} + \frac{-b-\sqrt{D}}{2a}$

$$= \frac{(-b)^2 + (\sqrt{D})^2}{4^2}$$

$$= \frac{b^2 - (\sqrt{b^2-4ac})^2}{4a^2} = \frac{b^2 - b^2 + 4ac}{4a^2} = \frac{\mathbf{c}}{\mathbf{a}}$$

so we have seen $\alpha + \beta = -\frac{b}{a}$

i.e., **sum of roots** is $= -\frac{\mathbf{b}}{\mathbf{a}}$

and $\alpha\,\beta = \frac{c}{a}$

i.e., **product of roots** $= \frac{c}{a}$

so simply by seeing the equation we can tell the sum and product of roots.
In equation $2x^2 + 5x + 1 = 0$

sum of roots is $\frac{-b}{a} = \frac{-5}{2}$

and product of roots $= \frac{c}{a} = \frac{1}{2}$

Now we will discuss how to make the equation when roots are given. Let us find a equation where sum of roots is 3 and product is 2

Let equation is $ax^2 + bx + c = 0$

as sum is given as 3 $\therefore$ $\frac{-b}{a} = 3$

$$b = -3a$$

and product is 2 $\therefore$ $\frac{c}{a} = 2$

$$c = 2a$$

equation (i) becomes

$$ax^2 - 3ax + 2a = 0$$

$$a(x^2 - 3x + 2) = 0$$

as **a** can not be equal to **0**

$\therefore$ **$x^2 - 3x + 2 = 0$ is the equation.**

So above we have seen that if **sum** and **product** of **roots** are known then **quadratic equation** having these roots is

x^2 – **(sum of roots)** x + **product of roots** = 0

$x^2 - Sx + P = 0$ is the equation

where, **S** is the **sum** and **P** is the **product**

Let us solve a simple example.

Example 16. Find an equation whose roots are 2 and 3

Sol : Sum of the roots i.e., $S = 2 + 3 = 5$

and **Product of roots i.e.,** $P = 2 \times 3 = 6$

so as discussed equation is **$x^2 - Sx + P = 0$**

i.e., $x^2 - 5x + 6 = 0$

We can cross check the equation

In $x^2 - 5x + 6 = 0$

$$\alpha = \frac{5 + \sqrt{1}}{2} = 3$$

$$\beta = \frac{5 - \sqrt{1}}{2} = 2$$

Now it is hoped that you can make a equation when answer is given i.e., its roots are given.

Factorisation by solving quadratic equation : If α and β are the **roots of**

$$ax^2 + bx + c = 0$$

then **$a(x - \alpha)(x - \beta)$** are the **factors of $ax^2 + bx + c = 0$**

now we understood it more clearly by following example

Example 17. Factorize $x^2 - 5x + 6 = 0$.

Sol : Ist method (by spliting)

$$x^2 - 5x + 6 = 0$$

$$x^2 - 3x - 2x + 6 = 0$$

$$x(x - 3) - 2(x - 3) = 0$$

$$(x - 3)(x - 2) = 0$$

IInd method (by solving quadratic equation)

$$x^2 - 5x + 6 = 0$$

$$D = 25 - 24 = 1$$

$$\alpha = \frac{5 + \sqrt{1}}{2} = \mathbf{3}$$

$$\beta = \frac{5 - \sqrt{1}}{2} = \mathbf{2}$$

Factors of $ax^2 + bx + c = 0$ are $(x - \alpha)(x - \beta)$

$\therefore$ Factors of $x^2 - 5x + 6 = 0$ are $(x - 3)(x - 2)$

Now It may be noted that here D is perfect square.

$\therefore$ Factor can be formed by spliting method.

It is not always possible to find factors by splitting method as discussed in following example.

Example 17. Factorize $x^2 + 4x + 2 = 0$.

Sol : This is not possible to factorize the above equation by spliting as **D** =16 – 8 = **8** is not a perfect square.

Now by **solving quadratic equation** its solution is given below

As **D > 0** and **not a perfect square**

$\therefore$ **roots** are **real**

and roots are $\alpha = \frac{-4 + \sqrt{8}}{2} = \frac{-4 - \sqrt{8}}{2}$

$$= \frac{-4 + 2\sqrt{2}}{2}$$

$$\alpha = -2 + \sqrt{2}$$

and similarly $\beta = -2 - \sqrt{2}$

$\therefore$ Factors are $1[x - (-2 + \sqrt{2})]\,[x - (-2 - \sqrt{2})]$

i.e., $(x + 2 - \sqrt{2})(x + 2 + \sqrt{2}) = 0$

[**Note:** If **D < 0** i.e., **negative then polynomial** can not be factorized into linear factors]

So It is hoped that you have understood **how to factorize a equation with the help of roots.**

So remember:

(*i*) $ax^2 + bx + c = 0$ is the **general** form of **quadratic**

(*ii*) $D = b^2 - 4ac$ is **discriminant**

(*iii*) If **D = 0 then roots** are **real** and **equal and** are $\frac{-b}{2a}$

(*iv*) If **D > 0** and **D** is a **perfect square** then **roots** are **rational, unequal and real**

(*v*) If **D < 0 then roots** are **imaginary**

(*vi*) If **one of the roots** of **equation** is $\boldsymbol{a} + \sqrt{\boldsymbol{b}}$ then **other** is $\boldsymbol{a} - \sqrt{\boldsymbol{b}}$

(*vii*) If **S** is the **sum of roots** and **P** is the **product of roots** then **quadratic equation** is $\boldsymbol{x^2 - Sx + p = 0}$

(*viii*) If α **and** β are **real roots** of an **equation** $\boldsymbol{ax^2 + bx + c = 0}$ then $\boldsymbol{ax^2 + bx + c = 0}$ can be **factorize** as $\boldsymbol{a(x - \alpha)(x - \beta)}$

Now we will discuss the types of question in quadratic equation.

TYPES OF QUESTIONS

Following are the types in which we have divided the questions of quadratic.

Type I : simple questions
Type II : questions related to nature of roots.
Type III : forming a equation when roots are given
Type IV : factorization with the help of roots
Type V : equation which are reducible to quadratic
Type VI : statement questions (where condition are reduced to quadratic)
Type VII : misc. questions.

TYPE I
(Simple Questions)

Ist type of question are simple question in which straight questions are asked.

Example 1. Solve $x^2 - 3x + 2 = 0$

Sol : $x^2 - 3x + 2 = 0$

is a quadratic equation

$$D = b^2 - 4ac$$
$$= 9 - 4 \times 2 = 9 - 8 = 1$$

$\therefore$ $a = 1, \quad b = -3, \quad c = 2$

$$\alpha = \frac{-b + \sqrt{b^2 - 4ac}}{2a}$$

$$\alpha = \frac{3+1}{2} = 2$$

$$\beta = \frac{3-1}{2} = \mathbf{1}$$

Example 2. Solve $x^2 - 3x - 18 = 0$

Sol : $x^2 - 3x - 18 = 0$

$$D = b^2 - 4ac$$
$$= 9 - 4x - 16$$
$$= 9 + 72$$
$$= 81$$

$$\alpha = \frac{3+\sqrt{81}}{2}$$

$$\alpha = \frac{3+9}{2} = 6$$

$$\beta = \frac{3-9}{2} = \frac{-6}{2} = \mathbf{-3}$$

TYPE II
(Question related to Nature of Roots)

Example 3. Find the value of p for which equation $px^2 + 4x + 1 = 0$ have real roots

Sol : For real roots

$$\begin{aligned} D &\geq 0 \\ b^2 - 4ac &\geq 0 \\ 16 - 4p &\geq 0 \\ 16 &\geq 4p \\ 4p &\leq 16 \\ \boldsymbol{p} &\leq \mathbf{4} \end{aligned}$$

Example 4. Find the value of p for which equation $x^2 - px + 4 = 0$ have equal roots

Sol : For equal roots

$$D = 0$$

i.e.,

$$\begin{aligned} b^2 - 4ac &= 0 \\ p^2 - 4 \times 4 \times 1 &= 0 \\ p^2 - 16 &= 0 \\ \boldsymbol{p} &= \pm\mathbf{4} \end{aligned}$$

Example 5. Find the value of p for which quadratic equation $px^2 - 2x + 2 = 0$, have real roots

Sol : For real roots

$$\begin{aligned} D &\geq 0 \\ b^2 - 4ac &\geq 0 \\ 4 - 8p &\geq 0 \\ 4 &\geq 8p \\ \frac{4}{8} &\geq p \\ \frac{1}{2} &\geq p \\ \boldsymbol{p} &\leq \frac{1}{2} \end{aligned}$$

Example 6. Find the value of p for which quadratic equation $px^2 - x - 2 = 0$, have equal roots

Sol : For equal roots

$$D = 0$$
$$b^2 - 4ac = 0$$
$$(1)^2 - 4 \times p \times -2 = 0$$
$$1 + 8p = 0$$
$$p = -\frac{1}{8}$$

TYPE III
(Forming a Equation when Roots are given)

Example 7. Form a equation whose roots are $2 + \sqrt{2}$ and $2 - \sqrt{2}$.

Sol : We know that equation is

$$x^2 - Sx + P = 0$$

here $S = 2 + \sqrt{2} + 2 - \sqrt{2} = 4$

$$P = (2+\sqrt{2})(2-\sqrt{2}) = 4 - 2 = 2$$

$\therefore$ $\mathbf{x^2 - 4x + 2 = 0}$ **is the equation**

Example 8. Form a equation root are 6 and –8

Sol : We know that equation is

$$x^2 - Sx + P = 0$$

here $S = 6 - 8 = -2$

$$P = 6x - 8 = -48$$

$\therefore$ **equation** is $\mathbf{x^2 + 2x - 48 = 0}$

TYPE IV
(Factorization with the Help of Roots)

Example 9. Factorize $x^2 - 6x + 6 = 0.$

Sol : $\alpha = \dfrac{6 + \sqrt{36 - 24}}{2}$

$$\alpha = \frac{6 + 2\sqrt{3}}{2} = \mathbf{3 + \sqrt{3}}$$

$$\beta = \mathbf{3 - \sqrt{3}}$$

Factor are a $(x - \alpha)(x - \beta)$

$$1(x - 3 - \sqrt{3})(x - 3 + \sqrt{3})$$

Example 10. Factorize

$$2x^2 - \sqrt{7} \times -7 = 0$$

Sol : $\alpha = \dfrac{\sqrt{7} + \sqrt{7 + 56}}{4}$

$$\alpha = \frac{-\sqrt{7}+3\sqrt{7}}{4} = \frac{2\sqrt{7}}{4} = \frac{\sqrt{7}}{2}$$

$$\beta = \frac{-\sqrt{7}-3\sqrt{7}}{4} = \frac{-4\sqrt{7}}{4} = -\sqrt{7}$$

$$2\left(x - \frac{\sqrt{7}}{2}\right)(x+\sqrt{7})$$

$$(x+\sqrt{7})(2x-\sqrt{7})$$

TYPE V
(Equation which are reducible to quadratic)

Some from degree equation which are reducible to quadratic are solved with this method.

A few examples are given below

Example 11. Solve $x^4 - 5x^2 + 6 = 0$

Sol : Put $x^2 = z$

$\therefore$ equation become $z^2 - 5z + 6 = 0$

$$z^2 - 5z + 6 = 0$$

$$z^2 - 3z - 2z + 6 = 0$$

$$z(z-3) - 2(z-3) = 0$$

$$z = 3 \qquad z = 2$$

But $z = x^2$

$\therefore$ $x^2 = 3 \qquad x^2 = 2$

$$x = \pm\sqrt{3} \qquad x = \pm\sqrt{2}$$

Example 12. Solve $2x + \frac{3}{x} = 5.$

Sol : $\frac{2x^2+3}{x} = 5$

$$2x^2 + 3 = 5x$$

$$2x^2 - 5x + 3 = 0$$

$$\alpha = \frac{5+1}{4} = \frac{6}{4} = \mathbf{\frac{3}{2}}$$

$$\beta = \frac{5-1}{4} = \frac{4}{4} = \mathbf{1}$$

Example 13. Solve $\sqrt{217-x} = x-7$

Sol : $\sqrt{217-x} = x-7$

squaring both sides

$$217 - x = (x-7)^2$$

$$217 - x = x^2 + 49 - 14x$$

$$x^2 - 14x + 49 - 217 + x = 0$$

$$x^2 - 13x - 168 = 0$$
$$x^2 - 21x + 8x - 168 = 0$$
$$x(x - 21) + 8(x - 21) = 0$$
$$x = 21 \qquad x = -8.$$

Example 14. Solve

$$2\left(x^2 + \frac{1}{x^2}\right) - 9\left(x + \frac{1}{x}\right) + 14 = 0$$

Sol : $$\left(x + \frac{1}{x}\right) = y$$

squaring both sides

$$x^2 + \frac{1}{x^2} + 2 = y^2$$
$$x^2 + \frac{1}{x^2} = y^2 - 2$$

$\therefore$ equation become $2(y^2 - 2) - 9y + 14 = 0$

$$2y^2 - 4 - 9y + 14 = 0$$
$$2y^2 - 9y + 10 = 0$$
$$2y^2 - 5y - 4y + 10 = 0$$
$$y(2y - 5) - 2(2y - 5) = 0$$
$$(2y - 5)(y - 2) = 0$$
$$y = \frac{5}{2} \qquad y = 2$$

$\therefore$ $$x + \frac{1}{x} = \frac{5}{2}$$
$$\frac{x^2 + 1}{x} = \frac{5}{2}$$
$$2x^2 + 2 = 5x$$
$$2x^2 - 5x + 2 = 0$$
$$2x^2 - 4x - x + 2 = 0$$
$$2x(x - 2) - 1(x - 2) = 0$$
$$(2x - 1)(x - 2) = 0$$
$$\boldsymbol{x = \frac{1}{2} \qquad x = 2}$$
$$x + \frac{1}{x} = 2$$

$$\frac{x^2+1}{x} = 2$$
$$x^2 + 1 = 2x$$
$$x^2 - 2x + 1 = 0$$
$$(x-1)^2 = 0$$
$$\mathbf{x = 1}$$

TYPE VI
(Question Regarding Statements)

Example 15. The sum of a number and its reciprocal is $\frac{10}{3}$. Find the number

Sol :
$$\text{let number} = x$$
$$\text{reciprocal} = \frac{1}{x}$$
$$\therefore \quad x + \frac{1}{x} = \frac{10}{3}$$
$$\frac{x^2+1}{x} = \frac{10}{3}$$
$$3x^2 + 3 = 10x$$
$$3x^2 - 10x + 3 = 0$$
$$3x^2 - 9x - x + 3 = 0$$
$$3x(x-3) - 1(x-3) = 0$$
$$(x-3)(3x-1) = 0$$
$$\mathbf{x = 3} \qquad \mathbf{x = \frac{1}{3}}$$

Example 16. The sides (in m) of a right triangle is $x - 1$, x and $x + 1$. Find the sides of triangle.

Sol : In right triangle ABC
$$AC^2 = AB^2 + BC^2$$
$$(x+1)^2 = (x)^2 + (x-1)^2$$
$$x^2 + 1 + 2x = x^2 + x^2 + 1 - 2x$$
$$x^2 + 1 + 2x = 2x^2 + 1 - 2x$$
$$2x^2 - 2x - x^2 - 2x = 0$$
$$x^2 = 4x$$
$$\mathbf{x = 4}$$

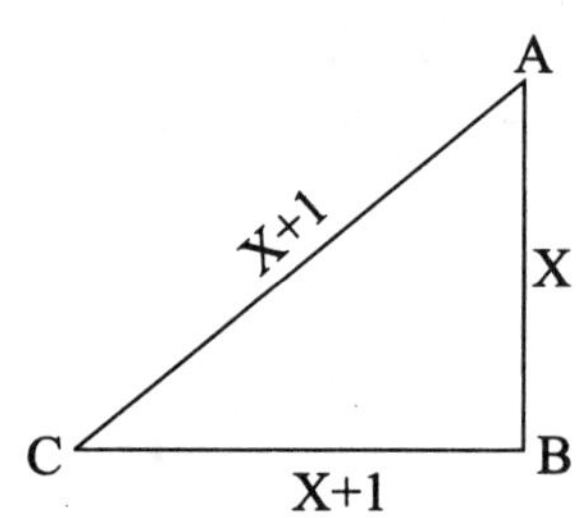

sides are $x - 1 = 4 - 1 = \mathbf{3m}$

$x = \mathbf{4m}$

and $x + 1 = 4 + 1 = \mathbf{5m}$

Example 17. The length of a rectangle is 5m more than its breadth. Its area is 336 m^2. Find its length.

Sol : let $b = x$

∴ $l = x + 5$

according to statement

$$x(x + 5) = 336$$
$$x^2 + 5x - 336 = 0$$
$$x^2 + 21x - 16x - 336 = 0$$
$$x(x + 21) - 16(x + 21) = 0$$
$$(x + 21)(x - 16) = 0$$
$$x = 16$$

∴ $b = 16$

and hence length $= 16 + 5 = \mathbf{21\ m}$

Example 18. Two numbers differ by 2 and sum of its squares is 100. Find the numbers.

Sol : Let Ist number $= x$

IInd no. $= x + 2$

According to statement

$$x^2 + (x + 2)^2 = 100$$
$$x^2 + x^2 + 4 + 4x = 100$$
$$2x^2 + 4x - 96 = 0$$
$$x^2 + 2x - 48 = 0$$
$$x^2 + 8x - 6x - 48 = 0$$
$$x(x + 8) - 6(x + 8) = 0$$
$$(x - 6)(x - 8) = 0$$
$$x = 6$$

∴ Ist number = **6m**

IInd number = **8 m**

Example 19. A passenger train takes 2 hour less for a journey of 300 km if its speed is increased by 5 km/h from its usual speed. What is the usual speed.

Sol : let usual speed $= k$

∴ increased speed $= k + 5$

we know that speed $= \dfrac{\text{distance}}{\text{time}}$

or Time $= \dfrac{\text{distance}}{\text{speed}}$

time taken by usual speed $= \dfrac{300}{x}$

$$\text{time taken by increased speed} = \frac{300}{x+5}$$

according to statement

$$\frac{300}{x} - \frac{300}{x+5} = \frac{2}{1}$$

$$\frac{300(x+5) - 300x}{x(x+5)} = \frac{2}{1}$$

$$300x + 1500 - 300x = 2(x)(x+5)$$

$$1500 = 2x(x+5)$$

$$x(x+5) = 750$$

$$x^2 + 5x - 750 = 0$$

$$x^2 + 30x - 25x - 750 = 0$$

$$x(x+30) - 25(x+30) = 0$$

$$(x-25)(x+30) = 0$$

$$x = 25$$

$\therefore$ usual speed = **25 km/h**

TYPE VII
(Misc. Question)

Example 20. If a and c are such that the quadratic equation $ax^2 - 5x + c = 0$ has 10 as the sum of the roots and as product of roots. Find a and c.

Sol : in $ax^2 - 5x + c = 0$

$$\text{Sum} = +\frac{5}{a} \text{ and product} = \frac{c}{a}$$

but $\frac{+5}{a} = 10$

$$a = \frac{5}{10} = \frac{1}{2}$$

$$\frac{c}{a} = 10$$

or $c = 10a = -10 \times \frac{1}{2} = \mathbf{+5}$

Example 21. If one of the roots of quadratic equation $2x^2 + px + 4 = 0$ is 2. Find the of value of p.

Sol : As 2 is the root of $2x^2 + px + 4 = 0$

$\therefore$ by putting $x = 2$ in equation we get the value of p

$\therefore$ $2(2)^2 + p \times 2 + 4 = 0$

$$8 + 2p + 4 = 0$$

$$2p = -12$$

$$\boldsymbol{p = -6}$$

Example 22. If α and β are roots of the quadratic equation $ax^2 + bx + c = 0$ then find the value of $\alpha^2 + \beta^2$.

Sol :

$$\alpha + \beta = \frac{-b}{a}$$

$$\alpha\beta = \frac{c}{a}$$

$$(\alpha + \beta)^2 = \alpha^2 + \beta^2 + 2\alpha\beta$$

$$\left(\frac{-b}{a}\right)^2 = \alpha^2 + \beta^2 + \frac{2c}{a}$$

$$\frac{b^2}{a^2} - \frac{2c}{a} = \alpha^2 + \beta^2$$

$$\frac{b^2 - 2ac}{a^2} = \alpha^2 + \beta^2$$

$$\alpha^2 + \beta^2 = \frac{\mathbf{b^2 - 2ac}}{\mathbf{a^2}}$$

Example 23. If sum of square of roots of the equation $x^2 + 2x - p = 0$ is 8. Then value of p is

Sol : let α and β be roots of the equation

$\therefore$ $\alpha^2 + \beta^2 = 8$ (...given)

and $\alpha + \beta = -2$

and $\alpha\beta = \frac{-p}{1}$

we know that

$$(\alpha + \beta)^2 = \alpha^2 + \beta^2 + 2\alpha\beta$$

$$(-2)^2 = 8 + 2 \times -p$$

$$4 = 8 - 2p$$

$$-2p = -4$$

$$p = 2$$

Now we will do the exercise of linear and quadratic.

EXERCISE

Q.1. Solution of x and y in the following equations is $x - y = -2$ and $2x + 3y = 36$

(a) 6, 8 (b) 8, 6 (c) 3, 5 (d) none

Q.2. The value of x and y in $x + y = 7$ and $3x - 2y = 11$ is

(a) 2, 5 (b) 5, 2 (c) 3, 4 (d) None

Q.3. The value of x and y in $3x + 2y = 11$ and $2x + 3y = 4$ is

(a) 2, 5 (b) 5, 2 (c) 5, –2 (d) none

Q.4. The graphs of the equation $3x + 2y = 5$ and $2x - y = 11$ are

(a) intersecting (b) parallel (c) concident (d) none

Q.5. The graphs of the equation $3x + 6y = 9$ and $9x + 18y = 27$ are

(a) intersecting (b) parallel (c) concident (d) none

Q.6. The graphs of the equation $4x - 5y = 7$ and $8x - 10y = 9$ are

(a) intersecting (b) parallel (c) concident (d) none

Q.7. The value of p for which graphs of $2x + py = 7$ and $4x + 2y = 14$ are concident

(a) 1 (b) 2 (c) 7 (d) none

Q.8. The system of equation $5x - 4y = 7$ and $3x - 2y = 15$ have

(a) unique solution (b) infinite solution (c) no solution (d) none

Q.9. The system of equation $9x - 17y = 34$ and $36x - 68y = 115$ have

(a) unique Solution (b) infinite Solution (c) no solution (d) none

Q.10. The price of 9 pencils and 5 pens is Rs 90. Whereas the price of 5 pencils and 4 pens is Rs. 61. The price of 6 pencils and 3 pens is

(a) Rs. 55 (b) Rs. 57 (c) Rs. 55 (d) none

Q.11. If 4 is added to the numerator of a fraction the fraction becomes equal to 1. If 1 is subtracted from the denominator, the fraction becomes equal to $\frac{1}{2}$. The fraction is

(a) $\frac{7}{3}$ (b) $\frac{3}{8}$ (c) $\frac{3}{7}$ (d) none

Q.12. The incomes of two persons are in the ratio 9:7 and their expenditures are in the ratio 4 : 3. If both of them save Rs. 200 per month. Find the monthly income of both

(a) Rs. 1700, Rs. 1200 (b) Rs. 1600, Rs. 1200

(c) Rs. 1800, Rs. 1400 (d) none

Q.13. If 4 is added to the numerator of a fraction the fraction becomes equal to $\frac{1}{2}$. If 5 is subtracted from the denominator. The fraction becomes equal to $\frac{1}{3}$. The fraction is

(a) $\frac{14}{5}$ (b) $\frac{3}{14}$ (c) $\frac{3}{11}$ (d) none

Q.14. There are two numbers. If we add one to each number their ratio becomes 2 : 3. If 1 be decreased from each no. their ratio become 1 : 2. The numbers are

(a) 3, 1 (b) 1, 3 (c) 1, 5 (d) none

Q.15. A father's age is equal to the ages of 5 children. In fifteen years, his age will be only half of their united age. Find his present age.

(a) 40 years (b) 45 years (c) 42 years (d) none

Q.16. The roots of equation $x^2 - 6x + 8 = 0$ are

(a) ! 2 (b) 4, 2 (c) 3, 1 (d) none

Q.17. The value of c for which the equation $2x^2 - 9x + c = 0$ have equal roots

(a) $\frac{81}{8}$ (b) $\frac{8}{81}$ (c) 9 (d) none

Q.18. The positive value of m for $6x^2 - mx + 5 = 0$ have roots in the ratio 1 : 2 is

(a) $15\sqrt{3}$ (b) $3\sqrt{15}$ (c) $\sqrt{15}$ (d) none

Q.19. The quadratic equation whose roots are $3 + \sqrt{5}$ and $3 - \sqrt{5}$ is

(a) $x^2 - 6x + 2 = 0$ (b) $x^2 - 4x + 6 = 0$ (c) $x^2 - 6x + 4 = 0$ (d) none

Q.20. The quadratic equation whose one of the roots is $6 + \sqrt{11}$

(a) $x^2 - 12x + 25=0$ (b) $x^2 - 25x + 12 = 0$ (c) $x^2 - 18x + 15 = 0$ (d) none

Q.21. Factor of $x^2 + 4\sqrt{2}\,x + 6$ are

(a) $(x+3\sqrt{2})(x+\sqrt{2})$ (b) $(x+\sqrt{2})(x+\sqrt{3})$

(c) $(x+2\sqrt{2})(x-2\sqrt{3})$ (d) none

Q.22. Factor of $x^2 + 2x - 6$ are

(a) $(x+1+\sqrt{2})(x+1+\sqrt{2})$ (b) $(x+1+\sqrt{2})(x+1-\sqrt{2})$

(c) $(x+2+\sqrt{2})(x-2-\sqrt{2})$ (d) none

Q.23. The roots in $x^4 - 26x^2 + 25 = 0$ are

(a) $\pm 1, \pm 5$ (b) $\pm 1, \pm 3$ (c) $\pm 2, \pm 5$ (d) none

Q.24. The roots of $2x + \frac{4}{x} = 9$ are

(a) 4, 2 (b) 4, 3 (c) 4, $\frac{1}{2}$ (d) none

Q.25. Value of x in $\sqrt{x} + 2x = 1$ is

(a) 4 (b) $\frac{1}{4}$ (c) 2 (d) none

Q.26. In $6\left(x^2 + \frac{1}{x^2}\right) - 25\left(x - \frac{1}{x}\right) + 12 = 0$. The value of x are

(a) 3, 4, 5, 6 (b) 3, $\frac{1}{3}$ 2, $\frac{1}{2}$ (c) 3, $\frac{1}{3}$ 2, $-\frac{1}{2}$ (d) none

Q.27. Product of Kush age five year ago to his age after 9 year his 51. The present age of Kush is

(a) 9 year (b) 8 year (c) 7 years (d) none

Q.28. The sides of a right triangle containing the right angle are $4x$ and $5x - 4$. If the area of triangle is 210 m^2. Find the sides of triangle

(a) 8, 15, 17 (b) 20, 21, 29 (c) 3, 4, 5 (d) None

Q.29. The sum of squares of two consecutive natural numbers is 841. The smaller number is

(a) 20 (b) 21 (c) 19 (d) none

Q.30. A fast train takes 3 hour less than a slow train for a journey of 600 km. If the speed of slow train is 10 km/h less then that of the fast train. Find the speed of fast train

(a) 20 (b) 30 (c) 40 (d) none

Q.31. In a cricket match Anil took one wicket more than thrice the number of wicket taken by Sachin. If the product of the number of wickets taken by both is 14. Find the number of wickets taken by Sachin.

(a) 3 (b) 7 (c) 2 (d) none

Q.32. If α and β are the roots of the quadratic equation $ax^2 + bx + c = 0$. The value of $\alpha^3 + \beta^3$ is

(a) $\frac{3abc - b^3}{a^3}$ (b) $\frac{3abc - a^3}{c^3}$ (c) $\frac{3ab - c^3}{a^3}$ (d) none

Q.33. If α and β are the roots of the equal square. $ax^2 + bx + c = 0$. Then the value of $a^2 + b^2$ is

(a) $\frac{b^2 - ac}{a^2}$ (b) $\frac{b^2 - 2ac}{a^2}$ (c) $\frac{b^2 - a^2}{a^2}$ (d) none

SOLUTIONS

1. (a)

$$x - y = -2 \quad \text{...(i)}$$
$$2x + 3y = 36 \quad \text{...(ii)}$$

multiply **(i)** by **3** and **add** to **(ii)**, we get

$$3x - 3y = -6$$
$$2x + 3y = 36$$
$$5x = 30$$
$$x = \mathbf{6}$$

now putting in **(ii)** we get $\boldsymbol{x = 6}$

$$2 \times 6 + 3y = 36$$

or $3y = 36 - 12 = 24$

or $y = \mathbf{8}$

2. (b)

$$x + y = 7 \quad \text{...(i)}$$
$$3x - 2y = 11 \quad \text{...(ii)}$$

multiply **(i)** by **2** and **adding** with **(ii),** we get

$$2x + 2y = 14$$
$$3x - 2y = 11$$
$$5x = 25$$
$$x = \mathbf{5}$$

put $\boldsymbol{x = 5}$ in **(i)** we get

$$5 + y = 7$$

or $y = 7 - 5 = \mathbf{2}$

3. (c)

$$3x + 2y = 11 \quad \text{...(i)}$$
$$2x + 3y = 4 \quad \text{...(ii)}$$

multiply **(i)** by **2** and **(ii)** by **3** and subtracting, we get

$$6x - 4y = 22$$
$$6x - 9y = 12$$
$$- \quad - \quad -$$
$$-5y = 10$$

$$y = -2$$

putting $\mathbf{y = -2}$

in **(i)** we get $3x + 2 \times -2 = 11$

or $3x = 11 + 4 = 15$

or $x = \frac{15}{5} = \mathbf{5}$

4. (a) $\frac{3}{2} \neq \frac{2}{-1}$

i.e., $\frac{a_1}{a_2} \neq \frac{b_1}{b_2}$

∴ **unique solution** hence **graphs** is **intersecting**

5. (c) $\frac{3}{9} = \frac{6}{18} = \frac{9}{27} = \frac{1}{3}$

∴ **solution** are **infinite** and **graphs** are **concident**

6. (b)

7. (a) for graphs to be concident

$$\frac{a_1}{a_2} = \frac{b_1}{b_2} = \frac{c_1}{c_2}$$

or $\frac{2}{4} = \frac{-4}{-2} = \frac{2}{1}$

or $\frac{p}{2} = \frac{1}{2}$

or $\mathbf{p = 1}$

8. (a) $\frac{a_1}{a_2} = \frac{5}{3}$

or $\frac{b_1}{b_2} = \frac{-4}{-2} = \frac{2}{1}$

∴ $\frac{a_1}{a_2} \neq \frac{b_1}{b_2}$ hence **unique solution**

9. (c) $\frac{a_1}{a_2} = \frac{9}{36} = \frac{1}{4}$

or $\frac{b_1}{b_2} = \frac{-p}{-68} = \frac{1}{4}$

or $\frac{c_1}{c_2} = \frac{34}{115}$

$\frac{a_1}{a_2} = \frac{b_1}{b_2} \neq \frac{c_1}{c_2}$ hence **no solution**

10. (b) let price of pencil $= x$

and price of pen $= y$

$\therefore$ $9x + 5y = 90$ ...(i)

$5x + 4y = 61$ ...(ii)

multiply **(i)** by **4** and **(ii)** by **5** and then subtracting, we get

$$\begin{array}{rcl} 36x + 20y &=& 360 \\ 25x + 20y &=& 305 \\ - \quad - & & - \\ \hline 11x &=& 55 \\ \hline x &=& \mathbf{5} \end{array}$$

for solving ***y*** we put value of ***x*** in **(*i*)**

i.e., $9 \times 5 + 5y = 90$

or $5y = 90 - 45$

or $y = 9$

we have to find $6x + 3y$

i.e., $6 \times 5 + 3 \times 9 = 30 + 27 = \mathbf{57}$

11. (c) let fraction $= \dfrac{x}{y}$

according to statement

$$\frac{x+4}{y} = 1$$

or $x + 4 = y$

or $x - y = -4$ ... (i)

according to statement

$$\frac{x}{y-1} = \frac{1}{2}$$

or $2x = y - 1$

or $2x - y = -1$ ...(ii)

Solving (i) and (ii) we get

$x = 3$ and $y = 7$

$\therefore$ fraction $= \dfrac{3}{7}$

12. (c) $9x - 4y = 200$

and $7x - 3y = 200$

solving above, we get $x = 200$ and $y = 400$

$\therefore$ income of 1st person = **Rs. 1800**

income of 2nd person = **Rs. 1400**

13. (b) let fraction $= \dfrac{x}{y}$

$\therefore$ $\dfrac{x}{y-1} = \dfrac{1}{2}$

or $2x + 8 = y$

or $2x - y = -8$ ...(i)

and $\dfrac{x}{y-5} = \dfrac{1}{3}$

or $3x = y - 5$

or $3x - y = -5$ ...(ii)

solving **(i)** and **(ii)** we get

$x = 3 \quad y = 14$

$\therefore$ fraction $= \mathbf{\dfrac{3}{14}}$

14. (d) let numbers are x and y

$$\frac{x+1}{y+1} = \frac{2}{3}$$

or $3x + 3 = 2y + 2$

or $3x - 2y = -1$ ...(i)

$$\frac{x-1}{y-1} = \frac{1}{2}$$

or $2x - 2 = y - 1$

or $2x - y = -1 + 2$

or $2x - y = 1$ ...(ii)

$$3x - 2y = -1$$
$$4x - 2y = 2$$
$$- \quad + \quad\quad -$$
$$-x = -3$$

and $2 \times 3 - y = 1$

$-y = 1 - 6$

or $-y = -5$

or $\mathbf{y = 5}$

15. (b) let father's present age $= x$

$\therefore$ children's present age $= x$

After fifteen years

father's age $= x + 15$

and five childrens age $= x + 5 \times 15 = x + 75$

according to question

$$x + 15 = \frac{1}{2}(x + 75)$$

or $2x + 30 = x + 75$

$2x - x = 45$

or $x = 45$

$\therefore$ father's present age $=$ **45 years**

16. (b)

$$D = \sqrt{36-32} = \sqrt{4} = 2$$

$$\therefore \quad \alpha = \frac{6+2}{2} = \frac{8}{2} = \mathbf{4}$$

and

$$\beta = \frac{6-2}{2} = \frac{4}{2} = \mathbf{2}$$

17. (a) for equal roots

$$D = 0$$

i.e.,

$$b^2 - 4ac = 0$$

or

$$81 - 8c = 0$$

or

$$81 = 8c$$

$$c = \mathbf{\frac{81}{8}}$$

18. (b)

let 1st root $= x$

$\therefore$ 2nd root $= 2x$

sum of roots $= -\dfrac{b}{a}$

i.e.,

$$x + 2x = \frac{m}{6}$$

$$3x = \frac{m}{6}$$

or

$$x = \frac{m}{18} \qquad \text{...(i)}$$

similarly products of roots $= \dfrac{c}{a}$

$$\therefore \quad x \times 2x = \frac{5}{6}$$

$$x^2 = \frac{5}{12}$$

or

$$x = \sqrt{\frac{5}{12}} \qquad \text{...(ii)}$$

from **(i)** and **(ii)**

$$\frac{m}{18} = \sqrt{\frac{5}{12}}$$

or

$$m = 18\sqrt{\frac{5}{12}}$$

or

$$m = 18 \times \frac{\sqrt{5}}{2\sqrt{3}}$$

$$m = \frac{9\sqrt{5}}{\sqrt{3}}$$

or

$$m = \frac{9\sqrt{5}}{\sqrt{3}} \times \frac{\sqrt{3}}{\sqrt{3}}$$

or

$$m = \frac{9\sqrt{15}}{3} = \mathbf{3\sqrt{15}}$$

19. (c)

$$S = 3+\sqrt{5}+3-\sqrt{5} = 6$$

and $$P = (3+\sqrt{5})(3-\sqrt{5}) = 9-5 = 4$$

∴ **equation** is $x^2 - Sx + p = 0$

i.e., $\mathbf{x^2 - 6x + 4 = 0}$

20. (a) If one root is $6+\sqrt{11}$ then other will be $6-\sqrt{11}$

∴ $$S = 6+\sqrt{11}+6-\sqrt{11} = 12$$

$$P = (6+\sqrt{11})(6-\sqrt{11}) = 36-11 = 25$$

∴ **equation** is $x^2 - Sx + p = 0$

i.e., $\mathbf{x^2 - 12x + 25 = 0}$

21. (a) in given equation $x^2 + 4\sqrt{2}\,x + 6$

$$\alpha = \frac{-4\sqrt{2}+\sqrt{(4\sqrt{2})^2-4\times 6}}{2} = \frac{-4\sqrt{2}\sqrt{32-24}}{2}$$

$$= \frac{-4\sqrt{2}+\sqrt{8}}{2} = \frac{-4\sqrt{2}+2\sqrt{2}}{2} = \frac{-2\sqrt{2}}{2} - \sqrt{2}$$

$$\beta = \frac{-4\sqrt{2}-\sqrt{8}}{2} = \frac{-4\sqrt{2}-2\sqrt{2}}{2} = \frac{-6\sqrt{2}}{2} = -\mathbf{3\sqrt{2}}$$

∴ factor are $\left[x-(-3\sqrt{2})\right]\left[(x-(-\sqrt{2})\right]$

$$= \mathbf{(x+3\sqrt{2})(x+\sqrt{2})}$$

22. (d) in given equation $x^2 + 2x - 6$

$$D = b^2 - 4ac = 4 - 4\times 1\times -6 = 4 + 24 = 28$$

as $D > 0$

∴ the given equation can be facterized into linear factors

now $$\alpha = \frac{-\sqrt{2}+\sqrt{28}}{2} = \frac{-2+2\sqrt{7}}{2} = \mathbf{-1} + \sqrt{7}$$

similarly $\beta = \mathbf{-1} - \sqrt{7}$

∴ factors of given equation are

$[(x-(-1+\sqrt{7})]\,[x-(-1-\sqrt{7})] = \mathbf{[(x+1-\sqrt{7})]\,[x+1+\sqrt{7})]}$

23. (a) given equation is

$$x^4 - 26x^2 + 25 = 0$$

now let $x^2 = z$

∴ given equation becomes

$$z^2 - 26z + 25 = 0$$

or $$z^2 - 25z - z + 25 = 0$$

$$z(z-25) - 1(z-25) = 0$$

or $$(z-25)(z-1) = 0$$

$$z = 25$$

or $$z = 1$$

put $z = x^2$

$\therefore$ $x^2 = 25$

$x^2 = 1$

$x = \pm 5$

$x = \pm \mathbf{1}$

24. (c) $2x + \frac{4}{x} = 9$

or $\frac{2x^2 + 4}{x} = 9$

$2x^2 + 4 = 9x$

or $2x^2 - 9x + 4 = 0$

or $2x^2 - 8x - x + 4 = 0$

$2x(x - 4) - 1(x - 4) = 0$

or $(x - 4)(2x - 1) = 0$

$x = \mathbf{4}$

or $x = \mathbf{\frac{1}{2}}$

25. (b) $\sqrt{x} + 2x = 1$

or $\sqrt{x} = 1 - 2x$

squaring both sides

$(\sqrt{x})^2 = (1 - 2x)^2$

or $x = 1 + 4x^2 - 4x$

$4x^2 - 5x + 1 = 0$

or $4x^2 - 4x - x + 1 = 0$

$4x(x - 1) - 1(x - 1) = 0$

or $(4x - 1)(x - 1) = 0$

$x = \mathbf{\frac{1}{4}}$

or $x = \mathbf{1}$

26. (c) $6\left(x^2 + \frac{1}{x^2}\right) - 25\left(x - \frac{1}{x}\right) + 12 = 0$ ...(i)

let $x - \frac{1}{x} = y$

squaring both sides

$x^2 + \frac{1}{x^2} - 2 = y^2$

or $x^2 + \frac{1}{x^2} = y^2 + 2$

putting these value in **(i)** we get

$6(y^2 + 2) - 25y + 12 = 0$

or $6y^2 + 12 - 25y + 12 = 0$

$$6y^2 - 25y + 24 = 0$$

or $$6y^2 - 16y - 9y + 24 = 0$$

$$2y(3y - 8) - 3(3y - 8) = 0$$

or $$(3y - 8)(2y - 3) = 0$$

$$y = \frac{8}{3} \quad \text{or} \quad y = \frac{3}{2}$$

$$x - \frac{1}{x} = \frac{8}{3} \qquad x - \frac{1}{x} = \frac{3}{2}$$

$$\frac{x^2 - 1}{x} = \frac{8}{3} \qquad \frac{x^2 - 1}{x} = \frac{3}{2}$$

$$3x^2 - 3 = 8x \qquad 2x^2 - 3x - 2 = 0$$

$$3x^2 - 8x - 3 = 0 \qquad 2x^2 - 4x + x - 2 = 0$$

$$3x^2 - 4x + x - 3 = 0 \qquad 2x(x - 2) + 1(x - 2) = 0$$

$$3x(x - 3) + 1(x - 3) = 0 \qquad (x - 2)(2x + 1) = 0$$

$$(x - 3)(3x + 1) = 0$$

or $x = \mathbf{3}$ and $x = -\frac{1}{3}$ $\qquad$ $x = \mathbf{2}$ and $x = -\frac{1}{2}$

27. (b) **5 years ago**

let Kush's age $= x$

present age of Kush $= x + 5$

Kush's age after 9 years $= x + 5 + 9$

according to question

$$x(x + 14) = 51$$

or $$x^2 + 14x - 51 = 0$$

$$x^2 + 17x - 3x - 51 = 0$$

or $$x(x + 17) - 3(x + 17) = 0$$

$$(x - 3)(x + 17) = 0$$

$\therefore$ $x = 3$

$\therefore$ Kush's present age $= 3 + 5 = \mathbf{8\ years}$

28. (b) area of right $\Delta = \frac{1}{2} \times b \times h$

$$210 = \frac{1}{2} \times 4x \times (5x - 4)$$

or $$420 = 4x \times (5x - 4)$$

$$20x^2 - 16x - 420 = 0$$

or $$4(5x^2 - 4x - 105) = 0$$

$$5x^2 - 4x - 105 = 0$$

or $$5x^2 - 25x + 21x - 105 = 0$$

$$5x(x - 5) + 21(x - 5) = 0$$

or $$(x - 5)(5x + 21) = 0$$

$\therefore$ $x = \mathbf{5}$

putting this value of x in sides, we get sides as **20cm, 21cm** and **29cm**

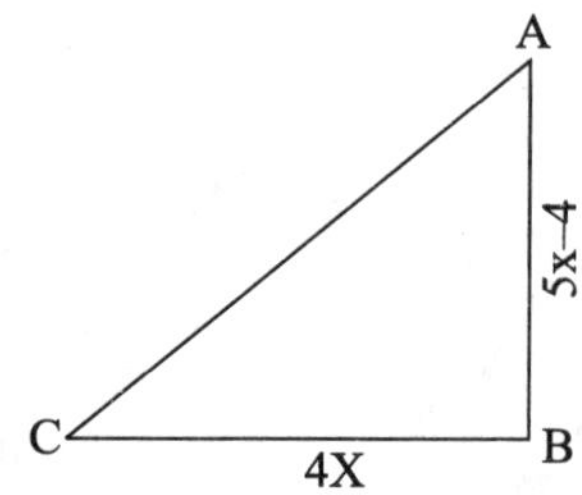

29. (a)

$$\text{let 1st number} = x$$
$$\therefore \quad \text{2nd number} = x + 1$$
$$x^2 + (x+1)^2 = 841$$
$$\text{or} \quad x^2 + x^2 + 2x + 1 = 841$$
$$2x^2 + 2x + 1 - 841 = 0$$
$$\text{or} \quad 2x^2 + 2x - 840 = 0$$
$$2(x^2 + x - 420) = 0$$
$$\text{or} \quad x^2 + x - 420 = 0$$
$$x^2 + 21x - 20x - 420 = 0$$
$$\text{or} \quad x(x+21) - 20(x+21) = 0$$
$$(x-20)(x+21) = 0$$
$$\text{or} \quad x = \mathbf{20}$$
$$\therefore \quad \text{smaller number} = \mathbf{20}$$

30 (a)

$$\text{let speed of slow train} = x \text{ km/h}$$
$$\therefore \quad \text{speed of fast train} = x + 10 \text{ km/h}$$

according to question

$$\frac{600}{x} - \frac{600}{x+10} = \frac{3}{1}$$

solving for x, we get $x = 40$

$$\therefore \quad \text{speed of fast train} = \mathbf{50\ km/h}$$

31. (c)

$$\text{let wickets taken by Sachin} = x$$
$$\therefore \quad \text{wickets taken by Anil} = 3x + 1$$

according to question

$$x(3x+1) = 14$$
$$3x^2 + x = 14$$
$$\text{or} \quad 3x^2 + 7x - 14 = 0$$
$$3x^2 + 7x - 6x - 14 = 0$$
$$\text{or} \quad x(3x+7) - 2(3x+7) = 0$$
$$(x-2)(x+7) = 0 \quad \text{or} \quad \therefore x = \mathbf{2}$$
$$\therefore \quad \text{wickets taken by Sachin} = \mathbf{2}$$

32. (a)

33. (b)

Day 23

Indices and Surds

Today we will do **indices and surds**. First we will do **indices**.

We know that $\mathbf{5^3} = 5 \times 5 \times 5 = \mathbf{125}$ and we say it as **five** raised to power **three** is $\mathbf{5 \times 5 \times 5}$ which equals to **125**.

Now here in $\mathbf{5^3}$, 5 is called **base** and 3 is called **power** or **exponent** or **indices.**

Now if we are to solve $\mathbf{5^3 \times 5^2}$ we will solve it as :

$$5^3 \times 5^2 = \underline{5 \times 5 \times 5} \times \underline{5 \times 5} = \mathbf{5^5} \text{ or can write it as}$$

$$5^3 \times 5^2 = 5^{3+2} = 5^5 \text{ means}$$

If we have same base we can add the powers.

This is also the first law of indices. We can generalise it as

$$a^m . a^n = a^{m+n}$$

Now we will define the laws of indices

LAWS OF INDICES

1. $a^m \times a^n = a^{m+n}$
2. $a^m \div a^n = \dfrac{a^m}{a^n} = a^{m-n}$
3. $(a^m)^n = a^{mn}$
4. $\left(\dfrac{a}{b}\right)^n = \dfrac{a^n}{b^n}$
5. $a^0 = 1$
6. $\sqrt[n]{a^m} = a^{\frac{m}{n}}$
7. $a^{-n} = \dfrac{1}{a^n}$.

These are very important laws and will be used in the questions to be followed.

Now comes the type of questions.

TYPES OF QUESTIONS

We have divided the question into following four types.

Type I : Simple Questions.

Type II : Question having algebric base.

Type III : Question requiring factorization into prime numbers.

Type IV : Miscellaneous Type

We will first do a few examples of all the above four types.

TYPE I
(Simple Questions)

Example 1. Find the value of $(2^3)^2$.

Sol : From law of indices we know that

$$(a^m)^n = a^{m\,n}$$
$$(2^3)^2 = 2^{3\times 2} = 2^6 = \mathbf{64}$$

Example 2. Find the value of $\sqrt[5]{32^4}$

Sol :
$$\sqrt[5]{32^4} = \sqrt[5]{2^{5\times 4}}$$
$$= \sqrt[5]{2^{20}}$$
$$= 2^{\frac{20}{5}}$$
$$= 2^4 = \mathbf{16}$$

Example 3. Solve $2^{2^3} - (2^2)^3$.

Sol :
$$2^{2^3} - (2^2)^3 = 2^8 - 2^6$$
$$= 2^6[2^2 - 1] = 2^6(4-1)) = 2^6[4-1]$$
$$= 64 \times 3 = \mathbf{192}$$

Example 4. Find the value of $2^{3^2} \div (2^3)^2$.

Sol :
$$2^{3^2} \div (2^3)^2 = \frac{2^{3^2}}{(2^3)^2}$$
$$= \frac{2^9}{2^6} = 2^{9-6} = 2^3 = \mathbf{8}$$

TYPE II
(Question having Algebric Base)

Example 5. Solve $\left(\frac{x^a}{x^b}\right)^{a+b} \times \left(\frac{x^b}{x^c}\right)^{b+c} \left(\frac{x^c}{x^a}\right)^{c+a}$.

Sol :
$$\left(\frac{x^a}{x^b}\right)^{a+b} \times \left(\frac{x^b}{x^c}\right)^{b+c} \left(\frac{x^c}{x^a}\right)^{c+a}$$

$$= x^{(a-b)(a+b)}\, x^{(b-c)(b+c)}\, x^{(c-a)(c+a)}$$

$$= x^{a^2-b^2}.x^{b^2-c^2}.x^{c^2-a^2}$$

$$= x^{a^2-b^2+b^2-c^2+c^2-a^2}$$

$$= x^0 = \mathbf{1}$$

Example 6. *Solve* $\left(\dfrac{x^a}{x^b}\right)^{a+b-c} . \left(\dfrac{x^b}{x^c}\right)^{b+c-a} . \left(\dfrac{x^c}{x^a}\right)^{c+a-b}$

Sol : $\left(\dfrac{x^a}{x^b}\right)^{a+b-c} . \left(\dfrac{x^b}{x^c}\right)^{b+c-a} . \left(\dfrac{x^c}{x^a}\right)^{c+a-b}$

$$= x^{(a-b)(a+b-c)}\, x^{(b-c)(b+c-a)}\, x^{(c-a)(c+a-b)}$$

$$= x^{a^2+ab-ac-ab-b^2+bc}\, x^{b^2+bc-ab-bc-c^2+ac}$$
$$x^{c^2+ac-bc-ac-a^2+ab}$$

$$= x^{a^2-ac-b^2+bc+b^2-ab-c^2+ac+c^2-bc-a^2+ab}$$

$$= x^0 = \mathbf{1}$$

Example 7. $\left(\dfrac{x^b}{x^c}\right)^{b^2+bc+c^2} \left(\dfrac{x^c}{x^a}\right)^{c^2+ac+a^2} . \left(\dfrac{x^a}{x^b}\right)^{a^2+ab+b^2}$

Sol : $\left(\dfrac{x^b}{x^c}\right)^{b^2+bc+c^2} \left(\dfrac{x^c}{x^a}\right)^{c^2+ac+a^2} \left(\dfrac{x^a}{x^b}\right)^{a^2+ab+b^2}$

$$= x^{(b-c)(b^2+bc+c^2)}.x^{(c-a)(c^2+ac+b^2)}$$
$$x^{(a-b)(a^2+ab+b^2)}$$

$$= x^{b^3-c^3}.x^{c^3-a^3}x^{a^3-b^3}$$

$$= x^{b^3-c^3+c^3-a^3+a^3-b^3}$$

$$= x^0 = \mathbf{1}$$

Example 8. Solve $\dfrac{2^{m+3}.3^{2m-n}.5^{m+n+3}.6^{n+1}}{6^{m+1}.10^{n+3}.15^m}$.

Sol : Here in this type we factorize the numbers into prime factor and the add the powers of same base and get result. so we will proceed as follows.

$$\frac{2^{m+3}.3^{2m-n}.5^{m+n+3}.6^{n+1}}{6^{m+1}.10^{n+3}.15^m} = \frac{2^{m+3}.3^{2m-n}.5^{m+n+3}.(2\times 3)^{n+1}}{(2\times 3)^{m+1}.(2\times 5)^{n+3}.(3\times 5)^m}$$

$$= \frac{2^{m+3}.3^{2m-n}.5^{m+n+3}.2^{n+1}.3^{n+1}}{2^{m+1}.3^{m+1}.2^{n+3}.5^{n+3}-3^{n}.5^{n}}$$

$$= 2^{m+3+n+1-m-1-n-3}.3^{2m-n+n+1-m-1-m}$$

$$5^{m+n+3-n-3-m}$$

$$= 2^0 . 3^0 . 5^0$$

$$= 1 . 1 . 1 = \mathbf{1}$$

Example 9. $\dfrac{3^{2m+3n}.5^{m-1}.10^{2n+1}.14^{m+1}}{6^{m-2}.7^{m+1}.12^{n+2}.15^{m+2n}}$.

Sol : $\dfrac{3^{2m+3n}.5^{m-1}.(2\times 5)^{2n+1}.(2\times 7)^{m+1}}{(2\times 3)^{m-2}.7^{m+1}.(2\times 2\times 3)^{n+2}.(3\times 5)^{m+2n}}$

$$= \frac{3^{2m+3n}.5^{m-1}.2^{2n+1}.5^{2n+1}.2^{m+1}.7^{m+1}}{2^{m-2}.3^{m-2}.7^{m+1}.2^{n+2}.2^{n+2}.3^{n+2}.3^{m+2n}.5^{m+2n}}$$

$$= 2^{2n+1+m+1-m+2-n-2-n-2}.3^{2m+3n-m+2-n-2-m-2n}$$

$$.5^{m-1+2n+1-m-2n}.7^{m+1-m-1}$$

$$= 2^0 . 3^0 . 5^0 . 7^0 = \mathbf{1}$$

Example 10. Solve $\dfrac{2^{m+1}.3^{2m-n}.5^{m+n+2}.6^{n}}{6^{m}.10^{n+1}.15^{m}}$

Sol : $\dfrac{2^{m+1}.3^{2m-n}.5^{m+n+2}.(2\times 3)^{n}}{(2\times 3)^{m}.(2\times 5)^{n+1}.(3\times 5)^{m}}$

$$= \frac{2^{m+1}.3^{2m-n}.5^{m+n+2}.2^{n}.3^{n}}{2^{m}.3^{m}.2^{n+1}.5^{n+1}.3^{m}.5^{m}}$$

$$= 2^{m+1+n-m-n-1}.3^{2m-n+n-m-m}.5^{m+n+2-n-1-m}$$

$$= 2^0 . 3^0 . 5^1$$

$$= 1 . 1 . 5 = \mathbf{5}$$

Example 11. $\dfrac{2^{n}.6^{m+1}.10^{m-n}.15^{m+n-2}}{4^{m}.3^{2m+n}.25^{m-1}}$

Sol : $\dfrac{2^{n}.(2\times 3)^{m+1}.(2\times 5)^{m-n}.(3\times 5)^{m+n-2}}{2^{2m}.3^{2m+n}.5^{2(m-1)}}$

$$= \frac{2^{n}.2^{m+1}.3^{m+1}.2^{m-n}.5^{m-n}.3^{m+n-2}.5^{m+n-2}}{2^{2m}.3^{2m+n}.5^{2m-2}}$$

$$= 2^{n+m+1+m-n-2m}.3^{m+1+m+n-2-2m-n}$$

$$.5^{m-n+m-n-2-2m+2}$$

$$= 2^1 . 3^{-1} . 5^0$$

$$= 2 \times \frac{1}{3} \times 1 = \mathbf{\frac{2}{3}}$$

Example 12. $\dfrac{2^n + 2^{n-1}}{2^{n+1} - 2^n}$

Sol :

$$\frac{2^n + 2^{n-1}}{2^{n+1} - 2^n} = \frac{2^n + 2^n.2^{-1}}{2^n.2^1 - 2^n}$$

$$= \frac{2^n\left[1 + 2^{-1}\right]}{2^n\left[2^1 - 1\right]} = \frac{1 + \frac{1}{2}}{2 - 1}$$

$$= \frac{\frac{3}{2}}{1} = \mathbf{\frac{3}{2}}$$

Example 13. Solve $9^x = \dfrac{9}{3^x}$

Sol : If $x^a = x^b$ then $a = b$

i.e., *if* bases are equal then power are equal

$$9^x = \frac{9}{3^x}$$

$$(3^2)^x = \frac{3^2}{3^x}$$

$$(3)^{2x} = 3^{2-x}$$

$$\therefore \quad 2x = 2 - x$$

$$3x = 2$$

$$x = \frac{2}{3}.$$

Example 14. $16^{x+1} = \dfrac{64}{4^x}$.

Sol :

$$16^{x+1} = \frac{64}{4^x}$$

$$(2^4)^{x+1} = \frac{2^6}{2^{2x}}$$

$$2^{4x+4} = 2^{6-2x}$$

$$\therefore \quad 4x + 4 = 6 - 2x$$

$$6x = 2$$

$$x = \frac{2}{6} = \frac{1}{3}$$

Now we will do the topic of surds

SURDS

In **surds** first we will define surds and its related terms (like **radical, order of surd radicand** etc.) then we will do **laws of surds, types of surds, rationalising factor of a surd**, rationalising of **binomial quadratic surds** etc. Then we will do **types of questions** followed by **exercise**.

SURD: If **a** is a **positive rational number** which cannot be expressed as the **nth power** of some **rational number**, then **irrational number** $\sqrt[n]{a}$ **or** $a^{\frac{1}{n}}$ *i.e.*, positive **nth root** of a is called a **surd**.

The symbool $\sqrt[n]{\ }$ is called radical sign, n is called **order of the surd** and a is called **radicand**.

We can understand the above statement by simple example $\sqrt[n]{a}$ is a surd because **3 is rational** and $\sqrt[n]{a}$ **is irrational.**

Now we compare

$\sqrt[n]{a}$	with	$\sqrt[5]{3}$
a or 3	is the	**radicand**
n or 5	is the	**order of surd**
$\sqrt[n]{\ }$ or $\sqrt[5]{\ }$	is	**radical sign.**

Now we will do **laws of surds.**

LAWS OF SURDS

1. $(\sqrt[n]{a})^n = a$
2. $\sqrt[n]{ab} = \sqrt[n]{a} \times \sqrt[n]{b}$
3. $\sqrt[m]{\left[\sqrt[n]{a}\right]} = a^{\frac{m}{n}}$
4. $(\sqrt[n]{a})^m = a^{\frac{m}{n}}$
5. $\sqrt[n]{\frac{a}{b}} = \frac{\sqrt[n]{a}}{\sqrt[n]{b}}$

Types of Surds

There are two types of surds

1. **Pure Surd**
2. **Mixed Surd.**

1. Pure Surd : The **surd** which has **unity** as its **rational factor** the **other factor** being **irrational** is called **pure surd.**

$\sqrt{6}, \sqrt{7}, \sqrt{8}$, are pure surd as their rational factor is 1 as $\sqrt{6}$ can be written as $1 \times \sqrt{6}$

$\sqrt{7}$ can be written as $1 \times \sqrt{7}$

and $\sqrt{8}$ can be written as $1 \times \sqrt{8}$

2. Mixed Surd: The **surds** which has a **rational factor other than unity** the other factor being **irrational** is called **mixed surd.**

e.g., $2\sqrt{7}, 2\sqrt{5}$ etc.

Now we will do rationalising of surds.

Rationalisation of Surds

If we multiply a surd $\sqrt{2}$ by $\sqrt{2}$ then we get the answer as **2**

i.e., $\sqrt{2} \times \sqrt{2} = 2$

Thus by multiplying above **two irrational numbers** we get a **rational number**. This each of the above two surds is called rationalising factor of the other. This process is called **rationalising of the** given **surd**.

We can further explain it by another example.

Let us suppose we are to find the **rationalising factor** of $\sqrt[3]{5}$. It is clear that there is a **deficency of two 5's** in the expression in order to make it a **rational number**. So $\sqrt[3]{5 \times 5}$ is **rationalising factor** of $\sqrt[3]{5}$ as

$$\sqrt[3]{5} \times \sqrt[5]{5 \times 5} = \sqrt[3]{5 \times 5 \times 5} = 5$$

which is a rational number.

So it is hoped that now you are clear about rationalising factor of a surd.

Rationalising of a Binomial Quadratic Surd

Surds of second order i.e., $\sqrt{2}$ are called **quadratic surds** and **quadratic surds** having **two terms** attached by a **sign + or –** are called **binomial quadratic surds.**

$\sqrt{3} + \sqrt{2}, \sqrt{5} - \sqrt{7}$ are example of **binomial quadratic surds**

The **binomial quadratic surds** which differ only in **sign + or –** connecting them are said to be **conjugate** to each other. Thus $\sqrt{5} + \sqrt{3}$ and $\sqrt{5} - \sqrt{3}$ are **conjugate** to each other and also **ratioinalising factor** of each other as by multiplying $\sqrt{5} + \sqrt{3}$ with $\sqrt{5} - \sqrt{3}$ we get

$(\sqrt{5}) - (\sqrt{3})^2 = 5 - 3 = 2$ which is a **rational number**.

Now it is hoped that you have understood the basic terminology related to surds. Now comes the types of question.

Types of Questions

We have divided the question into following types :

Type I : conversion of pure surds into mixed and vice versa.

Type II : comparison of surds by bringing them into same order.

Type III : simplifying the surds by combining the similar term.

Type IV : questions involving rationalisation of binomial quadratic surds.

We will do a few examples of all the types followed by exercise.

TYPE I
(Conversion of Pure Surds into Mixed and Vice Versa)

Example 1. Express following as pure surd

(i) $4\sqrt{3}$ (ii) $2^{3}\sqrt{3}$

(iii) $x\sqrt{y}$ (iv) $a^{3}\sqrt{b}$

Sol:

(i) $4\sqrt{3} = \sqrt{4^2}\sqrt{3} = \sqrt{4^2 \times 3} = \mathbf{48}$

(ii) $2\sqrt[3]{3} = \sqrt[3]{2^3}\sqrt[3]{3} = \sqrt[3]{2^3 \times 3} = \mathbf{\sqrt{24}}$

(iii) $x\sqrt{y} = \sqrt{x^2}\sqrt{y} = \sqrt{x^2 y}$

(iv) $a\sqrt[3]{b} = \sqrt[3]{a^3}\sqrt[3]{b} = \sqrt[3]{a^3 b}$

Example 2. Express following as a mixed surd in its simplest form.

(i) $\sqrt{80}$ (ii) $\sqrt[3]{144}$ (iii) $\sqrt[3]{75}$ (iv) $\sqrt[5]{320}$

Sol : (i)

$$\sqrt{80} = \sqrt{2\times 2\times 2\times 2\times 5}$$
$$= \sqrt{2^2 \times 2^2 \times 5}$$
$$= 2\times 2\times \sqrt{5} = \mathbf{4\sqrt{5}}$$

(ii)

$$\sqrt[3]{144} = \sqrt{2\times 2\times 2\times 2\times 3\times 3}$$
$$= \sqrt[3]{2^3 \times 2\times 3\times 3} = \mathbf{2\sqrt[3]{18}}$$

(iii)

$$\sqrt[3]{375} = \sqrt[3]{5\times 5\times 5\times 3}$$
$$= \sqrt[3]{5^3 \times 3} = \mathbf{5\sqrt[3]{3}}$$

(iv)

$$\sqrt[5]{320} = \sqrt[5]{2\times 2\times 2\times 2\times 2\times 2\times 5}$$
$$= a\sqrt[5]{2^5 \times 2\times 5} = \mathbf{2\sqrt[5]{10}}$$

TYPE II
(Reduced to Same Order for Comparison)

Example 3. Which is greater

(i) $\sqrt{3}$ or $\sqrt[3]{5}$ (ii) $\sqrt[3]{6}$ or $\sqrt[4]{8}$ (iii) $\sqrt[4]{4}$ or $\sqrt[3]{3}$

Sol:

(i) **LCM** of their **order** i.e., **2** and **3** is **6**

Thus $\sqrt{3} = \sqrt[2\times3]{3^3} = \sqrt[6]{27}$

$\sqrt[3]{5} = \sqrt[3\times2]{5^2} = \sqrt[6]{25}$

$$\sqrt{3} > \sqrt[3]{5}$$

(ii) **LCM** of **3** and **4** = **12**

$$\sqrt[3]{6} = \sqrt[3\times4]{6^4} = \sqrt[12]{1296}$$

$$\sqrt[4]{8} = \sqrt[4\times3]{8^3} = \sqrt[12]{512}$$

$$\therefore \quad \sqrt[3]{6} > \sqrt[4]{8}$$

(iii) **LCM** of **4** and **3** = **12**

$$\sqrt[4]{4} = \sqrt[4\times3]{4^3} = \sqrt[12]{64}$$

$$\sqrt[3]{3} = \sqrt[3\times4]{3^4} = \sqrt[12]{81}$$

$$\therefore \quad \sqrt[3]{3} > \sqrt[4]{4}$$

Example 4. Arrange in descending order

(i) $\sqrt[3]{4}, \sqrt[4]{5}, \sqrt{3}$ (ii) $\sqrt[3]{2}, \sqrt[6]{3}, \sqrt[9]{4}$

Sol :

(i) **LCM** of **3**, **4** and **2** is **12**

$$\sqrt[3]{4} = \sqrt[3\times4]{4^3} = \sqrt[3\times4]{4^3}$$

$$\sqrt[4]{5} = \sqrt[4\times3]{5^3} = \sqrt[12]{125}$$

$$\sqrt{3} = \sqrt[12]{3^6} = \sqrt[12]{729}$$

Arranging in descending order

$$\sqrt[12]{729} > \sqrt[12]{125} > \sqrt[12]{64}$$

$$\sqrt{3} > \sqrt[4]{5} > \sqrt[3]{4}$$

is the correct descending order.

(ii) **LCM** of **3, 6** and **9** is **18**

$$\sqrt[3]{2} = \sqrt[3\times6]{2^6} = \sqrt[18]{64}$$

$$\sqrt[6]{3} = \sqrt[6\times3]{3^3} = \sqrt[18]{27}$$

$$\sqrt[9]{4} = \sqrt[9\times2]{4^2} = \sqrt[18]{16}$$

$$\sqrt[18]{64} > \sqrt[18]{27} > \sqrt[18]{16}$$

$$\sqrt[3]{2} > \sqrt[6]{3} > \sqrt[9]{4}$$

is correct descending order.

TYPE III
(Simplifying the Surd into Simplest form by Combining Similar Terms)

Example 5. Simplfy by combining similar term

(i) $\sqrt{8}+\sqrt{32}-\sqrt{2}$ (ii) $4\sqrt{12}-\sqrt{75}-7\sqrt{48}$

(iii) $2\sqrt{40}-3\sqrt{625}-4\sqrt{320}$

Sol : (i)
$$\sqrt{8}+\sqrt{32}-\sqrt{2} = \sqrt{4\times 2}+\sqrt{16\times 2}-\sqrt{2}$$
$$= 2\sqrt{2}+4\sqrt{2}-\sqrt{2}$$
$$= 6\sqrt{2}-\sqrt{2}$$
$$= \sqrt{2}(6-1) = \mathbf{5\sqrt{2}}$$

(ii)
$$4\sqrt{4\times 3}-\sqrt{25\times 3}-7\sqrt{16\times 3} = 4\sqrt{4\times 3}-\sqrt{25\times 3}-7\sqrt{16\times 3}$$
$$= 4\times 2\sqrt{3}-5\sqrt{3}-7\times 4\sqrt{3}$$
$$= 8\sqrt{3}-5\sqrt{3}-28\sqrt{3}$$
$$= \sqrt{3}(8-5-28) = \mathbf{-25\sqrt{3}}$$

(iii)
$$2\sqrt[3]{40}-3\sqrt[3]{625}-4\sqrt[3]{320} = 2\sqrt[3]{2^3\times 5}-3\sqrt[3]{5^3\times 5}-4\sqrt[3]{4^3\times 5}$$
$$= 2+2\sqrt[3]{5}-3\times 5\sqrt[3]{5\times 5}-4\times 4\sqrt[3]{5}$$
$$= 4\sqrt[3]{5}-15\times\sqrt[3]{5}-16\sqrt[3]{5}$$
$$= \sqrt[3]{5}(4-15-16)$$
$$= \mathbf{\sqrt[3]{5}}(4-15-16) = -\mathbf{27\sqrt[3]{5}}$$

Example 6. Simplify and express result in simplest form

(i) $\sqrt{14}\times\sqrt{21}$ (ii) $\sqrt[3]{2}\times\sqrt[4]{3}$

(iii) $\sqrt[4]{28}\div\sqrt[3]{7}$

Sol :

(i)
$$\sqrt{14}\times\sqrt{21} = \sqrt{7\times 2}\times\sqrt{7\times 3}$$
$$= \sqrt{2}\sqrt{7}\times\sqrt{3}\sqrt{7}$$
$$= \sqrt{7}\sqrt{7}\times\sqrt{2}\sqrt{3}$$
$$= 7\sqrt{2}\times\sqrt{3} = \mathbf{7\sqrt{6}}$$

(ii) $\sqrt[3]{2}\times\sqrt[4]{3}$

LCM of 3 and 4 is 12

$$\sqrt[3]{2} = \sqrt[3\times 4]{2^4} = \sqrt[12]{16}$$
$$\sqrt[4]{3} = \sqrt[4\times 3]{3^3} = \sqrt[12]{27}$$
$$\sqrt[3]{2}\times\sqrt[4]{3} = \sqrt[12]{16}\times\sqrt[12]{27}$$

$$= \sqrt[12]{16 \times 27} = \sqrt[12]{\mathbf{432}}$$

(iii) $$4\sqrt{28} \div 3\sqrt{7} = \frac{4\sqrt{4 \times 7}}{3\sqrt{7}}$$

$$= \frac{4 \times 2\sqrt{7}}{3\sqrt{7}}$$

$$= \frac{8\sqrt{7}}{3\sqrt{7}} = \frac{\mathbf{8}}{\mathbf{3}}$$

TYPE IV
(Related to Rationalisation of Binomial Quadratic Surds)

Example 7. Rationalise the denominator of $\frac{1}{3+\sqrt{8}}$

Sol : Its rationalising factor is $3-\sqrt{8}$ we proceed as follows.

Let $$x = \frac{1}{3+\sqrt{8}}$$

So if we multiply and divide this expresstion by it will serve our purpose

$$\frac{1}{3+\sqrt{8}} = \frac{1}{3+\sqrt{8}} \times \frac{3-\sqrt{8}}{3-\sqrt{8}}$$

$$= \frac{3-\sqrt{8}}{9-8} = \mathbf{3-\sqrt{8}}$$

Example 8. Rationalise $\frac{1}{2+\sqrt{3}}$

Sol : Here Rationalisation mean to make the denominator of the expression rational.

Rationalising Factor of $2+\sqrt{3}$ is $2+\sqrt{3}$

$$\frac{1}{2+\sqrt{3}} = \frac{1}{2+\sqrt{3}} \times \frac{2-\sqrt{3}}{2-\sqrt{3}}$$

$$= \frac{2-\sqrt{3}}{4-3}$$

$$= \frac{2-\sqrt{3}}{4-3} = \mathbf{2-\sqrt{3}}$$

Example 9. If $x = 3+\sqrt{8}$ then find

(i) $x + \frac{1}{x}$ (ii) $x^2 + \frac{1}{x^2}$

Sol : $$x = 3-\sqrt{8}$$

$$\frac{1}{x} = \frac{1}{3+\sqrt{8}}$$

$$= \frac{1}{3+\sqrt{8}} \times \frac{3-\sqrt{8}}{3-\sqrt{8}}$$

$$= \frac{3-\sqrt{8}}{3-8} = 3-\sqrt{8}$$

$$x + \frac{1}{x} = 3+\sqrt{8}+3-\sqrt{8} = 6$$

now $$x + \frac{1}{x} = 6$$

Squaring both sides we get

$$\left(x + \frac{1}{x}\right)^2 = 6^2$$

$$x^2 + \frac{1}{x^2} + 2.x\frac{1}{x} = 36$$

$$x^2 + \frac{1}{x^2} = 36 - 2 = \mathbf{34}$$

Example 10. If $x = 5 + 2\sqrt{6}$. Find the value of $x^2 + \frac{1}{x^2}$

Sol.

$$x = 5 + 2\sqrt{6}$$

$$\frac{1}{x} = \frac{1}{5+2\sqrt{6}}$$

$$= \frac{1}{5+2\sqrt{6}} \times \frac{5 \times 2\sqrt{6}}{5-2\sqrt{6}}$$

$$= \frac{5-2\sqrt{6}}{25-24}$$

$$= \frac{5-2\sqrt{6}}{1} = 5-2\sqrt{6}$$

$$x + \frac{1}{x} = 5+2\sqrt{6}+5-2\sqrt{6} = 10$$

Squaring both sides we get

$$x^2 + \frac{1}{x^2} + 2.x\frac{1}{x} = 10^2$$

$$x^2 + \frac{1}{x^2} + 2 = 100$$

$$x^2 + \frac{1}{x^2} = 100 - 2 = \mathbf{98}$$

Example 11. If $x = 7 + 4\sqrt{3}$. Find the value of $x^3 + \frac{1}{x^3}$

Sol :

$$x = 7 + 4\sqrt{3}$$

$$\frac{1}{x} = \frac{1}{7+4\sqrt{3}}$$

$$= \frac{1}{7+4\sqrt{3}} \times \frac{7-4\sqrt{3}}{7-4\sqrt{3}}$$

$$= \frac{7-4\sqrt{3}}{(7)^2-(4\sqrt{3})^2}$$

$$= 7-4\sqrt{3}$$

$$\therefore \quad x+\frac{1}{x} = 7+4\sqrt{3} = 7-4\sqrt{3} = 14$$

Cubing both sides we get

$$\left(x+\frac{1}{x}\right)^3 = (14)^3$$

$$x^3+\frac{1}{x^3}+3x.\frac{1}{x}\left(x+\frac{1}{x}\right) = 2744$$

[$\because$ $(a+b)^3 = a^3 + b^3 + 3ab\,(a+b)$here $a = x$ and $b = 1/x$]

$$x^3+\frac{1}{x^3}+3x.\frac{1}{x}\left(x+\frac{1}{x}\right) = 2744$$

$$x^3+\frac{1}{x^3}+3.(14) = 2744$$

$$x^3+\frac{1}{x^3}+42 = 2744$$

$$x^3+\frac{1}{x^3}+3 = 2744-42 = \mathbf{2702}$$

Example 12. If $a = 2 - \sqrt{3}$ then find $a^4 + \frac{1}{a^4}$.

Sol :

$$a = 2-\sqrt{3}$$

$$\frac{1}{a} = \frac{1}{2-\sqrt{3}}$$

$$= \frac{1}{2-\sqrt{3}} \times \frac{2+\sqrt{3}}{2+\sqrt{3}}$$

$$= \frac{2+\sqrt{3}}{1}$$

$$a+\frac{1}{a} = 2-\sqrt{3}+2+\sqrt{3} = 4$$

now

$$a+\frac{1}{a} = 4$$

Squaring both sides we get

$$a^2 + \frac{1}{a^2} + 2a\frac{1}{a} = 4^2$$

$$a^2 + \frac{1}{a^2} + 2 = 4^2$$

$$a^2 + \frac{1}{a^2} = 16 - 2 = 14$$

Squaring both side we get

$$\left(a^2 + \frac{1}{a^2}\right)^2 = 14^4$$

$$a^4 + \frac{1}{a^4} + 2.a^2\frac{1}{a^2} = 196$$

$$a^4 + \frac{1}{a^4} + 2 = 196$$

$$a^4 + \frac{1}{a^4} = 196 - 2 = \mathbf{194}$$

EXERCISE

Q.1. The value of $(625)^{\frac{-3}{4}}$ is

(a) 125 (b) $\frac{1}{225}$ (c) $\frac{1}{125}$ (d) none

Q.2. The value of $\left(\frac{64}{729}\right)^{\frac{-2}{3}}$ is

(a) $\frac{64}{81}$ (b) $\frac{81}{164}$ (c) $\frac{81}{64}$ (d) none

Q.3. The value of $5^{3^2} \div (5^3)^2$

(a) 125 (b) 225 (c) 25 (d) none

Q.4. The value of $5\sqrt{\left(\frac{1}{32}\right)^{-3}}$

(a) 7 (b) 32 (c) 8 (d) none

Q.5. Simplifying $x^{a(b-c)}.x^{a(b-c)}.x^{c(a-b)}$ we get

(a) 0 (b) 1 (c) x (d) none

Q.6. Simplify $\left(\frac{x^l}{x^m}\right)^{l^2+lm+m^2}\left(\frac{x^m}{x^n}\right)^{m^2+mn+n^2}\left(\frac{x^n}{x^l}\right)^{n^2+nl+l^2}$

(a) 0 (b) 01 (c) x (d) none

Q.7. Simplify $\left(\dfrac{x^{\frac{1}{a}}}{x^{\frac{1}{b}}}\right)^{ab}\left(\dfrac{x^{\frac{1}{b}}}{x^{\frac{1}{c}}}\right)^{bc}\left(\dfrac{x^{\frac{1}{c}}}{x^{\frac{1}{b}}}\right)^{ca}$

(a) 0 (b) 1 (c) x (d) none

Q.8. On simplifying $\dfrac{2^{n-1}.6^{m+1}.10^{m-n}.15^{m+n}}{4^m.3^{2m+n+1}.25^m}$ we get

(a) 25 (b) 5 (c) 2 (d) none

Q.9. On simplifying $\dfrac{2^{m+1}\,3^{2m-n}\,5^{m+n}.6^n}{6^m\,10^{n+2}.15^m}$ we get

(a) 25 (b) $\dfrac{1}{500}$ (c) $\dfrac{1}{50}$ (d) none

Q.10. On simplifying $\dfrac{2^{m+3}\,3^{2m-n}\,5^{m+n+3}.6^{n+1}}{6^{m+1}\,10^{n+3}.15^m}$ we get

(a) 0 (b) 1 (c) 2 (d) none

Q.11. On simplifying $\dfrac{2^{2n}\,3^{3n}\,6^n}{2^{3n}\,3^{2n}\,30^n}$ we get

(a) $(.3)^n$ (b) 3^n (c) 1 (d) none

Q.12. On simplifying $\dfrac{5^{n+4}-122.5^{n+1}}{5^{n+2}-2.5^{n+1}}$ we get

(a) 0 (b) 1 (c) n (d) none

Q.13. The value of x in $4^{x+1} = 64^{x+3}$ is

(a) 1 (b) 3 (c) 2 (d) none

Q.14. The value of x in $\left(\dfrac{4}{5}\right)^{2x-3} = \left(\dfrac{5}{4}\right)^{x-3}$ is

(a) 1 (b) 3 (c) 2 (d) none

Q.15. Express following as pure surd

(i) $5\sqrt{7}$ (ii) $25\sqrt{5}$ (iii) $y\sqrt{x}$ (iv) $b^3\sqrt{a}$

Q.16. Express following as a mixed surd in its simplest form

(a) $\sqrt{120}$ (b) $\sqrt[3]{288}$ (c) $\sqrt{80}$ (d) $\sqrt[6]{128}$

Q.17. Which is greater

(i) $\sqrt{2}$ or $\sqrt[3]{9}$ (ii) $\sqrt[3]{2}$ or $\sqrt[4]{3}$ (iii) $\sqrt{5}$ or $\sqrt[3]{7}$

Q.18. The descending order of $\sqrt{5} > \sqrt[4]{3} > \sqrt[3]{2}$ is

(a) $\sqrt{5} > \sqrt[4]{3} > \sqrt[3]{2}$ (b) $\sqrt[4]{3} > \sqrt{5} > \sqrt[3]{2}$ (c) $\sqrt[3]{2} > \sqrt{5} > \sqrt[1]{3}$ (d) none

Q.19. The correct descending order of $\sqrt[3]{6} > \sqrt[4]{8} > \sqrt[6]{12}$ is

(a) $\sqrt[4]{8} > \sqrt[3]{6} > \sqrt[6]{12}$ (b) $\sqrt[3]{6} > \sqrt[4]{8} > \sqrt[6]{12}$ (c) $\sqrt[6]{12} > \sqrt[3]{6} > \sqrt[4]{8}$ (d) none

Q.20. Simplifying $\sqrt{12} + 2\sqrt{48} + 5\sqrt{147} - 45\sqrt{3}$ we get

(a) 1 (b) 2 (c) 3 (d) 0

Q.21. On simplifying $\sqrt[3]{16} + \sqrt[3]{54} + \sqrt[3]{128}$

(a) $\sqrt[3]{3}$ (b) $\sqrt[3]{2}$ (c) $3\sqrt[3]{2}$ (d) none

Q.22. If $x = 6 + \sqrt{35}$ then the value of $x^2 + \frac{1}{x^2}$

(a) 241 (b) 284 (c) 142 (d) none

Q.23. If $x = 19 - 6\sqrt{10}$ the value of is $x^2 + \frac{1}{x^2}$

(a) 1403 (b) 1442 (c) 1400 (d) none

Q.24. If $x = 3 + 2\sqrt{2}$ then value of $x^2 - \frac{1}{x^2}$ is

(a) $10\sqrt{2}$ (b) $24\sqrt{2}$ (c) $10\sqrt{2}$ (d) none

Q.25. If $a = 2 + \sqrt{3}$, $b = 2 - \sqrt{3}$ then value of $\frac{1}{a^3} + \frac{1}{b^3}$ is

(a) 104 (b) 52 (c) 88 (d) none

SOLUTIONS

1. (c)

$$(625)^{\frac{-3}{4}} = (5^4)^{\frac{-3}{4}}$$
$$= 5^{4\times-\frac{3}{4}}$$
$$= 5^{-3}$$
$$= \frac{1}{5^3} = \mathbf{\frac{1}{125}}$$

2. (d)

$$\left(\frac{64}{729}\right)^{\frac{-3}{2}} = \left(\frac{2^6}{3^6}\right)^{\frac{-2}{3}}$$
$$= \frac{2^{6\times-\frac{2}{3}}}{3^{6\times\frac{-2}{3}}}$$
$$= \frac{2^{-4}}{3^{-4}}$$

$$= \frac{3^4}{2^4} = \mathbf{\frac{81}{16}}$$

3. (a)

$$5^{3^2} \div (5^3)^2 = \frac{5^9}{5^6}$$

$$= 5^3 = \mathbf{125}$$

4. (c)

$$\sqrt[5]{\left(\frac{1}{2^5}\right)^{-3}} = \sqrt[5]{\left(\frac{1}{2^5}\right)^{-3}}$$

$$= \left(\frac{1}{2^5}\right)^{\frac{-3}{5}}$$

$$= (2^{-5})^{\frac{-3}{5}}$$

$$= 2^3 = \mathbf{8}$$

5. (b)

$$x^{a(b-c)}, x^{b(c-a)}, x^{c(a-b)} = x^{ab-ac}, x^{bc-ab}, x^{ca-bc}$$

$$= x^{ab-ac+bc-ab+ca-bc}$$

$$= x^0 = 1$$

6. (b) Same as exp. no. 7

7. (b)

$$\left(\frac{x^{\frac{1}{a}}}{x^{\frac{1}{b}}}\right)^{ab} \left(\frac{x^{\frac{1}{b}}}{x^{\frac{1}{b}}}\right)^{bc} \left(\frac{x^{\frac{1}{c}}}{x^{\frac{1}{a}}}\right)^{ac} = x^{\left(\frac{1}{a}-\frac{1}{b}\right)^{ab}} x^{\left(\frac{1}{b}-\frac{1}{c}\right)^{bc}} x^{\left(\frac{1}{c}-\frac{1}{a}\right)^{ac}}$$

$$= x^{\frac{(b-a)}{ab}xab} \; x^{\frac{(c-b)}{bc}xbc} \; x^{\frac{a-c}{ac}xac}$$

$$= x^{b-a} x^{c-b} x^{a-c}$$

$$= x^{b-a+c-b+a-c}$$

$$= x^0 = \mathbf{1}$$

8. (d)

$$\frac{2^{n-1}\, 6^{m+1}\, 10^{m-n}\, 15^{m+n}}{4^m \quad 3^{2m+n+1} \quad 25^m} = \frac{2^{n-1}\,(2\times 3)^{m+1}\,(2\times 5)^{m-n}\,(3\times 5)^{m+n}}{(2^2)^m \quad 3^{2m+n+1} \quad 5^{2m}}$$

$$= \frac{2^{n-1}\, x2^{m+1}\, 3^{m+1}\, 2^{m-n}\, 5^{m-n}\, 3^{m+n}\, 5^{m+n}}{2^{2m} \quad 3^{2m+n+1} \quad 5^{2m}}$$

$$= 2^{n-1+m+1+m-n-2m} \quad 3^{m+1+m+n-2m-n-1} \quad 5^{m-n+m+n-2m}$$

$$= \quad 2^0 \qquad 3^0 \qquad 5^0$$

$$= 1 \;.\; 1 \;.\; 1 = \mathbf{1}$$

9. (c)

$$\frac{2^{m+1}\, 3^{2m-n}\, 5^{m+n}\, 6^n}{6^m\, 10^{n+2}\, 15^m} = \frac{2^{m+1}\, 3^{2m-n}\, 5^{m+n}\,(2\times 3)^n}{(2\times 3)^m\,(2\times 5)^{n+2}\,(3\times 5)^m}$$

$$= \frac{2^{m+1}\, 3^{2m-n}\, 5^{m+n}\, 2^n\, 3^n}{2^m\, 3^m\, 2^{n+2}\, 5^{n+2}\, 3^m\, 5^m}$$

$$= 2^{m+1+n-m-n-2} \quad 3^{2m-n+n-m-m} \quad 5^{m+n-n-2-m}$$

$$= \quad 2^{-1} \quad 3^{0} \quad 5^{-2}$$

$$= \frac{1}{2} \times 1 \times \frac{1}{5^2} = \frac{1}{2 \times 25} = \mathbf{\frac{1}{50}}$$

10. (b)

$$\frac{2^{m+3}\,3^{2m-n}\,5^{m+n+3}\,6^{n+1}}{6^{m+1}\,10^{n+3}\,15^{m}} = \frac{2^{m+3}\,3^{2m-n}\,5^{m+n+3}\,(2\times 3)^{n+1}}{(2\times 3)^{m+1}\,(2\times 5)^{n+3}\,(3\times 5)^{m}}$$

$$= \frac{2^{m+3}\,3^{2m-n}\,5^{m+n+3}\,2^{n+1}\,3^{n+1}}{2^{m+1}\,3^{m+1}\,2^{n+3}\,5^{n+3}\,3^{m}\,5^{m}}$$

$$= 2^{m+3+n+1-m-1-n-3}\,3^{2m-n+n+1-m}5^{m+n+3-n-3-m}$$

$$= \mathbf{1\,.\,1\,.\,1}$$

$$= \mathbf{1}$$

11. (a)

$$\frac{2^{2n}3^{3n}6^{n}}{2^{3n}3^{2n}30^{n}} = \frac{2^{2n}3^{3n}(2\times 3)^{n}}{2^{3n}3^{2n}(2\times 3\times 5)^{n}}$$

$$= \frac{2^{2n}3^{3n}2^{n}3^{n}}{3^{3n}.3^{2n}.2^{n}.3^{n}.5^{n}}$$

$$= 2^{2n+n-3n-n}.3^{3n+n-2n-n}.5^{-n}$$

$$= 2^{-n}.3^{n}.5^{-n}$$

$$= \frac{3^{n}}{(2\times 5)^{n}} = \left(\frac{3}{10}\right)^{n} = \mathbf{(.3)^{n}}$$

12. (b)

$$\frac{5^{n+4} - 122.5^{n+1}}{5^{n+2} - 2.5^{n+1}} = \frac{5^{n+1}\left[5^{3} - 122\right]}{5^{n+1}\left[5^{1} - 2\right]}$$

$$= \frac{125 - 122}{5 - 2}$$

$$= \frac{3}{3} = 1$$

13. (d)

$$4^{x+1} = 64^{x+3}$$

$$4^{x+1} = (4^{3})^{x+3}$$

$$4^{x+1} = 4^{3x+9}$$

$$x + 1 = 3x + 9$$

$$3x - x = -1$$

$$2x = 8$$

$$x = \mathbf{4}$$

14. (c)

$$\left(\frac{4}{5}\right)^{2x-3} = \left(\frac{5}{4}\right)^{x-3}$$

$$\left(\frac{4}{5}\right)^{2x-3} = \left(\frac{4}{5}\right)^{-x+3}$$

$$2x - 3 = -x + 3$$

$$3x = 6$$

$$x = 2$$

15. (i) $5\sqrt{7} = \sqrt{5^2 7} = \sqrt{175}$

(ii) $25\sqrt{7} = 25\sqrt{7} = \sqrt{25^2 \times 7} = \sqrt{625 \times 5} = \sqrt{3125}$

(iii) $y\sqrt{x} = \sqrt{y^2}\sqrt{x} = \sqrt{y^2 x}$

(iv) $b^3\sqrt{a} = \sqrt[3]{b^3}\sqrt[3]{a} = \sqrt[3]{b^3 a}$

16. (i) $\sqrt{120} = \sqrt{2 \times 2 \times 5 \times 3}$

$= \sqrt{2^2 \times 2 \times 5 \times 3} = \mathbf{\sqrt[2]{30}}$

(ii) $\sqrt[3]{288} = \sqrt{2 \times 2 \times 2 \times 2 \times 2 \times 3 \times 3}$

$= \sqrt{2^3 \times 2 \times 2 \times 3 \times 3} = \mathbf{2\sqrt[3]{36}}$

(iii) $\sqrt{80} = \sqrt{2 \times 2 \times 2 \times 2 \times 5}$

$= \sqrt{2^2 \times 2^2 \times 5} = \mathbf{4\sqrt{5}}$

(iv) $\sqrt[6]{128} = \sqrt[6]{2 \times 2 \times 2 \times 2 \times 2 \times 2 \times 2}$

$= \sqrt[6]{2^6 \times 2} = \mathbf{2\sqrt[6]{2}}$

17. (i) LCM of 2 and 3 is 6

$$\sqrt{2} = \sqrt[6]{2^3} = \sqrt[6]{8}$$

$$\sqrt[3]{9} = \sqrt[3\times2]{9^2} = \sqrt[6]{81}$$

$$\sqrt[6]{81} > \sqrt[6]{8}$$

(ii) $\sqrt[3]{2}$ or $\sqrt[4]{3}$ LCM of 3 and 4 is

$$\sqrt[3]{2} = \sqrt[3\times4]{2^4} = \sqrt[12]{16}$$

$$\sqrt[4]{3} = \sqrt[4\times3]{3^3} = \sqrt[12]{27}$$

$$\sqrt[12]{127} > \sqrt[12]{16}$$

$$\sqrt[4]{3} > \sqrt[3]{2}$$

(iii) LCM of 2 and 3 is 6.

$$\sqrt{5} = \sqrt[6]{5^5} = \sqrt[6]{125}$$

$$\sqrt[3]{7} = \sqrt[3\times2]{7^2} = \sqrt[6]{49}$$

$$\sqrt[6]{125} > \sqrt[6]{49}$$

or $\sqrt{5} > \sqrt[3]{7}$

18. (a) In $\sqrt{5}, \sqrt[4]{3}, \sqrt[3]{2}$

LCM of 2, 3, 4 is 12

$$\sqrt{5} = \sqrt[2\times6]{5^6} = \sqrt[12]{5^6} = \sqrt[12]{15625}$$

$$\sqrt[4]{3} = \sqrt[4\times3]{3^3} = \sqrt[12]{27}$$

$$\sqrt[3]{2} = \sqrt[3\times4]{2^4} = \sqrt[12]{16}$$

Correct descending order is

$$\sqrt{5} > \sqrt[4]{3} > \sqrt[3]{2}$$

19. (b) In

$$\sqrt[3]{6} = \sqrt[3\times4]{6^4} = \sqrt[12]{1296}$$

$$\sqrt[4]{8} = \sqrt[4\times3]{8^3} = \sqrt[12]{512}$$

$$\sqrt[6]{12^2} = \sqrt[6\times2]{12^2} = \sqrt[12]{144}$$

$$\sqrt[3]{6} > \sqrt[4]{8} > \sqrt[6]{12}$$

is correct descending order.

20. (d)

$$\sqrt{12} + 2\sqrt{48} + 5\sqrt{147} - 45\sqrt{3} = \sqrt{4\times3} + 2\sqrt{2\times2\times2\times2\times3} + 5\sqrt{7\times7\times3} - 45\sqrt{3}$$

$$= 2\sqrt{3} + 8\sqrt{3} + 35\sqrt{3} - 45\sqrt{3}$$

$$= 45\sqrt{3} - 45\sqrt{3} = \mathbf{0}$$

21. (b)

$$\sqrt[3]{16} + \sqrt[3]{54} + \sqrt[3]{128} = \sqrt[3]{2\times2\times2\times2} + \sqrt[3]{2\times3\times3\times3} - \sqrt[3]{2\times2\times2\times2\times2\times2\times2}$$

$$= 2\sqrt[3]{2} + 3\sqrt[3]{2} - 4\sqrt[3]{2}$$

$$= \sqrt[3]{2}\ (2+3-4)$$

$$= \sqrt[3]{2}$$

22. (c)

$$x = 6 + \sqrt{35}$$

$$\frac{1}{x} = \frac{1}{6+\sqrt{35}}$$

$$= \frac{1}{6+\sqrt{35}} \times \frac{6-\sqrt{35}}{6-\sqrt{35}}$$

$$= \frac{6-\sqrt{35}}{36-36} = 6 - \sqrt{35}$$

$$x + \frac{1}{x} = 6 + \sqrt{35} + 6 - \sqrt{35}$$

$$= 12$$

Squaring both sides

$$\left(x + \frac{1}{x}\right)^2 = 12^2$$

$$x^2 + \frac{1}{x^2} + 2 = 144$$

$$x^2 + \frac{1}{x^2} = 144 - 2 = \mathbf{142}$$

23. (b)

$$\begin{aligned}
x &= 19-6\sqrt{10}\\
\frac{1}{x} &= \frac{1}{19-6\sqrt{10}}\\
&= \frac{1}{19-6\sqrt{10}}\\
&= \frac{1}{19-6\sqrt{10}}\times\frac{19+6\sqrt{10}}{19+6\sqrt{10}}\\
&= \frac{19+6\sqrt{10}}{(19)^2-(6\sqrt{10})^2}\\
&= \frac{19+6\sqrt{10}}{361-360}\\
&= 19+6\sqrt{10}\\
x+\frac{1}{x} &= 19+6\sqrt{10}+19-6\sqrt{10}\\
&= \mathbf{38}
\end{aligned}$$

Squaring both sides

$$\begin{aligned}
\left(x+\frac{1}{x}\right)^2 &= (38)^2\\
x^2+\frac{1}{x^2}+2\times\frac{1}{2} &= 1444\\
x^2+\frac{1}{x^2}+2 &= 1444\\
x^2+\frac{1}{x^2} &= 1444-2\\
&= 1442
\end{aligned}$$

24. (b)

$$\begin{aligned}
x &= 3+2\sqrt{2}\\
\frac{1}{x} &= \frac{1}{3+2\sqrt{3}}\times\frac{3-2\sqrt{2}}{3-2\sqrt{2}}\\
&= \frac{3-2\sqrt{2}}{9-8}\\
x+\frac{1}{x} &= 3+2\sqrt{2}+3-2\sqrt{2}\\
&= 6\\
x-\frac{1}{x} &= (3+2\sqrt{2})-(3-2\sqrt{2})\\
&= 3+2\sqrt{2}-3+2\sqrt{2}\\
&= 4\sqrt{2}\\
x^2-\frac{1}{x^2} &= \left(x+\frac{1}{x}\right)\left(x-\frac{1}{x}\right)\\
&= 6\times4\sqrt{2}=24\sqrt{2}
\end{aligned}$$

25. (b) $a = 2 + \sqrt{3}$ $b = 2 - \sqrt{3}$

We have to find
$$\frac{1}{a^3} + \frac{1}{b^3} = \frac{b^3 + a^3}{a^3 . b^3}$$
$$= \frac{b^3 + a^3}{a^3 b^3}$$
$$a + b = 2 + \sqrt{3} + 2 - \sqrt{3} = 4$$
$$ab = (2 + \sqrt{3})(2 - \sqrt{3})$$
$$4 - 3 = 1$$
$$(a + b)^3 = a^3 + b^3 + 3ab(a + b)$$
$$(4)^3 = a^3 + b^3 + 3 \times 1 \times 4$$
$$64 = a^3 + b^3 + 12$$
$$a^3 + b^3 = 64 - 12 = \mathbf{52}$$
$$(ab)^3 = (1)^3 = 1$$
$$\frac{b^2 + a^3}{a^3 b^3} = \frac{52}{1} = \mathbf{52}$$

Day 24

Sets

Today we will do the topic of **sets**. Now before going further we will first define what is a set?

A **Cricket team** consists of **players,** A **class** consists of **students** or we can say that a **cricket team** is **set of players** and a **class** is a **set** of **students.** So by set we mean collection of well defined objects.

So remember **set is a collection of well defined objects**. The objects are called **members** or **elements** of a set. **Sets** are denoted by capital letter **A, B, C, D, E** etc. and its **members** i.e., **elements** by small letters ***a, b, c, d, e,*** etc.

Now we will further discuss a few notations, if we say a is ***a*** member of set **A**. As stated the elements are denoted by **small letters** i.e., in this case by ***a*** and set by **capital letters** in this case by **A**. Now in language of sets we say **a belongs to A** and we write it as

$$a \in A$$

So the sign	$\in$	means **belongs to**
&	$\notin$	means **does not belongs to**.
so if we write		$a \in A$
		It means **a belong to A**.
and if we write		$a \notin A$
		It means **a does not belong to A**.

Now we will discuss **how to represents a set**?

Set can be represented in **two ways**

1. **Roaster Method**
2. **Set Builder Method.**

1. **Roaster Method :**

   In Roaster Method **each and every member of the set is listed but once**. Also, **every member** is enclosed in **brackets** i.e., { } and seprated by commas. As stated **each member is listed** but **is not written more than once**. We know that **English Vowels** are ***a, e, i, o***, and ***u***. So in roaster method **set** of **English vowels** can be written as **{*a, e, i, o, u,*}**.

   So each member is listed but once.

   Similarly set of **Natural Numbers** less than 8 is written as **{ 1, 2, 3, 4, 5, 6, 7, }.**

   So now it is hoped that you have understood the roaster method of representing a set.

Now comes the set builder form.

2. **Set Builder Method:**

   In set builder form **members** or **elements** are not listed but a **rule or property is written** which **defines all the elements of the set**.

   This can be more clearly understood by following examples.

   We have represented the set of **English vowels** in **Roaster method** as **{a, e, i, o, u}** Where all the elements are listed but here is set builder form it is written as

   A = **{x : x is a vowels in English alphabet}**

   Some more examples of sets in **Roaster method** and **Set builder** form are given below.

In **Roaster** **Method**	In **Set Builder** **Form**
{I, N, D, A}	$\{x : x$ is a letter of word INDIA$\}$
{2, 3}	$\{x : x$ is a prime no. and a divisor of 6$\}$
{1, 3, 5, 7, 9,}	$\{x : x$ is an odd natural number $< 10\}$
{2, 4, 6, 8,10}	$\{x : x$ is a positive even integer $\leq 10\}$

Now it is hoped that you have understood the method of representing a set by roaster method and set builder method.

Now we will define a few terms related to sets.

1. **Finite set and Infinite set :** Some sets contains finite numbers of elements and some contain Infinite numbers of elements. So simply sets **containing finite no. of elements are called finite sets** and **sets containing infinite no of elements called Infinite sets**.

   Examples of Infinite sets :

   (i) Set of **natural numbers,**

   (ii) Set of **odd numbers,**

   (iii) Set of **integers,**

   (iv) Set of **stars in sky.**

   Examples of Finite sets :

   (i) Set of **English vowels,**

   (ii) Set of **positive natural number** less than **10,**

   (iii) Set of **students** of **class x** of your **school,**

   (iv) Set of **ministers** in a **cabinet.**

2. **Equality of two sets :** Two **sets** are **equal** if every **element of one** is the other **element of other** and **vice versa** which means both must have **same elements**.

3. **Equivalent Sets :** Two sets are said to be **equivalent sets** if they contain **equal number** of **elements**. Which means

$$n\,(A) = n\,(B)$$

It is obvious that **two equal sets** are **equivalent** but **two equivalent sets** are **not necessary equal.**

4. **Empty Sets :** The set which does not contain any element is called **null set** or **empty set**. It is denoted by ϕ **or { }**

   e.g., (i) a set of **girls students** in a **boys school,**

   (ii) a set of **natural numbers** between **6 and 7**.

5. **Subset :**

   B is a **subset** of **A** means **B is contained in A.** It is represent as **B** $\subset$ **C** and pronounced as **B Subset A** Following example can explain it more clearly.

   consider a set A = {1,2,3,4,5,6,7,8,}

   B = {2,3,4,}

   It is clear that all the elements are B are from A.

   So **B is a subset of A**

6 **Super Set :**

   B is **subset** of **A** then **A is super set of B.**

7. **Cardinal number of a set :**

   The **number of elements** of a finite set is called its **cardinal number** and is represented by ***n*** **(A).**

   e.g., if A = { 2,3,5,7,9}

   Then n(A) = **5**

8. **Singleton Set :**

   A set consisting of a **single element** is called **singleton set.**

9. **Disjoint set :**

   Two sets **A** and **B** are **disjoint** if they **do not** have any **common element.**

   e.g., A = {1,3,5,7}

   B = {2,4,6,8}

   Above two sets are disjoint sets.

10. **Universal Set :**

   A set containing the **maximum number** of **elements** as compared to **other sets** is called **universal set**. The **other sets** are **subsets** of **universal set** and **universal set** is represented **by** $\cup$.

   e.g., $\cup$ = { 1,2,3,4,5,6,7}

   A = { 1,2}

   B = {3,4}

   In above examples A and B are **subsets** of **Universal set** $\cup$.

11. Complement of a set:

Consider a universal set

$$U = \{1, 2, 3, 4, 5, 6, 7\}$$

and $A = \{1, 2\}$

then **A's complement** which is denoted as **A′ or A^c**

$$A^c = \{3, 4, 5, 6, 7\}$$

So **complement set A** is the set of all those element of universal which are **not element of A**.

In above example

if $B = \{3, 4\}$

Then $B^c = \{1, 2, 5, 6, 7\}$

So it is hoped that you have clearly understood the above basic terminology.

Now we will do operation of sets

OPERATION OF SETS

We in our daily life do the operation of two numbers like addition, subtraction, multiplication, division. By doing any of the operation on two number we get a new number. Similarly in sets there are some operations which when performed on two sets gives rise to new set.

The two important operation are :

1. **Union of Sets**
2. **Intersection of Sets**

We will discuss of two above one by one.

1. Union of Sets :

Let A and B be any two sets A union B means the set which contain all the elements of A an all the elements of B. The common elements being takes only once.

We write the union of A and B as "$A \cup B$" and pronounce it as **A Union B**.

e.g., If $A = \{1,2,3,4,5,\}$

$B = \{2,3,4,5,6,\}$

then $A \cup B = \{1,2,3,4,5,6\}$

It is to be seen that common elements i.e., 2,3,4 and 5, are written only once.

2. Intersection of Sets :

Let A and B are two sets. A intersection B means the set which contain common elements of A & B.

We write the intersection of A and B as **$A \cap B$** and pronounce it is 'A intersection B'

e.g., If $A = \{1,2,3,4,5,6\}$

$B = \{2,3,4,5,6\}$

then $A \cap B = \{2,3,4,5\}$

So it is hoped that you have understood the operation of two sets.

Now we will do the venn diagram.

Venn Diagram :

Sets and their properties can be visulised by diagrams which are known as **'venn diagram'**.

In these diagrams universal set is represented by rectangle and its subsets are represented by circle, ellipse etc.

If is necessary we specifically mark the diagram

A few examples of venn diagram are shown below

1. In adjcent diagram

$\cup = \{1,2,3,4,5,6\}$

$A = \{3,4\}$

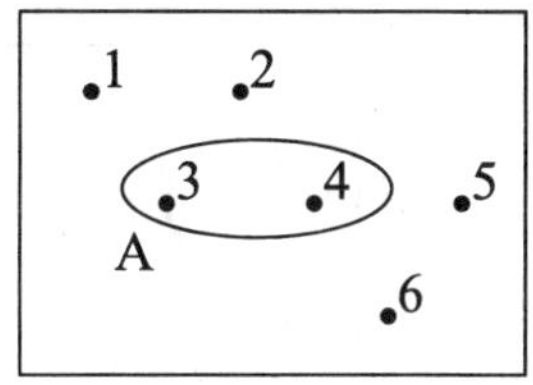

2. In adjcent diagram

$\cup = \{1,2,3,4,5,6,7,8,\}$

$A = \{5,6\}$

$B = \{5,7,8\}$

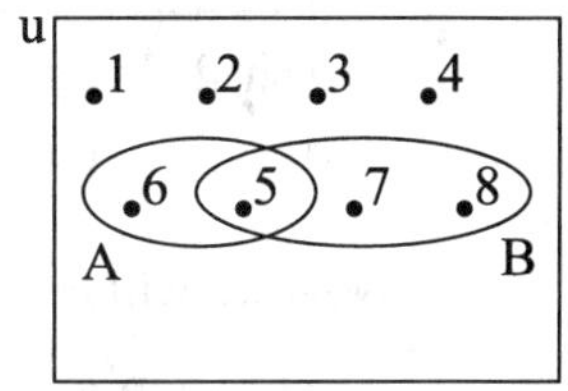

Some Standard Venn Diagrams are shown below:

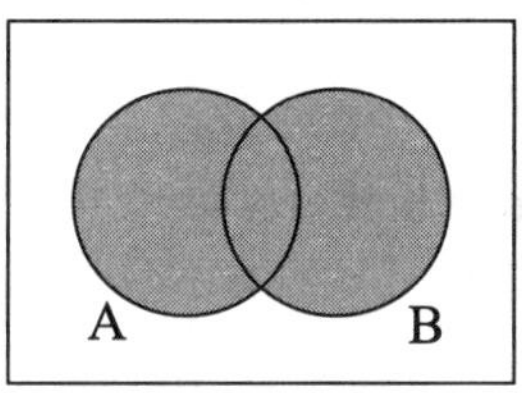

The shaded portion shows $\mathbf{A \cup B}$

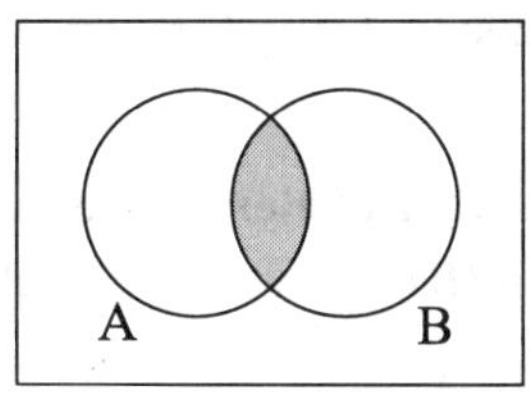

The shaded portion shows $\mathbf{A \cap B}$

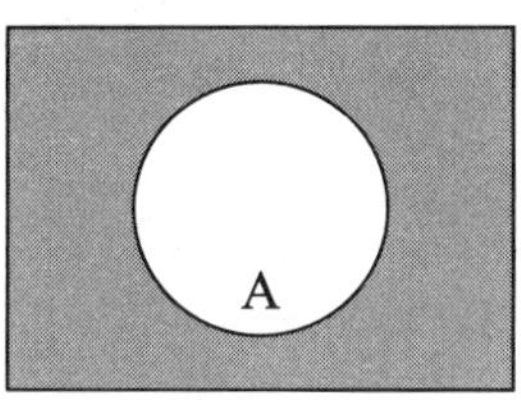

The shaded portion shows $\mathbf{A'}$

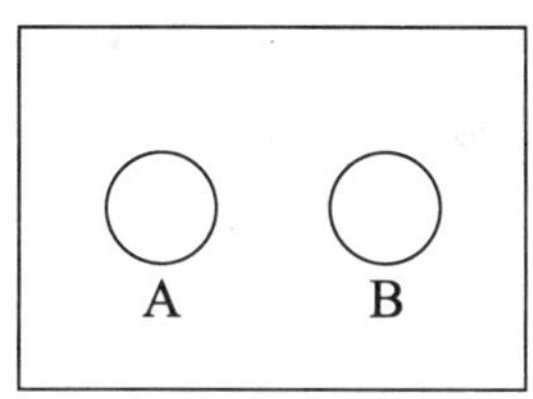

In above diagram A and B are disjoint sets

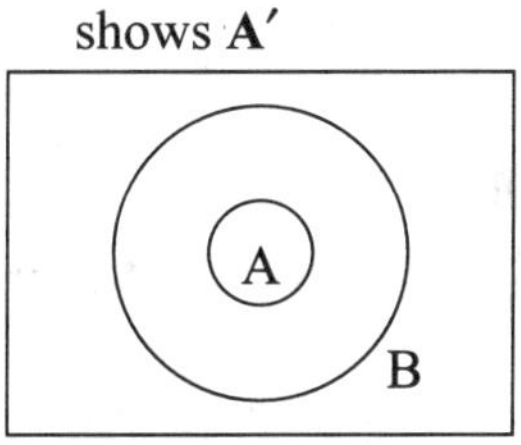

In above diagram A is a subset of B

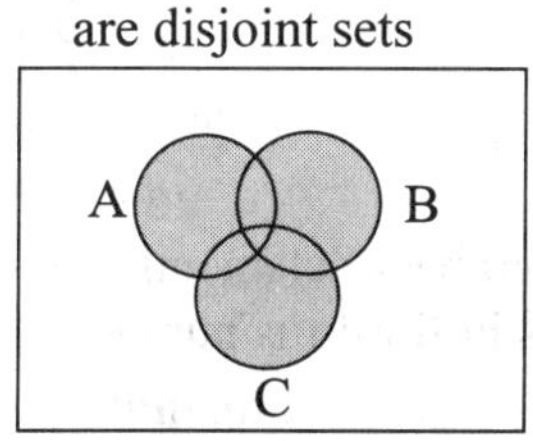

The shaded portion shows $A \cup B \cup C$

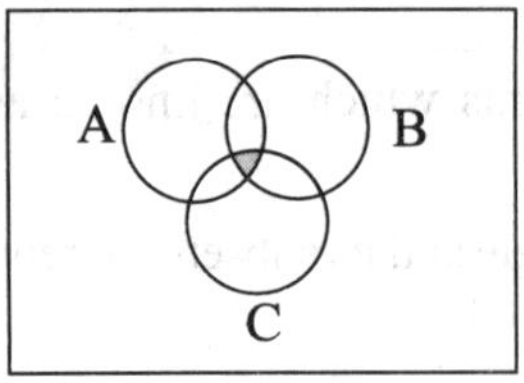

The shaded portion shows
$A \cap B \cap C$

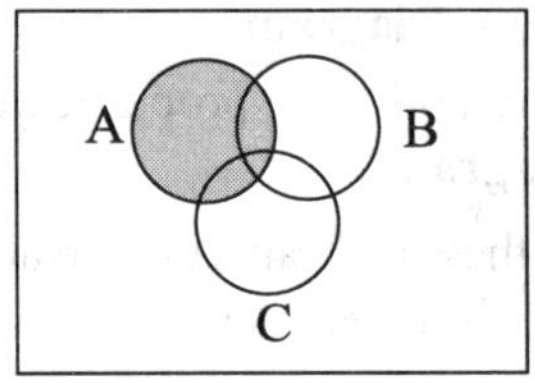

The shaded portion shows
$(A \cup B) \cap (A \cup C)$

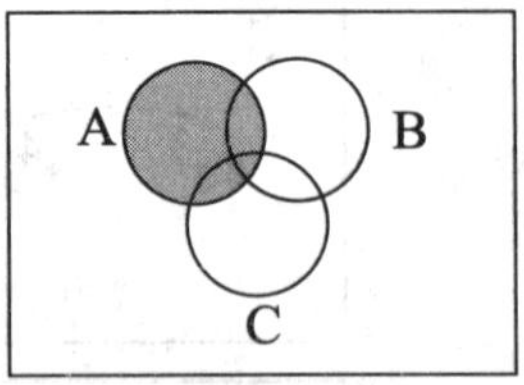

The shaded portion shows
$A \cup (B \cap C)$

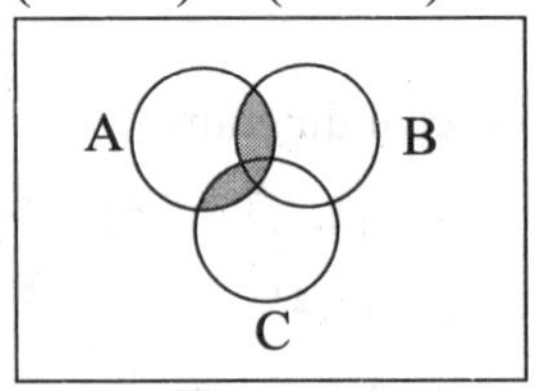

The shaded portion shows
$(A \cap B) \cap (A \cap C)$

Now we will do important application of operations of sets

Application of Operations of Sets :

Consider set $A = \{1,2,3,4\}$

$B = \{3,4,5,\}$

Number of elements in A are denoted by ***n*(A)** and are **4** in this case :

$\therefore$ ***n* (B)** i.e., **number of elements** in **B** are **3**

now $A \cup B = \{1,2,3,4,5\}$

so $n(A \cup B) = \mathbf{5}$

now $A \cap B = \{3, 4\}$

i.e., $n(A \cap B) = \mathbf{2}$

we observed that

$$n(A) = 4$$

$$n(B) = \mathbf{3}$$

$$n(A \cup B) = \mathbf{5}$$

$$n(A \cap B) = \mathbf{2}$$

i.e., $\mathbf{4 + 3 = 5 + 2}$

i.e., $n(A) + n(B) = n(A \cup B) + (A \cap B)$

or $\boldsymbol{n(A \cup B) = n(A) + n(B) - (A \cap B)}$

i.e., number of elements in **A union B** is **equal two** number of elements in **A plus** number of elements in **B minus** number of elements in **A intersection B**.

This is a **general formula** two **non disjoint** sets i.e., **intersecting** sets.

Now for three sets and A, B and C

$$\boldsymbol{n(A \cup B \cup C) = n(A) + n(B) + n(C) - n(A \cap B) - n(B \cap C) - n(C \cap A) + n(A \cap B \cap C)}$$

These two are **very important formula** and lot of question are asked on the basis of these two formula which will do it later part of the chapter.

Now it is hoped that you have understood the basic theory of sets.

Now we will do the types of question followed by exercise.

TYPES OF QUESTIONS

We have divided the questions into following four types

Type I : Question based on **basic terminology**

Type II : Based on **venn diagram**

Type III : Based on **application of operation of sets**

we will do a few examples of each of the above one by one followed by one exercise.

TYPE I
(Question based on basic Terminology)

Example 1. What is the cardinal number of following set

A = {x : x is two digit natural no. whose sum of digit is 6}

Sol : This set is in set builder method first we will write it in roaster method

which is A = {15, 24, 33, 42, 51, 60 }

cardinal number is no of element in a given set.

∴ here number of element is 6

∴ ***n* (A) = 6**

Example 2. which of the sets are equivelent {0}, (ϕ), ϕ.

Sol : {0} has one element i.e., 0

{ϕ} has an element i.e., ϕ

ϕ has no element

{0} {ϕ} are equivalent sets.

Example 3. Which of the following sets are equal

A = { x : x is a letter of word HIMACHAL}

B = { x : x is a letter of word HIMCAL}

C = { H, I, M, A, C, L }

Sol : All the elements of above three sets are contained in each other hence the above sets are equal.

Example 4. If $\cup$ = {1,2,3,4,5,6,7}

A = {1,2,3,4}

B = { 1,3,5,6}

then find $A^c \cap B^c$

Sol : A^c = {5,6,7}

A = { x : x is a letter of word HIMACHAL}

$$B^c = \{2, 4, 7\}$$
$$A^c \cap B^c = \{7\}$$

Example 5. $\cup = \{1, 2, 3, 4, 5, 6, 7\}$

$A = \{1, 2, 3, 4\}$

$B = \{1, 3, 5, 6\}$

find $(A \cap B)^c$

Sol : $A \cap B = \{1,3\}$

$(A \cap B)^c = \{2,4,5,6,7,\}$

Example 6. Define which of the following set is are finite or infinite.

(i) A = { men living in the different parts of world today}

(ii) A = { $x : x$ is a set of all natural number}

(iii) A = { x : x is a interger less than 20}

(iv) A = { $x : x$ is a set of odd natural number}

Sol :

(i) Finite set because as it is, we do not know show the numbers of elements in A but it is some definite number. May be quite a big number but **finite**

(ii) Infinite

(iii) This set can be rep. as { ...... –3, –2, –1, 0, –1 ......18, 19, 20}
Hence It is an **infinite** set

(iv) Infinite

Example 7. while the following is Roaster form

(i) A = { $x : x$ is a Integer and $-2 \le x < 6$ }

(ii) B = { $x : x$ is a natural no. less than 9 }

(iii) C = { $x : x$ is a two digit natural number such that it 3 is number of digits is 7}

(iv) D = {set of all letter of word ENGLAND}

(v) E = set of all letters of word SETISFY}

Sol :

(i) A = { –2, –1, 0, 1, 2, 3, 4, 5, 6 }

(ii) B = { 1, 2, 3, 4, 5, 6, 7, 8 }

(iii) C = { 16, 25, 34, 43, 52, 61, 70 }

(iv) D = { E, N, G, L, A, D }

(v) E = { S, E, T, I, F, Y }

Example 8. Find all the subset of { 2, 3 }

Sol : The subset of {2,3 } are as follows :

(i) {ϕ} As ϕ subset of every set.

(ii) { 2 }

(iii) { 3 }

(iv) { 2,3 } As every set is a subset of itself.

Example 9. Write all the possible subset of {1, { 2,3 }}.

Sol : The possible subsets are

(i) {φ} (ii) {1}

(iii) { {2, 3} } (iv) {1, {2,3} }

Example 10.

If A = {2,3,4,5}

B = {6,7,8,9,}

write $A \cup B$ and $A \cap B$

Sol :

(i) $A \cup B = \{2,3,4,5,6,7,8,9\}$

(ii) $A \cup B = \phi$ as there is no common element.

TYPE II
(Based on Venn's Diagram)

Example 11. The shaded area in the adjoining diagram is

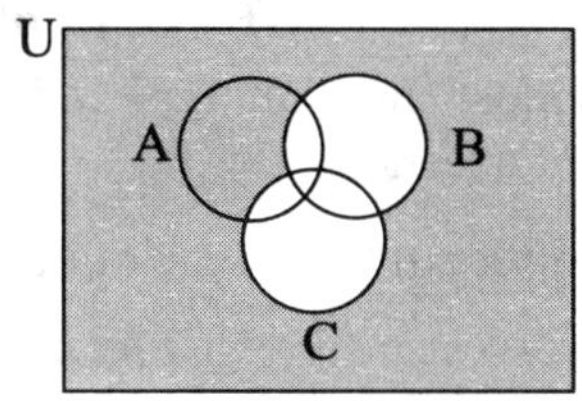

(i) $(A \cup B)^c$ (ii) $(B \cup C)$

(iii) $(B \cup C)^c$ (iv) $(B \cap C)$

Sol : The are is as (c) as the shaded region is excluding $B \cup C$

Example 12. The shaded region is the adjoining diagram

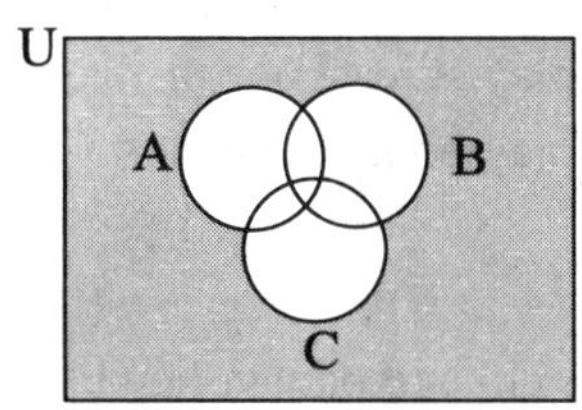

(a) $A \cup B \cup C$ (b) $A^c \cap B^c \cap C^c$

(c) $A^c \cup B^c \cap C$ (d) $A \cup B^c \cup C^c$

Sol : The answer is **(b)**.

Example 13 . The shaded area is adjoining diagram is

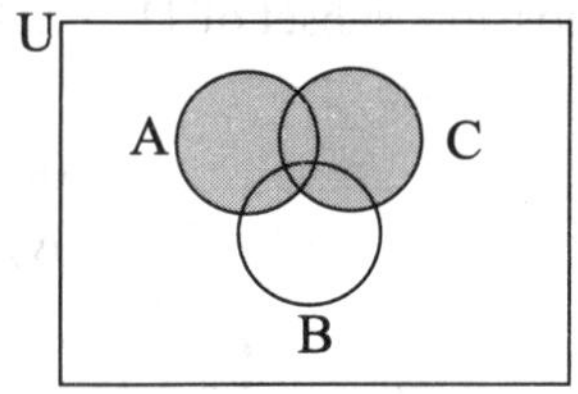

(a) $(A \cup C)^c$ (b) $A \cap C$

(c) $A \cup C$ (d) $A^c \cap C^c$

Sol : The ans. is (*c*)

TYPE III
(Based on Application of Operation of Sets)

In this type the question are based on the following formula.

1. $$\boldsymbol{n(A \cup B) = n(A) + n\,(B) - n\,A \cap B\,\}}$$

2. $$\boldsymbol{n(A \cup B \cup C) = n(A) + n(B) + n(C) - n(A \cap B) - n\,(B \cap C) - n(C \cap A) + n(A \cap B \cap C)}$$

As discussed earlier this is most important type and the help of venn diagram can also be taken while solving these type of questions.

Now we will do a few examples of these type of question.

Example 14. A and B are two sets such that A∪B has 20 elements, A has 18 elements while B has 7 elements, how many elements does $A \cap B$ have?

Sol . We know that

$$n(A \cup B) = n(A) + n(B) - n(A \cap B)$$
$$20 = 18 + 7 - n(A \cap B)$$
$$n(A \cap B) = 25 - 20 = \mathbf{5}$$

Example 15. In a class of 32 students, 20 likes to play volley ball, 15 like to play cricket. Also each of the student like to play at least one or two games?

How many student like to play both cricket and volley ball?

Sol : We know that

$$n(A \cup B) = n(A) + n\,(B) - n\,(A \cap B)$$

Let A be set of students who like to play volley ball

$\therefore$ $n(A) = 20$

similarly $n(B) = 15$

and $n(A \cup B) = 32$

$$32 = 20 + 15 - n\,(A \cap B)$$

$\therefore$ $n(A \cap B) = 35 - 32 = \mathbf{3}$

Example 16. In a group of 40 people, 25 speak English, 20 speak both Hindi and English. All the people speak at least one of the two languages. How many people speak Hindi ?

Sol:

$$\begin{aligned} n(E) &= 25 \\ n(E \cap H) &= 20 \\ n(E \cup H) &= 40 \\ n(E \cup H) &= n(E) + n(H) - n(E \cap H) \\ 40 &= 25 + n(H) - 20 \\ \therefore \quad n(H) &= 40 - 25 + 20 \\ &= \mathbf{35} \end{aligned}$$

Example 17. In a group 50 people 23 play cricket, 28 play football and 34 play badmination. Of these 16 play both cricket and football, 14 play both football and badmination and 11 play both cricket and badmination.

How many of these play all the three games.

Sol : Here

$$\begin{aligned} n(C) &= 23 \\ n(F) &= 28 \\ n(B) &= 34 \\ \text{also} \quad n(C \cap F) &= 16 \\ n(F \cap B) &= 14 \\ n(B \cap C) &= 11 \\ n(C \cup F \cup B) &= 50 \end{aligned}$$

we have to find $n\ C \cap F \cap B$

we know that

$$\begin{aligned} n(C \cup F \cup B) &= n(C) + n(F) + n(B) - n(C \cap F) - n(F \cap B) \\ &\quad - n(B \cap C) + n(C \cap F \cap B) \\ 50 &= 23 + 28 + 34 - 16 - 14 - 11 + n(C \cap F \cap B) \\ n(C \cap F \cap B) &= 50 - 23 - 28 - 34 + 16 + 14 + 11 \\ &= 50 + 41 - 85 \\ &= 91 - 85 = \mathbf{6} \end{aligned}$$

Example 18. In a group of 28 people, 14 drink coffee 14 like tea and 22 like juice. Of these 7 like both coffee and tea, 9 like both tea and juice and 11 like both juice and coffee. How many of these like all the three things?

$$A = \{x : x \text{ is a letter of word HIMACHAL}\}$$

Sol :

$$\begin{aligned} n(C) &= 14 \\ n(T) &= 14 \\ n(J) &= 22 \\ n(C \cap T) &= 7 \\ n(T \cap J) &= 9 \\ n(J \cap C) &= 11 \\ \text{also} \quad n(C \cup T \cup J) &= 28 \end{aligned}$$

we have to find $(C \cap T \cap J)$

we know that

$$
\begin{aligned}
n(C \cup T \cup J) &= n(C) + n(T) - n(C \cap T) - n(T \cap J) - n(J \cap C) \\
&\quad + n(C \cap T \cap J) \\
28 &= 14 + 14 + 22 - 7 - 9 - 11 + n(C \cap T \cap J) \\
n(C \cap T \cap J) &= 28 - 14 - 14 - 22 + 7 + 9 + 11 \\
&= 28 + 27 - 50 = \mathbf{5}
\end{aligned}
$$

Now we will do the exercise

EXERCISE

Q.1. If $A = \{a, b, c\}$

$B = \{c, d, 1, 3\}$

then cardinal number of $A \cup B$ is

(a) 3 (b) 7 (c) 6 (d) none

Q.2. $A = \{ \phi \}$ is

(a) null set (b) not a set (c) a singleton set (d) none of these

Q.3. If $A = \{x : x$ is a natural number less than $6\}$ the correct statement is

(a) $\{5\} \in A$ (b) $3 \in A$ (c) $7 \in A$ (d) $8 \in A$

Q.4. If A is a set of n distinct elements then the number of subsets of A is

(a) 2^{n-1} (b) $2n$ (c) 2^n (d) $2n - 1$

Q.5. If $A = \{1, 2, 3, 4, 5\}$

$B = \{1, 4, 7, 8\}$

$C = \{4, 7, 8, 9\}$

then $(A \cup B) \cap C$ is

(a) $\{4, 5, 7\}$ (b) $\{4, 8, 9\}$ (c) $\{4, 7, 8\}$ (d) none

Q.6. Let $\cup$ be the set of all triangles in a plane and if A is the set of all triangles with at least one different angle from 60° then A^c is

(a) a set of all right triangles (b) a set of all isosceles triangles

(c) a set of all equilateral triangles (d) none of these

Q.7. If $\cup = \{ a, b, c, d, e, f, g, h, i\}$

$A = \{b, d, f, h\}$

and $B = \{a, c, e, g, i\}$

then $(A \cup B)^c$ is

(a) $\{0]$ (b) $\{\phi\}$ (c) ϕ (d) none

Q.8. Which of the following is true if

$A = \{x : x$ is a letter of word CREATE$\}$

$B = \{x : x$ is a letter of word REACT$\}$

$C = \{x : x$ is a letter of ward ACTOR$\}$

(a) $A = B \neq C$ (b) $A \neq B = C$ (c) $A \neq B \neq C$ (d) $A = B = C$

Q.9. The number of subsets of set A = $\{a, b, c, d, e\}$ is

(a) 16 (b) 23 (c) 32 (d) none

Q.10. The number of all possible subsets of { a {b,c} } is

(a) 8 (b) 7 (c) 4 (d) none

Q.11. Which of the following statement is correct

(a) $\phi = 0$ (b) $\phi = \{0\}$ (c) $\phi = \{\phi\}$ (d) $\phi = \{\ \}$

Q.12. Which of the sets among ϕ, $\{0\}$ and $\{\phi\}$ are equivalent

(a) ϕ and $\{\phi\}$ (b) $\{0\}$ and $\{\phi\}$ (c) ϕ and $\{0\}$ (d) none of these

Q.13. $A \cap A^c$ is

(a) $\cup$ (b) ϕ (c) A^c (d) A

Q.14. $A \cup A^c$ is

(a) ϕ (b) A (c) A' (d) $\cup$

Q.15. $(A \cup B)^c$ is equal to

(a) $A^c \cup B^c$ (b) $A^c \cap B^c$ (c) A^c (d) B^c

Q.16. $(A^c \cup B^c)$ is equal to

(a) $A^c \cap B^c$ (b) $(A \cap B)^c$ (c) A^c (d) u

Q.17. The shaded region in the diagram is

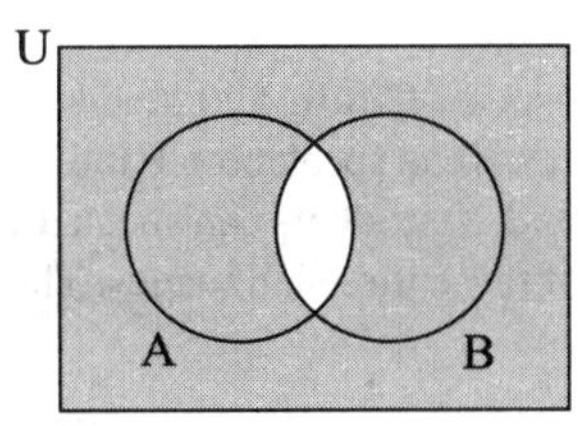

(a) $A \cap B$ (b) $(A \cap B)^c$ (c) $A^c \cap B^c$ (d) none

Q.18. The shaded region in the diagram is

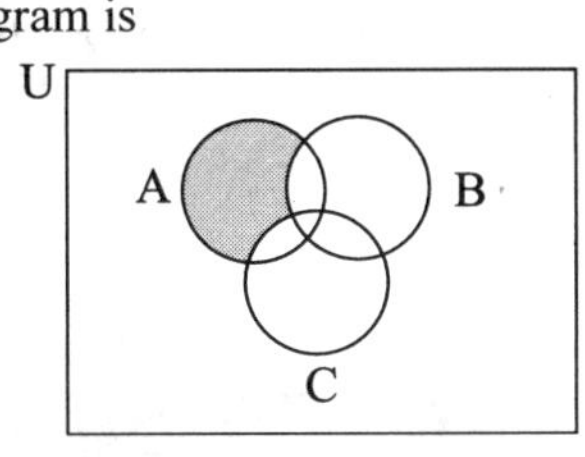

(a) $(A \cup B) \cap C$ (b) $(A \cup B) \cup C$ (c) $A \cap (B \cup C)^c$ (d) none

Q.19. The shaded region in the diagram is

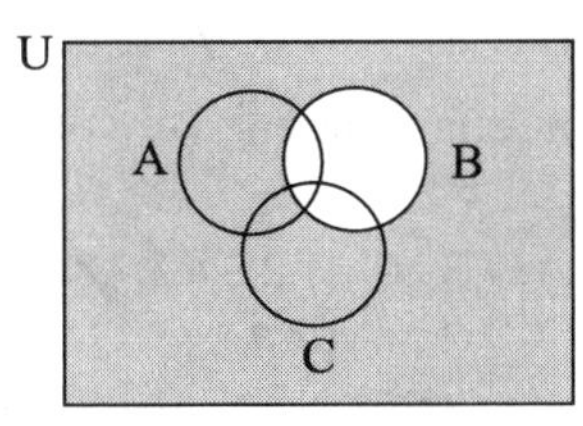

(a) $(A \cap B)^c \cap C^c$ (b) B^c (c) C^c (d) none of these

Q.20. If $n(A) = 20$ $n(B) = 25$

$n(A \cup B) = 30$ then $n(A \cap B)$ is

(a) 25 (b) 30 (c) 15 (d) none

Q.21. If $n(A) = 40$ $n(A \cup B) = 60$

and $n(A \cap B) = 10$ then $n(B)$ is

(a) 30 (b) 40 (c) 50 (d) none

Q.22. In a group of 100 people, 70 can speak Hindi and 45 speak English, then no. of people what can speak English only and not Hindi

(a) 30 (b) 40 (c) 50 (d) none

Q.23. In a class of 80 students, 60 offered maths, 21 both math and physics then number of students enrolled in physics is

(a) 48 (b) 52 (c) 49 (d) none

Q.24. Among 27 players, 17 like cricket, 9 likes both cricket and football, if each player like at least one game than number of players who like football and not cricket is

(a) 8 (b) 9 (c) 10 (d) none

Q.25. In above question the number of player who like cricket only and not football is

(a) 8 (b) 9 (c) 10 (d) none

Q.26. In a school 500 students can speak **Hindi**, 590 students can speak **English** & 410 can speak **Marathi.** Out of these 100 students can speak both **Hindi and English**, 160 **English and Marathi** and 120 **Hindi and Marathi** and 20 students can speak **all the three languages**. Each student of the school speak at least one of these three languages. The total number of student in the school is

(a) 1140 (b) 1100 (c) 1150 (d) none

Q.27. In a class of 111 students, 48 students took **physics** 45 took **chemistry** and 52 took **maths** out of these 8 took both **physics and chemistry** 11 took **chemistry and maths** and 16 took **physics and math.** If each student of class took at least one of the three subjects then how many took all the three subjects?

(a) 10 (b) 2

(c) 1 (d) none of these

SOLUTIONS

1. **(c)** $A \cup B = \{a, b, c, d, 1, 3\}$

   $\therefore$ cardinal number is 6

2. **(c)**
3. **(b)** a is not correct as {5} is a set not element
4. **(c)**
5. **(c)** $A \cup B = \{1,2,3,4,5,7,8\}$

   $C = \{4,7,8\}$

$\therefore \quad (A \cup B) \cap C = \{4,7,8\}$

6. (c)

7. (c) $(A \cup B)^c = \phi$ a null set.

8. (a)

9. (c) Number of subsets of a set contaning n elements is equal to 2^n in the case $n = 5$

$\therefore$ number of subsets $= 2^5 = \mathbf{32}$

10. (c) Same as example no 9

11. (d)

12. (b) Each in a singleton set

13. (b) It is obvious that $A \cap A^c$ will not have any common element

14. (d) It will be a universal set.

15. (b) Can be proved by taking simple example.

16. (d) This can be also be proved by taking simple example.

17. (b) The shaded region shows the portion which is complement of $A \cap B$.

18. (c)

19. (b)

20. (c)

$$n(A \cup B) = n(A) + n(B) - n(A \cap B)$$
$$30 = 20 + 25 - n(A \cap B)$$
$$\therefore \quad n(A \cap B) = 45 - 30 = \mathbf{15}$$

21. (a)

$$n(A \cup B) = n(A) + n(B) - n(A \cap B)$$
$$60 = 40 + n(B) - 10$$
$$n(B) = 60 - 40 + 10 = \mathbf{30}$$

22. (a)

$$n(H \cup E) = 100, \quad n(H) = 70, \quad n(E) = 45$$
$$\therefore \quad n(H \cup E) = n(H) + n(E) - n(H \cap E)$$
$$\therefore \quad n(H \cap E) = 115 - 100 = \mathbf{15}$$

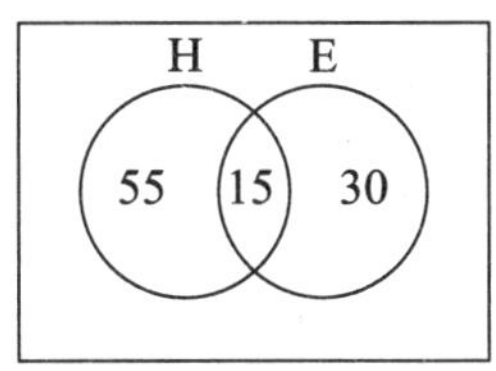

now no. of people who can speak Eng but not Hindi are

$$n(E) - n(H \cap E) = 45 - 15 = \mathbf{30}$$

The venn dig. of this sol. is also show.

23. (d)

$$n(M) = 60$$
$$n(M \cap P) = 21$$
$$n(M \cup P) = 80$$
$$\therefore \quad n(M \cup P) = n(M) + n(P) - n(M \cap P)$$
$$80 = 60 + n(P) - 21$$
$$n(P) = 80 - 60 + 21 = \mathbf{41}$$

24. (c)

$$n(C) = 17$$
$$n(C \cap F) = 9 \text{ and } n(C \cup F) = 27$$
$$\therefore \quad n(C \cup F) = n(C) + n(F) - n(C \cap F)$$
$$27 = 17 + n(F) - 9$$
$$n(F) = 27 - 17 + 9 = 19$$

$\therefore$ no of players who play football but not cricket $= n(F) - n(C \cap F) = 19 - 9 = \mathbf{10}$

25. (a) number of player who play cricket only $= n(C) - n(C \cap F) = 17 - 9 = \mathbf{8}$

26. (a)

$$\begin{aligned} n(H \cup E \cup M) &= n(H) + n(E) + n(M) - n(H \cap E) - n(E \cap M) \\ &\quad - n(H \cap M) + n(H \cap E \cap M) \\ &= 500 + 590 + 410 - 100 - 160 - 120 + 20 \\ &= 1520 - 380 = \mathbf{1140} \end{aligned}$$

27. (c) same as above

Section IV

Test Yourself

1. $n^3 - 7n + 3$ is divisible by k for all values of n, where k and n are natural numbers. Then k is... **(NIFT UG 2017)**

   (a) -3 (b) 13 (c) 7 (d) 3

2. In an international school, 80% of the students know French, 85% know German and 75% know both French and German. If 40 students don't know both, then find total number of students. **(SNAP 2017)**

   (a) 400 (b) 350 (c) 460 (d) 450

3. If A is the set of odd Integers less than 50, and B is the set of the squares of integers in A, how many elements does the intersection of A and B contain? **(SNAP 2017)**

   (a) 4 (b) 5 (c) 3

   (d) Cannot be determined

4. If $x + y = 10$ and $xy = 4$, then what is the value of $(x^4 + y^4)$? **(SSC CGL Tier-II 2018)**

   (a) 84646 (b) 8432 (c) 7478 (d) 6218

5. If $x^2 - 16x + 59 = 0$, then what is the value of $(x-6)^2 + \left[\frac{1}{(x-6)^2}\right]$?**(SSC CGL Tier-II 2018)**

   (a) 14 (b) 18₹ (c) 16 (d) 20

6. If $x^3 - 4x^2 + 19 = 6(x - 1)$, then what is the value of $\left[x^2 + \frac{1}{x-4}\right]$? **(SSC CGL Tier-II 2018)**

   (a) 3 (b) 5 (c) 6 (d) 8

7. Given that $1^2 + 2^2 + 3^2 + \ldots\ldots + 10^2 = 385$, the value of $2^2 + 4^2 + 6^2 + \ldots.. + 20^2 =$ **(SSC CGL Tier-II 2018)**

   (a) 770 (b) 1540 (c) 1155 (d) $(385)^2$

8. If $\left[a + \frac{1}{a}\right]^2 = 3$, then the value of $\left[a^2 + \frac{1}{a^2}\right] = 3$ Will be: **(SSC CGL Tier-I 2017)**

   (a) 0 (b) 1 (c) 2 (d) 3

9. If $\frac{4\sqrt{3} + 5\sqrt{2}}{\sqrt{48} + \sqrt{18}} = a + b\sqrt{6}$, then the values of a and b are respectively **(SSC CGL Tier-I 2011)**

   (a) $\frac{9}{15}, -\frac{4}{15}$ (b) $\frac{3}{11}, \frac{4}{33}$ (c) $\frac{9}{10}, \frac{2}{5}$ (d) $\frac{3}{5}, \frac{4}{15}$

10. If $\left[x+\frac{1}{x}\right]=5$, then find the value of $\frac{6x}{x^2+x+1}$ **(SSC CGL Tier-II 2014)**

(a) 3 (b) 2 (c) 1 (d) 0

11. If $(2+\sqrt{3})a=(2-\sqrt{3})b=1$, then the value of $\frac{1}{a}+\frac{1}{b}$ is **(SSC CGL Tier-I 2017)**

(a) 1 (b) 2 (c) $2\sqrt{3}$ (d) 4

12. If $a-b=2$ and $ab=24$, then what is the value of (a^3-b^3)? **(SSC CHSL Tier-I 2017)**

(a) 280 (b) 124 (c) 140 (d) 152

13. If xy = 48 and $x^2+y^2=100$, then (x + y) is **(SSC CHSL Tier-I 2017)**

(a) 12 (b) 16 (c) 18 (d) 14

14. factories : $x^2+5x-14$ **(SSC CHSL Tier-I 2017)**

(a) (x – 7) (x + 2) (b) (x – 1) (x + 14) (c) (x + 7) (x – 2) (d) (x – 14) (x + 1)

15. If sum of the roots of a quadratic equation is 1 and product of the roots is – 20. Find the quadratic equation **(SSC CHSL Tier-1 2017)**

(a) $x^2-x-20=0$ (b) $x+x+20=0$ (c) $x^2+x-20=0$ (d) $x^2-x+20=0$

16. What is the value of $\frac{(a^2+b^2)(a-b)-(a-b)^3}{a^2b-ab^2}$ **(SSC CAPF Delhi Police Exam 2017)**

(a) 0 (b) 1 (c) –1 (d) 2

17. If $3x+\left[\frac{1}{(5x)}\right]=7$, then what is the value of $\frac{5x}{(15x^2+15x+1)}$? **(SSC CHSL Tier-I 2017)**

(a) $\frac{1}{5}$ (b) $\frac{1}{10}$ (c) $\frac{2}{5}$ (d) 10

18. If $x^2+x=19$ then what is the value of $(x+5)^2+\left[\frac{1}{(x+5)^2}\right]$? **(SSC CHSL Tier-I 2017)**

(a) 77 (b) 79 (c) 81 (d) 83

19. If $x^3-y^3=81$ and x – y = 3, what is the value of x^2+y^2? **(SSC CGLTier-I 2017)**

(a) 18 (b) 21 (c) 27 (d) 36

20. If $\left(\frac{x^2}{yz}\right)+\left(\frac{y^2}{zx}\right)+\left(\frac{z^2}{xy}\right)=3$, what is the value of $(x+y+z)^3$? **(SSC CHSL Tier-I 2017)**

(a) 0 (b) 1 (c) 2 (d) 3

21. x, y and z are real numbers .$x^3+y^3+z^3=13$, x + y + z = 1, then what is the value of xy + yz + zx? **(SSC CGLTier-II 2018)**

(a) –1 (b) 1 (c) 3 (4 – 3

22. If 3x + 5y + 7z = 49 and 9x + 8y + 21z =126, then what is the value of y? **(SSC CGLTier-II 2018)**

(a) 4 (b) 2 (c) 3 (d) 5

23. If the expression $px^3 - 2x^2 - qx + 18$ is completely divisible by $(x^2 - 9)$, what are the ratio between p and q respectively? **(SSC CGLTier-I 2017)**

(a) 1:9 (b) 1:3 (c) 3:1 (d) 9:1

24. A man went to the Reserve Bank of India with ₹2000. He asked the cashier to give him ₹10 and ₹20 notes only in return. The man got 150 notes in all. How many notes of ₹10 did he receive? **(CLAT 2014)**

(a) 100 (b) 150 (c) 50 (d) 70

25. If the product of zeroes of the polynomial $ax^2 - 6x - 6$ is 4, find the value of 'a' **(CLAT 2016)**

(a) 1/2 (b) –1/2 (c) –3/2 (d) 3/2

26. If $x + \frac{1}{2x}$ find the value of $8x^3 + \frac{1}{x^3}$ **(SSC CHSL 2016)**

(a) 48 (b) 88 (c) 40 (d) 44

27. If $x^2 + y^2 - 4x - 4y + 8 = 0$, then the value of $x - y$ is **(SSC CHSL 2016)**

(a) 4 (b) – 4 (c) 0 (d) 8

28. If $a^3 - b^3 - c^3 - 3abc = 0$, then **(SSC CHSL 2016)**

(a) $a = b = c$ (b) $a + b + c = 0$ (c) $a + c = b$ (d) $a = b + c$

29. Ravi has ₹ 3 more than Ramu, but then Ramu wins on the horses and trebles his money, so that he now has ₹ 2 more than the original amount of money that the two boys had between them. How much money did Ravi and Ramu have between them before Ramu's win? **(MAT Sept. 2016)**

(a) ₹ 11 (b) ₹ 9 (c) ₹ 13 (d) ₹ 15

30. In a school of 500 students, 102 students can read Hindi and Tamil both, 200 students can read only Hindi. The students who can read Tamil are **(CLAT 2013)**

(a) 98 (b) 402 (c) 302 (d) 300

31. If x and y are real numbers, then the minimum value of $x^2 + 4xy + 6y^2 - 4y + 4$ is **(XAT 2010)**

(a) –4 (b) 0 (c) 2 (d) 4

(e) None of the above

32. If $(t - 7)$ is a factor of $t^2 - dt - 14 = 0$, then d = ? **(ATMA 2010)**

(a) 18 (b) 3 (c) 22 (d) 5

33. Which of the following equations has a root common with $x^2 + 7x + 12 = 0$? **(ATMA 2010)**

(a) $x^2 + 1 = 37$ (b) $x^2 + 11x + 28 = 0$ (c) $x^2 - 8x + 15 = 0$ (d) $3x^2 - 42 = 33$

34. What is the relation between x and y based on the following quadratic equations ? **(ATMA 2010)**

I. $x^2 + 14x + 48 = 0$

II. $y^2 + 20y + 96 = 0$

(a) $x > y$ (b) $x < y$ (c) $x \geq y$ (d) $x \leq y$

35. If a and b are negative, and c is positive, which of the following statement/s is/are true? **(SNAP 2009)**

I. $a - b < a - c$

II. if $a < b$, then $a/c < b/c$

III. $1/b < 1/c$

(a) I only (b) II only (c) III only (d) II and III only

DIRECTION ***for questions (36-37):*** Based on the information given below.

The Venn diagram given below shows the estimated readership of 3 daily newspapers (X, Y & Z) in a city. The total readership and advertising cost for each of these papers is as below.

Newspaper	Readership (lakhs)	Advertising cost (Rs. per sq. cm)
X	8.7	6000
Y	9.1	6500
Z	5.6	5000

The total population of the city is estimated to be 14 million. The common readership (in lakhs) is indicated in the given venn diagram

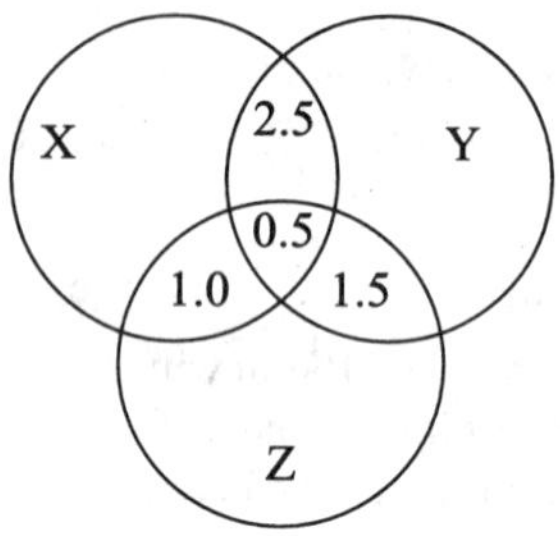

36. The number of people (in lakhs) who read at least one newspaper is :

(a) 4.7 (b) 11.9 (c) 17.4 (d) 23.4

37. The number of people (in lakhs) who read only one newspaper is :

(a) 4.7 (b) 11.9 (c) 17.4 (d) 23.4

38. There are 240 second year students in a B-School. The Finance area offers 3 electives in the second year. These are Financial Derivatives, Behavioural Finance, and Security Analysis. Four students have taken all the three electives, and 48 students have taken Financial Derivatives. There are twice as many students who study Financial Derivatives and Security Analysis but not Behavioural Finance, as those who study both Financial Derivatives and Behavioural Finance but not Security Analysis, and 4 times as many who study all the three. 124 students study Security Analysis. There are 59 students who could not muster courage to take up any (if these subjects). The group of students who study both Financial Derivatives and Security Analysis but not Behavioural Finance, is exactly the same as the group made up of students who study both Behavioural Finance and Security Analysis. How many students study Behavioural Finance only? **(XAT 2011)**

(a) 29 (b) 30 (c) 32 (d) 35

(e) None of the above options

39. In a certain B-School, of 110 students only 30 students have enrolled in Computer Science; 35 students have enrolled in Marketing; and only 20 students have enrolled in the courses in the first term. How many students of 110, have not enrolled in any of these two courses? **(ATMA (2008))**

(a) 45 (b) 65 (c) 15 (d) 25

40. If the first term of an Arithmetic Progression is 5 and the fourth term is 17, then the tenth term is **(ATMA 2008)**

(a) 22 (b) 37 (c) 41 (d) 45

SOLUTIONS

1. (4) As k and n are natural numbers. So option 1 can not be the answer.

Putting $n = 3$ we get

$$\begin{aligned} x^2 - 7x + 3 &= (3)^3 - 7 \times 3 + 3 \\ &= 27 - 21 + 3 \\ &= 9 \end{aligned}$$

Which is disisible by 3.

Hence 4th option is the answer.

2. (1)

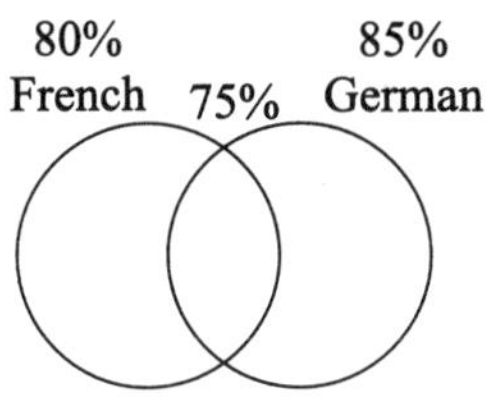

Number of persons knowing at least on language

$$= 80 + 85 - 75 = 90\%$$

$\therefore$ Not knowing any language

$$= 100\% - 90\% = 10\%$$

$$\therefore \quad 10\% \text{ of Total} = 40$$

$$\therefore \quad \text{Total} = \frac{40}{10\%} = \frac{40}{10\%} \times 100$$

$$= 400$$

3. (1)

$$\begin{aligned} \text{Set A} &= \{1, 3, 5, 7 \ldots 49\} \\ \text{Set B} &= \{1, 9, 25, 49\} \\ \therefore \quad \text{AAB} &= \{1, 9, 25, 49\} \end{aligned}$$

So have four commons elements.

4. (2)

$$\begin{aligned} (x + y)^2 &= x^2 + y^2 + 2xy \\ (10)^2 &= x^2 + y^2 + 2 \times 4 \\ 100 &= x^2 + y^2 + 8 \\ x^2 + y^2 &= 92 \end{aligned}$$

angain squaring both sides

$$\begin{aligned} (x^2 + y^2)^2 &= 92^2 \\ x^4 + y^4 + 2x^2y^2 &= 8464 \end{aligned}$$

$$x^4 + y^4 + 2(4)^2 = 8464$$

$$\therefore \quad x^4 + y^4 = 8464 - 32 = 8432$$

5. (2)

$$x^2 - 16x + 59 = 0$$

$$x^2 - 12x + 36 - 4x + 24 - 1 = 0$$

$$(x-6)^2 - 4(x-6) - 1 = 0$$

$$(x-6)^2 - 1 = 4(x-6)$$

$$(x-6) - \frac{1}{(x-6)} = 4$$

Squaring both sidus

$$(x-6)^2 + \frac{1}{(x-6)^2} - 2 = 16$$

$$\therefore \quad (x-6)^2 + \frac{1}{(x-6)^2} = 18$$

6. (3)

$$x^3 - 4x^2 + 19 = 6(x-1)$$

$$x^3 - 4x^2 + 1 = 6x - 6 - 18$$

$$x^2(x-4) + 1 = 6x - 24$$

$$x^2(x-4) + 1 = 6(x-4)$$

$$\frac{x^2(x-4)}{(x-4)} + \frac{1}{(x-4)} = 6$$

$$x^2 + \frac{1}{(x-4)} = 6$$

7. (2)

$$1^2 + 2^2 + 3^2 \ldots 10^2 + 385$$

$$2^2 + 4^4 + 6^2 + 20^2$$

$$= 2^2[1^2 + 2^2 + 3^2 \ldots 10^2]$$

$$= 4 \times 385 = 1540$$

8. (2)

$$\left(a + \frac{1}{a}\right)^2 = 3$$

$$a^2 + \frac{1}{a^2} 2a \times \frac{1}{a} = 3$$

$$a^2 + \frac{1}{a^2} + 2 = 3$$

$$a^2 + \frac{1}{a^2} = 3 - 2 = 1$$

9. (4)

$$\frac{4\sqrt{3} + 5\sqrt{2}}{\sqrt{48} + \sqrt{18}} = \frac{4\sqrt{3} + 5\sqrt{2}}{4\sqrt{3} + 3\sqrt{2}}$$

Rationlising we get

$$\frac{4\sqrt{3}+5\sqrt{2}}{4\sqrt{3}+3\sqrt{2}} \times \frac{4\sqrt{3}-3\sqrt{2}}{4\sqrt{3}-3\sqrt{2}}$$

$$= \frac{16\times 3-12\sqrt{6}+20\sqrt{6}-30}{(4\sqrt{3})^2-(3\sqrt{2})^2}$$

$$= \frac{48+8\sqrt{6}-30}{48-18} = \frac{18+8\sqrt{6}}{30}$$

$$= \frac{18}{30}+\frac{8\sqrt{6}}{30} = \frac{3}{5}+\frac{4\sqrt{6}}{15}$$

Now $\frac{3}{5}+\frac{4\sqrt{6}}{15} = a+b\sqrt{6}$

$\Rightarrow \quad a=\frac{3}{5}$ and $b=\frac{4}{15}$

10. (3) $x+\frac{1}{x} = 5$

Now $$\frac{6x}{x^2+x+1} = \frac{6x}{x\left(x+1+\frac{1}{x}\right)}$$

$$= \frac{6}{\left(x+\frac{1}{x}+1\right)} = \frac{6}{5+1} = \frac{6}{6} = 1$$

11. (4) $(2+\sqrt{3})a = (2-\sqrt{3})b = 1$

So $a=\frac{1}{2+\sqrt{3}} \Rightarrow \frac{1}{a} = 2+\sqrt{3}$

$b=\frac{1}{2+\sqrt{3}} \Rightarrow \frac{1}{b} = 2-\sqrt{3}$

$\therefore \quad \frac{1}{a}+\frac{1}{b} = 2+\sqrt{3}+2-\sqrt{3}$

$= 4$

12. (4)

$$(a-b)^3 = a^3-b^3-3ab\,(a-b)$$
$$(2)^3 = a^3-b^3-3\times 24\times 2$$
$$8 = a^3-b^3-144$$

$\therefore \quad a^3-b^3 = 144+8 = 152$

13. (4)

$$(x+y)^2 = x^2+y^2+2xy$$
$$= 100+2\times 48$$
$$= 100+96$$

$$= 196$$

$$\therefore \quad x + y = 14$$

14. (3)
$$x^2 + 5x - 14 = x^2 + 7x - 2x - 14$$
$$= x(x + 7) - 2(x + 7)$$
$$= (x + 7)(x - 2)$$

15. (1) The quadratic equation

$$x^2 - (\text{sum})x + p = 0$$

$$\therefore \quad x^2 - x - 2 = 0$$

16. (4)

$$\frac{(a^2 + b^2)(a - b) - (a - b)^3}{a^2b - ab^2}$$

$$= \frac{(a - b)[(a^2 + b^2) - (a - b)^2]}{ab(a - b)}$$

$$= \frac{a^2 + b^2 - a^2 - b^2 + 2ab}{ab}$$

$$= \frac{2ab}{ab} = 2$$

17. (2)
$$3x + \frac{1}{5x} = 7$$
$$15x^2 + 1 = 35x$$

Now
$$\frac{5x}{15x^2 + 15x + 1} = \frac{5x}{15x^2 + 1 + 15x}$$
$$= \frac{5x}{35x + 15x} = \frac{5x}{50x}$$
$$= 1/10$$

18. (2)
$$x^3 + x = 19$$
$$x^2 + 2 \times x \times 5 + 5^2 - 9x$$
$$= 19 + 5^2$$
$$(x + 5)^2 - 9x = 44$$
$$(x + 5)^2 - 9x - 45 = 44 - 45$$
$$(x + 5)^2 - 9(x - 5) = -1$$
$$(x + 5) - 9 = \frac{-1}{(x + 5)}$$
$$(x + 5)\frac{+1}{(x + 5)} = 9$$
$$\Rightarrow \quad (x + 5)^2 \frac{+1}{(x + 5)^2} = 81 - 2 = 79$$

19. (2)
$$(x-y)^3 = x^3 - y^3 - 3xy(x-y)$$
$$(3)^3 = 81 - 3xy \times 3$$
$$27 = 81 - 9xy$$
$$9xy = 81 - 27 = 54$$
$$xy = 6$$
Now
$$(x-y)^2 = x^2 + y^2 - 2xy$$
$$(3)^2 = x^2 + y^2 - 2 \times 6$$
$$9 = x^2 + y^2 - 12$$
$$\Rightarrow \quad x^2 + y^2 = 9 + 12 = 21$$

20. (1)
$$\frac{x^2}{yz} + \frac{y^2}{zx} + \frac{z^2}{xy} = 3$$
$$\frac{x^3 + y^3 + z^3}{xyz} = 3$$
$$x^3 + y^3 + z^3 = 3xyz$$
$$\Rightarrow \quad (x + y + z) = 0$$
$$\therefore \quad (x + y + z)^3 = 0$$

21. (4)
$$x^3 + y^3 + z^3 - 3xyz$$
$$= (x + y + z)(x^2 + y^2 + z^2 - xy - yz - zx)$$
$$\Rightarrow \quad x^3 + y^3 + z^3 - 3xyz = (x + y + z)\{(x + y + z)^2 1 - 3(xy + y^2 + 2x)\}$$
$$\Rightarrow \quad 13 - 3 \times 1 = 1\{1 - 3(xy + yz + 2xy)\}$$
$$10 = 1 - 3(xy + yz + 2x)$$
$$3(xy + yz + 2x) = 1 - 10 = -9$$
$$\therefore \quad xy + yz + yz = \frac{-9}{3} - 3$$

22. (3)
$$3x + 5y + 7z = 49 \quad \text{...(i)}$$
$$9x + 8y + 21^2 = 126 \quad \text{...(ii)}$$
Multiple test equation by 3 and subtractry equation (ii) from it weget
$$y = \frac{21}{7} = 3$$

23. (1) Since $x^2 - 9$ i.e.,

$(x + 3)(x - 3)$ is a fraction of
$$p(x) = px^2 - 2x^2 - qx + 18$$
$\therefore$ $p(-3)$ and $p(3)$ should be equal to zero.
$$\therefore \quad p(-3)^2 - 2(-3)^2 - q(-3) + 18 = 0$$
$$-27p - 18 + 3q + 18 = 0$$
$$27p = 3q$$

$$\Rightarrow \quad \frac{p}{q} = \frac{3}{27} = \frac{1}{9}$$

24. (a) Let the number of ₹10 and ₹20 notes be x and y respectively.

$$\therefore \quad x + y = 150 \text{ and } 10x + 20y = 2000$$

Solving the above equation, we get $x = 100$.

25. (d) The product of zeroes of the polynomial $ax^2 - 6x - 6$ is 4.

$$\text{It means } \alpha\beta = (-6)/a$$
$$4 = -6/a$$

We get $\quad a = -3/2$

26. (c) We have $x + \dfrac{1}{2x} = 2$

$$\Rightarrow \quad 2x + \frac{1}{x} = 4 \quad \text{... (i)}$$

Now, $\quad \left(2x + \dfrac{1}{x}\right)^3 = 64$

$$\Rightarrow \quad 8x^3 + \frac{1}{x^3} + 3 \times 2x \times \frac{1}{x}\left(2x + \frac{1}{x}\right) = 64$$

$$\Rightarrow \quad 8x^3 + \frac{1}{x^3} + 6 \times 4 = 64$$

$$\Rightarrow \quad 8x^3 + \frac{1}{x^3} = 40.$$

27. (c) We have $\quad x^2 + y^2 - 4x - 4y + 8 = 0$

$$\Rightarrow \quad x^2 - 4x + 4 + y^2 - 4y + 4 = 0$$

$$\Rightarrow \quad (x-2)^2 + (y-2)^2 = 0$$

$\Rightarrow$ $x - 2 = 0$ and $y - 2 = 0 \Rightarrow x = 2$ and $y = 2$. Hence $x - y = 2 - 2 = 0$.

28. (d) If $a = b + c$, then $a^3 = b^3 + c^3 + 3bc(b + c)$

$$\Rightarrow \quad a^3 = b^3 + c^3 + 3bca \text{ (as } a = b + c)$$

$$\Rightarrow \quad a^3 - b^3 - c^3 - 3abc = 0$$

Hence if $a^3 - b^3 - c^3 - 3abc = 0$ then $a = b + c$.

29. (c) Let Ramu has = ₹ x

Ravi = $x + 3$

According to question,

$$3x = x + x + 3 + 2$$
$$30 = 2x + 5$$
$$\Rightarrow \quad x = 5$$

$\Rightarrow$ Total money with Ramu and Ravi before Ramu's win = $x + x + 3$

$$= 2x + 3 = 2 \times 5 + 3 = ₹13$$

30. (d)

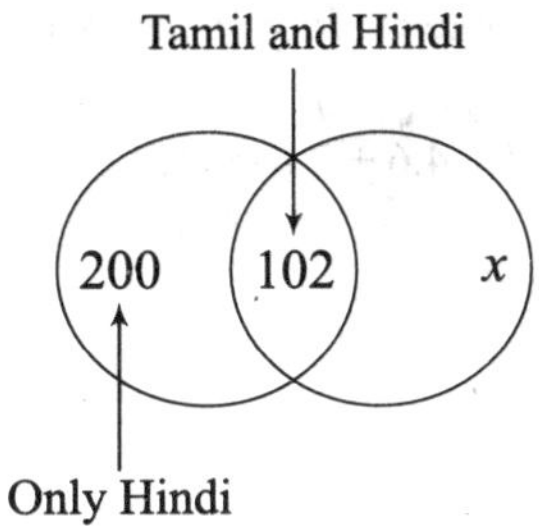

$$500 = 200 + 102 + x$$

$$\Rightarrow \quad x = 198$$

198 can read only Tamil

So students who can read Tamil = 198 + 102 = 300

31. (c) $x^2 + 4xy + 6y^2 - 4y + 4$

$= x^2 + 4y^2 + 4xy + 2y^2 + 2 - 4y + 2$

$= (x + 2y)^2 + (\sqrt{2}y - \sqrt{2})^2 + 2)$

As minimum value of $(x + 2y)^2$ and $(\sqrt{2}y - \sqrt{2})^2$ can be zero. Hence the minimum value of above expression is 2.

32. Since $t - 7$ is a factor of $t^2 - dt - 14 = 0$ so $t = 7$ must satisfy it. On putting the value of t we get $d = 5$.

33. (b) On solving $x^2 + 7x + 12 = 0$ we find $x = -3, -4$, On checking the options we find that $x = -4$ satisfies option (b).

34. (c) On solving the first equation we get

$x = -6, -8$

On solving the 2nd equation

We get $y = -12, -8$

So $x \geq y$.

Hence option (c).

35. (d) In Statement 1 we donot know about the magnitudes of a,b,c therefore we cannot tell that $a - b < a - c$ is true or not. But in statement 2 $a < b$ is given And c is a positive number so on dividing by c on both sides the sign of inequality will not change. Hence statement 2 is true. In statement 3 the LHS is negative and the RHS is positive hence statement 3 is also true. So from the options we find that option (d) is correct.

36-37

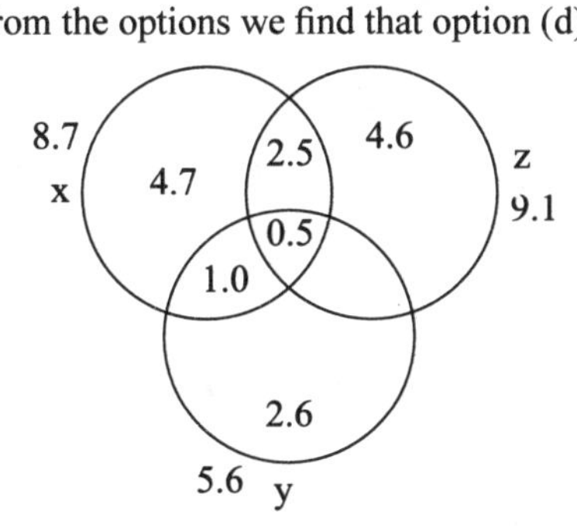

36. (c) As can be seen from the above diagram the number of people reading atleast one newspaper is = 4.7 + 4.6 + 2.6 + 2.5 + 1.5 + 1 + 0.5 = 17.4 lakhs.

37. (b) The number of people who read only one newspaper = 4.7 + 4.6 + 2.6 = 11.9 lakhs.

38. (a)

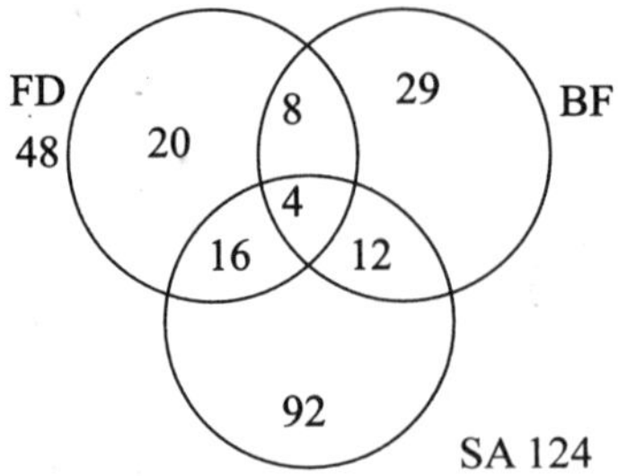

Since 59 students do not study any subject, so the total students = 181. Let total students who study BF only = x. So x + 20 + 8 + 16 + 12 + 4 + 92 = 181. ⇒ x = 29.

39. (b)

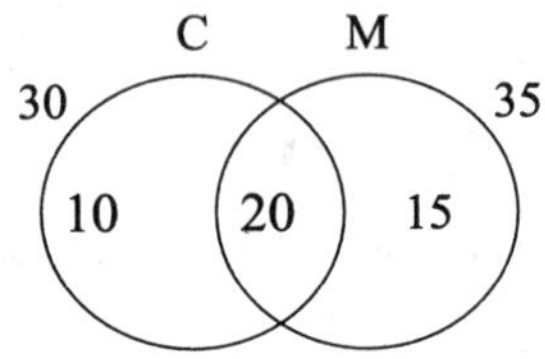

No. of students enrolled

$$10 + 20 + 15 = 45$$

Hence not enrolled $= 110 - 45 = \mathbf{65}$

40. (c)

$$a + 3d = 17$$
$$3d = 12$$
$$d = 4$$

Hence 10th term $= a + 9d$

$$= 5 + 9 \times 4 = \mathbf{41}$$

SECTION-V

ADVANCED MATHS

Day 25

Trigonometry–I

The word **trigonometry** is derived from the greek word **'trignon'** and **'metria'**. **Trignon** means **triangle** and **metria** means **to measure**. So the basic task of trigonometry is the solution of triangles means finding unknown quantities of triangle from given values of other of its quantities.

Now trigonometry plays an important role in the field of engineering, navigation, astronomy etc. So it is very useful in practical life and for a person to be successful in practical life the understanding of this topic is must.

trigonometry is very vast topic but here we have summarised and simplified it in simple language so that it may be understood by a layman and should cover all the contents from competitive point of view. So we will do this topic in two days i.e., **day 25** and **day 26**. On **day 25** we will discuss in detail the **basic terminology, formulae, their measure** etc. and simple application of trigonometry i.e., **height and distance** and on **day 26** we will discuss the **typical implementation of formulae**.

Simply, we have divided the topic of **trigonometry** into two parts i.e., **basic** and **advance**. On day 25 we will do **part I** i.e., basic and day 26 we will do **IInd part** i.e., **advance part**.

So we start with day 25 i.e., **basic part**. In this part we will discuss the **basic terminology** and **simple examples** to **master** the basics of trigonometry and simple application i.e., **height and distance**. So we start with part I i.e., basic.

PART–I [BASIC]

First of all in trigonometry if we have **an angle** and **a side** then we can find the **other sides** with the help of some ratio's called **trignometric ratio's** or **t- ratio's**. This is explained below.

Explanations :

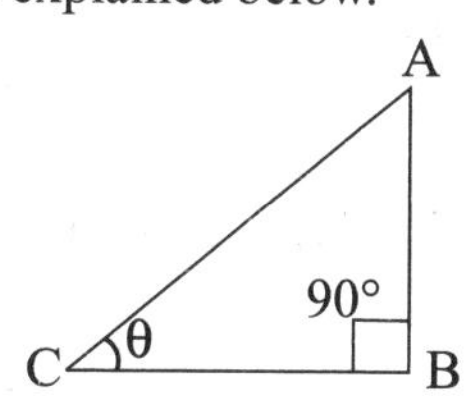

Consider a right angle triangle Where $\angle$**B = 90°**

Side **opposite to right angle** (in this case **AC**) is called **hypotenuse**.

Now we take sides with respect to $\angle$C i.e., θ

Side between **hypotenuse** and **the give angle** is called **base** and this case is **BC**.

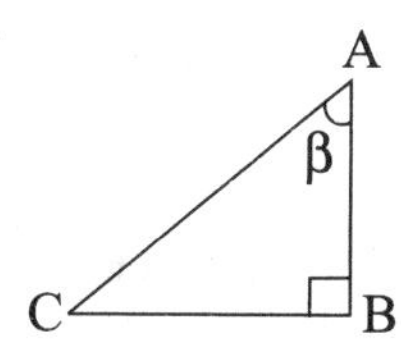

Side **opposite** to **given angle** i.e., θ is called **Perpendicular** and in this case it is **AB**.

Note: If we had taken the side in accordance with $\angle$**A** i.e., β instead of $\angle$C i.e., θ then **AB** would have been the **base** and **side opposite** to β i.e., **BC** would have been the **perpendicular**.

So it is hoped that you can now understand **hypotenuse**, **base** and **perpendicular** of the right angle triangle **with respect** to any **angle**.

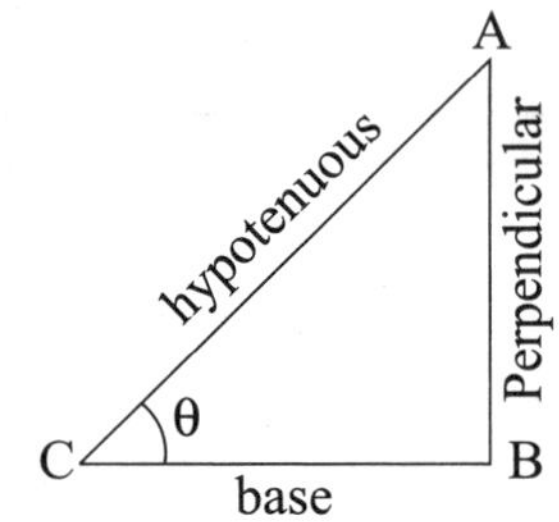

Now there are certain relations between the three sides i.e., base, hypotenuse and perpendicular of the right angle. triangle. These relations as known as **trignometric ratio's** or **t-ratio's**.

Ratio of **perpendicular** to the **hypotenuse** of right angle triangle for an angle (i.e., in this case) is called **sine of the angle** and is written as **sin.**

This **ratio** is **same** for any particular value of θ how big the triangle may be

For example in following **triangles θ is 30°** and **ratio** of **perpendicular to hypotenuse** in each case is $\frac{1}{2}$ irrespective of **size of triangle**.

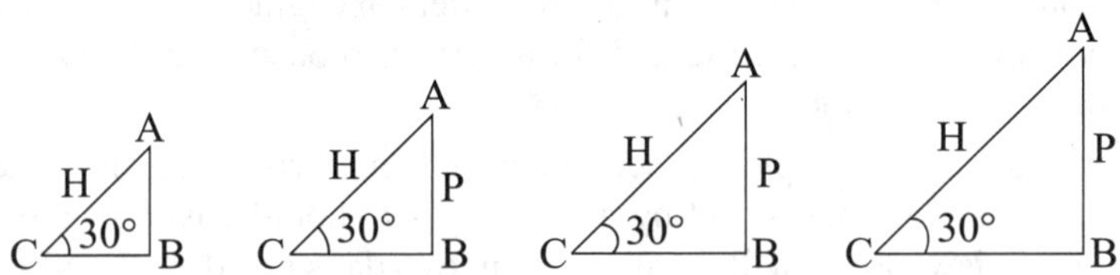

$\frac{P}{H}=\frac{1}{2}$ $\frac{P}{H}=\frac{1}{2}$ $\frac{P}{H}=\frac{1}{2}$ $\frac{P}{H}=\frac{1}{2}$

So for **30°** value of **sine** is $\frac{1}{2}$ i.e., **sin 30°** $= \frac{1}{2}$.

Similarly certain names have been given to other ratio's as **base is to hypotenuse** is called **cosine of the angle**.

So in this case

$$\frac{\text{Base}}{\text{Hypotenuse}} = \frac{BC}{AC} = \text{cosine } \theta = \cos\theta$$

Similarly $\frac{\text{Perpendicular}}{\text{Base}}$ is named as **tangent** of that angle i.e., **tan θ**.

Now we will define all the trignometric ratio's for any angle.

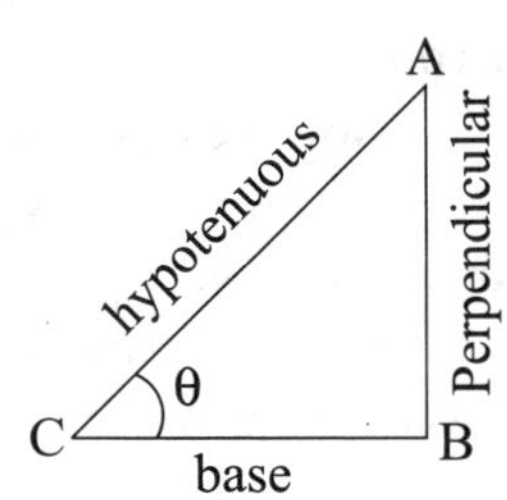

1. $\sin\theta = \frac{\text{Perpendicular}}{\text{Hypotenuse}} = \frac{AB}{AC}$

2. $\cos\theta = \frac{\text{Base}}{\text{Hypotenuse}} = \frac{BC}{AC}$

3. $\tan\theta = \frac{\text{Perpendicular}}{\text{Base}} = \frac{AB}{BC}$

Now the reciprocal of **sin θ** is **cosec θ**

The reciprocal of **cos θ** is called **secent** or **sec θ**.

The reciprocal of **tangent θ** is called **cotangent θ** or **cot θ**.

4. **cosec θ** $= \dfrac{1}{\sin\theta} = \dfrac{\text{Hypotenuse}}{\text{Perpendicular}} = \dfrac{AC}{AB}$

5. **sec θ** $= \dfrac{1}{\cos\theta} = \dfrac{\text{Hypotenuse}}{\text{Base}} = \dfrac{AC}{BC}$

6. **cot θ** $= \dfrac{\mathbf{1}}{\tan\theta} = \dfrac{\textbf{Base}}{\textbf{Perpendicular}} = \dfrac{\mathbf{BC}}{\mathbf{AB}}$

These are six basic trignometric ratio. Which have to be understood and remembered in order to do the topic of trigonometry.

*Here very popular poem is given to learn the above **t-ratio's** by heart. This poem will help you to remember the six **basic t-ratio's** even years after when you are out of touch of the topic. So understand this very **amusing and interesting poem**, which is as follows:*

Pandit पंडित	**B**adri बद्री	**P**rasad प्रसाद
Har हर	**H**ar हर	**B**ole बोले
Sona सोना	**C**handi चांदी	**T**ole तोले

Taking out the first letter of each word of above poem i.e., **P** *from* **P**andit, **B** *from* **B**adri *and* **so on** *we get the following table*

$\left[\dfrac{P}{H}\right]$	$\left[\dfrac{B}{H}\right]$	$\left[\dfrac{P}{B}\right]$
S	C	T
↓	↓	↓
(sin)	(cos)	(tan)
sin $= \dfrac{\mathbf{P}}{\mathbf{H}}$	**cos** $= \dfrac{\mathbf{B}}{\mathbf{H}}$	**tan** $= \dfrac{\mathbf{P}}{\mathbf{B}}$

P : perpendicular

B : base

H : Hypotenuse

and once these three are found their reciprocal i.e., **cosec θ, sec θ** *and* **cot θ** *can be easily found. How interesting it is. Isn't it.*

After learning the basic t-ratio's we will define a few basic term which we must understand to do the topic of trigonometry.

1. **Angles:**

   As angle we have discuss in detail in geometry here we will define positive angle and negative angle which plays an important role in trigonometry.

(i) **Positive angle:** When ray **OA** moves from its **initial position** to its **final position OB** in anti clock wise direction a **positive angle** is formed.

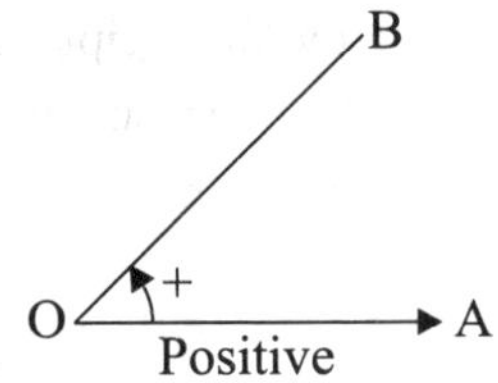

Positive

(ii) **Negative angle:** When ray **OA** moves from its **initial postion** to its **final position OB** in clock wise direction a **negative angle** is formed.

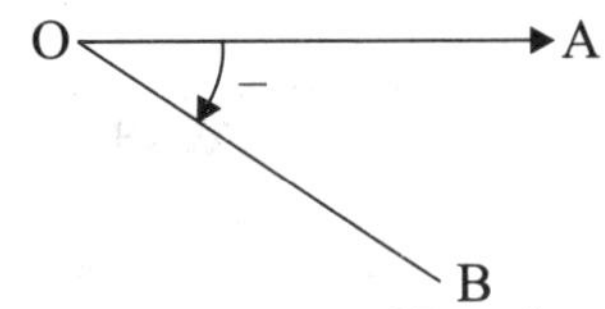

Negative

2. Quadrants:

Quadrants means **one fourth**. So as shown in figure **X′OX** and **YOY′** be two lines perpendicular to each other.

X′ OX is called as **X axis** and **YOY′** is called as **Y axis**.

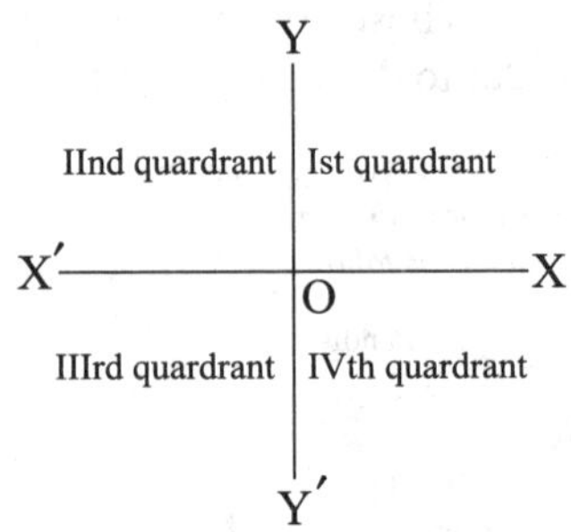

Four equal parts are made by intersection of these two perpendicular lines. Each part is known as a quadrants as shown in figure.

So XOY is Ist quadrants

YOX′ is IInd quadrants

X′OY is IIIrd quadrants

Y′OX is IVth quadrants

An angle is said to be in particular quadrants if its final sides lies in that particular quadrants.

3. Radian:

Radian is an angle subtanded at the center of a circle by an arc whose length is equal to the radius of the circle.

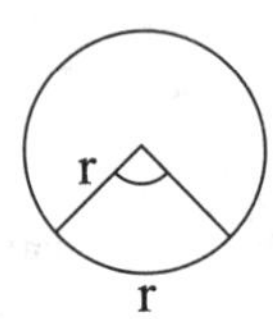

and Radian = 180°

Value of *t*-ratio's : While defining *t*-ratio's we have seen that **sin 30°** $= \frac{1}{2}$.

Now we will define the value of trignometric ratio's of angles between 0° and 90°. (0°, 30°, 45°, 60° and 90°) in the following table :

	0°	30°	45°	60°	90°
sin θ	0	$\frac{1}{2}$	$\frac{1}{\sqrt{2}}$	$\frac{\sqrt{3}}{2}$	1

cos θ	1	$\frac{\sqrt{3}}{2}$	$\frac{1}{\sqrt{2}}$	$\frac{1}{2}$	0
tan θ	0	$\frac{1}{\sqrt{3}}$	1	$\sqrt{3}$	not defined or ∝

The above table shows the value of **sin θ, cos θ** and **tan θ.** The value of **cosecθ, secθ** and **cotθ** can be found by reversing the value sin, cos and tan.

$$\text{for example } \sin 30° = \frac{1}{2} \qquad \therefore\ \mathbf{Cosec\ 30° = \frac{2}{1}}$$

$$\cos 30° = \frac{\sqrt{3}}{2} \qquad \therefore\ \mathbf{Sec\ 30°} = \frac{2}{\sqrt{3}}$$

$$\tan 30° = \frac{1}{\sqrt{3}} \qquad \therefore\ \mathbf{Cot\ 30° = \sqrt{3}}$$

So you can now find the remaining values of **cosec θ, sec θ** and **cot θ** in similar method.

Note : It may be noted that any number divided by zero is infinite or not defined.

$$\sin 0° = 0 \qquad \therefore\ \text{cosec } 0° = \frac{1}{0} = \propto \text{ or not defined.}$$

and

$$\cos 90° = 0 \qquad \therefore\ \sec 0° = \frac{1}{0} = \propto \text{ or not defined.}$$

You should learn the value of above *t* ratio's by heart as these will be very helpful in solving the question.

4. Trignometric Identities :

We know that in the given right angle triangle

$$\sin \theta = \frac{P}{H} \qquad ...(i)$$

$$\cos \theta = \frac{B}{H} \qquad ...(ii)$$

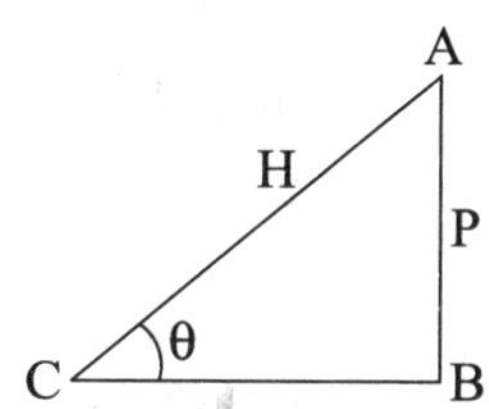

dividing (i) by (ii)

$$\frac{\sin\theta}{\cos\theta} = \frac{\frac{P}{H}}{\frac{B}{H}} = \frac{P}{B} = \tan\theta$$

$$\therefore \qquad \mathbf{\frac{\sin\theta}{\cos\theta} = \tan\theta}$$

The above equation is a trignometric identity. Now trignometric equation is an identity if it is true for all values of the angle i.e. θ

Following are the three basic identities which will help us in deriving the other identities.

(i) $\mathbf{\sin^2\theta + \cos^2\theta = 1}$ **...(i)**

It can also be written as

$$\sin^2\theta = 1 - \cos^2\theta$$

or $$\cos^2\theta = 1 - \sin^2\theta$$

(ii) $\mathbf{1 + \tan^2\theta = \sec^2\theta}$

It can also be written as

$$\tan^2\theta = \sec^2\theta - 1$$

or $$\sec^2\theta - \tan^2\theta = 1$$

(iii) $\mathbf{1 + \cot^2\theta = \text{cosec}^2\theta}$

It can also be written as

$$\cot^2\theta = \text{cosec}^2\theta - 1$$

or $$\text{cosec}^2\theta - \cot^2\theta = 1$$

Sine +ve	All +ve
Tangent +ve	Cosine +ve

The above which are basics identities. All the other identities are derived from the above identities.

5. Sign of T Ratio:

The adjcent figure shows the sign of various *t* ratio in different quadrants.

In **Ist quadrant** All **six ratios** are + **ve.**

In **IInd quadrant** only **sin** θ(and its reciprocal **cosec** θ) is + **ve.**

In **IIIrd quadrant** **cos** θ (and its reciprocal **sec** θ) is + **ve.**

In **IVth quadrant** **tan** θ (and its reciprocal **cot** θ) is + **ve.**

So remember

I	II	III	IV
All	sin	tan	cos

As we had an intresting poem to remember basic t-ratio's there is also a very easy way to remember the Sign of different t-ratio's in difference quadrants which is as follows:

After School **To** College

After school to college we take the first letter of above four words i.e., **A, S, T** and **C.** A means all, S means sine, T means tan and C means cos.

∴ In **Ist** quadrant **A** mean **all** are + **ve**

In **IInd** quadrant **S** mean **sin** (and its reciprocal **cosec** θ) are + **ve**

In **IIInd** quadrant **T** mean **tan** θ (and its reciprocal **cot** θ) are +**ve**

In **IVth** quadrant **C** mean **cos** θ (and its reciprocal **sec** θ) are +**ve**

So It is hoped that you have understood the above.

Now this completes our basic part. So before going to IInd part we will do a few simple examples so that we may master the basics.

Examples

Example 1. If in ΔABC, $\angle B = 90°$ and $\cos A = \frac{3}{5}$. Find sin A and tan A.

Sol : In $\Delta ABC = \angle B = 90°$

and $\cos A = \frac{3}{5}$

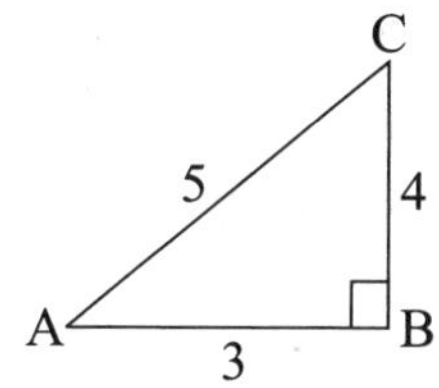

$$\text{means } \cos A = \frac{B}{H} = \frac{3}{5}$$

$\therefore$ $AB = 3, AC = 5$

$$AB^2 + BC^2 = AC^2$$
$$3^2 + BC^2 = 5^2$$
$$BC^2 = 5^2 - 3^2 = 25 - 9 = 16$$
$$BC^2 = 4^2$$
$$BC = 4$$

now AB is Base, BC is perpendicular and AC is Hypotenuse.

$$\sin A = \frac{P}{H} = \frac{BC}{AC} = \mathbf{\frac{4}{5}}$$

$$\tan A = \frac{P}{B} = \mathbf{\frac{4}{3}}$$

Example 2. If in ΔABC, $\angle B = 90°$ and $\sin A = \frac{15}{17}$. Find other *t*-ratio's.

Sol : $\sin A = \frac{15}{17}$

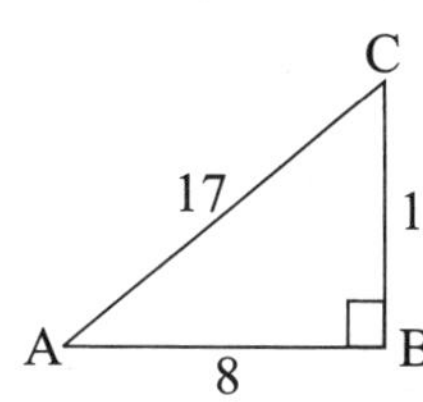

$$\text{means } \frac{P}{H} = \frac{15}{17}$$

i.e., $\frac{BC}{AC} = \frac{15}{17}$

$\therefore$ $BC = 15, AC = 17$

In Δ ABC

$$AC^2 = AB^2 + BC^2$$
$$17^2 = AB^2 + 15^2$$
$$AB^2 = 289 - 225 = 64$$
$$AB^2 = 8^2$$

$$AB = 8$$

now we are to find other *t*-ratio's means cos A, tan A, sec A, cosec A and cot A

$$\cos A = \frac{B}{H} = \frac{AB}{AC} = \frac{\mathbf{8}}{\mathbf{17}}$$

$$\tan A = \frac{P}{B} = \frac{BC}{AB} = \frac{\mathbf{15}}{\mathbf{8}}$$

$$\text{cosec } A = \frac{H}{P} = \frac{AC}{BC} = \frac{\mathbf{17}}{\mathbf{15}}$$

$$\sec A = \frac{H}{B} = \frac{AB}{AB} = \frac{\mathbf{17}}{\mathbf{8}}$$

$$\cot A = \frac{B}{P} = \frac{AB}{BC} = \frac{\mathbf{8}}{\mathbf{15}}$$

Example 3. If in ΔABC, $\angle B = 90°$ $AB = 6$ cm and $BC = 8$ cm. Find $\sin\alpha$, $\cos\alpha$, $\tan\alpha$, $\sin\beta$, $\cos\beta$, $\tan\beta$.

Sol :

$$AC^2 = AB^2 + BC^2$$
$$= 6^2 + 8^2$$
$$AC^2 = 36 + 64$$
$$AC^2 = 100$$
$$AC = 10 \text{ cm}$$

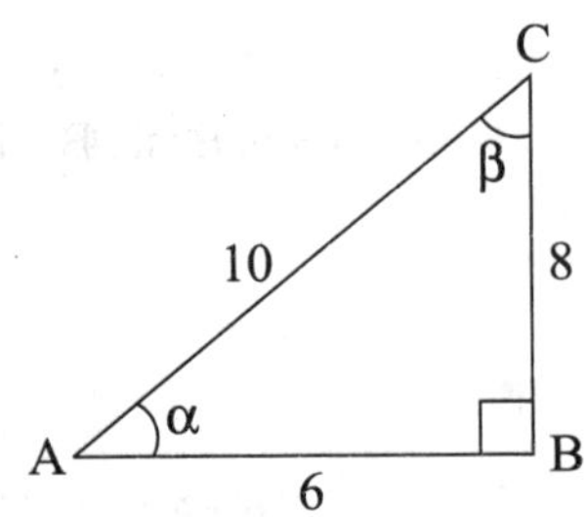

now Ist we evaluate w.r.t. angle α

now for angle α

AB is base

BC is perpendicular

and AC is Hypotenuse

$\therefore$

$$\sin\alpha = \frac{P}{H} = \frac{BC}{AC} = \frac{8}{10}$$

$$\cos\alpha = \frac{B}{H} = \frac{AB}{AC} = \frac{6}{10}$$

$$\tan\alpha = \frac{P}{B} = \frac{BC}{AB} = \frac{8}{6}$$

now for angle β

BC is Base

AB is perpendicular

and AC is hypotenuse

$$\therefore \qquad \sin\beta = \frac{P}{H} = \frac{AB}{AC} = \frac{\mathbf{6}}{\mathbf{10}}$$

$$\cos\beta = \frac{B}{H} = \frac{BC}{AC} = \frac{\mathbf{8}}{\mathbf{10}}$$

$$\tan\beta = \frac{P}{B} = \frac{AB}{BC} = \frac{\mathbf{6}}{\mathbf{8}}$$

Example 4. If $\sin A = \frac{1}{2}$ then find the value of $3\cos A - 4\cos^3 A$

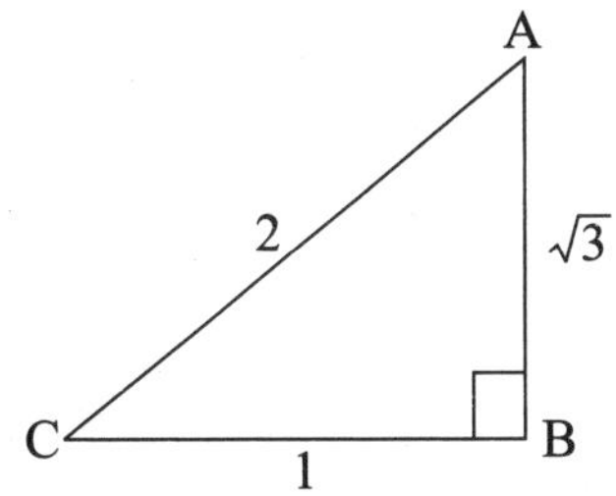

Sol : $\sin A = \frac{1}{2}$

So in given Δ $\frac{BC}{AC} = \frac{1}{2}$

i.e., $BC = 1$ and $AC = 2$

∴ By Pythagoras

$$AB^2 + BC^2 = AC^2$$

$$AB^2 + 1^2 = 2^2$$

$$AB^2 = 4 - 1$$

$$AB^2 = 3$$

$$AB = \sqrt{3}$$

$$\cos A = \frac{B}{H} = \frac{AB}{AC} = \frac{\sqrt{3}}{8}$$

$$\therefore \qquad 3\cos A - 4\cos^3 A = 3 \times \frac{\sqrt{3}}{2} - 4\left(\frac{\sqrt{3}}{2}\right)^3$$

$$= \frac{3\sqrt{3}}{2} - 4 \times \frac{3\sqrt{3}}{8}$$

$$= \frac{3\sqrt{3}}{2} - \frac{3\sqrt{3}}{2} = \mathbf{0}$$

Example 5. If $\sin A = \frac{1}{2}$ then find $\cot A + \frac{\sin A}{1 + \cos A}$

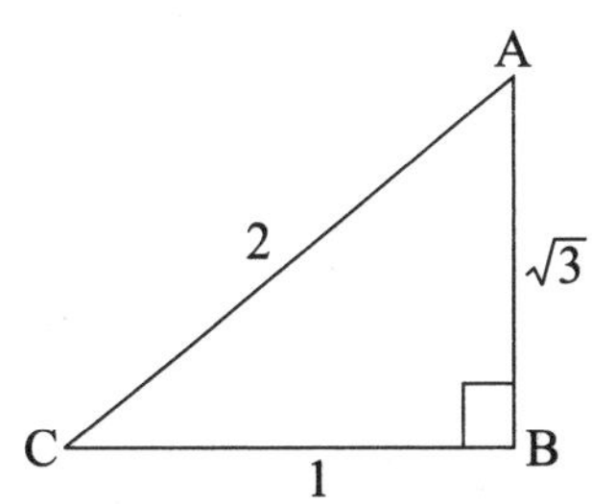

Sol : As in previous example

if $\sin A = \frac{1}{2}$

then $AB = \sqrt{3}, BC = 1$

and $AC = 2$

$$\therefore \quad \cot A + \frac{\sin A}{1+\cos A} = \frac{\sqrt{3}}{1} + \frac{\frac{1}{2}}{1+\frac{\sqrt{3}}{2}}$$

$$= \frac{\sqrt{3}}{1} + \frac{\frac{1}{2}}{\frac{2+\sqrt{3}}{2}}$$

$$= \frac{\sqrt{3}}{1} + \frac{1}{2} \times \frac{2}{2+\sqrt{3}}$$

$$= \frac{\sqrt{3}}{1} + \frac{1}{2+\sqrt{3}}$$

$$= \frac{\sqrt{3}(2+\sqrt{3})+1}{2+\sqrt{3}}$$

$$= \frac{2\sqrt{3}+3+1}{2+\sqrt{3}}$$

$$= \frac{2(2+\sqrt{3})}{2+\sqrt{3}} = \mathbf{2}$$

Example 6. Evaluate tan 60° cosec² 45° + sec² 60° cot 45°

Sol : As discussed, we know that

$$\tan 60° = \sqrt{3}$$

$$\text{cosec } 45° = \sqrt{2}$$

$$\sec 60° = 2$$

$$\cot 45° = 1$$

$\therefore$ tan 60° Cosec² 45° + Sec² 60° Cot 45°

$$= \sqrt{3} \times (\sqrt{2})^2 + (2)^2 \times 1$$

$$= 2\sqrt{3} + 4$$

Example 7. Evaluate sin 30° cos 60° + cos 30° sin 60°

Sol : sin 30° cos 60° + cos 30° sin 60°

$$= \frac{1}{2} \times \frac{1}{2} + \frac{\sqrt{3}}{2} \times \frac{\sqrt{3}}{2}$$

$$= \frac{1}{4} + \frac{3}{4}$$

$$= \frac{1+3}{4} = \frac{4}{4} = \mathbf{1}$$

Example 8. Evaluate $\sec^2 60° \sin^2 45° - \text{cosec } 30°$

Sol : $\sec^2 60° \sin^2 45° - \text{cosec } 30°$

$$= (2)^2 \times \left(\frac{1}{\sqrt{2}}\right)^2 - \frac{2}{1}$$

$$= 4 \times \frac{1}{2} - 2$$

$$= 2 - 2 = \mathbf{0}$$

Example 9. Evaluate $= \dfrac{\tan 60° + \tan 30°}{1 + \tan 60° \times \tan 30°}$.

Sol : $\dfrac{\tan 60° + \tan 30°}{1 + \tan 60° \times \tan 30°}$

$$= \frac{\sqrt{3} + \frac{1}{\sqrt{3}}}{1 + \sqrt{3} \times \frac{1}{\sqrt{3}}}$$

$$= \frac{\frac{3+1}{\sqrt{3}}}{1+1} = \frac{\frac{4}{\sqrt{3}}}{\frac{2}{1}}$$

$$= \frac{4}{\sqrt{3}} \times \frac{1}{2} = \frac{2}{\sqrt{3}}$$

Example 10. Evaluate $= \dfrac{\cos 30° + \sin 60°}{1 + \sin 30° + \cos 60°}$.

Sol : $$\frac{\cos 30° + \sin 60°}{1 + \sin 30° + \cos 60°} = \frac{\frac{\sqrt{3}}{2} + \frac{\sqrt{3}}{2}}{1 + \frac{1}{2} + \frac{1}{2}}$$

$$= \frac{\frac{2\sqrt{3}}{2}}{2} = \frac{\sqrt{3}}{2}$$

Example 11. Convert $\dfrac{3\pi}{2}$ radian into degree

Sol : $$\pi \text{ radian} = 180°$$

$$1 \text{ radian} = \frac{180°}{\pi}$$

$$\frac{3\pi}{2} \text{ radian} = \frac{180°}{\pi} \times \frac{3\pi}{2} = \mathbf{270°}$$

Example 12. Convert 75° into radian

Sol :

$$180° = \pi \text{ radian}$$

$$1° = \frac{\pi}{180°}$$

$$75° = \frac{\pi}{180°} \times 75$$

$$= \frac{5\pi}{12} \text{ radian}$$

Example 13. Convert following radian into degree

(i) $\frac{\pi}{12}$ (ii) $\frac{\pi}{4}$ (iii) $\frac{2\pi}{3}$ (iv) $\frac{5\pi}{6}$ (v) $\frac{35\pi}{36}$ (vi) $\frac{23\pi}{12}$

Sol :

(i)

$$\pi \text{ radian} = 180°$$

$$1 \text{ radian} = \frac{180°}{\pi}$$

$$\frac{\pi}{12} \text{ radian} = \frac{180°}{\pi} \times \frac{\pi}{12} = \mathbf{15°}$$

(ii)

$$\pi \text{ radian} = 180°$$

$$1 \text{ radian} = \frac{180°}{\pi}$$

$$\frac{\pi}{4} \text{ radian} = \frac{180°}{\pi} \times \frac{\pi}{4} = \mathbf{45°}$$

(iii)

$$\pi \text{ radian} = 180°$$

$$1 \text{ radian} = \frac{180°}{\pi}$$

$$\frac{2\pi}{3} \text{ radian} = \frac{180°}{\pi} \times \frac{2\pi}{3} = \mathbf{120°}$$

(iv)

$$\pi \text{ radian} = 180°$$

$$1 \text{ radian} = \frac{180°}{\pi}$$

$$\frac{5\pi}{6} \text{ radian} = \frac{180°}{\pi} \times \frac{5\pi}{6} = \mathbf{150°}$$

(v)

$$\pi \text{ radian} = 180°$$

$$1 \text{ radian} = \frac{180^\circ}{\pi}$$

$$\frac{35\pi}{36} \text{ radian} = \frac{180^\circ}{\pi} \times \frac{35\pi}{36} = \mathbf{175^\circ}$$

(vi)
$$\pi \text{ radian} = 180^\circ$$

$$1 \text{ radian} = \frac{180^\circ}{\pi}$$

$$\frac{23\pi}{12} \text{ radian} = \frac{180^\circ}{\pi} \times \frac{23\pi}{12} = \mathbf{345^\circ}$$

Example 14. Convert following degree into radian

(i) 105° (ii) 135° (iii) 165° (iv) 190° (v) 270° (vi) 360°

Sol :

(i)
$$180^\circ = \pi \text{ radian}$$

$$1^\circ = \frac{\pi}{180^\circ}$$

$$105^\circ = \frac{\pi}{180^\circ} \times 105^\circ$$

$$= \frac{7\pi}{12} \text{ radian}$$

(ii)
$$180^\circ = \pi \text{ radian}$$

$$1^\circ = \frac{\pi}{180^\circ}$$

$$135^\circ = \frac{\pi}{180^\circ}$$

$$= \frac{\pi}{180^\circ} \times 135^\circ \text{ radian}$$

(iii)
$$180^\circ = \pi \text{ radian}$$

$$1^\circ = \frac{\pi}{180^\circ}$$

$$165^\circ = \frac{\pi}{180^\circ} \times 165^\circ$$

$$= \frac{11\pi}{12} \text{ radian}$$

(iv)
$$180^\circ = \pi \text{ radian}$$

$$1^\circ = \frac{\pi}{180^\circ}$$

$$190^\circ = \frac{\pi}{180^\circ} \times 190^\circ$$

$$= \frac{19\pi}{18} \text{ radian}$$

(v)

$$180^\circ = \pi \text{ radian}$$

$$1^\circ = \frac{\pi}{180^\circ}$$

$$270^\circ = \frac{\pi}{180^\circ} \times 270^\circ$$

$$= \frac{3\pi}{2} \text{ radian}$$

(vi)

$$180^\circ = \pi \text{ radian}$$

$$1^\circ = \frac{\pi}{180^\circ}$$

$$360^\circ = \frac{\pi}{180^\circ} \times 360^\circ$$

$$= 2\pi \text{ radian}$$

Example 15. $\dfrac{\sin\theta}{1+\cos\theta}$ is equal to

(a) $\cot\theta$ (b) $\dfrac{\cos\theta - 1}{\sin\theta}$ (c) $\dfrac{1-\cos\theta}{\sin\theta}$ (d) none

Sol : (c)

$$\frac{\sin\theta}{1+\cos\theta} = \frac{\sin\theta}{1+\cos\theta} \times \frac{1-\cos\theta}{1-\cos\theta}$$

$$= \frac{\sin\theta\,(1-\cos\theta)}{1-\cos^2\theta}$$

$$= \frac{\sin\theta\,(1-\cos\theta)}{\sin^2\theta} \qquad [\because\ 1-\cos^2\theta = \sin^2\theta]$$

$$= \frac{1-\cos\theta}{\sin\theta}$$

Example 16. $\dfrac{1+\sec\theta}{\sec\theta}$ is equal to

(a) $1+\sec\theta$ (b) $1+\cos\theta$ (c) $1-\cos\theta$ (d) none

Sol : (b) $$\frac{1+\sec\theta}{\sec\theta} = \frac{1+\dfrac{1}{\cos\theta}}{\dfrac{1}{\cos\theta}}$$

$$= \frac{\cos\theta+1}{\cos\theta}\times\frac{\cos\theta}{1}$$

$$= \mathbf{1 + \cos\theta}$$

Example 17. $\tan^2\theta + \cot^2\theta + 2$ equal to

(a) $\sin^2\theta\cos^2\theta$ (b) $\sec^2\theta\,\text{cosec}^2\theta$

(c) $\sec^2\theta\cot\theta$ (d) none

Sol : (b) $$\tan^2\theta + \cot^2\theta + 2 = \tan^2\theta + \frac{1}{\tan^2\theta} + 2$$

$$= \frac{\tan^4\theta + 1 + 2\tan^2\theta}{\tan^2\theta}$$

$$= \frac{(1+\tan^2\theta)^2}{\tan^2\theta}$$

$$= \frac{(\sec^2\theta)^2}{\tan^2\theta}$$

$$= \frac{(\sec^2\theta)^2}{\tan^2\theta} \qquad [\because\ 1+\tan^2\theta = \sec^2\theta]$$

$$= \frac{\sec^2\theta.\sec^2\theta}{\tan^2\theta}$$

$$= \frac{1}{\cos^2\theta}\,\frac{\sec^2\theta}{\dfrac{\sin^2\theta}{\cos^2\theta}}$$

$$= \frac{\sec^2\theta}{\sin^2\theta}$$

$$= \mathbf{\sec^2\theta\,\text{cosec}^2\theta}$$

HEIGHT AND DISTANCE

This is the practicle application of trigonometry. Till now we have learnt that **if one side** and **an angle** of a **triangle** is known then we can find **the other side**.

for example in adjcent triangle

$$BC = 3$$

and $$\angle C = 30°$$

We can find AB as

$$\frac{AB}{3} = \tan 30°$$

$$\frac{AB}{3} = \frac{1}{\sqrt{3}}$$

$$AB = \frac{3}{\sqrt{3}} = \sqrt{3}$$

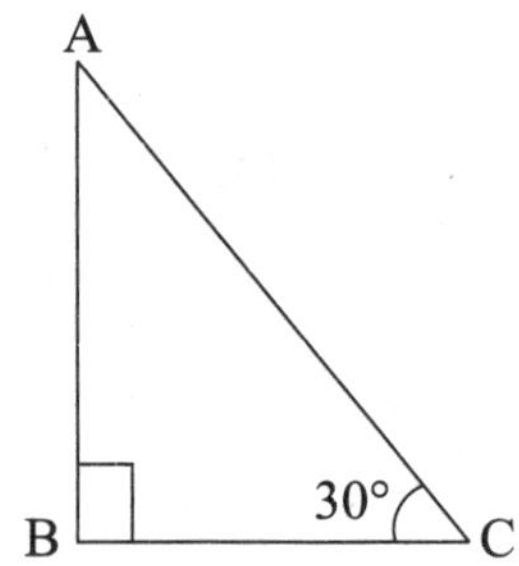

So now imagine **AB as a tree** and we are to find the height of it.

So practically it is very easier if from some **horizental distance** from the **foot of tree** say **BC** we can measure the **angle ACB**. Then we can find the height of tree as discussed above.

(**Angle ACB** is measured with the help of an instrument called **sextant**).

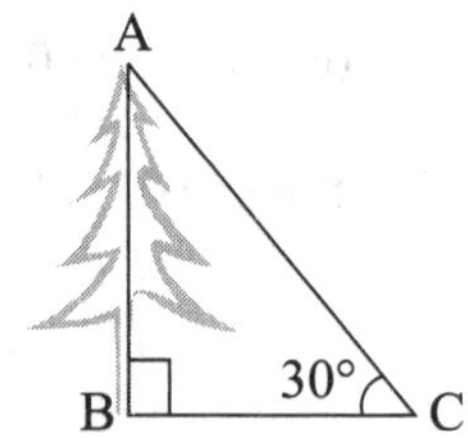

So the same case is applied to **high building**, **mountain** etc. So this was the very **basic concept** of the problem of **height and distance**.

Now before going further we will understand the basic terminology that is used while solving the question or height and distance.

1. Angle of elevation
2. Angle of depression

1. Angle of Elevation: When we view an object the **line of sight** or **line of vision** is a straight line from the object we are viewing. If the object is **above the horizontal** from the **eye** means it is at the **higher level** than our **eyes** (as shown in figure) than obviously, we have to **lift up** our **head** to **view the object**. In this process, our eye moves through an angle.

This angle is called **angle of elevation** of object.

2. Angle of Depression: If the object is at a **lower level** then we have to turn our **head down** to view the object. In this process our eyes moves **through an angle** in **downward direction**. This angle is called **angle of depression** of the object as shown in figure.

It may be noted that angle of elevation or depression of the same object is different at different positions.

As shown in figure angle of elevation of object A is different point C and D respectively.

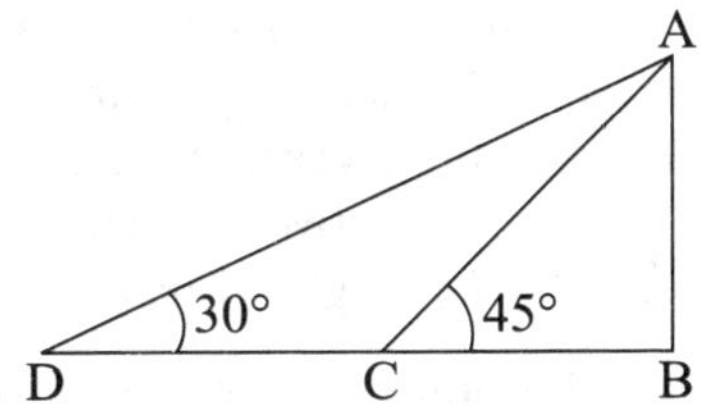

Now It's hoped that you are clear about the basic idea of height and distance problem. Now we will do a few simple examples of height and distance.

Example 1. A tower stands vertically on the ground. At the point on the ground, 20m away from the foot of tower, the angle of elevation of the top of tower is 45°. What is the height of tower.

Sol : Let A is the top and B is the foot of tower and P is point on ground at which the angle of elevation is 45°.

$$\text{then } PB = 20\text{m}$$

$$\angle APB = 45°$$

So height of tower is height AB i.e., **h**

Now in ABP

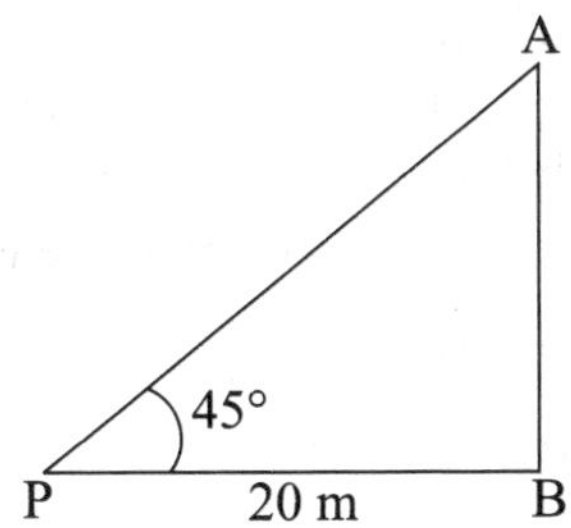

$$\tan \angle APB = \frac{AB}{PB}$$

$$\tan 45° = \frac{h}{20}$$

$$1 = \frac{h}{20}$$

$$h = 20 \times 1 = 20 \text{ m}$$

Example 2. At a point 30 m away from the foot of tower the angle of elevation at the top of tower is 30°. Find the height of tower.

Sol : Let A is the top and B is the foot of tower and P is point of observation.

$$\therefore \quad PB = 30 \text{ m}$$

$$\angle APB = 30°$$

So height of tower is height AB i.e., h

now in ABP

$$\frac{h}{30} = \tan 30°$$

$$\frac{h}{30} = \frac{1}{\sqrt{3}}$$

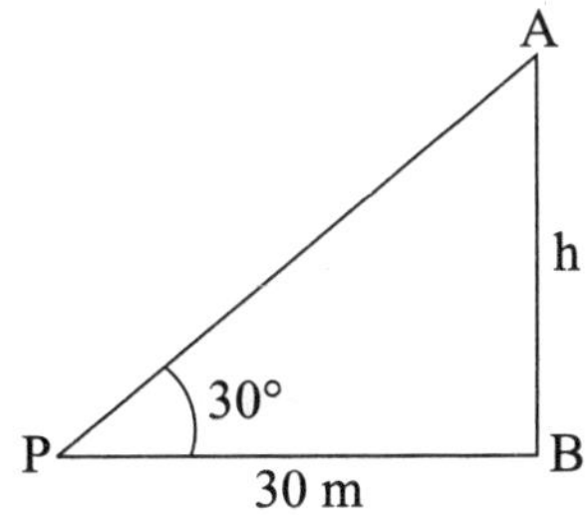

$$h = \frac{30}{\sqrt{3}} = \frac{30}{\sqrt{3}} \times \frac{\sqrt{3}}{\sqrt{3}}$$

$$= \frac{30\sqrt{3}}{3} = 10\sqrt{3} = \mathbf{17.32 \text{ m}}$$

Now some times it is **not possible** to find the **horizontal distance** from the **foot of tower or tree** (of which height is to be found). For example if we are to find the **height of a tree** which is **across the river** and we are **standing** on the **other bank**. In this case we take the **angle of elevation** from **two diffrent points** (at a few meters apart) and then using **different t ratios** measure the **height of tree**. This can be **clearly understood** by following example.

Example 3. A tall tree stand vertically on a bank of river. At a point on the other bank directly opposite the tree, the angle of elevation at the top of tree is 60° & at a point 45m behind this point on the same bank the angle of elevation is of the top of tree is 30°. Find the height of tree.

Sol : As shown in fig.

Let h is the height of tree. P is the point on other bank opposite to tree so that BP is the width of river say x.

As shown in fig. from point Q which is at 45 m behind the P the angle of elevation is 30°.

Now we will solve it as follow :

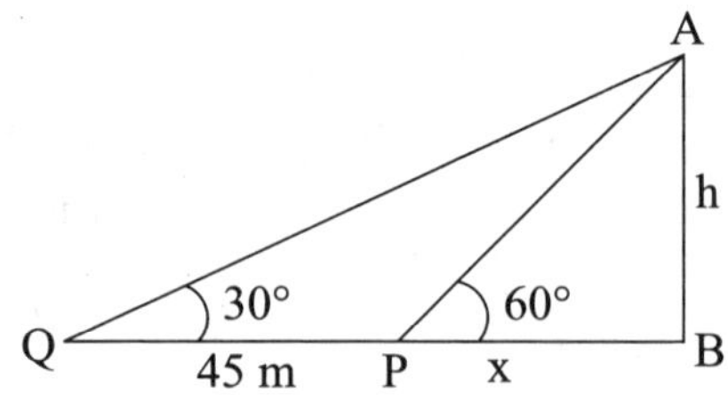

In ΔABP
$$\frac{h}{x} = \tan 60°$$
$$\frac{h}{x} = \sqrt{3}$$
$$h = \sqrt{3}x \quad ...(i)$$

In Δ ABQ
$$\frac{h}{BQ} = \tan 30°$$
$$\frac{h}{45 + x} = \frac{1}{\sqrt{3}}$$
$$h = \frac{45 + x}{\sqrt{3}} \quad ...(ii)$$

from (i) and (ii)
$$\sqrt{3}x = \frac{45 + x}{\sqrt{3}}$$
$$3x = 45 + x$$
$$3x - x = 45$$
$$2x = 45$$
$$x = \frac{45}{2} = 22.5 \text{ m}$$

now from (i)
$$h = \sqrt{3}\, x$$
$$= \sqrt{3} \times 22.5$$
$$= 1.732 \times 22.5$$
$$= 38.97 \text{ m} = \mathbf{39\ m}$$

Example 4. The angle of elevation of top of a tower from a point on the ground is 30°. After walking 15m towards the tower the angle of elevation is 45°. Find the height of tower.

Sol : As shown in fig.

Let $AB = h$ is the height of tower.

$QP = 15$ m.

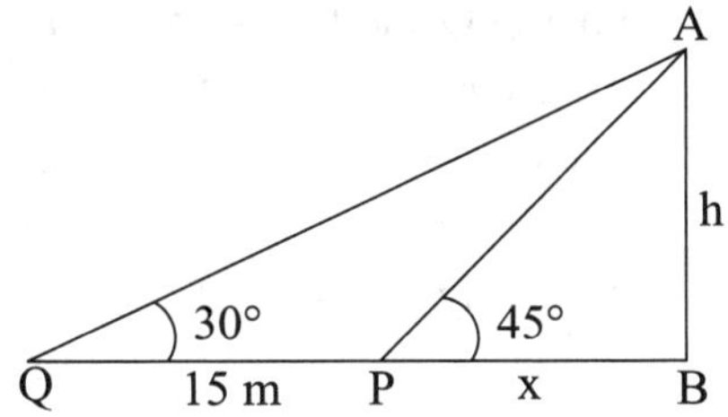

In ΔABP

$$\frac{h}{x} = \tan 45°$$

$$\frac{h}{x} = 1$$

$$h = x \quad \text{...(i)}$$

In ΔABQ

$$\frac{h}{BQ} = \tan 30°$$

$$\frac{h}{15 + x} = \frac{1}{\sqrt{3}}$$

$$h = \frac{15 + x}{\sqrt{3}} \quad \text{...(ii)}$$

from (i) and (ii)

$$x = \frac{15 + x}{\sqrt{3}}$$

$$\sqrt{3}x = 15 + x$$

$$\sqrt{3}x - x = 15$$

$$x(\sqrt{3} - 1) = 15$$

$$x = \frac{15}{\sqrt{3} - 1}$$

$$= \frac{15}{\sqrt{3} - 1} \times \frac{\sqrt{3} + 1}{\sqrt{3} + 1}$$

$$= \frac{15(\sqrt{3} + 1)}{3 - 1} = \frac{15}{2}(\sqrt{3} + 1)$$

now from (i)

$$h = x$$

$$\therefore \quad h = \frac{15}{2}(\sqrt{3} + 1)$$

$$= 7.5 \times 2.73 = \mathbf{20.47m}$$

Example 5. As observed from the top of 50m tall light house, the angle of depression of two ships approaching it are 30° and 45°. If one ship is directly behind the other. Find the distance between the two ships.

Sol : In fig. AB is light house which is of 50m and let d is a distance between two ships. Ist ship is at Q and IInd ship is at P.

So $\angle CAQ$ is the angle of depression of Ist ship which is of 30°.

but $\angle CAQ = \angle AQB$ (alternate angle)

$\therefore$ $\angle AQB = 30°$

Similarly,

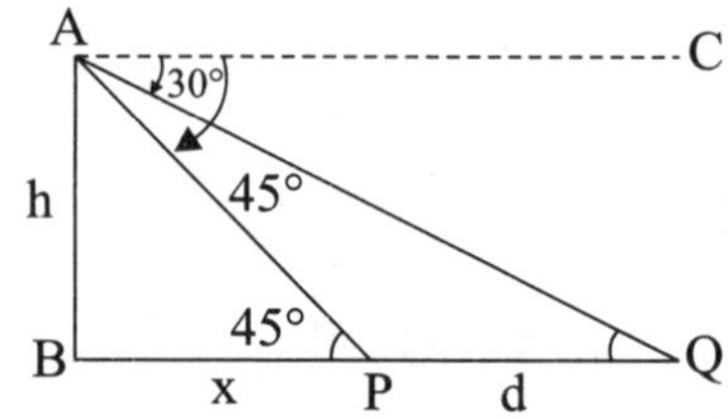

$\angle APB = 45°$

now in ΔAPB

$$\frac{h}{x} = \tan 45°$$

$$\frac{50}{x} = 1$$

$$x = 50\text{m}$$

In ΔABQ

$$\frac{50}{d+x} = \tan 30°$$

$$\frac{50}{d+x} = \frac{1}{\sqrt{3}}$$

$$50 = \frac{d+x}{\sqrt{3}}$$

$$d+x = 50\sqrt{3}$$

$$d+50 = 50\sqrt{3}$$

$$d = 50(\sqrt{3}-1)$$

$$= 50\, x\, .73 = \mathbf{36.5m}$$

EXERCISE

Q.1. In ΔABC, $\sin A = \frac{21}{29}$ then the value of cot A and cosec A are respectively

(a) $\frac{20}{21}, \frac{29}{21}$ (b) $\frac{20}{21}, \frac{21}{29}$ (c) $\frac{20}{21}, \frac{20}{29}$ (d) none

Q.2. In ΔABC $\angle B = 90°$ and $\tan A = \frac{8}{15}$ then the value of cot C

(a) $\frac{8}{15}$ (b) $\frac{15}{8}$ (c) $\frac{8}{17}$ (d) none

Q.3. In ΔABC cosec $A = \frac{5}{4}$. Find the value of $4\sin^2 A - \frac{11}{5}\cos A$ is

(a) $\frac{53}{25}$ (b) $-\frac{53}{25}$ (c) $\frac{43}{25}$ (d) none

Q.4. In ΔABC $\angle B = 90°$ and $\cos C = \frac{7}{25}$ then the value of $\cos^2 A - \sin A$

(a) $\frac{570}{21}$ (b) $\frac{572}{21}$ (c) $\frac{569}{21}$ (d) none

Q.5. In $\sin\theta = \frac{3}{5}$. Find the value of $\tan^2\theta + \sin\theta\cos\theta + \cot\theta$

(a) $\frac{2852}{1200}$ (b) $\frac{2851}{1200}$ (c) $\frac{2800}{1201}$ (d) none

Q.6. Convert following radian into degree

(i) $\frac{2\pi}{3}$ (ii) $\frac{23\pi}{12}$ (iii) $\frac{7\pi}{12}$ (iv) $\frac{19\pi}{18}$

(v) $\frac{3\pi}{2}$

Q.7. Convert following degree into radian

(i) 90° (ii) 105° (iii) 165° (iv) 175°

(v) 345°

Q.8. Evaluate $\frac{\sin 30° + \cos 30°}{2\sin 30° \times \cos 30°}$.

(a) $\frac{1}{2}$ (b) $\frac{\sqrt{3}}{2}$ (c) $\frac{1}{\sqrt{3}}$ (d) none

Q.9. Evaluate $\frac{\tan 60° + \tan 30°}{\cot 60° + \cot 30°}$

(a) 0 (b) 2 (c) 1 (d) none

Q.10. The value of $3\tan^2\frac{\pi}{6} + \frac{4}{3}\cos^2\frac{\pi}{6} - \frac{1}{2}\cot^3\frac{\pi}{4} - \frac{2}{3}\sin^2\frac{\pi}{3} + \frac{1}{8}\sec^4\frac{\pi}{3}$.

(a) 3 (b) $\frac{1}{2}$ (c) 8 (d) none

Q.11. $\cos^4\theta - \sin^4\theta$ is equal to

(a) $\cos^2\theta - \sin^2\theta$ (b) $\cos^2\theta + \sin^2\theta$ (c) $\cos^3\theta - \sin\theta$ (d) none

Q.12. $\frac{1+\sin\theta}{\cos\theta} + \frac{\cos\theta}{1+\sin\theta} = ?$

(a) $2\cos\theta$ (b) $2\sec\theta$ (c) $\sec\theta$ (d) none

Q.13. $\dfrac{\cos\theta}{1+\sin\theta} = ?$

(a) $\dfrac{1-\cos\theta}{\sin\theta}$ (b) $\dfrac{1-\sin\theta}{\cos\theta}$ (c) $1-\tan\theta$ (d) none

Q.14. $\dfrac{1-\sin\theta}{1+\sin\theta} = ?$

(a) $(\cos\theta-\tan\theta)^2$ (b) $(\sec\theta-\tan\theta)^2$ (c) $(\sec\theta-\operatorname{cosec}\theta)^2$ (d) none

Q.15. $\sqrt{\dfrac{1-\sin\theta}{1+\sin\theta}} = ?$

(a) $\sec\theta-\tan\theta$ (b) $\sec\theta+\tan\theta$ (c) $\cos\theta+\tan\theta$ (d) none

Q.16. $\dfrac{\cot A+\tan B}{\cot B+\tan A} = ?$

(a) cot B tan A (b) cot A tan A (c) cot A tan B (d) none

Q.17. The value θ if $\dfrac{\cos\theta}{1-\sin\theta}+\dfrac{\cos\theta}{1+\sin\theta} = 4$ is

(a) 45° (b) 30° (c) 60° (d) none

Q.18. Find the quadrant in which θ lies if

(*i*) cosec θ is +ve but tan θ is –ve (*ii*) sec θ is +ve but cot θ is –ve

(*iii*) tan θ is +ve but sin θ is –ve (*iv*) cos θ is +ve but sin θ is –ve

(*v*) Both sec θ and cosec θ are negative (*vi*) Both sin θ and tan θ are positive.

Q.19. A tower stands vertically on the ground. At the point on the ground 10 m away from the foot of tower the angle of elevation of the top of tower is 30°. Find the height of tower.

(a) $10\sqrt{3}$ (b) $\dfrac{10}{3}\sqrt{3}$ (c) $\sqrt{3}$ (d) none

Q.20. At a point 35 m away from the tower the angle elevation at the top of tower is 60°. Find the height of tower.

(a) 35 m (b) $\dfrac{35}{\sqrt{3}}$ m (c) $35\sqrt{3}$ m (d) none

Q.21. The angle of elevation of the top of a hill at the foot of a tower is 60° and angle of elevation of the top of the tower at the foot the hill is 30°. If the tower is 100 m tall what is the height of hill.

(a) 30 m (b) 300 m (c) $300\sqrt{3}$ m (d) none

Q.22. The angle of elevation at the top of a rock at the top and foot of a 40 m high tower are respectively 30° and 45°. Find the height of rock.

(a) $10\sqrt{3}+1$ (b) $20(\sqrt{3}+1)$ (c) $20\sqrt{3}$ (d) none

Q.23. The angle of elevation of an artificial satellite as measured from two earth station are 30° and 60°. If two distance between two earth station is 2000 k.m. Find for height of satellite

(a) 5760 m (b) 3460 m (c) 2730 m (d) none

Q24. An electric pole of 10 m high. A steel wire as tied to the top of the pole is affixed at a point on the ground to keep pole upright. If the steel wire makes an angle of 45° with the horizontal through the foot of the pole. Find the length of the steel wire.

(a) 10 m (b) 10 m (c) 15 am (d) none

SOLUTIONS

1. (a)

$$\sin A = \frac{21}{29} = \frac{P}{H}$$

$\therefore$ $BC = 21$

$AC = 29$

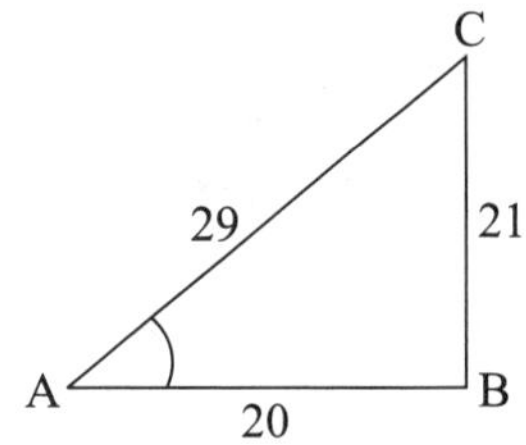

and by Pythagoras $AB = 20$

$\therefore$ $\cot A = \frac{B}{B} = \frac{20}{21}$

and $\text{cosec } A = \frac{H}{P} = \frac{29}{21}$

2. (a)

$$\tan A = \frac{P}{B} = \frac{BC}{AB} = \frac{8}{15}$$

$\therefore$ $BC = 8$

$AB = 15$

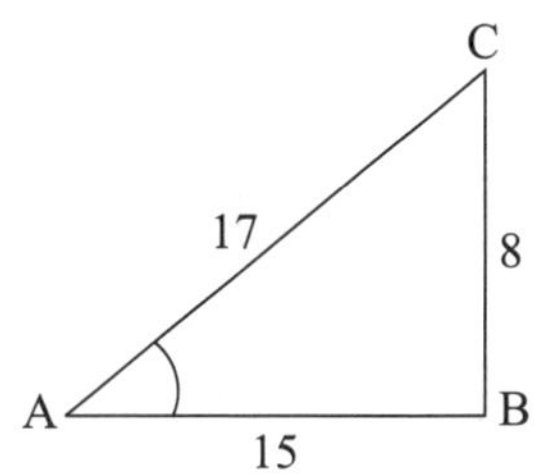

So by Pythagoras $AC = 17$

$\therefore$ $\cot C = \frac{B}{P} = \frac{8}{15}$

3. (a)

$$\text{cosec } A = \frac{5}{4} = \frac{H}{P} = \frac{AC}{BC}$$

$\therefore$ $AC = 5$

$BC = 4$

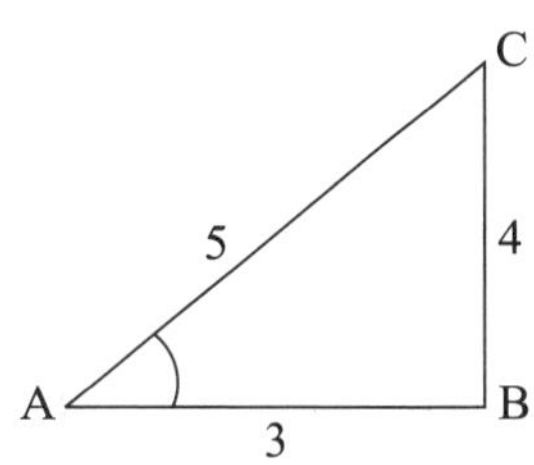

and by Pythagoras $AB = 3$

$\therefore$ $\sin A = \frac{P}{H} = \frac{4}{5}$

and $\cos A = \frac{B}{H} = \frac{3}{5}$

$$\therefore \quad 4 \sin^2 A - 3 \cos A = 4 \times \frac{16}{25} - \frac{3}{5} \times \frac{11}{15}$$

$$= \frac{64}{25} - \frac{11}{25}$$

$$= \frac{64 - 11}{25} = \frac{53}{25}$$

4. (c) $\cos C = \frac{7}{25} = \frac{B}{H}$

$\therefore$ $BC = 7$

$AC = 25$

by Pythagoras $AB = 24$

now $\cos A = \frac{B}{H} = \frac{24}{25}$

and $\sin A = \frac{P}{H} = \frac{7}{25}$

$$\therefore \quad \cos^2 A - \text{Sin } A = \left(\frac{24}{25}\right)^2 - \frac{7}{25}$$

$$= \frac{576}{25} - \frac{7}{25}$$

$$= \frac{576}{25} - \frac{7}{25} = \frac{569}{25}$$

5. (b) Same as above.

6. (i) 120° (ii) 345° (iii) 105° (iv) 190° (v) 270°

(**Hint :** Same as example no. 13)

7. (i) $\frac{\pi}{2}$ (ii) $\frac{7\pi}{12}$ (iii) $\frac{11\pi}{12}$ (iv) $\frac{35\pi}{36}$ (v) $\frac{23\pi}{12}$

(**Hint :** Same as example no. 14)

8. (d)
$$\frac{\sin 30° + \cos 30°}{2 \sin 30° \cos 30°} = \frac{\frac{1}{2} + \frac{\sqrt{3}}{2}}{2 \times \frac{1}{2} \times \frac{\sqrt{3}}{2}} = \frac{1+\sqrt{3}}{2} \times \frac{2}{\sqrt{3}} = \frac{1+\sqrt{3}}{\sqrt{3}}$$

9. (c)
$$\frac{\sqrt{3} + \frac{1}{\sqrt{3}}}{\frac{1}{\sqrt{3}} + \sqrt{3}} = 1$$

10. (d)
$$3 \times \left(\frac{1}{\sqrt{3}}\right)^2 + \frac{4}{3}\left(\frac{\sqrt{3}}{2}\right)^2 - \frac{1}{2} \times 1 - \frac{2}{3} \times \left(\frac{1}{2}\right)^2 + \frac{1}{8} \times (2)^4$$

$$= 1 + 1 - \frac{1}{2} - \frac{1}{6} + 2 = 4 - \frac{1}{2} - \frac{1}{6}$$

$$= \frac{24 - 3 - 1}{6} = \frac{20}{6} = 3\frac{1}{3}$$

11. (a)
$$\cos^4 \theta - \sin^4 \theta = (\cos^2 \theta + \sin^2 \theta)(\cos^2 \theta - \sin^2 \theta)$$

$$= 1 \times (\cos^2\theta - \sin^2\theta)$$

$$= \cos^2\theta - \sin^2\theta$$

12. (b) $$\frac{1+\sin\theta}{\cos\theta} + \frac{\cos\theta}{1+\sin\theta} = \frac{1+\sin^2\theta + 2\sin\theta + \cos^2\theta}{\cos\theta\,(1+\sin\theta)} = \frac{2+2\sin\theta}{\cos\theta\,(1+\sin\theta)}$$

$$= 2\ \text{Sec}\theta$$

13. (b) $$\frac{\cos\theta}{1+\sin\theta} = \frac{\cos\theta}{1+\sin\theta} \times \frac{1-\sin\theta}{1-\sin\theta}$$

$$= \frac{\cos\theta\,(1-\sin\theta)}{1-\sin^2\theta} = \frac{1-\sin\theta}{\cos\theta}$$

14. (b) $$\frac{1-\sin\theta}{1+\sin\theta} = \frac{1-\sin\theta}{1+\sin\theta} \times \frac{1-\sin\theta}{1-\sin\theta}$$

$$= \frac{(1-\sin\theta)^2}{1+\sin^2\theta} = \frac{(1-\sin\theta)^2}{\cos^2\theta}$$

$$= \left(\frac{1-\sin\theta}{\cos\theta}\right)^2 = \left(\frac{1}{\cos\theta} - \frac{\sin\theta}{\cos\theta}\right)$$

$$= (\sec\theta - \tan\theta)^2$$

15. (a) $$\sqrt{\frac{1-\sin\theta}{1+\sin\theta}} = \sqrt{\frac{1-\sin\theta}{1+\sin\theta}}$$

$$= \sec\theta - \tan\theta$$

16. (c) $$\frac{\cot A + \tan B}{\cot B + \tan A} = \frac{\dfrac{\cos A}{\sin A} + \dfrac{\sin B}{\cos B}}{\dfrac{\cos B}{\sin B} + \dfrac{\sin A}{\cos A}}$$

$$= \frac{\dfrac{\cos A\cos B + \sin A\ \sin B}{\sin A\cos B}}{\dfrac{\cos A\cos B + \sin A\sin B}{\cos A\sin B}}$$

$$\frac{\cos A\sin B}{\sin A\cos B} = \cot A\,.\,\tan B$$

17.(c) $$\frac{\cos\theta}{1-\sin\theta} + \frac{\cos\theta}{1+\sin\theta} = 4$$

$$\frac{\cos\theta\,(1+\sin\theta) + \cos\theta\,(1-\sin\theta)}{(1-\sin\theta)\,(1+\sin\theta)} = 4$$

$$\frac{\cos\theta + \cos\theta\sin\theta + \cos\theta - \cos\theta\sin\theta}{1-\sin^2\theta} = 4$$

$$\frac{2\cos\theta}{\cos^2\theta} = 4$$

$$\cos\theta = \frac{1}{2}$$

$$\theta = \mathbf{60°}$$

18. (i) IInd quadrant (ii) IVth quadrant (iii) IIIrd quadrant (iv) IVth quadrant
(v) IIIrd quadrant (vi) Ist quadrant

19. **(b)** In ΔAPB

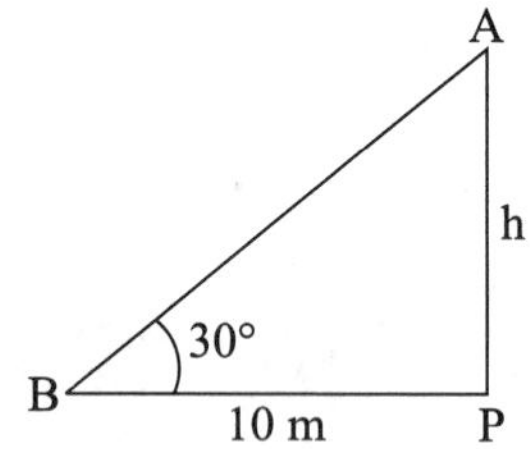

$$\frac{h}{10} = \tan 30°$$

$$\frac{h}{10} = \frac{1}{\sqrt{3}}$$

$$h = \frac{10}{\sqrt{3}} = \frac{10}{\sqrt{3}} \times \frac{\sqrt{3}}{\sqrt{3}}$$

$$= \frac{10\sqrt{3}}{3}$$

20. **(c)** In Δ APB

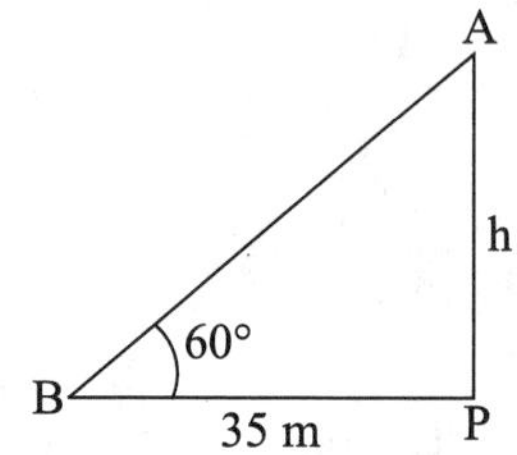

$$\frac{h}{35} = \tan 60°$$

$$h = 35\sqrt{3}$$

21. **(b)** Let PQ is tower and AB is hill

In ΔPQB

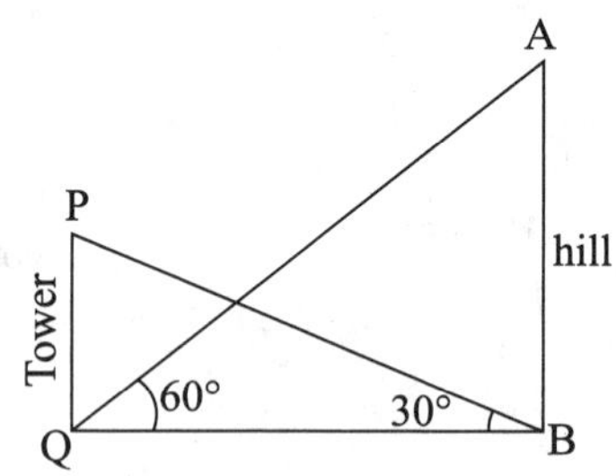

$$\frac{PQ}{QB} = \tan 30°$$

$$\frac{100}{QB} = \frac{1}{\sqrt{3}}$$

$$QB = 100\sqrt{3} \quad \text{...(i)}$$

In ΔAQB

$$\frac{AB}{QB} = \tan 60°$$

$$QB = \frac{AB}{\sqrt{3}} \quad \text{...(ii)}$$

from (i) and (ii)

$$\frac{AB}{\sqrt{3}} = 100\sqrt{3}$$

$$AB = 100\sqrt{3} \times \sqrt{3} = \mathbf{300\ m}$$

22. (b) In ΔBRP

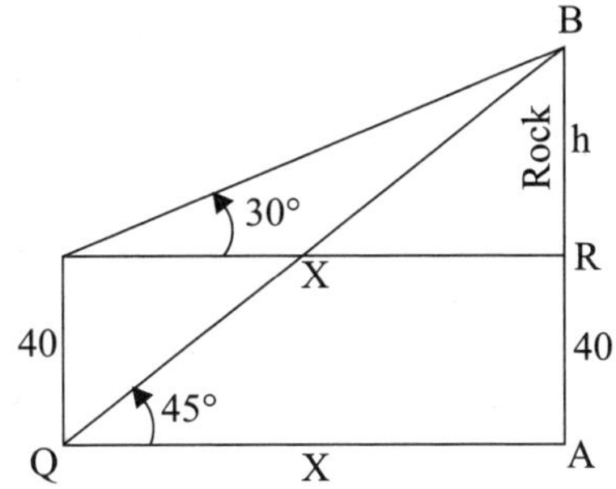

$$\frac{h}{x} = \tan 30^\circ$$

$$\frac{h}{x} = \frac{1}{\sqrt{3}}$$

$$x = h\sqrt{3} \quad ...(i)$$

In ΔBQA

$$\frac{h+40}{x} = \tan 45^\circ \quad ...(ii)$$

from (i) and (ii)

$$h + 40 = h\sqrt{3}$$

$$h(\sqrt{3}-1) = 40$$

$$h = \frac{40}{\sqrt{3}-1} = \frac{40}{\sqrt{3}-1} \times \frac{\sqrt{3}+1}{\sqrt{3}+1}$$

$$= \frac{40(\sqrt{3}+1)}{2} = 20(\sqrt{3}+1)$$

23. (c) Let satellite is at D and two earth station are at A and B respectively and AB = 2000 mt now in ΔDCB

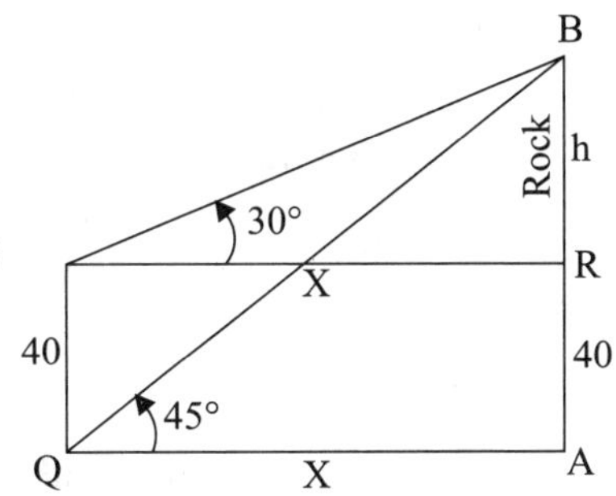

$$\frac{h}{x} = \tan 45^\circ$$

$$h = x \quad ...(i)$$

now in ΔDCA

$$\frac{h}{x+2000} = \tan 30^\circ$$

$$h = \frac{x+2000}{\sqrt{3}} \quad ...(ii)$$

from (i) and (ii)

$$h = x = \mathbf{2730\ m.}$$

24. (d) Let AC length of steel wire

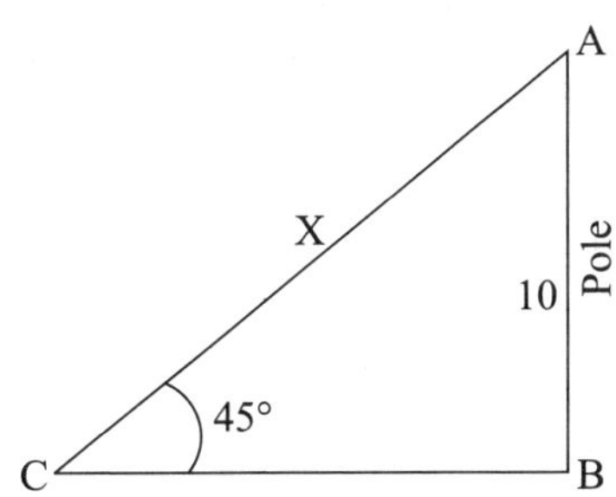

$$\frac{10}{x} = \sin 45^\circ$$

$$\frac{10}{x} = \frac{1}{\sqrt{2}}$$

$$x = 10\sqrt{2}$$

$$= 10 \times 1.42$$

$$= \mathbf{14.2\ m}$$

Day 26

Trigonometry–II

Today we will do **part II** i.e., **advanced part of Trigonometry.** In this part Ist we will do the t ratio's of allied angles (i.e., 90° + θ, 90° – θ, 180° – θ, 180° – θ, 270° + θ, 270° – θ etc)

Then we will do a few important formulae which include sum and difference formulae tangent formula, t-ratio of multiple angles, sub-multiple angles etc.

So first we will do t-ratio's of allied angles.

1. **t-ratio of allied angles:** t-ratio of allied angles means angles in terms of **90°– θ, 90° + θ, 180°– θ, 180° + θ, 270° + θ, 360° + θ, 360°– θ.**

   First we will explain t-ratio in terms of (90° – θ) followed by t-ratio's of all other allied angle.

   (*i*) **t -ratio's of (90° – θ):** For this consider a triangle ABC as shown in fig.

Here $\angle \mathbf{B} = \mathbf{90°}$

and $\angle \mathbf{C} = \theta$

But we know that

$$\angle A + \angle B + \angle C = 180°$$
$$\angle A + 90° + \theta = 180°$$
$$\angle \mathbf{A} = 180° - 90° - \theta = \mathbf{90° - \theta}$$

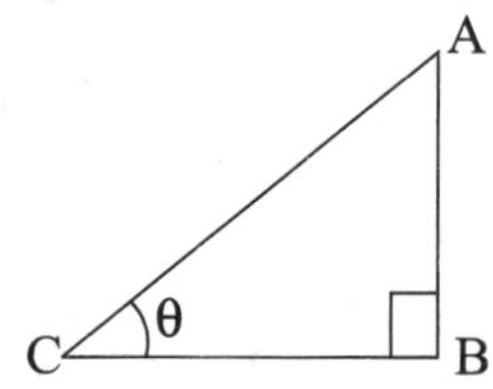

Now in ΔABC If we take t-ratio's according to θ

then **AB** is **perpendicular, BC is base** & **AC** is **hypotenuse**

$$\sin\theta = \frac{P}{H} = \frac{AB}{AC} \qquad \text{...(i)}$$

Now if we take t-ratio's in terms of ∠A i.e., (90° – θ)

than **AB** is **base, BC** is **perpendicular** and **AC** is **hypotenuse**

Here $$\frac{AB}{AC} = \cos A = \cos\,(\mathbf{90° - \theta}) \qquad \text{...(ii)}$$

∴ from (i) and (ii)

$$\mathbf{\sin\theta = \cos\,(90° - \theta)}$$

Similarly

$$\cos\theta = \sin\,(\mathbf{90° - \theta})$$
$$\tan\theta = \cot\,(\mathbf{90° - \theta})$$
$$\cot\theta = \tan\,(\mathbf{90° - \theta})$$
$$\sec\theta = \text{cosec}\,(\mathbf{90° - \theta})$$
$$\text{cosec}\,\theta = \sec\,(\mathbf{90° - \theta})$$

and so on

Here all **signs** are taken as **+ ve** because **(90° – θ)** lies in **Ist quadrant** and in Ist quadrant all the ratios are + ve.

So t-ratio's of **(90° – θ)** in terms of θ are

$$\sin(90^\circ - \theta) = \cos\theta$$
$$\cos(90^\circ - \theta) = \sin\theta$$
$$\tan(90^\circ - \theta) = \cot\theta$$
$$\sec(90^\circ - \theta) = \text{cosec}\,\theta$$
$$\text{cosec}(90^\circ - \theta) = \sec\theta$$
$$\cot(90^\circ - \theta) = \tan\theta$$

Now we will define the t-ratio of **(90° + θ)**

t-ratio's of (90° + θ):

$$\sin(90^\circ + \theta) = \cos\theta$$
$$\cos(90^\circ + \theta) = -\sin\theta$$
$$\tan(90^\circ + \theta) = -\cot\theta$$
$$\cot(90^\circ + \theta) = -\tan\theta$$
$$\sec(90^\circ + \theta) = -\text{cosec}\,\theta$$
$$\text{cosec}(90^\circ + \theta) = \sec\theta$$

As shown above in t-ratio's of allied angles **two parameters** are to be taken care of. **One** is the **sign** and **another** is the **t-ratio**, which will come on **right hand side**. For example if we are to find the t-ratio's of **cos (90° + θ)**. As shown above **cos (90° + θ) = – sin θ.**

Now two things will come in your mind.

(i) How the **–ve sign** has come on **R.H.S.**?

(ii) How **sin θ** has come on **R.H.S.**?

We will discuss each of the above one by one

(i) How the **–ve sign** has come on **R.H.S.**?

For this we will follow the rule of finding the sign of different t-ratio in different quadrant.

That we have discussed in details on day 25. Here we will once again revise that in brief.

[Sign of T Ratio: The adjcent figure shows the sign of various t-ratio's in different quadrant.

Sine + ve	All + ve
Tangent + ve	Cosine + ve

We have a very easy way to remember the Sign of different t-ratio's in differenc quadrant which is as follows:

After **S**chool **T**o **C**ollege

After school to college we take the first letter of above four words i.e., **A, S, T** and **C. A** means all, **S** means sine, **T** means tan & **C** means cos.

$\therefore$	In **Ist** quadrant	**A** mean **all** are +ve
	In **IInd** quadrant	**S** mean **sin** (& **cosec** θ) are +ve
	In **IIInd** quadrant	**T** mean **tan** θ (& **cot** θ) are +ve
	In **IVth** quadrant	**C** mean **cos** θ (& **sec** θ) are +ve]

Now for finding the sign of t-ratio's first we will determine in which quadrant our allied angle lies. As shown in figure 90° + θ lies in IInd quadrant and as discussed. In **IInd** quadrant only **sin** (& **cosec** θ) are **+ve** and rest are negative. So cos(90° + θ) is negative in IInd quadrant.

$\therefore$ negative sign comes on R.H.S.

(ii) How **sin** θ has come on **R.H.S**?

Now to find t-ratio's of allied angles. There are two rules

(a) For allied angles which are odd multiples of 90° [i.e., (90° – θ), (90° + θ), (270° – θ), (270° + θ).

(b) For allied angles which are even multiples of 90° [i.e., (180° – θ), (180° + θ), (360° – θ), (360° + θ).

As we are discussing **cos (90° – θ)** = – **sin** θ and 90° – θ is a odd multiple of 90°. So first we will discussed t-ratio's of allied angles which are odd multiples of 90°.

(*a*) For allied angles which are odd multiples of 90° [i.e., (90° – θ), (90° + θ), (270° – θ), (270° + θ):

(i) For allied angles which are odd multiple of 90°

sine	is expressed in term of **cosine**
cosine	is expressed in term of **sine**
tangent	is expressed in term of **co-tangent**
secant	is expressed in term of **cosecant**

cosecant is expressed in term of **secant**

If we minutely observe the above then we can see that we have prefix co if it is not there and omitted co if it is there.

For example

(i) **in sine** it is expressed in term of **cosine.** We have prefixed co with sine and made it cosine as co is not there (with sine)

(ii) **in cosine** it is expressed in term of **sine**

we have omitted co from cosine as it is there and made it sine.

This is the **aid to memory** which we once again defined as:

for t-ratio's of allied angles which are odd multiples of 90° **Prefix co if it is not there and omit co if it is there and assign the proper sign according to the quadrant in which the angle alies.**

So **cos (90° + θ)** = – **sin** θ can be easily understood by the above theory.

To understand the above concept more clearly we will do a few examples

(i) sin (90° + θ) = ?

Sol: from above aid to memory i.e., **Prefix co if it is not there,** as in this case **co is not there.**

$\therefore$ sin (90° + θ) becomes cosine θ

and (90° + θ) lies in IInd quadrant in which sign of sine is taken as positive.

$\therefore$ **sin (90° + θ) = cosine θ**

(ii) cos **(270° – θ)** = ?

Sol: from above aid to memory i.e., **omit co if it is there.** As in this case **co is there.**

$\therefore$ cosine (270° – θ) becomes sin θ

and (270° – θ) lies in IIInd quadrant in which sign of cosine is taken as negative.

$\therefore$ cos **(270° – θ) = – sin θ**

Now it is hoped that you yourself can write all the t-ratio's allied angles which are odd multiples of 90° i.e., **(90° – θ), (90° + θ), (270° – θ), (270° + θ).**

Here we have summerised all the t-ratio's of allied angle which are odd multiples of 90°.

1. t-ratio's of (90° - θ)

$$\sin(90° - \theta) = \cos\theta \qquad \text{cosec}(90° - \theta) = \sec\theta$$
$$\cos(90° - \theta) = +\sin\theta \qquad \sec(90° - \theta) = \text{cosec}\,\theta$$
$$\tan(90° - \theta) = \cot\theta \qquad \cot(90° - \theta) = \tan\theta$$

2. t-ratio's of (90° + θ)

$$\sin(90° + \theta) = \cos\theta \qquad \text{cosec}(90° + \theta) = \sec\theta$$
$$\cos(90° + \theta) = -\sin\theta \qquad \sec(90° + \theta) = -\text{cosec}\,\theta$$
$$\tan(90° + \theta) = -\cot\theta \qquad \cot(90° + \theta) = -\tan\theta$$

3. t-ratio's of (270° – θ)

$$\sin(270° - \theta) = -\cos\theta \qquad \text{cosec}(270° - \theta) = -\sec\theta$$
$$\cos(270° - \theta) = -\sin\theta \qquad \sec(270° - \theta) = -\text{cosec}\,\theta$$
$$\tan(270° - \theta) = \cot\theta \qquad \cot(270° - \theta) = \tan\theta$$

4. t-ratio's of (270° + θ)

$$\sin(270° + \theta) = -\cos\theta \qquad \text{cosec}(270° + \theta) = -\sec\theta$$
$$\cos(270° + \theta) = \sin\theta \qquad \sec(270° + \theta) = \text{cosec}\,\theta$$
$$\tan(270° + \theta) = -\cot\theta \qquad \cot(270° + \theta) = -\tan\theta$$

So these were the t-ratio's of allied angles which are odd multiples of 90°. Here once again we will revise the **aid to memory "for *t* ratios that are odd multiple of 90° prefix co if it is not there and omit co if it is there and assign the proper sign according to the quadrant in which the angle lies".**

Now we will discussed the t-ratio's of allied angles which are even multiple of 90° i.e., [(180° – θ), (180° + θ), (360° – θ), (360° + θ)]

(b) For allied angles which are even multiples of 90° [i.e., (180° – θ), (180° + θ), (360° – θ), (360° + θ)]

In this case t-ratio's are not changed in form i.e.,

sine	is expressed in term of **sine**
cosine	is expressed in term of **cosine**
tangent	is expressed in term of **tangent**
secant	is expressed in term of **secant**
cosecant	is expressed in term of **cosecant**

and

proper sign according to the quadrant in which the angle alies.

To understand the above concept more clearly we will do a few examples

(i) $\sin \mathbf{(180° + \theta)} = ?$

Sol : As **180° + θ** lies in IIIrd quadrant and in IIIrd quadrant only tan and cot are +ve rest all are negative.

$\therefore$ $\mathbf{\sin (180° + \theta) = -\sin \theta}$

(ii) $\cos \mathbf{(360° - \theta)} = ?$

Sol : As **360°** – θ lies in IVth quadrant and in IVth quadrant cos and sec are +ve rest all are negative.

$\therefore$ $\mathbf{\cos (360° - \theta) = \cos \theta}$

So keeping in mind the above discussed rules we will summarised the t-ratio's of all the allied angles which are even multiple of 90°.

1. t-ratio's of (180° – θ)

180° – θ **lies in IInd quadrant. Here sin and cosec are +ve rest are –ve. So**

$\sin (180° - \theta) = \sin \theta$ $\quad$ $\text{cosec} (180° - \theta) = \text{cosec}\, \theta$

$\cos (180° - \theta) = -\cos \theta$ $\quad$ $\sec (180° - \theta) = -\sec \theta$

$\tan (180° - \theta) = -\text{tant}\, \theta$ $\quad$ $\cot (180° - \theta) = -\cot \theta$

2. t-ratio's of (180° + θ)

180° + θ **lies in IIInd quadrant. Here tan and cot are +ve rest are –ve. So**

$\sin (180° + \theta) = -\sin \theta$ $\quad$ $\text{cosec} (180° + \theta) = -\text{cosec}\, \theta$

$\cos (180° + \theta) = -\cos \theta$ $\quad$ $\sec (180° + \theta) = -\sec \theta$

$\tan (180° + \theta) = \tan \theta$ $\quad$ $\cot (180° + \theta) = \cot \theta$

3. t-ratio's of (360° – θ)

360° – θ lies in IVth quadrant. Here cos and sec are +ve rest are –ve. So

$\sin (360° - \theta) = -\sin \theta$ $\quad$ $\text{cosec} (360° - \theta) = -\text{cosec}\, \theta$

$\cos (360° - \theta) = \cos \theta$ $\quad$ $\sec (360° - \theta) = \sec \theta$

$\tan (360° - \theta) = -\tan \theta$ $\quad$ $\cot (360° - \theta) = -\cot \theta$

[Note: It may be noted that final position of revolving line for the angle 360° – θ **occupy the same position as for same angle (– θ). Hence t-ratio of** 360° – θ **have the same value as of – θ]**

4. t-ratio's of (360° + θ)

360° + θ lies in Ist quadrant. Here all +ve. So

$$\sin(360° + \theta) = \sin\theta \qquad \text{cosec}(360° + \theta) = \text{cosec}\,\theta$$

$$\cos(360° + \theta) = \cos\theta \qquad \sec(360° + \theta) = \sec\theta$$

$$\tan(360° + \theta) = \tan\theta \qquad \cot(360° + \theta) = \cot\theta$$

Now it is hoped that you have understood that above comcept. Now let us have simple test to check your understanding.

Test I

(i) sin (90° + θ)............ (ii) cos (90° – θ)............

(iii) tan (180° – θ)............ (iv) cot (180° + θ)............

(v) cosec (270° – θ)............ (vi) sec (270° + θ)............

(vii) cot (360° + θ)............ (viii) tan (180° + θ)............

(ix) sin (– θ)............ (x) cos (360° + θ)............

(xi) tan (360° + θ)............ (xii) sec (270° – θ)............

ANSWERS

(i) cos θ (ii) sin θ (iii) – tan θ (iv) cot θ

(v) – sec θ (vi) cosec θ (vii) cot θ (viii) tan θ

(ix) – sin θ (x) cos θ (xi) tan θ (xii) – cosec θ

If your answers to above test are correct then only you should proceed to next part of chapter other wise revise the above chapter once again.

Now we will do a few examples based as the above concepts.

Example 1. Find the value of cos 135°.

Sol :

$$\cos 135° = \cos(180° - 45°) = -\cos 45° = -\frac{1}{\sqrt{2}}$$

Example 2. Find the value of sin 150°.

Sol :

$$\sin 150° = \sin(90° + 60°) = \cos 60° = \frac{\sqrt{3}}{2}$$

Example 3. Find the value of cot 225°.

Sol :

$$\cot 225° = \cot(180° + 45°) = \cot 45° = 1$$

Example 4. Find the value of sin 330°.

Sol :
$$\begin{aligned}\sin 330° &= \sin(360° - 30°)\\ &= -\sin 30°\\ &= -\frac{1}{2}\end{aligned}$$

Example 5. Find the value of sin 405°.

Sol :
$$\begin{aligned}\sin 405° &= \sin(360° + 45°)\\ &= \sin 45°\\ &= \frac{1}{\sqrt{2}}\end{aligned}$$

Example 6. Find the value of Cosec 240°

Sol :
$$\begin{aligned}\text{cosec } 240° &= \text{cosec }(270° - 30°)\\ &= -\sec 30°\\ &= -\frac{2}{\sqrt{3}}\end{aligned}$$

Example 7. Find the value of $\frac{\cos 32°}{\sin 58°}$.

Sol :
$$\begin{aligned}\frac{\cos 32°}{\sin 58°} &= \frac{\cos(90° - 58°)}{\sin 58°}\\ &= \frac{\sin 58°}{\sin 58°}\\ &= \mathbf{1}\end{aligned}$$

Example 8. Find sin 40° sec 50° + cos 40° cosec 50°

Sol : sin 40° sec 50° + cos 40°. cosec 50°
$$\begin{aligned}&= \sin(90° - 50°)\sec 50° + \cos(90° - 50°)\text{ cosec } 50°\\ &= \cos 50° \sec 50° + \sin 50° \text{ cosec } 50°\\ &= \cos 50° \times \frac{1}{\cos 50°} + \sin 50° \times \frac{1}{\sin 50°}\\ &= 1 + 1 = 2\end{aligned}$$

Example 9. $\sin^2 59 + \sin^2 41°$.

Sol :
$$\begin{aligned}\sin^2 59° + \sin^2 41° &= \sin^2 59° + \sin^2(90° - 59°)\\ &= \sin^2 59° + \cos^2 59° = 1\end{aligned}$$

Example 10. $\frac{\tan 43°}{\cot 47°}$

Sol :
$$\frac{\tan 43°}{\cot 47°} = \frac{\tan(90° - 47°)}{\cot 47°}$$

$$= \frac{\cot 47^\circ}{\cot 47^\circ} = \mathbf{1}$$

Example 11. Find the value of tan 1° tan 2° tan 3° ....tan 45° .... tan 89°

Sol :

tan 1° tan 2° tan 3° .... tan 45° .... tan 89°

= (tan 1° tan 89°) (tan 2° tan 88°) .... (tan 44° tan 46°) ...... tan 45°

= [tan 1° tan (90° – 1°)] [tan 2° tan (90° – 2°)]

[tan 44° tan (99° – 44°)] ..... tan 45°

= [tan 1°. cot 1°] [tan 2°. cot 2°] [tan 44°. cot 44°] 1

= 1 . 1 . 1 . 1

= **1**

Example 12. Find the value of $\frac{\cot(360° - \theta).\sec(180° - \theta).\sin(-\theta)}{\sin(180° + \theta).\tan(90° - \theta)\ \text{cosec}(90° + \theta)}$

Sol : $\frac{\cot(360° - \theta).\sec(180° - \theta).\sin(-\theta)}{\sin(180° + \theta).\tan(90° - \theta)\ \text{cosec}(90° + \theta)}$

$$\frac{-(\cot\theta).(-\sec\theta)(-\sin\theta)}{(-\sin\theta).(\cot\theta).(\sec\theta)} = \mathbf{-1}$$

Now we will do the important formulae which will helps us to have better understanding of this topic which is followed by few examples based on these formulae.

1. (i) $\mathbf{\sin(A + B) = \sin A \cos B + \cos A \sin B}$

(ii) $\mathbf{\sin(A - B) = \sin A \cos B - \cos A \sin B}$

(iii) $\mathbf{\cos(A + B) = \cos A \cos B - \sin A \sin B}$

(iv) $\mathbf{\tan(A + B) = \dfrac{\tan A + \tan B}{1 - \tan A \tan B}}$

(v) $\mathbf{\tan(A - B) = \dfrac{\tan A - \tan B}{1 + \tan A.\tan B}}$

These are basic formulae below given formulae can be derived from these

(i) $\mathbf{\sin 2A = 2 \sin A \cos A}$

[sin (A + A) = sin A cos A + cos A sin A = 2 sin A cos A]

(ii) $\mathbf{\cos 2A = \cos^2 A - \sin^2 A}$

(iii) $\mathbf{\tan 2A = \dfrac{2\tan A}{1 - \tan^2 A}}$

(iv) $\mathbf{\sin 3A = 3 \sin A - 4 \sin^3 A}$

(v) $\mathbf{\cos 3A = 4 \cos^3 A - 3 \cos A}$

(vi) $\mathbf{\sin x = 2 \sin \dfrac{x}{2} \cos \dfrac{x}{2}}$

$$\left[\sin x = \sin\left(\frac{x}{2}+\frac{x}{2}\right) = \sin\frac{x}{2}.\cos\frac{x}{2} + \cos\frac{x}{2}\sin\frac{x}{2} = 2\sin\frac{x}{2}\cos\frac{x}{2}\right]$$

(vii) $$\cos x = \cos^2\frac{x}{2} - \sin^2\frac{x}{2} = 2\cos^2\frac{x}{2} - 1 = 1 - 2\sin^2\frac{x}{2}$$

(viii) $$\tan x = \frac{2\tan^2\frac{x}{2}}{1-\tan^2\frac{x}{2}}$$

A part from the above we have four formulae (known as **C**, **D**, formulae) which are very important and will help us to solve large numbers of questions.

(i) $$\sin C + \sin D = 2\sin\left(\frac{C+D}{2}\right).\cos\left(\frac{C-D}{2}\right)$$

(ii) $$\sin C - \sin D = 2\cos\left(\frac{C+D}{2}\right).\sin\left(\frac{C-D}{2}\right)$$

(iii) $$\cos C + \cos D = 2\cos\left(\frac{C+D}{2}\right).\cos\left(\frac{C-D}{2}\right)$$

(iv) $$\cos C - \cos D = -2\sin\left(\frac{C+D}{2}\right).\sin\left(\frac{C-D}{2}\right)$$

Now you should have all the above formulae on your finger tips so before starting the questions you should learn these formulae by heart and should devote at least 10 - 15 minutes for formula revision

REVISE THE FORMULAE

It is hoped that you have gone through all the formulae. Now let us have a simple test to check your revision.

Test II

Fill in the blanks

1. $\sin(A+B) =$ .... $+ \cos A \sin B$
2. $\cos(A+B) = \cos A \cos B -$ ....
3. .... $= \dfrac{\tan A - \tan B}{1 + \tan A. + \tan B}$

4. $\tan(A+B) = \dfrac{\tan A + \tan B}{1 - ....}$

5. .... $= \cos^2 A - \sin^2 A$

6. $\sin 3A =$ ....

7. $\cos 3A =$ .... $- 3\cos A$

8. .... $= \cos^2 \dfrac{x}{2} - \sin^2 \dfrac{x}{2}$

9. $\sin C + \sin D =$ ....

10. $\cos C + \cos D =$ ....

11. $\cos C - \cos D =$ ....

ANSWERS

1. $\sin A \cos B$	**2.** $\sin A \sin B$	**3.** $\tan(A-B)$	**4.** $\tan A \tan B$
5. $\cos 2A$	**6.** $3\sin A - 4\sin^3 A$	**7.** $4\cos^3 A$	**8.** $\sin x \cos x$

9. $2\sin\left(\dfrac{C+D}{2}\right).\cos\left(\dfrac{C-D}{2}\right)$ **10.** $2\cos\left(\dfrac{C+D}{2}\right).\cos\left(\dfrac{C-D}{2}\right)$

11. $2\sin\left(\dfrac{C+D}{2}\right).\sin\left(\dfrac{C-D}{2}\right)$

If all the answer in above exercise are correct then you should do the next part of the chapter, otherwise you should revise the formulae again.

Now we will do a few examples based on above formulae.

Example 1. Find the value of

(i) $\sin 15°$ (ii) $\cos 75°$ (iii) $\tan 15°$ (iv) $\tan 105°$

(v) $\sin 135°$ (vi) $\tan 150°$

Sol : (i)

$$\sin 15° = \sin(45° - 30°)$$
$$= \sin 45° \cos 30° - \cos 45° \sin 30°$$

[sin (A – B) = sin A cos B – cos A sin B]

$$= \frac{1}{\sqrt{2}} \times \frac{\sqrt{3}}{2} - \frac{1}{\sqrt{2}} \times \frac{1}{2}$$
$$= \mathbf{\frac{\sqrt{3}-1}{2\sqrt{2}}}$$

(ii)

$$\cos 75° = \cos(45° - 30°)$$
$$= \cos 45° \cos 30° - \sin 45° \sin 30°$$
$$= \frac{1}{\sqrt{2}} \times \frac{\sqrt{3}}{2} - \frac{1}{\sqrt{2}} \times \frac{1}{2}$$
$$= \frac{\sqrt{3}}{2\sqrt{2}} - \frac{1}{2\sqrt{2}}$$

$$= \frac{\sqrt{3}-1}{2\sqrt{2}}$$

(iii) $\tan 15° = \tan(45° - 30°)$

$$= \frac{\tan 45° - \tan 30°}{1 + \tan 45°.\tan 30°} = \frac{1 - \frac{1}{\sqrt{3}}}{1 + \frac{1}{\sqrt{3}}}$$

$$= \frac{\sqrt{3}-1}{\sqrt{3}+1}$$

(iv) $\tan 105° = \tan(60° + 45°)$

$$= \frac{\tan 60° + \tan 45°}{1 - \tan 60°.\tan 45°}$$

$$= \frac{\sqrt{3}+1}{1-\sqrt{3}} = \frac{1+\sqrt{3}}{1-\sqrt{3}}$$

(v) $\sin 135° = \sin(90° + 45°) = \cos 45° = \frac{1}{\sqrt{2}}$

(vi) $\tan 150° = \tan(180° - 30°) = -\tan 30° = -\frac{1}{\sqrt{3}}$

Example 2. Find the value of

(i) $\sin 78° \cos 18° - \cos 78° \sin 18°$

(ii) $\cos 68° \cos 38° + \sin 68° \sin 38°$

(iii) $\cos 40° \cos 20° - \sin 40° \sin 20°$

(iv) $\sin 33° \cos 12° + \cos 33° \sin 12°$

Sol :

(i) $\sin 78° \cos 18° - \cos 78° \sin 18° = \sin(78° - 18°)$

[sin A cos B – cos A sin B = sin (A – B)]

$$= \sin 60° = \frac{\sqrt{3}}{2}$$

(ii) $\cos 68 \cos 38° + \sin 68° \sin 38° = \cos(68° - 38°)$

[cos A cos B + sin A sin B = cos (A – B)]

$$= \cos 30° = \frac{\sqrt{3}}{2}$$

(iii) $\cos 40° \cos 20° - \sin 40° \sin 20° = \cos(40° + 20°) = \cos 60°$

$$= \frac{1}{2}$$

(iv) $\sin 33° \cos 12° + \cos 33° \sin 12° = \sin(33° + 12°) = \sin 45°$

$$= \frac{1}{\sqrt{2}}$$

Example 3. Find the value of

(*i*) sin 18° (*ii*) cos 18° (*iii*) cos 36° (*iv*) sin 54°

Sol :

(i) Let $\theta = 18°$

then $5\theta = 18° \times 5 = 90°$

$$2\theta = (5\theta - 3\theta)$$

$$\sin 2\theta = \sin (5\theta - 3\theta)$$

$$\sin 2\theta = \sin (90° - 3\theta) = \cos 3\theta$$

$$\sin 2\theta = \cos 3\theta$$

$$2 \sin\theta \cos\theta = 4 \cos^3 \theta - 3 \cos\theta$$

$$2 \sin\theta \cos\theta = \cos\theta (4 \cos^2 \theta - 3)$$

$$2 \sin\theta = 4 \cos^2 \theta - 3$$

$$2 \sin\theta = 4 (1 - \sin^2 \theta) -3$$

$$2 \sin\theta = 4 - 4 \sin^2 \theta - 3$$

$$2 \sin\theta = - 4 \sin^2 \theta + 1$$

$$4 \sin^2 \theta + 2 \sin\theta - 1 = 0$$

solving by quadratic equation method, we get

$$\sin \theta = \frac{-1 \pm \sqrt{5}}{4}$$

but we will take the positive value of sinθ given below as

$$\sin \theta = \frac{-1 + \sqrt{5}}{4}$$

[as θ lies in Ist quadrant and in Ist quadrant sin θ is +ve]

therefore $\sin \theta \text{ or } \sin 18° = \dfrac{-1 + \sqrt{5}}{4}$

(ii) $\cos 18° = \sqrt{1 - \sin^2 18°}$

$$= \sqrt{1 - \left(\frac{-1 + \sqrt{5}}{4}\right)^2}$$

$$= \sqrt{1 - \left(\frac{1 + 5 - 2\sqrt{5}}{16}\right)}$$

$$= \sqrt{1 - \left(\frac{6 - 2\sqrt{5}}{16}\right)}$$

$$= \sqrt{\frac{16 - 6 + 2\sqrt{5}}{16}}$$

$$= \sqrt{\frac{10+2\sqrt{5}}{16}}$$

$$= \frac{\sqrt{10+2\sqrt{5}}}{4}$$

(iii) We know that

$$\cos 2A = \cos^2 A - \sin^2 A$$
$$= (1 - \sin^2 A) - \sin^2 A$$
$$= 1 - 2\sin^2 A$$

Now we apply the above formula to cos 36°

$$\cos 36° = 1 - 2\sin^2 18°$$

$$= 1 - 2\left(\frac{-1+\sqrt{5}}{4}\right)^2$$

$$= 1 - 2\left(\frac{1+5-2\sqrt{5}}{16}\right)$$

$$= 1 - \left(\frac{6-2\sqrt{5}}{8}\right)$$

$$= \frac{8-6+2\sqrt{5}}{8}$$

$$= \frac{2+2\sqrt{5}}{8}$$

$$= \frac{2(1+\sqrt{5})}{8}$$

$$= \mathbf{\frac{1+\sqrt{5}}{4}}$$

(iv)

$$\sin 54° = \sin(90° - 36°) = \cos 36°$$

$$= \frac{1+\sqrt{5}}{4}$$

Note: We have seen in above example that we have to find the value of sin18° rest cos18°, cos36°, sin36°, sin54°, cos54°, etc. can be derived from it. A few have been derived in examples rest we will derive in exercise.

EXERCISE

1. The value of cos 330° is

(a) $\frac{\sqrt{3}}{2}$ (b) $\frac{1}{2}$ (c) 1 (d) none

2. The value of tan (– 210°) is

(a) $\frac{-1}{\sqrt{3}}$ (b) $\frac{1}{\sqrt{3}}$ (c) $\sqrt{2}$ (d) none

3. The value of sin 225° is

(a) $\frac{-1}{\sqrt{2}}$ (b) $\frac{1}{2}$ (c) $\frac{\sqrt{3}}{2}$ (d) none

4. The value of cosec 330° is

(a) $\frac{\sqrt{3}}{2}$ (b) $\frac{-2}{\sqrt{3}}$ (c) $\sqrt{2}$ (d) none

5. The value of sec 240° is

(a) $\frac{\sqrt{3}}{2}$ (b) $\frac{-2}{\sqrt{3}}$ (c) $\sqrt{2}$ (d) none

6. The value of cos 28° – sin 62° is

(a) 1 (b) 2 (c) 0 (d) none

7. The value of $\sin^2 51° + \sin^2 39°$ is

(a) 0 (b) 1 (c) 2 (d) none

8. The value of $\frac{\tan 53°}{\cot 37°}$ is

(a) 0 (b) 1 (c) 2 (d) none

9. The value of $\frac{\sin (90° - \theta).\text{cosec}\,(180° - \theta)}{\sin (270° + \theta).\sec (270° + \theta)}$ is

(a) 0 (b) 1 (c) 2 (d) none

10. The value of $\frac{\cos 41°}{\sin 49°} - \frac{\sin 72°}{\cos 18°}$ is

(a) 0 (b) –1 (c) 1 (d) none

11. The value of $\sin^2 41° + \sin^2 49°$ is

(a) 0 (b) 1 (c) 2 (d) none

12. The value of $\frac{\cos(90° - \theta).\cos(360° - \theta).\tan(180° - \theta)}{\cos(90° + \theta).\sin(90° + \theta).\cot(90° + \theta)}$ is

(a) –1 (b) 1 (c) 2 (d) none

13. The value of sin 15° is

(a) $\frac{\sqrt{3}+1}{2\sqrt{2}}$ (b) $\frac{\sqrt{3}-1}{2\sqrt{2}}$ (c) $\frac{\sqrt{5}-1}{4}$ (d) none

14. The value of cos 75° is

(a) $\frac{\sqrt{3}+1}{2\sqrt{2}}$ (b) $\frac{\sqrt{3}-1}{2\sqrt{2}}$ (c) $\frac{\sqrt{5}-1}{4}$ (d) none

15. The value of cos 15° is

(a) $\frac{\sqrt{3}+1}{2\sqrt{2}}$ (b) $\frac{\sqrt{3}-1}{2\sqrt{2}}$ (c) $\frac{\sqrt{5}-1}{4}$ (d) none

16. The value of $\sin 75°$ is

(a) $\frac{\sqrt{3}+1}{2\sqrt{2}}$ (b) $\frac{\sqrt{3}-1}{2\sqrt{2}}$ (c) $\frac{\sqrt{5}-1}{4}$ (d) none

17. The value of $\sin 105°$ is

(a) $\frac{\sqrt{3}+1}{2\sqrt{2}}$ (b) $\frac{\sqrt{3}-1}{2\sqrt{2}}$ (c) $\frac{\sqrt{5}-1}{4}$ (d) none

18. The value of $\sin 18°$ is

(a) $\frac{\sqrt{5}-1}{4}$ (b) $\frac{\sqrt{10+2\sqrt{5}}}{4}$ (c) $\frac{\sqrt{10-2\sqrt{5}}}{4}$ (d) $\frac{\sqrt{5}+1}{4}$

19. The value of $\cos 18°$ is

(a) $\frac{\sqrt{5}-1}{4}$ (b) $\frac{\sqrt{10+2\sqrt{5}}}{4}$ (c) $\frac{\sqrt{10-2\sqrt{5}}}{4}$ (d) $\frac{\sqrt{5}+1}{4}$

20. The value of $\cos 36°$ is

(a) $\frac{\sqrt{5}-1}{4}$ (b) $\frac{\sqrt{10+2\sqrt{5}}}{4}$ (c) $\frac{\sqrt{10-2\sqrt{5}}}{4}$ (d) $\frac{\sqrt{5}+1}{4}$

21. The value of $\sin 36°$ is

(a) $\frac{\sqrt{5}-1}{4}$ (b) $\frac{\sqrt{10+2\sqrt{5}}}{4}$ (c) $\frac{\sqrt{10-2\sqrt{5}}}{4}$ (d) $\frac{\sqrt{5}+1}{4}$

22. The value of $\sin 22\frac{1}{2}^{\circ}$ is

(a) $\frac{\sqrt{2-\sqrt{2}}}{2}$ (b) $\frac{\sqrt{2+\sqrt{2}}}{2}$ (c) $\sqrt{2}-1$ (d) none

23. The value of $\cos 22\frac{1}{2}^{\circ}$ is

(a) $\frac{\sqrt{2-\sqrt{2}}}{2}$ (b) $\frac{\sqrt{2+\sqrt{2}}}{2}$ (c) $\sqrt{2}-1$ (d) none

24. The value of $\tan 22\frac{1}{2}^{\circ}$ is

(a) $\frac{\sqrt{2-\sqrt{2}}}{2}$ (b) $\frac{\sqrt{2+\sqrt{2}}}{2}$ (c) $\sqrt{2}-1$ (d) none

25. $\frac{\sin\theta+\sin x}{\cos\theta+\cos x}$ is equal to

(a) $\tan\left(\frac{\theta+x}{2}\right)$ (b) cot (c) $\tan x$ (d) none

26. $\frac{\cos 2C-\cos 2D}{\sin 2D+\sin 2C}$

(a) tan D (b) tan C (c) tan (C + D) (d) none

27. The value of $\frac{1+\tan 15°}{1-\tan 15°}$ is

(a) $\sqrt{3}$ (b) $\frac{1}{\sqrt{3}}$ (c) $-\sqrt{3}$ (d) none

28. If $\tan 43° = \frac{a}{b}$ then value of $\tan 47°$ is

(a) $\frac{\sqrt{5}-1}{4}$ (b) $\frac{b}{a}$ (c) $a\,b$ (d) none

29. $\sin(45° + \theta) - \sin(45° - \theta)$ is equal to

(a) $\sqrt{2}\cos\theta$ (b) $\sqrt{2}\sin\theta$ (c) $\frac{1}{\sqrt{2}}\sin\theta$ (d) none

30. The value of $\sin 7\frac{1°}{2}\cos 37\frac{1°}{2}$ is

(a) $\frac{\sqrt{2}-1}{4}$ (b) $\frac{\sqrt{2}-1}{2}$ (c) $\frac{1-\sqrt{2}}{2}$ (d) none

31. The maximum value of $\sin x + \cos x$ is

(a) $\sqrt{2}$ (b) 1 (c) $\frac{1}{2}$ (d) none

32. the minimum value of $\sin x + \cos x$ is

(a) $\sqrt{2}$ (b) 1 (c) $\frac{1}{2}$ (d) none

33. The maximum value of $\sin x \cos x$ is

(a) $\sqrt{2}$ (b) 1 (c) $\frac{1}{2}$ (d) none

34. If $\cos\alpha\cos B = 1 + \sin\alpha\sin B$ then value of $\tan x \cot B$ is

(a) $\sqrt{2}$ (b) 1 (c) $\frac{1}{2}$ (d) none

35. If $\sin A = \frac{4}{5}$ $\cos B = \frac{5}{13}$ and $0 < A < 90°$, $0 < B < 90°$ then $\sin(A + B)$ is

(a) $\frac{2}{13}$ (b) $\frac{51}{65}$ (c) $\frac{53}{65}$ (d) none

36. If $\sin x = \frac{1}{2}$ then $\tan 2x$ is equal to

(a) $\sqrt{2}$ (b) $\frac{1}{\sqrt{3}}$ (c) $2\sqrt{2}$ (d) none

37. $\sqrt{2+\sqrt{2+2\cos 4\theta}} = ?$

(a) $2\sec\theta$ (b) $2\cos\theta$ (c) $2\tan\theta$ (d) none

38. The value of θ If $3\tan\theta + \cot\theta = 5\,\text{cosec}\,\theta$ when $0 < \theta < 90°$

(a) 45° (b) 60° (c) 90° (d) none

39. The value of cos1, cos 2, cos 3, ... cos 88, cos 89, cos 90°

(a) 0 (b) 1 (c) 2 (d) 3

40. The value of sin 1°, sin 2°, sin 3°, ... sin 171°, sin 180°

(a) 0 (b) 1 (c) 2 (d) 3

41. The value of sin 1470° × cos 750° is

(a) $\sqrt{3}$ (b) $\frac{1}{\sqrt{3}}$ (c) 2 (d) none

42. The value of $\frac{\tan 40° + \tan 20°}{1 - \tan 40° \tan 20°}$ is

(a) $\sqrt{3}$ (b) $\frac{1}{\sqrt{3}}$ (c) 2 (d) none

43. If A + B = 45° then value of (1 + tan A) (1 + tan B) is equal to

(a) 2 (b) 4 (c) 6 (d) none

44. The value of $\cos^2 5 + \cos^2 10 + \cos^2 15 + \cos^2 20 + \cos^2 70 + \cos^2 75 + \cos^2 80 + \cos^2 85$ is

(a) 2 (b) 4 (c) 6 (d) none

SOLUTIONS

1. (a)
$$\cos 330° = \cos(360° - 30°) = \cos 30° = \frac{\sqrt{3}}{2}$$

2. (a)
$$\tan(-210°) = -\tan 210° = -\tan(180° + 30°) = -\tan 30° = \frac{-1}{\sqrt{3}}$$

3. (a)
$$\sin 225° = \sin(180° + 45°) = -\sin 45° = \frac{-1}{\sqrt{2}}$$

4. (d)
$$\text{cosec } 330° = \text{cosec}(360° - 30°) = -\text{cosec } 30° = \mathbf{-2}$$

5. (d)
$$\sec 240° = \sec(270° - 30°) = -\text{cosec } 30° = \mathbf{-2}$$

6. (c)
$$= \cos 28° - \sin 62° = \cos(90° - 62°) - \sin 62° = \sin 62° - \sin 62° = \mathbf{0}$$

7. (b)
$$\sin^2 51° + \sin^2 39° = \sin^2(90° - 39°) + \sin^2 39 = \cos^2 39 + \sin^2 39 = \mathbf{1}$$

8. (b)
$$\frac{\tan 53°}{\cot 37°} = \frac{\tan(90° - 37°)}{\cot 37°} = \frac{\cot 37°}{\cot 37°} = \mathbf{1}$$

9. (d)
$$\frac{\sin(90° - \theta)\text{cosec}(180° - \theta)}{\sin(270° - \theta)\sec(270° - \theta)} = \frac{\cos\theta \text{ cosec}\theta}{-\cos\theta \text{ cosec}\theta}$$

$$= \mathbf{-1}$$

10. (a)

$$\frac{\cos 41^\circ}{\sin 49^\circ} - \frac{\sin 72^\circ}{\cos 18^\circ} = \frac{\cos(90^\circ - 49^\circ)}{\sin 49^\circ} - \frac{\sin(90^\circ - 18^\circ)}{\cos 18^\circ}$$

$$= \frac{\sin 49^\circ}{\sin 49^\circ} - \frac{\cos 18^\circ}{\cos 18^\circ}$$

$$= 1 - 1 = \mathbf{0}$$

11. (b)

12. (b)

$$\frac{\sin\theta.\cos\theta.\tan\theta}{(-\sin\theta).\cos\theta.(-\tan\theta)} = \mathbf{1}$$

13. (b) Same as example 1 (i)

14. (b) Same as example 1 (ii)

15. (a)

$$\cos 15^\circ = \cos(45^\circ - 30^\circ)$$

$$= \cos 45^\circ \cos 30^\circ + \sin 45^\circ \sin 30^\circ$$

$$= \frac{1}{\sqrt{2}}\frac{\sqrt{3}}{2} + \frac{1}{\sqrt{2}} \times \frac{1}{2}$$

$$= \frac{\sqrt{3}}{2\sqrt{2}} + \frac{1}{2\sqrt{2}}$$

$$= \frac{\sqrt{3}+1}{2\sqrt{2}}$$

16. (b)

$$\sin 75^\circ = \sin(90^\circ - 15^\circ)$$

$$= \cos 15^\circ$$

$$\cos 15^\circ = \frac{\sqrt{3}+1}{2\sqrt{2}}$$

17. (b)

$$\sin(105^\circ) = \sin(60^\circ + 45^\circ)$$

$$= \sin 60^\circ \cos 45^\circ + \cos 60^\circ \sin 45^\circ$$

$$= \frac{\sqrt{3}}{2}\frac{1}{\sqrt{2}} + \frac{1}{2}\frac{1}{\sqrt{2}}$$

$$= \frac{\sqrt{3}+1}{2\sqrt{2}}$$

18. (a) Same as example 2 (i)

19. (b) Same as example (ii)

20. (d) Same as example 2 (iii)

21. (c)

22. (a) Same as example 3 (i)

23. (b)

$$\cos 22\frac{1^\circ}{2} = \sqrt{1 - \sin^2 22\frac{1^\circ}{2}}$$

Solving we get (b) as answer.

24. (c) Same as example 3 (ii)

25. (a)

$$\frac{\sin\theta + \sin x}{\cos\theta + \cos x} = \frac{2\sin\left(\frac{\theta+x}{2}\right).\cos\left(\frac{\theta-x}{2}\right)}{2\cos\left(\frac{\theta+x}{2}\right).\cos\left(\frac{\theta-x}{2}\right)}$$

$$= \tan\left(\frac{\theta + x}{2}\right)$$

26. (d)

27. (a)

$$\frac{1+\tan 15^\circ}{1-\tan 15^\circ} = \frac{\tan 45^\circ + \tan 15^\circ}{1 - \tan 45^\circ \tan 15^\circ}$$

$$= \tan(45^\circ + 15^\circ)$$

$$= \tan 60^\circ = \sqrt{3}.$$

28. (b) $\tan 47^\circ = \tan(90^\circ - 43^\circ) = \cot 43^\circ = \frac{b}{a}$

29. (b)

$$\sin(A+B) - \sin(A-B) = 2\cos A \sin B$$

$$\sin(45^\circ + \theta) - \sin(45^\circ - \theta) = 2\cos 45^\circ \sin\theta$$

$$= \sqrt{2}\sin\theta$$

30. (a)

31. (a) The maximum value of A sin θ + B cos θ is $+\sqrt{A^2+B^2}$

Here $A = 1$

and $B = 1$

∴ Maximum value in this case is $\sqrt{1^2+1^2} = \sqrt{2}$

32. (d) The minimum value of A sin θ + B cos θ is $-\sqrt{A^2+B^2}$

Here $A = 1$

and $B = 1$

∴ Minimum value in this case is $-\sqrt{1^2+1^2} = -\sqrt{2}$

33. (c)

$$\sin x \cos x = \frac{1}{2}(2\sin x \cos x)$$

$$= \frac{1}{2}\sin 2x$$

Now we know that sin θ lies between – 1 and + 1 i.e., maximum value of sin (angle) is + 1

maximum value of sin x cos x is $\frac{1}{2}$

34. (d)

$$\cos\alpha\cos B = 1 + \sin\alpha\sin B$$

$$\cos\alpha\cos B - \sin\alpha\sin B = 1$$

i.e., $\cos(\alpha + B) = 1$

we know that $\cos 0 = 1$

$$\alpha + B = 0$$

$$\sin(\alpha + B) = \sin 0 = 0$$

i.e., $\sin\alpha\cos B + \cos\alpha\sin B = 0$

$$\sin\alpha\cos B = -\cos\alpha\sin B$$

or $\frac{\sin\alpha\cos B}{\cos\alpha\sin B} = -1$

$$\tan\alpha\cot\beta = -1$$

35. (d) By Pythagoras theorem

If $\sin A = \frac{4}{5}$

then $\cos A = \frac{3}{5}$

similarly if $\cos B = \frac{5}{13}$

then $\sin B = \frac{12}{13}$

now $\sin(A+B) = \sin A \cos B + \cos A \sin B$

$$= \frac{4}{5} \times \frac{5}{13} + \frac{3}{5} \times \frac{12}{13} = \frac{56}{65}$$

36. (d) By Pythagoras theorem

$$\sin x = \frac{1}{2}$$

$$\tan x = \frac{1}{\sqrt{3}}$$

now $\tan 2x = \frac{2\tan x}{1-\tan^2 x}$

$$= \frac{2 \times \frac{1}{\sqrt{3}}}{1-\left(\frac{1}{\sqrt{3}}\right)^2}$$

$$= \frac{\frac{2}{\sqrt{3}}}{1-\frac{1}{3}}$$

$$= \frac{2}{\sqrt{3}} \times \frac{3}{2}$$

$$= \sqrt{3}$$

37. (b) $\sqrt{2+\sqrt{(1+\cos 4\theta)}} = \sqrt{2+\sqrt{2 \times 2\cos^2 2\theta}}$

$$= \sqrt{2+2\cos 2\theta}$$

$$= \sqrt{2(1+\cos 2\theta)}$$

$$= \sqrt{2.2\cos^2 \theta}$$

$$= 2\cos\theta$$

38. (b) $3\tan\theta + \cot\theta = 5\operatorname{cosec}\theta$

$$\frac{3\sin\theta}{\cos\theta} + \frac{\cos\theta}{\sin\theta} = \frac{5}{\sin\theta}$$

$$3\sin^2\theta + \cos^2\theta = 5\cos\theta$$

$$3(1-\cos^2\theta) + \cos^2\theta = 5\cos\theta$$

$$2\cos^2\theta + 5\cos\theta - 3 = 0$$

$$2\cos^2\theta + 6\cos\theta - \cos\theta - 3 = 0$$

$$2\cos\theta(\cos\theta+3)-1(\cos\theta+3) = 0$$
$$(2\cos\theta-1)(\cos\theta+3) = 0$$
$$2\cos\theta = 1$$
$$\cos\theta = \frac{1}{2}$$
$$\theta = 60°$$

39. (a) As $\cos 90° = 0$

Product of whole expression is zero

40. (a) As $\sin 180° = 0$

Product of whole expression is zero

41. (d)
$$\sin 1470° = \sin(360° \times 4 + 30°)$$
$$= \sin 30° = \frac{1}{2}$$
$$\cos 750° = \cos(360° \times 2 + 30°) = \cos 30° = \frac{\sqrt{3}}{2}$$
$$\sin 1470° \times \cos 750° = \frac{1}{2} \times \frac{\sqrt{3}}{2} = \frac{\sqrt{3}}{4}$$

42. (a)
$$\frac{\tan 40° + \tan 20°}{1-\tan 40° \tan 20°} = \tan(40°+20°) = \tan 60° = \sqrt{3}$$

43. (a)
$$A + B = 45°$$
$$\tan(A+B) = \tan 45° = 1$$
$$\frac{\tan A + \tan B}{1-\tan A \tan B} = 1$$
$$\tan A + \tan B = 1 - \tan A \tan B$$
$$\tan A + \tan B + \tan A \tan B = 1$$

Add 1 on both sides, we get
$$1 + \tan A + \tan B + \tan A \tan B = 1 + 1$$
$$1 + \tan A + \tan B(1 + \tan A) = 2$$
$$(1+\tan A)(1+\tan B) = \mathbf{2}$$

44. (b) $(\cos^2 5° + \cos^2 85°) + (\cos^2 10° + \cos^2 80°) + (\cos^2 15° + \cos^2 75°) + \cos^2 20° + \cos^2 70°$

$= [\cos^2 5° + \cos^2(90° - 5°)] + [\cos^2 10° + \cos^2(90° - 10°)] + [\cos^2 15° + \cos^2(90° - 15°)] + [\cos^2 20° + \cos^2(90° - 20°)]$

$= (\cos^2 5^2 + \sin^2 5°) + (\cos^2 10° + \sin^2 10°) + (\cos^2 15° + \sin^2 15°) + (\cos^2 20° + \sin^2 20°)$

$= 1 + 1 + 1 + 1$

$= 4$

Day 27

Combination and Permutation

In this topic we are going to understand the basic difference between the application of **Combination** and **Permutation** i.e. what is **Combination** and **Permutation** & how to use **Combination** and **Permutation** in solving the questions.

To make this concept clear let us consider the two experiments. Let first experiment is selecting a cricket team from given set of players and second experiment is sending the batsmen in certain order to play the game.

Let us consider the first experiment. In this we will select two different teams from given set of players.

for e.g. we can select teams as follows.

Let Ist team = **Sachin, Saurav, Mongia, Sehwag.**

Let 2nd team = **Dravid, Srinath, Ajay, Vijay.**

By seeing above teams one can say that the teams are different as the players in each team is not same.

Now to consider the second experiment in which we will be sending players to play the game.

(i) We can send : **Sachin, Saurav, Mongia, Sehwag.**

(ii) We can send : **Mongia, Sachin, Saurav, Sehwag**

Or any other order which wc like.

So we see that all the batting orders are different, although the players are same.

So in first experiment we have selected two different teams out of total given players and in second experiment we send the players in two different orders to play the game.

So in Ist experient there was selection of players from given set of players and different and in second experiment the players are same but the order of batting hasbeen changed.

Therefore the difference between two experiment is difference between **Combination** and **Permutation**.

Hence **Combination** is ways of selection of objects or persons out of total objects or persons.

Permutation is arrangement of objects or persons out of total objects or persons.

REMEMBER in Permutation the order or the ways of arrangment persons or objects is important i.e., we can have different arrangments for same set of different persons or objects.

In **Combination** the selection of objects or persons is important so we will not have different Combination if the objects selected are same.

After understanding the concept now we will do some important results and formulae which will help us to solve the questions. First we will do some important definitions

DEFINITIONS

(i) Factorial

It is the mathematical operation denoted by symbol **!** or **L**

For e.g., 5! or L5 (read as 5 factorial) means $5 \times 4 \times 3 \times 2 \times 1$

i.e., we go on reducing the given number (here 5) by one, multiplying these numbers and it is done until we reach 1 at the last.

$$\therefore \text{L}n \text{ or } n! = (n)(n-1)(n-2)(n-3) \ldots\ldots\ldots 1$$

Also $$\text{L}0 \text{ or } 0! = 1$$

(ii) Permutation

The ways of arrangements of r different objects or persons taken one or more than one at a time out of given total objects n, is known as Permutation. It is given by

$$^nP_r = \frac{n!}{(n-r)!}$$

WHERE ^nP_r read as Permutations or arrangement **of '*r*'** different object out of total **'*n*'**.

For e.g., If we have to arrange five persons in five vacant seats, taking all the persons at a time then it is given by

$$^5P_5 = \frac{5!}{(5-5)!} = \frac{5!}{0!} = 5 \times 4 \times 3 \times 2 \times 1 = 120$$

(iii) Combination

The ways of selecting or chosing **'*r*'** different persons or objects out of given (total) n objects or person taken one or more than one at a time is known as **Combination.**

It is given by

$$^nC_r = \frac{n!}{(n-r)!\,r!}$$

where ^nC_r is read as Combination of r objects or persons out of total n objects or persons.

For e.g., If we have to select one or more then one person at a time from given five persons then it is given as

selection of one out of five $= {}^5C_1 = \frac{5!}{4!1!} = 5$

selection of two out of five $= {}^5C_2 = \frac{5!}{3!2!} = \frac{5 \times 4}{2} = 10$

selection of three out of five $= {}^5C_3 = \frac{5!}{2!3!} = 10$

selection of four out of five $= {}^5C_4 = \frac{5!}{4!1!} = 5$

selection of five out five $= {}^5C_5 = \dfrac{5!}{0!5!} = 1$

(iv) Fundamental Principle of Counting

There are two fundamental laws of counting given below :

(a) Law of Multiplication

It is again a mathematical result according to it if any event **A** can happen in ***m*** ways and another event **B** can happen in ***n*** ways then simultaneous occurance of both the events **A and B** can happen in **(*m* × *n*)** ways.

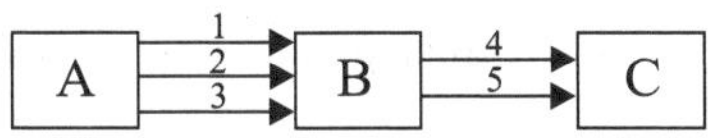

e.g., if there are 3 ways of moving from **A** to **B** and 2 ways of moving from **B** to **C**, then **A** to **C** can be reached in **3 × 2 = 6** ways.

(b) Law of Addition

It is again a mathematical result according to it if any event **A** can happen in ***m*** ways and another event **B** can happen in ***n*** ways then either of them can happen in ***m* + *n*** ways.

e.g., if there are 3 ways of moving from **A** to **B** and 2 ways of moving from **B** to **C**, then one can go from **B** to **A** or **B** to **C** in 3 + 2 = 5 ways.

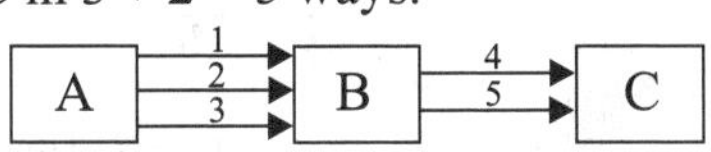

Important Results

Permutations

(i)

(a) The numbers of ways of arranging n different objects taken all at a time when clockwise and anticlockwise ways are different in circle is **$(n-1)!$**. Fix any one as reference point, other $(n-1)$ things can be arranged in $(n-1)!$ ways. Three persons around a circular table can be arranged in 2 ways, *i.e.,* $(3-1)!$ ways

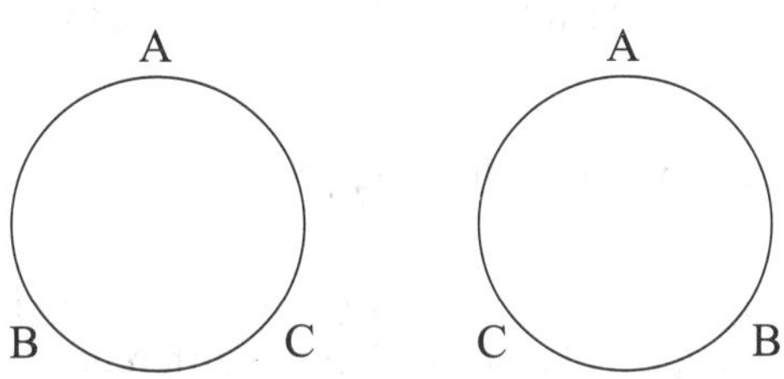

(b) If in circle we can see the things from behind for example beads in the necklace then the half the cases are obtained in front and half in the back. The same concept applies in the case of floral wreath or a garland. In such cases number of arrangements is $\dfrac{(n-1)!}{2}$

(ii)

(a) The number of ways in which n different things can be divided into r groups of sizes $k_1, k_2, ..., k_r$

$$= \frac{n!}{k_1!\, k_2! \ldots k_r!}$$

But if the groups are named, then number of ways

$$= \left(\frac{n!}{k_1!\, k_2! \ldots k_r!}\right) r!$$

Clearly $\qquad n = k_1 + k_2 + \ldots + k_r$

(b) If $k_1 = k_2 = \ldots = k_r = k$ *i.e.,* if the r groups are of equal size,

then the number of ways of dividing in r groups $= \dfrac{n!}{(k!)^r\, r!}$

But if the groups are named, then the number of ways $= \dfrac{n!}{(k!)^r}$

(c) The number of ways in which n different things can be distributed among r persons such that they get $k_1, k_2, \ldots, k_r$ things

$$= \frac{n!}{k_1!\, k_2! \ldots k_r!} . r!$$

But if $\qquad k_1 = k_2 \ldots k_r = k$, then number of ways

$$= \frac{n!}{(k!)^r\, r!} . r! = \frac{n!}{(k!)^r}$$

The above concept can be further understood by the following examples.

Example. In how many ways can 15 soldiers be divided into 3 groups equally?

Sol : As the groups are not named therefore required number of ways $= \dfrac{15!}{3!(5!)^3}$.

Example. In how many ways can 15 soldiers be drafted into 3 battalion **(ALPHA, BETA, GAMMA).**

Sol : As the groups are named therefore required number of ways $= \dfrac{15!}{3!(5!)^3} \times 3! = \dfrac{15!}{(5!)^3}$.

(iii) The number of ways of arrangements of n objects out which P are of one type, Q are of one type and R are of one type and rest all other are different is given by

$$\frac{\mathbf{n!}}{\mathbf{P!Q!R!}}$$

Mostly we use this result in alphabets related problems in which a word is given, having some of its alphabets repeated and we are asked to find that how many different words we can make using alphabets of given word.

Example. How many different arrangements can be made with the letters of the word **"MISSISSIPPI"**.

Sol : As MISSISSIPPI contains 4 S, 4 I and 2 P

$\therefore$ Number of arrangements $= \dfrac{11!}{4!\ 4!\ 2!}$

Combination

(i) The total number of ways of selection of one or more than one objects or persons out of total n objects or persons is $2^n - 1$

Because if we are allowed to select one or two or three or upto n persons or objects then the ways of selections for these are given as follows:

selection of 1 out of $n = {}^nC_1$

or selection of 2 out of $n = {}^nC_2$

or selection of 3 out of $n = {}^nC_3$

or .... upto n out of $n = {}^nC_n$

$\therefore$ **total ways** $= {}^nC_1 + {}^nC_2 + {}^nC_3 + \ldots\ldots + {}^nC_n = 2^n - 1$

The above concept is explained with the help of example given below.

Example. Every one of the 10 available lamps can be switched on to illuminate certain Hall. The total number of ways in which, the hall can be illuminated is :

Sol : The total number of ways in which the lamp can be illustrated

$$= {}^{10}C_1 + {}^{10}C_2 + \ldots\ {}^{10}C_{10} = 2^{10} - 1 = 1023.$$

(ii) If n points are given in a plane then

(a) Total number of triangles we can make by joining these n points is equal to ${}^nC_3$

Beacuse to make a triangle we need three points at least.

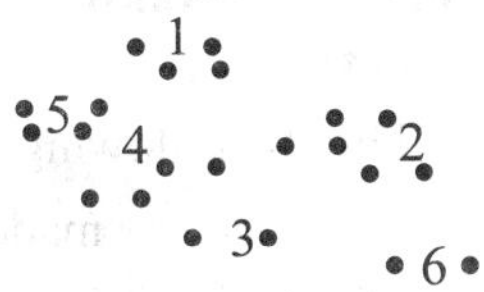

As shown in figure we can select any three points for *e.g.* in figure we can join points 1, 2, and 3 and points 1, 4, and 5 to make triangles. So ways of selections of any 3 points is equal to number of triangles formed which is equal to ${}^nC_3$.

(b) Total numbers of lines = ${}^nC_2$

because to make a line we need to join atleast two points so if n points are given then number of lines is equal to the numbers of ways of selecting two points out of total n points.

(c) If m points out of total n given points are collinear (when the points lies in straight line they are collinear) Then number of lines will be ${}^nC_2 - {}^mC_2 + 1$

here ${}^nC_2$ = total number of lines (as we get a line by joining any two selected points.)

${}^mC_2$ = number of lines made by using m points if they are not collinear or scattered but here m point are collinear so ${}^mC_2$ is number lines that are lost due to collinearity of m points.

Also here we have added 1, as we can get **one** line by joining m collinear points

∴ Exact number of lines = total lines – lines lost due to collinearty of some given points + 1.

(d) If n sided figure is there then number of diagonals of this figure is equal to $^nC_2 - n$

Consider a seven sided figure as shown below

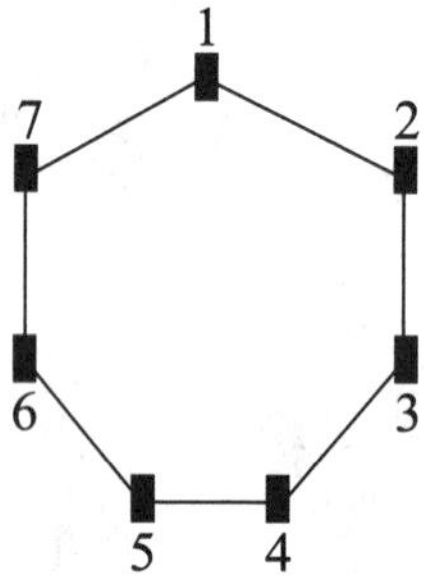

Seven sides mean seven vertices or seven points, shown as points (1, 2, 3, 4, 5, 6, and 7). Now to make a line we need atleast 2 points and we can join these two selected points and get a line.

So total lines that can be made using 7 pts = $^7C_2$

But out these $^7C_2$ lines 7 are sides of this figure.

∴ We have to subtract these 7 sides from total $^7C_2$ lines to get number of diagonals. (shown by dotted lines)

Hence number of diagonals = total lines made by using given n points - number of sides in that figure = $^7C_2 - 7$

(iii) Sum of the numbers formed by taking all the given n digits = (*Sum* of all *n* digits) *x* (*n* – 1)! *x* (111..upto n times)

These can be clearly understood by the following example

Example. How many 4 digit numbers can be made using the digits 1, 2, 3, 4 if repetition of digits is not allowed and what is the sum of these numbers.

Sol : 4 digit numbers that can be made using 1, 2, 3, 4 = 4 ! = 24

Now second part the question is we have to find the sum of these 24 numbers this can be found by the above given formula.

Hence the sum of the numbers is = 3! (1 + 2 + 3 + 4). (1111)

$$= 6 \times 10 \times 1111 = \mathbf{66660}$$

(iv) $^nC_r + {}^{n+1}C_r = {}^{n+1}C_{r+1}$

(v) $^nC_r = {}^nC_{n-r}$

Note : In all the above result we have not given any proof of them but we have explained these results by giving examples so that it may be easier for the student to learn and understand them properly. So without going into the detail of proofs for these results it is advisable that you must learn these results and as far as application of these results are concerned your are provided with fully explained solved examples. and then exercises which too are solved.

Types of Questions

Type I Simple Questions regarding arrangements or selection of object out of given objects without any condition or based on fundamental principal of counting.

Type II Arrangement of alphabets or persons or object keeping one or more than one, together or separate or according to given condition

Type III Numbers of triangles, lines, squares, parallelograms can be made by using given n points.

Type IV Distribution of different objects to different people.

Type V Based on circular arrangements.

TYPE I

Example 1. How many 4 digits numbers can be formed by using the digits 2,3,4,5,6 using each digit only once.

Sol : The given digits = 2, 3, 4, 5, 6.

Now the question is of permutation as we shall get different numbers just by arranging the given digits in different ways for example 2345, 6543, etc .

So total four digits numbers = $^5P_4$ (as here $n = 5, r = 4$)

$$= \frac{5!}{(5-4!)} = \frac{5!}{1!} = 5 \times 4 \times 3 \times 2 \times 1 = 120$$

Example 2. The number of ways of selecting 6 players out of 11 players.

Sol : Here we have to select persons therefore this question is of combination

Here $n = 11, \quad r = 6$

$\therefore$ number of ways of selecting of 6 players and of total 11

$$= {}^{11}C_6 = \frac{11!}{5!6!}$$

$$= \frac{11 \times 10 \times 9 \times 8 \times 7}{5 \times 4 \times 3 \times 2 \times 1}$$

$$= \mathbf{11 \times 2 \times 3 \times 7 = 462}$$

Example 3. 10 students compete for a race, the number of ways in which first three places can be taken

Sol : As the first three places can be taken by any of given 10 students so this is the question of permutation or ways of arrangement of students in first 3 places. Also we can not chose or select that which student will take Ist or IInd or IIIrd place.

$\therefore$ The first 3 places can be taken as

$$= {}^{10}P_3 = \frac{10!}{7!} = 10 \times 9 \times 8 = \mathbf{720\ ways}$$

Another method

Ist place IInd place IIIrd place

In first place any one from given 10 students can come. So ways of filling

1st place = 10

Similarly IInd place can be won by remaining 9 students. So ways of filling the

IInd place = 9

Also 3rd place can be won by remaining 8 student.So the ways of filling the

IIIrd place = 8

Whole event or first 3 places can be filled in

$$10 \times 9 \times 8 = \mathbf{720\ ways}$$

Example 4. There are 6 ways go from place A to place B and 5 ways to go from place B to place C. In how many ways a person can go from place A to place C.

Sol : A —— B —— C (6 ways from A to B, 5ways from B to C)

By **Fundamental Principle of Counting**

Ways of going from A to C is $6 \times 5 =$ **30 ways**

TYPE II

Example 5. How many different words can be made by using the letters of word **"MAXIMUM"**?

Sol : Given Word = M A X I M U M

total alphabets = 7, M = 3 times,

The question is of permutation as we can get different words by arranging the given alphabets in different places.

Here we cannot apply combination as for each word we make from the alphabets of word MAXIMUM the selection of alphabets remain the same.

The number of different words = $\frac{7!}{3!}$

$= 7 \times 6 \times 5 \times 4 =$ **840**

Example 6. How many different words can be made from the alphabets of word 'TEACHING' in which vowels are kept together.

Sol : Word = TEACHING

total alphabets = $n = 8$

vowels = A, E, I

To keep vowels together we will consider three vowels as single unit and rest alphabets separate

| AEI | TCHNG

So this will constitute 6 objects (5 consonants and three vowels)

Now if we arrange these six objects the vowels will always remain together

The arrangment = ${}^6P_6 \times {}^3P_3 = 4320$

Where ${}^6P_6$ = ways of arrangements of six objects out of six objects.

and ${}^3P_3$ = ways of arrangement of three vowels in a box.

Example 7. There are 8 true - false questions in a paper. The number of ways in which the paper can be answered if all the questions are compulsaroy.

Sol : Here it is very confusing to judge that whether to use combination or permutation. To avoid this we use fundamental principal of counting.

$\underset{\wedge}{Q_1}\ \underset{\wedge}{Q_2}\ \underset{\wedge}{Q_3} - - - - - \underset{\wedge}{Q_8}$ [^, symbol represents two choices]

Given eight question and each question has two choices as shown above

$\therefore$ Q_1 can be attempted in 2 ways

Similarly Q_2, Q_3 .... Q_8 also can be attempted in two ways each.

Hence simultaneously eight questions or whole paper can be attempted in

$$2 \times 2 \times 2 \times 2 \times 2 \times 2 \times 2 \times 2 = 2^8 \text{ ways}$$

Example 8. If $2\,{}^nP_3 = {}^{n+1}P_3$ then the value of n is

Sol : Given $\quad 2\,{}^nP_3 = {}^{n+1}P_3$

We know $\quad {}^nP_r = \dfrac{n!}{(n-r)!}$

$\therefore \quad 2.\dfrac{n!}{(n-3)!} = \dfrac{(n+1)!}{(n+1-3)!}$

or $\quad 2.\dfrac{n!}{(n-3)!} = \dfrac{(n+1)!}{(n-2)!}$

or $\quad 2.\dfrac{(n)!}{(n-3)!} = \dfrac{(n+1)(n)!}{(n-2)(n-3)!}$

or $\quad 2(n-2) = n+1$

or $\quad 2n-4 = n+1$

or $\quad n = 5$

Example 9. How many 5 digits numbers can be made by using digits 0, 1, 2, 3, 4, 5, 6, 7.

Sol : 5 digits number is like

___	___	___	___	___
10 thousands place	thousands place	hundreads place	tens place	unit place

Now first place i.e., 10 thousands place can be filled in 7 ways as 0 can not come at these place.

Now second place i.e., thousands place can also be filled in 7 ways as 0 can come at these place but the number that is already used at 10 thousands place can not be placed here.

Similarly next three places can be filled in 6, 5 and 4 ways respectively.

Hence total number of 5 digit numbers are = $7 \times 7 \times 6 \times 5 \times 4 =$ **5880**

TYPE III

Example 10. The number of triangles that can be made with 7 points on the circle.

Sol :

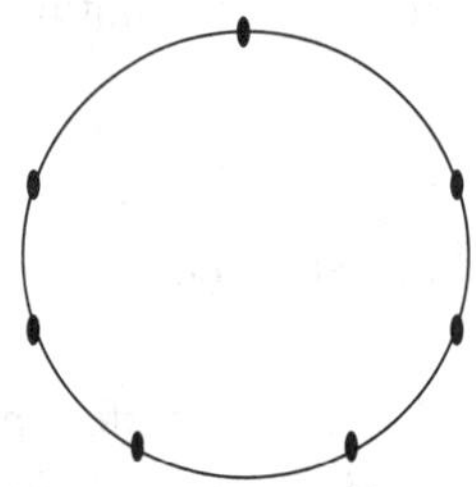

The question of combination as we cannot arrange the points to get different triangles. Instead have to select different points on the circle to form the triangles.

So we can select any three points and join them to get a triangle *i.e.*, number of triangles formed is equal to the number ways of choosing any three points on the circle.

$\therefore$ The number of triangles formed $= {}^7C_3 = \dfrac{7!}{4!3!} = \dfrac{7 \times 6 \times 5}{3 \times 2} = 35$

Example 11. The number of diagonals of polygon having 8 sides?

Sol :

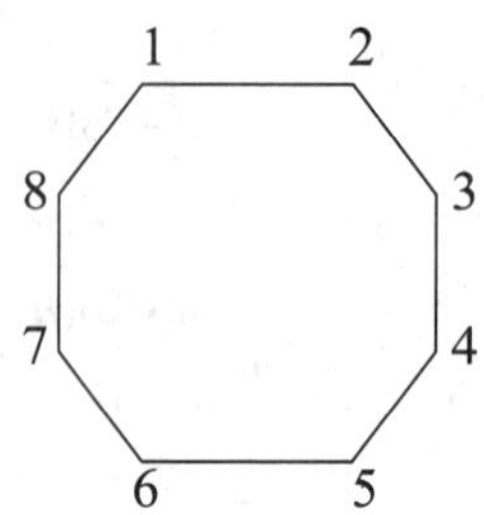

As it is clear from the figure that we have 8 different points in the figure.

Again like previous question this question is of combination as we will select any two points out these 8 points join them and get a line.

Therefore number of lines formed is equal to ways selecting any two points in the figure. But out of total lines formed, we have to subtract 8 sides to get the total number of diagonals.

$$\text{Total numbers diagonals} = {}^8C_2 - 8$$

$$= \frac{8!}{6!2!} - 8 = \frac{8 \times 7}{2} - 8 = 28 - 8 = 20$$

Example 12. There are 15 points in a plane no three of them which are in straight line except 6, all of which are in straight line. The number straight lines which can be drawn.

Sol : Again the question is of combination as ex. 11 and ex. 12. Here also we will select any two points to get a line.

If all 15 points are scattered then total number lines = ${}^{15}C_2$

But 6 points are in straight line if these 6 points are also scattered then number of lines formed by joining these 6 points

$$ {}^{6}C_2 = \frac{6!}{2!4!} = \frac{6 \times 5}{2} = 15 $$

So 15 lines are lost out of total ${}^{15}C_2$ lines but we can make one line by joining these 6 points as it is given in the question that these 6 points are in straight line.

total straight lines formed by 15 points in a plane out of which 6 are in straight line.

$$ = {}^{15}C_2 - {}^{6}C_2 + 1 = 91 $$

TYPE IV

Example 13. In how many ways can three rings be worn in 4 fingers if any finger can accommodate any number of rings?

Sol : Answer to the above question can be either $\mathbf{4^3}$ or $\mathbf{3^4}$. To avoid this confusion always remember a basic principle **"thing to be distributed always come in power."** So in this question as rings are to be distributed and they are 3 number so 3 will come in power and fingers that are 4 in numbers will come as base. So answer of these question is $\mathbf{4^3}$.

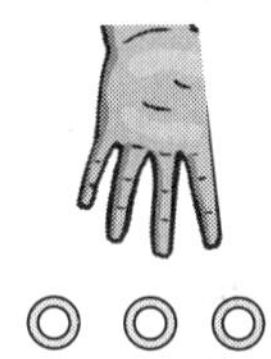

Fingers

Rings

Now we will do a few more important examples of the above said type.

Example 14. Five persons entered the lift cabin on ground floor of a 8-floor house. Suppose each of then can leave the cabin independently at any floor beginning with the first, the total number of ways in which each of the five persons can leave the cabin at any of the 7 floors is.

Sol : As things to be distributed are the **5 persons** that are to be settled or distributed in **7 floors** (as we are considering from the first floor). So **5** will come in **power** and **7** as **base** so the required number of ways = $\mathbf{7^5}$.

Example 15. A person has 6 friends to be invited for dinner through invitation cards, and he has 3 servants. In how many ways can be extend the invitation card?

Sol : As things to be distributed are the **cards** that are **6** in number so **6** will come in **power** and the **servants** that are **3** in number will come in **base**.

Required number of ways = $\mathbf{3^6}$.

Example 16. In how many ways. We can give 7 toys to 3 children. When each child is eligible for all the toys.

Sol : It may be now clear that things to be distributed are **7** in numbers so required number of ways = $\mathbf{3^7}$.

TYPE V

Example 17. The ways of arrangement of 4 men and 4 women around the circular table if no two women may sit together.

Sol :

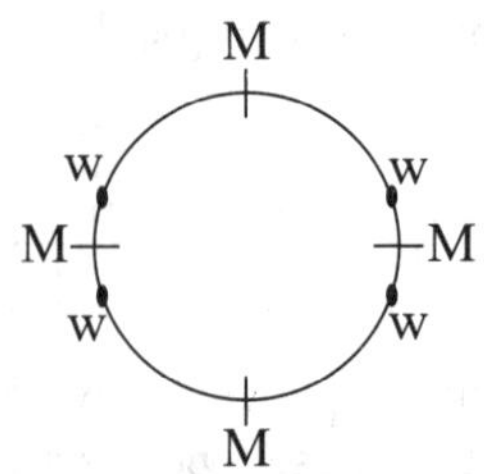

As shown in figure we can arrange the men first and then women or women first and then men.

Ways of arranging the men = (4 – 1)! = 3! = 3 × 2 = 6

In between them we can arrange women but now that arrangement cannot be considered as circular but linear.

∴ Ways of arranging women $^4P_4 = 4! = 4 \times 3 \times 2 \times 1$

$= \mathbf{24}$

Hence total number of ways of arrangment of 4 men and 4 women in circle = 24 × 6 = 144.

Example 18. Find the number of ways in which 10 different beads can be arranged to form a necklace.

Sol : As discussed earlier in case of necklace we can observe the things from both sides therefore

Required number of ways = $\frac{(n-1)!}{2} = \frac{\mathbf{9!}}{\mathbf{2}}$

EXERCISE

Q.1. How many ways we can direct 4 letters in 4 envelope in wrong ways ?

(a) 23 (b) 25 (c) 30 (d) 45

Q.2. Each of the three questions in a multiple choice test has four alternatives. The number ways to attempt the paper?

(a) 24 (b) 64 (c) 36 (d) 52

Q.3. From group 10 batsmen and 10 bowlers a team of 11 players is to be selected so that there may be 6 batsmen and 5 bowlers.

(a) 210 (b) 320 (c) 715 (d) $^{10}C_6 \times {}^{10}C_5$

Q.4. How many panel of 5 judges can be formed from 6 men and 4 women contaning at least two women.

(a) 156 (b) 148 (c) 186 (d) 110

Q.5. The number of ways in which 5 boys and 4 girls can be seated in a row so that each girl is between two boys?

(a) 1440 (b) 2880 (c) 1536 (d) 1584

Q.6. How many different words can be made from word PANPHONE if

(1) vowels are kept together (2) If vowels are not kept together.

(a) 1120,1020 (b) 1080, 9000 (c) 14700, 15810 (d) none

Q.7. How many different numbers can be made by using the digits 1, 2, 4, 5, 6, 7 using one or more digits at a time.

(a) 1859 (b) 1956 (c) 1942 (d) 1998

Q.8. There are 3 copies each of 4 different books. In how many ways they can be arranged on a shelf.

(a) $^{12}P_3$ (b) $\frac{12!}{3!3!3!3!}$ (c) $^{12}C_3$ (d) none

Q.9. Out of 10 blue and 8 white balls, 5 blue and 4 white balls can be drawn in how many ways.

(a) $^{10}C_5 \times {}^8C_4$ (b) 1420 (c) 1520 (d) 1480

Q.10. If $^nP_r = 720.\ ^nC_r$ then n is equal to

(a) 8 (b) 18 (c) 6 (d) none

Q.11. The number of lines that can be formed by joining 10 points out of which 5 are in straight line.

(a) 49 (b) 50 (c) 36 (d) 25

Q.12. The number of 5 digit telephone numbers having at least one of their digits repeated is

(a) 69760 (b) 58960 (c) 69850 (d) none

Q.13. In a class a candidate is required to pass in 5 different subjects. The number of ways in which he can fail.

(a) 28 (b) 29 (c) 31 (d) 25

Q.14. How many words can be formed from the letters of word **PROTECT** which are starting as well as ending with T.

(a) 140 (b) 120 (c) 165 (d) none

Q.15. How many 4 digit number can be made by using digits 0,1, 2, 3, 4, 5.

(a) 65 (b) 52 (c) 56 (d) 300

Q.16. A person has 12 friends out of which 7 are politicans and 5 are doctors. In how many ways he can invite 6 friends such that at least 3 politicians and atleast one doctor should be there in party.

(a) 596 (b) 582 (c) 800 (d) 805

Q.17. A man has five different work to do. In how many ways he can choose to finish one or more work at a time.

(a) 30 (b) 26 (c) 31 (d) none

Q.18. Given 5 different green papers and four different yellow papers. How many combination of letter pads can be made taking at least one green and one yellow paper.

(a) 465 (b) 456 (c) 650 (d) none

Q.19. In how many ways 7 boys be seated at a round table so that 2 particular boys are together

(a) 240 (b) 120 (c) 480 (d) none

Q.20. In how many ways 7 boys be seated at a round table so that 2 particular boys are separated ?

(a) 240 (b) 120 (c) 480 (d) none

Q.21. In how many ways can 12 things be divided in three sets having 4 things each?

(a) $\frac{12!}{(4!)^3 (3!)}$ (b) $\frac{12!}{(4!)^3}$ (c) $\frac{12!}{(4!)^3 (4!)}$ (d) none

Q.22. In how many ways can 12 different things be divided equally among 3 persons

(a) $\frac{12!}{(4!)^3 (3!)}$ (b) $\frac{12!}{(4!)^3}$ (c) $\frac{12!}{(4!)^3 (4!)}$ (d) none

Q.23. In how many ways can 4 boys and 4 girls can be seated in a row so that boys and girls are alternate

(a) 1152 (b) 576 (c) 2304 (d) none

SOLUTIONS

1. (a) Let letters and envelopes be as shown below

Letters L_1 L_2 L_3 L_4

Envelopes E_1 E_2 E_3 E_4

This is question of permutation because we can direct or arrange the letters (L_1, L_2, L_3, L_4) into four envelopes (E_1, E_2, E_3, E_4) in $^4P_4$ ways *i.e.*, $4! = 4 \times 3 \times 2 \times 1 = 24$ ways. But out these 24 ways only one way is correct *i.e.*, L_1, into E_1, L_2 into E_2, L_3 in E_3 and L_4 into E_4, rest 23 ways are wrong.

$\therefore$ Wrong ways = $^4P_4 - 1 = 24 - 1 =$ **23 ways**

2. (b) Now for each of three questions we have four way to attempt each equation

$\therefore$ to attempt the paper we have $4 \times 4 \times 4 = 4^3$ ways

Note : (Here we have applied the fundamental principle of counting)

3. (d) The given question is of combination as we have to select persons (players)

Now 6 batsmen can be selected from 10 batsmen in $^{10}C_6$ ways.

And 5 bowlers can be selected from 10 bowlers

in $^{10}C_5$ ways.

So team of 11 players can be selected in

$$= {}^{10}C_6 \times {}^{10}C_5 \text{ Ways.}$$

4. (c) Again the question is of selection and the possible combinations can be

(i) 2 women and 3 men

(ii) 3 women and 2 men

(iii) 4 women and 1 men

The ways of selection in (i) is $^4C_2 \times {}^6C_3 = 6 \times 20 = 120$

The ways of selection in (ii) posibility = $^4C_3 \times {}^6C_2 = 4 \times 15 = 60$

The ways of selection in (iii) posibility = $^4C_4 \times {}^6C_1 = 1 \times 6 = 6$

Hence total ways = $120 + 60 + 6 =$ **186 ways**

5. (b) This is question of arrangement as we have to arrange 5 boys and 4 girls in a row.

The possible arrangement are B G B G B G B G B

As shown above the 5 boys can be arranged in ${}^5P_5$ ways or 5!

It is also clear from figure that we have 4 spaces in between 5 boys.

∴ The ways of arrangement of 4 girls in these 4 places = ${}^4P_4 = 4!$

Hence total ways of arranging 5 boys and 4 girls = 5! × 4!

= 120 × 24 = **2880**

6. (b) This question is of permutation as we can get different words (Whether meaningful or not) just by placing given letters at different positions though the letters used are same.

(i) No. of vowels = A, O, E

Number of consonants = 5

Consider three vowels as single unit

i.e., (A, O, E) PNPHN we have 6 different items to be arranged

$$\therefore \quad \text{Their arrangement} = \frac{6!}{2!2!} \times {}^3P_3 = 1080$$

As we have 2N and 2P therefore we have divided 6! by 2! and 2! and multiplied by ${}^3P_3$ (are the ways to arrange vowels (in a box) themselves.)

(ii) If vowels are not kept together we have

Total arrangements without any condition - number of arrangements in which vowels are kept together.

$$\frac{8!}{2!2!} - \frac{6!}{2!2!} \times {}^3P_3 = 9000$$

7. (b) The question is of permutation or ways of arrangement. As we can make one digit numbers or two digit numbers, or 3 digit numbers or four digit numbers, or 5 digit numbers or 6 digit numbers.

$$\text{1 digit numbers} = {}^6P_1 = \frac{6!}{5!} = 6$$

$$\text{2 digit numbers} = {}^6P_2 = \frac{6!}{4!} = 6 \times 5 = 30$$

$$\text{3 digit numbers} = {}^6P_3 = \frac{6!}{3!} = 6 \times 5 \times 4 = 120$$

$$\text{4 digit numbers} = {}^6P_4 = \frac{6!}{2!} = 6 \times 5 \times 4 \times 3 = 360$$

$$\text{5 digit numbers} = {}^6P_5 = \frac{6!}{1!} = 720$$

$$\text{6 digit numbers} = {}^6P_6 = \frac{6!}{0!} = 720$$

$$\therefore \quad \text{total number} = 6 + 30 + 120 + 360 + 720 + 720 = \mathbf{1956}$$

8. (b) Total books = 12

out of which 4 different sets of 3 identical books are there

$$\text{way of arrangement} = \frac{12!}{3!3!3!3!}$$

9. (a) The question is of selection or combination.

Selecting 5 blue balls out of total ten blue balls = $^{10}C_5$ selecting 4 white balls out 8 = $^{8}C_4$

$$\text{total ways of selection} = {}^{10}C_5 \times {}^{8}C_4$$

10. (c)

$$^{n}P_r = 720\ {}^{n}C_r$$

$$\frac{n!}{(n-r)!} = 720 \times \frac{n!}{(n-r)!r!}$$

or $r! = 720$

or $r! = 6 \times 5 \times 4 \times 3 \times 2$

or $r! = 6!$

$\therefore$ $\boldsymbol{r = 6}$

11. (c) The question is of combination as we can select any two points from set of given points, join them and we will get a straight line.

If all the 10 points are scattered, the number of lines = $^{10}C_2$

But out of these 10 points 5 are in straight line, so if these 5 points are also scattered then number of lines = $^{5}C_2$

But these 5 points are in straight line so $^{5}C_2$ number of lines will be lost. As these 5 points are in straight line we will get one line by joining them.

$\therefore$ $$\text{total lines} = {}^{10}C_2 - {}^{5}C_2 + 1 = \mathbf{36}.$$

12. (a) We have 10 natural numbers or digits i.e., from 0 to 9

The question is of permutations as we can get different telephone numbers by just arranging the given digits at different places.

If all the digit are repeated then number of 5 digit telephone numbers

$$= 10 \times 10 \times 10 \times 10 \times 10 = 10^5$$

(As we can fill each place in 10 ways.)

But if digit are not repeated then ways to form telephone numbers

$$= 10 \times 9 \times 8 \times 7 \times 6,$$

As for first place we have 10 ways, for the second place or digit we have 9 ways to fill and so on.

$\therefore$ The number of ways to form telephone numbers with atleast one digit repeated

$$= 10^5 - 10 \times 9 \times 8 \times 7 \times 6 = \mathbf{69760}$$

13. (c) The candidate is fail if he is fail in one or two or three or four subjects or all the five subjects.

The question of combination as the candidate cannot arrange the subject to fail himself instead he is fail in any one, or two, or three, or four or five selected subjects out of total given subjects.

$\therefore$ Number of ways in which he can fail is $^{5}C_1 + {}^{5}C_2 + {}^{5}C_3 + {}^{5}C_4 + {}^{5}C_5 = 31$
where $^{5}C_1$ ways of selection of one subject out of five

$^{5}C_2$ ways of selection of two subject out of five

$^{5}C_3$ ways of selection of three subject out of five

$^5C_4$ways of selection of four subject out of five

$^5C_5$ways of selection of five subject out of five

14. (b) As the words have to start and end with ***T***. We will fix the ***T*** at the start and at the end as shown below.

T _ _ _ _ _ T

So we have 5 letters and 5 spaces for them.

$\therefore$ The ways of arrangement of 5 letters in 5 spaces = $^5P_5$ = **120**

(The question was of arrangement or permutation as one can get different words just by arranging the remain letters at given places or spaces.)

15. (d) Required number of 4 digit numbers = 5 × 5 × 4 × 3 = **300**.

16. (d) Possible selections or combinations are as follows

	Politicians	Doctors
(i)	3	3
(ii)	4	2
(iii)	5	1

$\therefore$ total ways of inviting six friends is

$$^7C_3 \times {}^5C_3 + {}^7C_4 \times {}^5C_2 + {}^7C_5 \times {}^5C_1 = 805$$

17. (c) He can select 1, 2, 3, 4, or 5 work at a time.

$\therefore$ Number of ways of completing the task by selecting 1, 2, 3, 4, or 5 tasks respectively is

$$= {}^5C_1 + {}^5C_2 + {}^5C_3 + {}^5C_4 + {}^5C_5$$

$$= 5 + 10 + 10 + 5 + 1 = 31$$

18. (a) One or more green paper can be chosen or selected out of 5 different green papers in

$$= {}^5C_1 + {}^5C_2 + {}^5C_3 + {}^5C_4 + {}^5C_5$$

$$= \text{ways} \qquad ...(1)$$

Similarly one or more yellow paper can be chosen out of 4 different yellow paper in

$$^4C_1 + {}^4C_2 + {}^4C_3 + {}^4C_4 = 2^4 - 1 = 15 \qquad ...(2)$$

$\therefore$ By fundamental principal of counting Both (1) and (2) can happen simultaneously or we can make letter pad in 31 × 15 = **465 ways**.

19. (a) Let the 2 particular boys be taken together as one unit.

Then the number of units will be 6.

They can sit around the table in 5! ways.

For each of this arrangement, the 2 can be interchanged in 2! ways.

$\therefore$ The total number of arrangement = 5! 2! = **240**.

20. (c) The arrangements that the 2 persons are separated = 6! – 5! 2! = **480**.

21. (a) As sets are not named therefore required number of ways = $\frac{\mathbf{12!}}{\mathbf{(4!)^3\ 3!}}$.

22. (b) As sets are named therefore required number of ways = $\frac{\mathbf{12!}}{\mathbf{(4!)^3}}$.

23. (a) The 4 boys and 4 girls can be seated in 4! 4! ways. There are further two cases when the arrangement starts with boy or girls. Thus the required no. of arrangements is 2 × 4! × 4! = **1152**.

Day 28

Probability

Today we **will discuss and understand the concept of Probability,** for this let us once again consider the simple case.

But first we should keep in Mind that we calculate or find the **probability of an event, Phenomenon or an experiment therefore in Mathematics Probability alone means nothing.**

Let us **consider an example** in which a person is given **10 bullets to hit the given target**. He fires the shots one by one. He finds that only **5 bullets have hit the target**

Now he again **fire** the another set of 10 bullets to the target again he finds that only **five** bullets have hit the target so his target hitting accuracy or chance of hitting the target is 50%, as he can hit five out of ten times.

So **again** if he tries to hit the target with 10 bullets we can very well say that most **probably** he will have five hits at the target out of ten **So this probably word stated above in mathematics is known as Probability**.

Here the event **was hitting the target**. He was given **total ten chances** and he managed to **hit only five shots accurately at the target** i.e., his accuracy in percentage was $\mathbf{\frac{5}{10} \times 100 = 50\%}$ or simply **Probability** of his hitting the target is 5 /10 i.e., 1/2

So, for example if this person is given 20 bullets, we can say that most **probably** he will fire 10 shots accurately or if he is given 100 bullets, most **Probably** he will fire 50 shots accurately. **But practically, one may suppose that it may not happen always that this person can hit with 50% accuracy. This we are talking in terms of chances only**. It may be possible that he may hit **4 out 10 or** 6 or 7 out of 10 times. But again the **most probably** seeing his record one can **say his accuracy is approximately 50%.**

Now if we have followed the above event or case or an **experiments then we can define and try to understand the probabilty mathematically** and also some of basic terms related to this **topic**.

(a) Probability (of any given event A) = P(A), is ratio of chances favourable to given event (A) to total chances of happening of given event (A). It is denoted by P(A) read as **Probability of (A)**

$$\text{or} \qquad \mathbf{P(A) = \frac{chances\ in\ favour\ of\ (A)}{total\ number\ of\ equally\ likely\ cases}}$$

Remember that to find the **probability of any event there must be some definite and unbiased outcomes** (results). Also there must not be **unknown external factors**

affecting the outcomes or result or given event or events. Otherwise **we may not arrive at correct answer mathematically as well as in real application of Probability.**

(b) Experiment: it is process which can be repeated and in any trial we can expect fair unbaised and definite outcome or result, though we may not be knowing the value outcome or what that outcomes would be unless. It is calculated or shown.

For e.g., **if coin is tossed**, we know that **either head** or tail, will come but what actually has come, can be known after seeing the tossed coin. so there was certain and fair outcome, no matter it could be either head or tail **Therefore tossing of a coin was an Experiment**.

(c) Sample space : It is the **set or collection of total fair possible outcomes of any given experiment.** Usually the set of outcomes is written with in bracket { }. For e.g., if a coin is tossed there can be only two outcomes i.e., head or tail.

$$\therefore \text{ Sample space} = \underset{\text{elements}}{\{\underline{HT}\}} \text{ or } \underset{\text{elements}}{\{\underline{TH}\}}$$

The elements of sample space are called sample points i.e., **(H,T)**

(d) Simple events : When there is **single possible outcome of an experiment it is a simple event** for e.g., if we toss a coin then **getting a tail is simple or** in case of tossing of two coins getting **only two tails** or **only two heads is simple event.**

(e) Compound events: When there are possibilities of more than one outcome for an event then it is a compound event. For e.g., in a toss of two coins the **probability** of getting at least one head is compound as we have two possibilities for this (1) one head, one tail (2) two heads i.e., it consists of two simple events.

(f) Exhaustive events : When there are **fair, unbiased and certain possible outcomes, events are exhaustive**. for e.g., in tossing of a coin, outcome will either be head or tail. Also there can be no other possible outcome

(g) Mutually exclusive events : The events are said to be Mutually exclusive events if only one and one can occur, not each of them can occur simultaneously for e.g., in a toss of coin either head will come or tail will come i.e., both head and tail cannot come up simultaneously

(h) Equally likely events : Events are equally likely if the probability of each possible outcome is equal. For e.g., in a toss of a coin, the **probability** of getting a head or a tail is same or equal.

(i) Independent events : Two events are said to be Independent if the outcome of first is not affecting or related to or having any outcome in common to second and vice-versa for e.g., consider a **tossing of coin and throwing a dice,** we get either a head or a tail in a toss and any number from 1 to 6 in throw of a dice. In both the cases outcome of each event is not having any effect on each other at all i.e., outcome of toss is not affecting the outcome of throw of dice. Therefore the events are independent.

(j) Dependent events : Two events are said to be dependent when the outcome is first is related to second or have outcomes in common or vice-versa.

For e.g., consider a **draw of king or spade from deck of 52 playing cards.** In deck we have one card which is king of spade i.e., a card which is common to both the events. So drawing of king or spade is **dependent** event.

(k) **Complementry events : Two mutually exclusive events are said to be complementry if the outcome of one is just opposite of other.** In other words one is about happening of that event and second is about not happening of that event. So if one is denoted by A (say) than other is denoted by $\bar{A}$ (read as $\bar{A}$) For e.g. in a toss of coin if **Probability of getting head** is asked, then **probability** of not getting head is complementry to getting a head.

(l) **Sum of events :** For two events **A** and **B**, **the sum of these two, which is compound means either event A or event B** occur i.e., atleast one of A or B occurs. It is denoted by **A + B** (read as **A** or **B**)

(m) **Product of events : For two events A and B the product of these two means that event A and B both will occur.** It is denoted by **AB** (read as **A** and **B**)

(n) **If Odds in favour** of A is a : b then it means, **ratio of chances that event will happen to chances that event will not happen**

$$\therefore \qquad \mathbf{P(A)} = \frac{a}{a+b} \text{ and } \mathbf{P}(\bar{A}) = \frac{b}{a+b} \text{ where a + b are total chances.}$$

If odd against **A** is a : b then, it is ratio of not happening of A to happening of A

$$\therefore \qquad \mathbf{P(A)} = \frac{a}{a+b} \text{ and } \mathbf{P}(\bar{A}) = \frac{b}{a+b}$$

When $a + b$ are total chances.

(o) For any event **A**, $0 \le$ **P(A)** ≤ 1

i.e., **Probability** of any event A is always greater than or equal to 0 and it is always less than or equal to 1

(p) (i) Total cases when n coins are tossed once or one coin is tossed n times = 2^n

(ii) Total cases when n dice are thrown once or one dice is thrown n times = 6^n

Note: In addition to the above results there are few other important theorems like addition theorem of probability, conditional probability, bayes rule etc. These are explained with the questions in the different types of questions categorised below.

Types of Questions

Type I **Simple questions based on concepts of permutation and combinations.**

Type II **Based on addition theorem of probability.**

Type III **Based on conditional probability.**

Type IV **Based on Bayes theorem.**

Type V **Based on Binomial distribution.**

TYPE I
(Simple questions)

Example 1. A dice is thrown what is the probability of getting a six.

Sol : Total cases = 6 [i.e., {1,2,3,4,5,6 As any number from 1 to 6 can come in throw of a dice.]

Favourable cases = 1 [as only one six as a number is there on dice.]

Probability (of getting 6) = $\frac{1}{6}$

Example 2. A coin is tossed, what is probability of getting a head

Sol : When a coin is tossed, total outcomes = 2 i.e., {H,T} favourable outcome for given event = {H} i.e., one outcome.

$\therefore$ P(getting a head) = $\frac{1}{2}$

Example 3. Two coins are tossed, what is the probability of getting two tails only.

Sol : Sample space, when two coin are tossed

= { H,H}, {T,T}, {T,H} {H,T}

$\therefore$ Total outcomes = 2^2 or 4

As it clear from the sample space, we have 1 outcome only in which there are exactly two tails.

$\therefore$ P(getting two tails) = $\frac{1}{4}$

Example 4. Two dices are thrown, then what is the probability of getting a total of 8.

Sol : Total possible outcomes when two dice are thrown = $6^2 = 36$

Let **A denotes the event of getting the sum of 8** Outcomes **favourable to** A are **{2,6}, {6,2}, {4,4}, {5,3}, {3,5}** i.e., 5 favourable outcomes.

$\therefore$ **P(A)** = $\frac{5}{36}$

Example 5. A bag contains 4 Red, 6 white and 7 blue balls. Two balls are drawn at random, what is the probability that a both are red.

Sol : Total balls = 4 + 6 + 7 = 17

Total cases = Ways of selecting two balls

$$= {}^{17}C_2 = \frac{17!}{15!2!}$$

Let A be event if drawing two red balls.

Also ways of selecting 2 red balls out of 4 red, which is also number of favourable outcomes for this question.

$$= {}^{4}C_2 = \frac{4!}{2!2!} = \frac{4\times 3}{2} = 6$$

$\therefore$ **P(A)** i.e., of drawing 2 red balls) = $\frac{6}{136} = \frac{3}{68}$

Example 6. A card is drawn at random from deck of 52 cards. What is the **probability** of getting.

(i) an ace

(ii) a red card

(iii) a diamond

Sol : Since total cards are 52

$\therefore$ total no. of outcomes = 52

(i) Favourable outcomes = 4 **[as 4 aces are there in of 52 cards]**

$\therefore$ **P** (an ace) = $\frac{4}{52} = \mathbf{\frac{1}{13}}$

(ii) Favourable outcomes = 26 **[as there are 26 blackand 26 red cards are in a pack of cards]**

$\therefore$ **P(A** red card) = $\frac{26}{52} = \mathbf{\frac{1}{2}}$

(iii) Favourable outcomes = 13 [as there are 13 cards each of 4 different suits
13 of diamond
13 of club
13 of Heart
13 of spade]

$\therefore$ **P** (a diamond) = $\frac{13}{52} = \mathbf{\frac{1}{4}}$

Example 7. Two cards are drawn at random from the pack of 52 cards. What is the **probability** of getting **a king or Queen**.

Sol : The drawing of king or Queen is an independent event, because there is no card in common between king and Queen or i.e., there is no card on which king and Queen are present or printed simultaneously.

Therefore by **additive rule** of probability.

P (King or Queen) = P (King) + P (Queen)

$$= \frac{4}{52} + \frac{4}{52} = \mathbf{\frac{2}{13}}$$

[as these are 4 kings and 4 queens in a pack of 52 cards]

Example 8. What is the chance that a leap year selected at random will have 53 sundays.

Sol : We know that a leap year has 366 days.

Also Number of sundays is equal to number of weeks in a year.

Therefore number of weeks = $7\overline{)366}$ (quotient 52)

35

16

14

2

So we have 52 complete weeks plus two days more Now these 2 days can have, any of the total (7) following possibilities as follows.

(i) Sunday, Monday

(ii) Monday, Tuesday

(iii) Tuesday, Wednesday

(iv) Wednesday, Thursday

(v) Thursday, Friday

(vi) Friday, Saturday

(vii) Saturday, Sunday

Favourable cases for required **probability** = 2 [i.e., **Possibilityno. (i) and no. (vii)**]

$$\therefore \quad \text{P\{leap year will have 53 sundays)} = \frac{2}{7}$$

Example 9. Two coins are tossed, then the **probability** of getting at least one head?

Sol :

$$\text{P (At least 1 head)} = 1 - \text{P (both tails)}$$

$$= 1 - \frac{1}{4} = \frac{3}{4}$$

As it is clear from sample space in the start of solution that we have one case in which having both the outcomes are tails out of total 4 outcomes.

$$\therefore \quad \text{P [both tails)} = \frac{1}{4}$$

Example 10. 3 coins are tossed together. Find the probability of getting

(i) Exactly one head (ii) At least 1 head (iii) At most 1 head.

Sol : Total number of cases when three coins are tossed together = $2^3 = 8$. Now we will solve each of the above questions one by one.

(i) The cases of exactly one head are {(HTT) (TTH) (THT)}

$$\therefore \quad \text{Probability of getting exactly one head} = \frac{3}{8}.$$

(ii) Probability of getting atleast one head = 1 – probability of getting no head

$$= 1 - \frac{1}{8} = \frac{7}{8}$$

(iii) The cases of at most one head are {(T T T) (HTT) (TTH) (THT)}

$$\therefore \quad \text{Probability of getting at most one head} = \frac{4}{8} = \frac{1}{2}.$$

Example 11. 3 dice are thrown together. Find the probability that total score is at most 15.

Sol : Total number of cases when 3 dice are thrown together = $6^3 = 216$.

Cases in which the score is more than 15 are {(4, 6, 6) (6, 4, 6) (6, 6, 4) (5, 5, 6) (5, 6, 5) (6, 5, 5) (6, 6, 6) (5, 6, 6) (6, 5, 6) (6, 6, 5)} = 10.

∴ Probability of getting score at most 15 $= 1 - \frac{10}{216} = \frac{206}{216} = \mathbf{\frac{103}{108}}$.

Example 12. From a group of a six **professors** and 3 lecturers. What is probability of **forming a team of 3**, having at least one lecturer.

Sol : Total persons = 6 + 3 = 9

Total cases = Ways of selecting 3 persons from 9 = $^9C_3$

Now at least one lecturer means we can select one two or three lecturers. So different possibilites are :

(i) 1 lecturer and 2 professors

(ii) 2 lecturers and 1 professor

(iii) 3 lecturers

∴ **P (team with at least one lecturer)** $= \frac{^3C_1 \times {}^6C_2 + {}^3C_2 \times {}^6C_1 + {}^3C_3}{^9C_3} = \mathbf{\frac{16}{21}}$

Example 13. In a box containing 100 blades,10 are defective. What is the probability that out of sample of 4 blades, none is defective.

Sol : Total blades = 100

No. of defective blades = 10

$$\textbf{P (a defective blade)} = \frac{\text{no. of defective blades}}{\text{total no. of blades}} = \frac{10}{100} = \frac{1}{10} \quad ...(i)$$

$$\textbf{P (not getting a defective blade)} = \textbf{1 – P (getting a defective blades)}$$

$$= 1 - \frac{1}{10} = \frac{9}{10} \quad ...(ii)$$

[As (ii) is complementry of (i), therefore P (ii) = 1– P (i)]

∴ **P (all blades are not defective)** $= \frac{9}{10} \times \frac{9}{10} \times \frac{9}{10} \times \frac{9}{10} = \mathbf{\left(\frac{9}{10}\right)^4}$.

Example 14. A card drawn from the pack of 50 cards numbered 1 to 50. The **probability** of drawing a number, which is a square.

Sol : Total possible outcomes = 50 [as there are 50 cards]

Favourable outcomes = Number of cards on which square of numbers from 1 to 50 are there.

Those can be [1, 4, 9, 16, 25, 36, 49] i.e., seven favourable outcomes.

∴ **P (drawing a number which is square)** $= \mathbf{\frac{7}{50}}$.

Example 15. What is **probability** of selecting a consonant from the word **PROBABIL-ITY.**

Sol : Total letters = 11 = Total cases.

No. of consonants = 7 = no. of favourable outcomes.

P(a consonant is selected) = $\frac{7}{11}$.

Example 16. If there are 3 children in a family, then the **probability** that there is one girl in the family is.

Sol : Total possibilites or sample space, whether the child is boy or girl in family of 3 children is {BBB, **BBG, BGB, GBB,** GGG, GGB, GBG, BGG}

So out of 8 possibility, there are 3 favourable cases [shown by underline] in which there is one girl.

$\therefore$ Required **probability = 3/8.**

TYPE II
(Based on addition theorem of probability)

Before doing this type we will first understand a few notations followed by the addition theorem of probability.

Notations

(i) P(A ∪ B) = Probability of **either A or B**

or

Probability of **atleast one of A or B**.]

(ii) P(A ∪ B ∪ C) = Probability of **either A or B or C**

or

Probability of **atleast one of A, B, C**.

(iii) P(A ∩ B) = Probability of **both A and B.**

(iv) P(A ∩ B ∩ C) = Probability of **simultaneous occurrence of A, B, C.**

(v) P(A ∩ $\bar{B}$) = Probability of **A but not B.**

(vi) P($\bar{A} \cap \bar{B}$) = Probability of **neither A nor B.**

Now we will do addition theorems of probability.

Theorem 1 : If A and B be any two events in a sample space S, then the probability of occurrence of at least one of the events A and B is given by

$$\boxed{P(A \cup B) = P(A) + P(B) - P(A \cap B)}$$

Proof : From set theory, we know that

$$n(A \cup B) = n(A) + n(B) - n(A \cap B)$$

Dividing both sides by n(S), we get

$$\frac{n(A \cup B)}{n(S)} = \frac{n(A)}{n(S)} + \frac{n(B)}{n(S)} - \frac{n(A \cap B)}{n(S)}$$

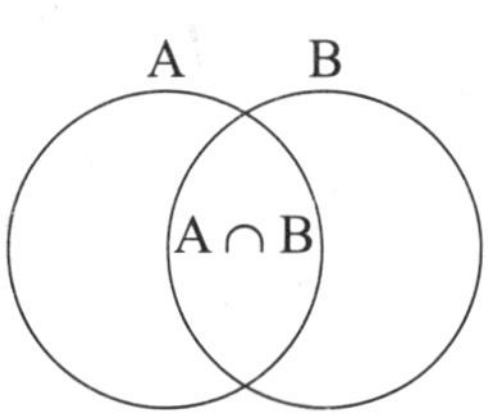

or

$$\boxed{P(A \cup B) = P(A) + P(B) - P(A \cap B)}$$

Note : (i) If A and B are mutually exclusive, events, then A ∩ B = φ and hence P(A ∩ B) = 0

$$\therefore \qquad P(A \cup B) = P(A)\frac{1}{8} + P(B)$$

(ii) Two events A and B are mutually exclusive if and only if

$$P(A \cup B) = P(A) + P(B)$$

(iii) $1 = P(S) = P(A \cup A') = P(A) + P(A')$ $[\because A \cap A' = \phi]$

or $$P(A') = 1 - P(A)$$

Theorem 2 : If A, B and C are any three events in a sample space S, then

$$\mathbf{P(A \cup B \cup C) = P(A) + P(B) + P(C) - P(A \cap B) - P(B \cap C)}$$
$$\mathbf{- P(C \cap A) + P(A \cap B \cap C)}$$

Now we will understand the above theorems with the help of following examples.

Example 17. If **P(A)** = 1/4, **P(B)** = 1/2, **P(A ∪ B)** = $\frac{1}{8}$ then **P(A ∩ B)** is

Sol : We know **P(A ∪ B) = P(A) + (B) – P(A ∩ B)**

$$\therefore \qquad \mathbf{P(A \cap B) = \frac{1}{4} + \frac{1}{2} - \frac{1}{8} = \frac{5}{8}}$$

Example 18. Twenty tickets are numbered from 1 to 20 and one of them is drawn at random, the **probability** that number is divisible by 2 or 5 is.

Sol : Total possible outcomes = 20

Let **A** denote the event that number drawn is multiply of 2 and these no. are (2, 4, 6, 8, 10, 12, 14, 16, 18, 20) ∴ $n(A) = 10$.

B denote the event that number drawn is multiple of 5 and these number are {5, 10, 15, 20}. ∴ $n(B) = 4$. So we see the from the sample space for multiple of 2 and 5 that two cases (multiple i.e., 10) are common i.e., **(A ∩ B)** = {10, 20}. ∴ $n(A \cap B) = 2$.

Therefore the events are not independent.

$$\textbf{Required probability} = \mathbf{P(A) + P(B) - P(A \cap B)}$$
$$= \frac{10}{20} + \frac{4}{20} - \frac{2}{20} = \frac{12}{20} = \frac{3}{5}$$

Example 19. One ticket is drawn at random out of 30 tickets numbered from 1 to 30. Find the probability that the number on the ticket is a multiple of 5 or 7.

Sol : Here random experiment is : Drawing one ticket out of 30 tickets numbered from 1 to 30.

Let S = the sample space

A = event of drawing a ticket having a number which is a multiple of 5

B = event of drawing a ticket having a number which is a multiple of 7.

Then
$$S = \{1, 2, 3, ..., 30] \qquad \therefore \quad n(S) = 30$$
$$A = \{5, 10, 15, 20, 25, 30] \quad \therefore \quad n(A) = 6$$
$$B = \{7, 14, 21, 28\} \qquad \therefore \quad n(B) = 4$$

Clearly $A \cap B = \phi$, hence events A and B are mutually exclusive.

Now $P(A \cup B) = P(A) + P(B) = \frac{6}{30} + \frac{4}{30} = \frac{1}{3}$.

Example 20. Two dice are tossed once. Find the probability of getting on even number on first dice, or a total of 8.

Sol : Here random experiment is : Tossing of two dice

Let S be the sample space. Then, $n(S) = 36$.

Let A = the event of getting an even number on first die

and, B = the event of getting a total of 8.

Then, $A \cup B$ = the event of getting an even number on first dice and a total of 8.

Now, $A = \{(2, 1), (2, 2), (2, 3), (2, 4), (2, 5), (2, 6), (4, 1), (4, 2), (4, 3), (4, 4), (4, 5), (4, 6), (6, 1), (6, 2), (6, 3), (6, 4), (6, 5), (6, 6)\}$;

and $B = \{(2, 6), (6, 2), (3, 5), (5, 3), (4, 4)\}$

$\therefore$ $A \cap B = \{(2, 6), (6, 2), (4, 4)\}$.

$\therefore$ $n(A) = 18$, $n(B) = 5$ and $n(A \cap B) = 3$.

Now $P(A) = \frac{n(A)}{n(S)} = \frac{18}{36} = \frac{1}{2}$; $P(B) = \frac{n(B)}{n(S)} = \frac{5}{36}$

and $P(A \cap B) = \frac{n(A \cap B)}{n(S)} = \frac{3}{36} = \frac{1}{12}$.

Now **required probability** $P(A \cup B) = P(A) + P(B) - P(A \cap B)$

$$= \left(\frac{1}{2} + \frac{5}{36} - \frac{1}{12}\right) = \frac{20}{36} = \frac{5}{9}.$$

Example 21. The probability of a horse A winning a race is $\frac{1}{3}$ and that of a horse B winning the same race is $\frac{1}{5}$, find the probability that (i) either of them will win, (ii) none of them will win.

Sol : Let E_1 = the event that horse A wins the race

and E_2 = the event that horse B wins the race

The $E_1 \cup E_2$ = the event that either of horses A and B will win the race

$(E_1 \cup E_2)'$ = the event that none of the horses A and B will win the race.

According to question $P(E_1) = \frac{1}{3}$, $P(E_2) = \frac{1}{5}$

Since both the horses A and B take part in the same race, therefore, both of them cannot win at a time.

Hence E_1 and E_2 are mutually exclusive events

$$\therefore \qquad P(E_1 \cup E_2) = P(E_1) + P(E_2) = \frac{1}{3} + \frac{1}{5} = \frac{8}{15}$$

$$\text{and} \qquad P(E_1 \cup E_2)' = 1 - P(E_1 \cup E_2) = 1 - \frac{8}{15} = \frac{7}{15}$$

TYPE III
(Based on conditional probability)

First we will define what the conditional probability is then we will understand it with the help of related examples.

Multiplication Theorem of Probability or Theorem of Compound Probability

If A and B are any two events, then

$P(A \cap B) = P(B) \,.\, P\left(\frac{A}{B}\right)$, where $B \neq \phi$ and $P\left(\frac{A}{B}\right)$ denotes the probability of occurrence of event A when B has already occurred.]

If A and B are independent events, then probability of occurrence of event A is not affected by occurrence or nor occurrence of event B, therefore

$$P\left(\frac{A}{B}\right) = P(A)$$

Hence, $\qquad P(A \cap B) = P(B) \,.\, P(A)$

Thus $\qquad P(A \cap B) = P(A) \,.\, P(B)$

where A and B are independent events.

Note : **1.** $P(B/A) = \dfrac{P(A \cap B)}{P(A)}$

or $P(A \cap B) = P(A) \,.\, P(B/A)$

2. If A and B are independent events, then $P\left(\frac{B}{A}\right) = P(B)$

3. Two events A and B are independent if and only if

$P(A \cap B)$ or $P(AB) = P(A) \,.\, P(B)$

4. If A, B and C are any three independent events, then

$$P(A \cap B \cap C) = P(A) \,.\, P(B) \,.\, P(C)$$

Now we will do a few examples to understand the above theorems.

Example 22. A and B are any two events such that $P(A) = 0.4$, $P(B) = 0.8$, $P\left(\frac{B}{A}\right) = 0.6$. Find the probability of at least one of A or B.

Sol : $\qquad P(A \cap B) = P(A) \,.\, P\left(\frac{B}{A}\right) = 0.4 \times 0.6 = 0.24$

Also $\qquad P(A \cup B) = P(A) + P(B) - P(A \cap B)$

$$= 0.4 + 0.8 - 0.24 = \mathbf{0.96}.$$

Example 23. In an exams 20% of the students failed in maths, 10% failed in physics, 5% in both maths and physics. A student is selected randomly find the probability that

(i) He failed in maths it is known that he had already failed in physics

(ii) He failed in physics it is known that he had already failed in maths.

Sol : $P(A) = 0.2$, $P(B) = 0.1$, $P(A \cap B) = 0.05$

(i) $$P\left(\frac{B}{A}\right) = \frac{P(A \cap B)}{P(B)} = \frac{0.05}{0.1} = \mathbf{0.5}$$

(ii) $$P\left(\frac{B}{A}\right) = \frac{P(A \cap B)}{P(A)} = \frac{0.05}{0.2} = \mathbf{0.25}$$

TYPE IV
(Based on bayes theorem)

Bayes theorem is used when outcome is given and we have to find the probability of the outcome in a particular case. This can be clearly understood by the following examples.

Example 24. In a bolt factory, machines A, B and C manufacture 25%, 35% and 40% respectively. Of the total of their output 5%, 4% and 2% are defective. A bolt is drawn and is found to be defective. What are the probabilities that it was manufactured by the machine A.

Sol : Here one bolt has been drawn and its result is known that the bolt drawn is defective, therefore Bayes theorem should be used.

In this case we will first find the total probability of a bolt to be defective and then find what is the probability that this defective bolt has been manufactured by machine A. The defective bolt should have been manufactured by either machine A or machine B or machine C. If this bolt is from machine A and it is also defective then this case can be mathematically be represented as 0.25×0.05 as first the probability of a bolt to be from A is 25% that is 0.25 and further the probability of a bolt that is manufactured by machine A to be defective is 0.05 similarly if it is from B then the probability is 0.35×0.04 and if it is from C then its probabilities 0.40×0.02 hence total probability of a bolt to be defective is $0.25 \times 0.05 + 0.35 \times 0.04 + 0.40 \times 0.02$. We will put this value in the denominator and our favourable case that it is manufactured by machine A whose value is 0.25×0.05 is to be placed in numerator.

$$\text{Hence required probability} = \frac{0.25 \times 0.05}{0.25 \times 0.05 + 0.35 \times 0.04 + 0.40 \times 0.02} = \mathbf{\frac{25}{69}}$$

Example 25. Bag X contains 2 white and 3 red balls and bag Y contains 4 white and 5 red balls. One ball is drawn at random from one of the bags and is found to be red. Find the probability that it was from bag Y.

Sol : Here result is known that the ball drawn is red, therefore Bayes theorem should be used.

$$\text{Required probability} = \frac{\frac{1}{2}\cdot\frac{5}{9}}{\frac{1}{2}\cdot\frac{5}{9}+\frac{1}{2}\cdot\frac{3}{5}} = \mathbf{\frac{25}{52}}$$

Example 26. A man is known to speak the truth 3 out of 4 times. He throws a dice and reports that it is a six. Find the probability that it is actually a six.

Sol : As the outcome is given so Bayes theorem is used

$$\text{Required probability} = \frac{\frac{1}{6}\cdot\frac{3}{4}}{\frac{1}{6}\cdot\frac{3}{4}+\frac{5}{6}\cdot\frac{1}{4}} = \mathbf{\frac{3}{8}}$$

TYPE V
(Based on Binomial distribution)

This concept we will understand with the help of following examples.

Example 27. A coin is tossed 6 times. what is the **probability** of getting head three times.

Sol : In this question, we will not follow the simple method to find required probability which is

$$= \frac{\textbf{number of favourable cases for given event}}{\textbf{total no. of cases}}$$

because the sample space is very large. Instead of this method, we will adopt another method given by Bernoulli it is also known as finding the **probability** by Binomial distribution method, which is

P[of getting r success out of n trails] = $^nC_r p^r q^{n-r}$

Where p and q denotes the success and failure in single trial of that event respectively.

Remember that usually we use this result when the number of trials are large.

Here $n = 6$ $r = 3$ [as 3 success are required]

$p = 1/2$ $q = 1/2$ [since prob. of getting a head or a tail is 1/2]

$$\therefore \quad \text{P (of getting 3 head]} = {}^6C_3\left(\frac{1}{2}\right)^3\left(\frac{1}{2}\right)^3 = \mathbf{\frac{5}{16}}$$

Example 28. 10 coins are tossed what is the probability that exactly 5 heads appear?

Sol : Let E = the event of occurrence of head on one coin

Then $p = P(E) = \frac{1}{2}$ and $q = 1 - p = \frac{1}{2}$

Here $n = 10$(since 10 coins have been tossed)

Now $P(r)$ = probability that event E will occur exactly r times in 5 trials

$= {}^nC_r\, p^r\, q^{n-r}$

$$\therefore \text{ Required probability} = {}^{10}C_5\, p^5\, q^5 = \frac{10!}{5!5!}\cdot\left(\frac{1}{2}\right)^5\left(\frac{1}{2}\right)^5 = \frac{63}{256}$$

Example 29. 10 coins are tossed. Find the probability of getting at least 8 heads.

Sol : Required probability $= P(8) + P(9) + P(10)$

$$= {}^{10}C_8\, p^8 q^2 + {}^{10}C_9\, p^9 q^1 + {}^{10}C_{10}\, p^{10} q^0$$

$$= 45\left(\frac{1}{2}\right)^8 \cdot \left(\frac{1}{2}\right)^2 + 10\left(\frac{1}{2}\right)^9 \cdot \frac{1}{2} + 1 \cdot \left(\frac{1}{2}\right)^{10}$$

$$= 56 \cdot \left(\frac{1}{2}\right)^{10} = \frac{7}{2^7} = \mathbf{\frac{7}{128}}$$

EXERCISE

Q.1. A pair of dice is thrown, then the **Probability** of not getting the sum of 9.

(a) $\frac{2}{9}$ (b) $\frac{8}{9}$ (c) $\frac{1}{9}$ (d) $\frac{7}{9}$

Q.2. Two persons A and B appear for an interview if the **probability** of their selection is $\frac{1}{3}$ and $\frac{1}{5}$ respectively then **probability** that none of them will be selected is

(a) $\frac{25}{41}$ (b) $\frac{9}{16}$ (c) $\frac{8}{15}$ (d) $\frac{8}{15}$

Q.3. The **probability P(ϕ)** and **P(s)** is equal to.

(a) 0,1 (b) 1, $\frac{1}{2}$ (c) $\left(\frac{1}{4}\right)^2 \times \left(\frac{1}{2}\right)^3$ (d) none

Q.4. If the odds in favour of eve nt **A** are 3:8 then $\mathbf{P(\bar{A})}$

(a) $\frac{8}{11}$ (b) $\frac{3}{11}$ (c) $\frac{1}{11}$ (d) $\frac{7}{11}$

Q.5. A card is drawn from a pack of 52 cards at random. what is **probability** that it is either a heart or a jack.

(a) $\frac{16}{52}$ (b) $\frac{7}{52}$ (c) $\frac{4}{51}$ (d) $\frac{17}{18}$

Q.6. Three coins are tossed,what is the **probability** of getting at the most two heads.

(a) $\frac{7}{8}$ (b) $\frac{1}{9}$ (c) $\frac{2}{19}$ (d) $\left(\frac{1}{4}\right)^2 \times \left(\frac{1}{2}\right)^3$

Q.7. Two cards are drawn at random from the pack of 52 cards, then what is the **probability** of drawing two aces if the chosen card is not replaced.

(a) $\frac{7}{8}$ (b) $\frac{4}{52} \times \frac{3}{51}$ (c) $\frac{2}{19}$ (d) $\left(\frac{1}{4}\right)^2 \times \left(\frac{1}{2}\right)^3$

Q.8. A dice is so formed that probability of getting any number its proportional to that number. Find the probability of getting an even number when dices thrown

(a) $\frac{4}{7}$ (b) $\frac{4}{21}$ (c) $\frac{1}{2}$ (d) none

Q.9. Twenty tickets are numbered from 1 to 20 and one of them is drawn at random, the **probability** that number is divisible by 3 or 5 is

(a) $\frac{1}{5}$ (b) $\frac{9}{20}$ (c) $\frac{3}{5}$ (d) $\frac{4}{5}$

Q.10. In a single throw of pair of dice what is **probability** of getting a doublet.

(a) $\frac{3}{7}$ (b) $\frac{1}{2}$ (c) $\frac{2}{5}$ (d) $\frac{1}{6}$

Q.11. For independent events A and B.

(a) $\mathbf{P(A \cap B) = P(A) \times P(B)}$

(b) $\mathbf{P(A \cup B) = P(A) + P(B) - P(A \cap B)}$

(c) $\mathbf{P(B/A) = \frac{P(A \cap B)}{P(B)}}$

(d) $\mathbf{P(A/B) = \frac{P(A \cap B)}{P(B)}}$

Q.12. A bag contains 5 brown and 4 white shirts A man pulls out 2 shirts. The **probability** that they are of the same colour.

(a) $\frac{19}{36}$ (b) $\frac{7}{8}$ (c) $\frac{2}{7}$ (d) $\frac{4}{9}$

Q.13. In a single throw of pair of dice. The **probability** of getting a total of at least 10 is

(a) $\frac{1}{4}$ (b) $\frac{1}{5}$ (c) $\frac{1}{6}$ (d) $\frac{5}{6}$

Q.14. A single letter is selected at random from word **"PROBABILITY"** the probability that it is vowel is

(a) $\frac{3}{11}$ (b) $\frac{2}{11}$ (c) $\frac{4}{11}$ (d) 0

Q.15. In a class of 10 students 4 are boys and the rest are girls. The probability that a student selected will be a girl is

(a) $\frac{1}{5}$ (b) $\frac{2}{3}$ (c) $\frac{3}{5}$ (d) $\frac{4}{5}$

Q.16. A speaks truth in 75% cases and B speaks truth in 80% cases. The **probability** that they contradict each other in a statement is

(a) $\frac{2}{27}$ (b) $\frac{1}{19}$ (c) $\frac{7}{20}$ (d) $\frac{7}{18}$

Q.17. Two persons A and B throw a dice alternately till one of them gets a 5 and wins the game. Find their respective **probability** of winning the game if A starts the game.

(a) $\frac{1}{11}, \frac{1}{10}$ (b) $\frac{6}{11}, \frac{5}{10}$ (c) $\frac{6}{11}, \frac{5}{10}$ (d) none

Q.18. Three prize are to be distributed among three persons at random the chances that none of persons gets all the prizes is.

(a) $\frac{210}{216}$ (b) $\frac{19}{37}$ (c) $\frac{1}{7}$ (d) $\frac{170}{210}$

Q.19. Three numbers are chosen from 1 to 20 the **probability** that they are consecutive is

(a) $\frac{190}{311}$ (b) $\frac{210}{216}$ (c) $\frac{3}{190}$ (d) $\frac{1}{8}$

Q.20. A pair of dice is thrown once and total of 8 has appeared. what is the chance that odd no. appear of each dice.

(a) $\frac{1}{9}$ (b) $\frac{2}{10}$ (c) $\frac{1}{7}$ (d) $\frac{2}{5}$

Q.21. From first 1000 natural numbers a natural number is selected randomly. Find the probability that it is either a perfect square or a perfect cube or a perfect 4th power.

(a) $\frac{19}{1000}$ (b) $\frac{38}{500}$ (c) $\frac{19}{500}$ (d) $\frac{2}{5}$

SOLUTIONS

1. (b) When two dice are thrown total possible outcomes = $6^2 = 36$

Possible cases of getting total of 9 = {(6,3), (3,6), (5,4), (4,5)}

i.e., 4 cases.

P(of getting the total of 9) = $\frac{4}{36} = \frac{1}{9}$

∴ **P(not getting the total of 9)**

= **1 – P(of getting the total of 9)**

$= 1 - \frac{1}{9} = \frac{8}{9}$

2. (c)

$$\mathbf{P(A)} = \frac{1}{3}.\mathbf{P(B)}\frac{1}{5},$$

$$\mathbf{P(\bar{A})} = 1 - \frac{1}{3} = \frac{2}{3}$$

$$\mathbf{P(\bar{B})} = 1 - \frac{1}{5} = \frac{4}{5}$$

Required probability = **P(neither is selected)**

$$= \mathbf{P(\bar{A} \times \bar{B}) = P(\bar{A}) \times P(\bar{B})} = \frac{2}{3} \times \frac{4}{5} = \frac{8}{15}$$

As the events are independent. Because selection or rejection of A or B is not going to effect each other at all.

3. (a) Where,

P(ϕ) = Probability of nothing or no event

P(S) = happening of all the events.

4. (a) Odds or chances in favour of event **A** = 3:8

$$\therefore \quad P(A) = \frac{3}{11} \text{ and } P(\bar{A}) = \frac{8}{11}$$

5. (a) Let **A = event of drawing a jack and B = event of drawing the heart.**

Here the events **A** and **B** are not independent as we have a jack which is of heart therefore we have to subtract this probability of drawing this common card from required **probability**.

$\therefore$ **P(heart or Jack) = P(heart) + P(Jack) – P(Jack of heart)**

$$= \frac{13}{52} + \frac{4}{52} - \frac{1}{52} = \frac{16}{52}$$

6. (a) Total possible outcomes when 3 coins are tossed = $2^3 = 8$

Here at the most two heads means, we can get no head on all the three coins tossed, or one head and rest two tails on all the three coins or two heads and one tail on the three tossed coins.

$\therefore$ Required probability = P(at the most two heads)

= P(no head) + P (one head) + P(two head) ...(1)

= **1 – P(3 heads)** ...(2)

$$= 1 - \frac{1}{8} = \frac{7}{8}$$

7. (b) Required **probability** $= \frac{4}{52} \times \frac{3}{51}$

Because at first we have four aces out of 52 cards and after drawing one ace, we have 3 aces out of 51 cards.

8. (a)

$$P(x) \propto x$$

$$P(x) = k(x)$$

$$P(1) + P(2) + P(3) + P(4) + P(5) + P(6) = 1$$

$$k + 2k + 3k + 4k + 5k + 6k = 1$$

$$k = \frac{1}{21}$$

$\therefore$ Probability of getting an even number $= P(2) + P(4) + P(6) = 2k + 4k + 6k = 12k$ $= \frac{12}{21} = \frac{4}{7}$.

9. (b) Total possible outcomes = 20

Let **A** denote the event that number drawn is multiply of 3 and these no. are (3, 6, 9, 12, 15, 18)

B denote the event that no drawn is multiple of 5 and these no. are {5, 10, 15, 20} So we see the from the sample space for multiple of 3 and 5 that one case (multiple i.e., 15) is common i.e., **(A $\cap$ B)** = {15} therefore the events are not independent.

Required probability = P(A) + P(B) – P(A $\cap$ B)

$$= \frac{6}{20} + \frac{4}{20} - \frac{1}{20} = \frac{9}{20}$$

10. (d) Total possible outcomes = $6^2 = 36$

Favourable cases = {(1, 1) (2, 2) (3, 3) (4, 4) (5, 5) (6, 6)}

$= 6$ cases

(doublet means the number on each dice is same.)

Required probability $= \frac{6}{36} = \frac{1}{6}$

11. (a)

12. (a) Total cases = Ways of drawing two shirts out of 9 shirts.

$= {}^9C_2 = \frac{9!}{7!2!} = \frac{9 \times 8}{2} = \mathbf{36}$

Since the shirts should be of same colour, it can be either 2 brown shirts or 2 white shirts.

∴ Required probability = **P(2 brown shirts) + P(2 white shirts)**

$= \frac{{}^5C_2}{36} + \frac{{}^4C_2}{36} = \frac{4}{9}$

13. (c) Total cases or possible out comes when pair of dice is thrown

$= 6^2 = 36$

Here getting the total of at least ten means we can have total of 10 or total of 11 or total of 12 (i.e., ≥ 10)

Required probability = **P(total of 10) + P(total of 11) + P(total of 12)**

Sample space for total of 10 = { (5,5), (6,4) (4,6)} = 3 cases

Sample spaces or outcomes favourable for the total of 11

= {6,5} (5,6} = 2 cases

Sample space or outcomes favourable for total of 12

= {6,6} = 1 case

∴ Required probability $= \frac{3}{36} + \frac{2}{36} + \frac{1}{36} = \frac{6}{36} = \mathbf{\frac{1}{6}}$

14. (c) As there are 4 vowels out of 11 words therefore required probability $= \frac{4}{11}$

15. (c)

16. (c) Let α = event that A speaks truth and $\bar{\alpha}$ = event that A tells lie.

β = event that B speaks truth and $\bar{\beta}$ = event that B tells lie.

∴ Both A and B will contradict each other if A speaks truth and B tells lie or B speaks truth and A tells the lie. Since A speaks truth in 75% cases

Therefore $\mathbf{P(\alpha)} = \frac{75}{100}$ and $P(\bar{\alpha}) = \mathbf{1} - P(\alpha) = \frac{25}{100}$

$\mathbf{P(\beta)} = \frac{80}{100}$ and $P(\bar{\beta})$

$= \mathbf{1 - P(\beta)} = \frac{20}{100}$

Required probability = P(α and $\bar{\beta}$) + P($\bar{\alpha}$ and β) $= \frac{75}{100} \times \frac{20}{100} + \frac{25}{100} \times \frac{80}{100} = \mathbf{\frac{7}{20}}$

17. (d) Let **E** be event that **A** gets 5 and F be event that B gets 5

$\therefore$ **P(E)** $= \frac{1}{6}$, $\mathbf{P(\bar{E})} = 1 - \frac{1}{6} = \frac{5}{6}$

and **P(F)** $= \frac{1}{6}$, $\mathbf{P(\bar{F})} = 1 - \frac{1}{6} = \frac{5}{6}$

If A starts the game then A can win in 1st game or 3rd game, or 5th game and so on. B can get a chance to win in 2nd try or 4th try or 6th try and so on.

$$\mathbf{P(A)} = \mathbf{P(E) + P(\bar{E}\,\bar{F}\,E) + P(\bar{E}\,\bar{F}\,\bar{E}\,\bar{F}\,E)}$$

$$= \frac{1}{6} + \frac{5}{6}\cdot\frac{5}{6}\cdot\frac{1}{6} + \left(\frac{5}{6}\right)^4 \cdot \frac{1}{6} + \ldots$$

$$= \frac{1}{6}\left[1 + \left(\frac{5}{6}\right)^2 + \left(\frac{5}{6}\right)^4 + \ldots\right]$$

$$= \frac{1}{6} \times \left[\frac{1}{1 - \left(\frac{5}{6}\right)^2}\right]$$ [The series is infinite G.P. in which

$$S_\infty = \frac{a}{1-r}, \text{ here } a = 1, \quad \left[r + \frac{5}{6}\right]$$

$$= \frac{1}{6} \times \frac{1}{1 - \frac{25}{36}}$$

$$= \frac{1}{6} \times \frac{1}{\frac{11}{36}} = \frac{1}{6} \times \frac{36}{11} = \frac{6}{11}$$

$$\mathbf{P(B)} = \mathbf{1 - P(A)} = 1 - \frac{6}{11} = \mathbf{\frac{5}{11}}$$

18. **(a)** **Req. prob.** $=$ **1 – P(all the prizes go to one person)**

$$= 1 - \frac{6}{216} = \mathbf{\frac{210}{216}}$$

Since total no. of ways to distribute 3 prizes = 6 × 6 × 6 = 216

Also way of distributing the 3 prizes to same person = 6

19. **(c)**Total cases or outcomes = ways of selecting any three balls out of 20 balls. = $^{20}C_3$ [Because when we have to select any no. of objects from total given objects at once or all at a time,then we have to use the method of combination for this. And if we have select objects one by one for *e.g.*, in the above question the total cases will be = $\frac{1}{20} \times \frac{1}{20} \times \frac{1}{20} = \frac{1}{80,00}$

Favourable cases or outcomes can be = (1, 2, 3) (2, 3, 4), (3, 4, 5).......(18, 19, 20) = 18 cases

Req. prob. $= \frac{18}{^{20}C_3} = \frac{3}{190}$

20. **(d)** Let A be event showing total of 8. B be event showing odd no. on each dice.

we have to find **P(B/A)**

we know $$\mathbf{P(B/A)} = \frac{\mathbf{P(A \cap B)}}{\mathbf{P(A)}}$$

Sample space or outcome for A i.e., total of 8 = {2, 6}, (6, 2), (4, 4), (5, 3), (3, 5) = 5

Sample space or outcomes for B i.e., showing odd no. each dice = {(3, 5)(5, 3)}

$$\therefore \quad \boldsymbol{n}\mathbf{(A \cap B)} = 2$$

$$\textbf{Req. prob.} = \frac{\frac{2}{36}}{\frac{5}{36}} = \frac{2}{5}$$

21. (c) As there are 31 perfect squares in first 1000 natural numbers therefore $n(A) = 31$.

Similarly, there are 10 perfect cubes in first 1000 natural numbers therefore $n(B) = 10$.

Similarly, there are 5 fourth power perfect four power in first 1000 natural numbers therefore $n(C) = 5$.

As 1 is present in all the above cases therefore $n(A \cap B \cap C) = 1$.

$$\therefore \quad P(A \cup B \cup C) = P(A) + P(B) + P(C) - P(A \cap B) - P(B \cap C) - P(C \cap A) + P(A \cap B \cap C)$$

$$= \frac{31}{1000} + \frac{10}{1000} + \frac{5}{1000} - \frac{3}{1000} - \frac{1}{1000} - \frac{5}{1000} + \frac{1}{1000}$$

$$= \frac{38}{1000} = \mathbf{\frac{19}{500}}$$

Day 29

Sequences and Series

Today we will do the topic of **Arithmetic progressions, Geometric progressions** and **a few more types of sequences of numbers** having **special property or nature attached to them.**

In English **progression** means **going forward** but in **Mathematics** when some **special property is added to series of number,** it may result into **A. P or G. P or any other form of number series** we will discuss each or other series **{A P or G P or any other series}** one by one and in detail.

Lets as now consider a **few examples** which will help us to understand this topic in a better way

Example 1. Let a man starts **saving** some money. He deposits **1 rupee** on first day, **2 rupees** on second day, **3 rupees** on third day and so on **upto two hunderd days**. So if we try to calculate his savings after two hundred days and that too by **normal method of addition** i.e. by adding amount of each day upto two hundred days. It will become very time consumming task.

To make this calculation easy we have one formula on the basis of which we can calculate his saving at the end of two hundred days is given as follows:

$$1 + 2 + 3 + 4 + 5 + \ldots\ldots\ldots\ldots + \text{upto 200 terms}$$

$$= \frac{200}{2}[1 + (200 - 1) \times 1] = 20100$$

So his savings at the end of two hundred days is **rupees twenty thousand and one hundred only**. In this we have applied the formula for sum of n terms of A. P which we will also discuss later on in this chapter

Example 2. Let a man save rupees 2 on first day, rupees 4 on second day, rupees 8 on third day and so on. Now his saving after two hundred days is as follows:

$$2 + 4 + 8 + 16 + \ldots\ldots\ldots + \text{upto 200 terms}$$

$$= \frac{2(2^{200} - 1)}{2 - 1} = 2.(2^{200} - 1) \text{ rupees only}$$

Which is very large in terms of amount of money. But instead of applying this formula if we had tried to find out this sum by normal addition, it would have been a very time consuming exercise. So here we have used the formula of sum of *n*-terms of G.P. We will also discuss this formula later on this chapter.

Now it is clear that by these two examples that how these concepts are useful in solving very time consuming problems in simpler ways. **So after seeing the importance of these concepts in solving above examples now we will learn the mathematical at concepts** behind these type of number series which follows some definite pattern.

In this topic we will discuss **Five type of series**

1. **Arithmetic Progression or A. P.**
2. **Geometric Progression or G. P.**
3. **Harmonic Progression or H.P.**
4. **Arithmetico-Geometric Progression or A. G. P.**
5. **Miscellaneous type of series**

First we will start with an **A. P.** series

PART I
Arithmetic Progressions

(i) Consider the sequences of numbers **1, 3, 5, 7, 9, 11, 13......** . We see that the **difference between any two consecutive** terms is same

i.e., $3 - 1 = 5 - 3, = 7 - 5, = 9 - 7 .....$

the difference is same and is equal to 2

(Therefore the sequences of numbers in which the difference between any two consecutive terms is same is known as Arithmetic Progression or an A.P.)

Generally we denote **this difference,** which is also known as common difference by **'*d*'**

So if **'*a*'** is first term of any **A.P.** and **'*d*'** be its common difference then the series will be written as

$$a, a + d, a + 2d, a + 3d, a + 4d, ... a + (n - 1) d.$$

Where

First term $T_1 = a$

second term $T_2 = a + d$

Third term $T_3 = a + 2d$

and so on upto $T_n = a + (n - 1) d.$

This T_n is also known as **last term** or **general term** or **nth term** of the given **A. P.**

$\therefore$ **Last term** or **nth term** or **general term**

$$T_n = a + (n - 1) d$$

Where **'*a*'** is **first term** and **'*d*'** is **common difference**. ($d = T_2 - T_1$ or $T_3 - T_2$ or $T_4 - T_3$ and so on)

(ii) if we are asked to find the sum of first ten or twenty terms of an **A. P.** then anybody can find it by simple addition.

i.e., $1 + 3 + 5 + 7 + 9 + 11 + 13 + 15 + 17 + 19 = 100$

But what if we are asked to find out the sum of 199 or 500 or 1000 terms of any A.P.

So for this we have the **formula** to find out the sum upto any terms or upto ***n* terms** where ***n*** can take any value.

The sum of ***n* terms** of any **A.P.** is given by

$$S_n = \frac{n}{2} [2a + (n - 1) d]$$

where n = no of terms in the series

a = first term of the series

d = common difference of series

There is another **Formula** also to find out the sum of given **A. P.** series if First term **'*a*'** and last term (**'*l*'**) then it is given as.

$$S_n = \frac{n}{2}[a + l]$$

$$= \frac{n}{2}[\text{first term} + \text{last term}]$$

Remember Formula for $\mathbf{S_n}$ is sum of ***n* terms** is applicable for the terms starting from begining.

(iii) If **'*a*'** and **'*b*'** are any two numbers then **Arithmetic mean** between them is given by

$$\text{Arithmetic mean} = \frac{a + b}{2}$$

Remember: we can find as many as Arithmetic means between any two given numbers and if we put all calculated arithmetic means of any two given numbers, between given two numbers. The resulting sequence will also be in A. P.

(iv)

(a) If we have to suppose three numbers in **A. P,** these will be $a - d, a, a + d$

(b) And we have to suppose four numbers in A. P. these will be $a - 3d, a - d, a + d, a + 3d$

(v) If the same number is **added** or **subtracted** from each term of given **A. P.**, the resulting sequence will also be in **A. P.** with same common difference

(vi) If each term is **multiplied** or **divided** by same number then the resulting sequence is also an **A. P.** with common difference **K** times or $\frac{1}{K}$ times the original common difference respectively

PART II
Geometric Progressions

(i) Consider the sequence of numbers **3, 6, 12, 24, 48, 96, ...**

We find that the ratio between any two successive terms *i.e.* the ratio.

$$\frac{6}{3} = \frac{12}{6} = \frac{24}{12} = \frac{48}{24}$$

is equal to 2, which is same and constant throughout the series. This ratio is known as **common ratio** and is denoted by **'*r*'**

∴ **The series in which the ratio between any two successive terms is same is known as Geometric Progression.**

If **'*a*'** is the first term and **'*r*'** be the **common** ratio then the **G. P.** in general is written as

$$a, ar, ar^2, ar^3 \text{ ...... } ar^{n-1}$$

where

first term $T_1 = a$

second term $T_2 = ar$

third term $T_3 = ar^2$

and so on upto ***n* terms**

Here **nth term = last term = general term = ar^{n-1}**

$\therefore \quad T_n = ar^{n-1}$

Here we can put $n = 1, 2, 3, 4$ .... to get **first term (T_1), second term (T_2), third term (T_3)** and so on respectively.

Also $\frac{T_2}{T_1}, \frac{T_3}{T_2}, \frac{T_4}{T_3}$ etc. is common ratio of this series.

(ii) The sum of n terms of **Geometric Progression** is denoted by S_n and is given by

$$S_n = a + ar + ar^2 + ... + ar^{n-1}$$

$$= \frac{a(1-r^n)}{1-r} \quad \text{if} \quad r < 1$$

$$= \frac{a(r^n - 1)}{r-1} \quad \text{if} \quad r > 1$$

where n is numbers of terms and 'a' and 'r' is **first term** and **common ratio** respectively.

(iii) The sum of Infinite terms of **G.P.** is denoted by $S\infty$ and given by

$$S\infty = a + ar + ar^2 + ... + \infty$$

$$= \frac{a}{1-r} \quad r < 1$$

Where a = first term of this series

r = common ratio of this series

PART III
Harmonic Progression

(i) Consider any **A.P.** series 1, 3, 5, 7, 9, 11 if we reverse each term of this series we get $\frac{1}{1}, \frac{1}{3}, \frac{1}{5}, \frac{1}{7}, \frac{1}{9}, \frac{1}{11}$ **then** resulting sequence will be in **Harmonic Progression**.

Since nth term of **A.P.** is $T_n = a + (n-1)d$

$\therefore$ ***n*th or last term or general terms of H.P. is**

$$T_n = \frac{1}{a + (n-1)d}$$

Where 'a' is first term and 'd' is common difference of corresponding **A.P.**

(ii) Harmonic mean between any two numbers 'a' and 'b' is given by

$$H = \frac{2ab}{a+b}$$

Remember: while solving the question of H.P. usually we do not use the standard result for finding nth terms or sum of n terms or total number of terms of H.P. instead of this we reverse the each term of H.P. series to make it an A.P. series. Then we use the result of A.P. series and reverse the answer to get the final answer in H.P.

(iii) Let 'a' and 'b' be two numbers

their $\quad \text{A.M.} = A = \dfrac{a+b}{2}$

$\therefore \quad \mathbf{a+b = 2A}$

their $\quad \text{G. M.} = G = \mathbf{(ab)^{1/2}}$

$\therefore \quad \mathbf{ab = G^2}$

So if 'a' and 'b' are roots of any equation then **Quadratic Equation** with roots 'a' and 'b' is given by

$$\mathbf{x^2 - 2Ax + G^2 = 0}$$

Where **2A** is sum of the roots 'a' and 'b', $\mathbf{G^2}$ is product of roots 'a' and 'b'. If you are having any problem in concept of Quadratic Equation then go through the chapter of Quadratic Equation.

PART IV
Arithmetico – Geometric Series (A. G. P.)

Consider the sequences

(i) 1, 3, 5, 7, 9, 11, 13 ...... an **(A.P.)**

(ii) $x\, x^2, x^3, x^4, x^5, x^6$ .... a **(G.P.)**

We see that **(i)** is an **A.P.** series and **(ii)** is a **G.P.** series if we multiply the corresponding terms of **(i)** and **(ii)** i.e.,

First term of AP with first term of GP, second term of AP with second term of GP and so on we get

$$1x + 3x^2 + 5x^3 + 7x^4 + 9x^5 \ldots.$$

The above series we get is known as **A.G.P.**

So A.G.P. is obtained by multiplying first term of A.P. with the first term of G.P., second term of A.P. with second term of G.P. and so on.

Now to find out the sum of this series or nth term of this series we do not use the standard results of **A.G.P.** instead we use other techniques which we will discuss later on this chapter.

PART V
Miscellaneous Series

Consider the series

(i) $1 + 2 + 3 + 4 + 5 + \ldots\ldots.$ upto 500 terms

(ii) $1^2 + 2^2 + 3^2 + 4^2 + \ldots\ldots..$ upto 30 terms

(iii) $1^3 + 2^3 + 3^3 + 4^3 + \ldots\ldots..$ upto 30 terms

Now to find out the sum of the series of (i), (ii) or (iii) we have a few **results** which will help us to find the sum of these type of series in simpler and shorter way.

For series (*i*) we have $\sum_{k=1}^{n} K = 1 + 2 + 3 + ... + n$

$= \frac{(n)(n+1)}{2}$, Where $\sum_{k=1}^{n} K$ is read

"as summation K" in which we can put

K = 1, 2, 3 ...... upto n and add them

Now to find out the sum of (*i*) we put

$n = 500$

therefore (*i*) sum of $= \frac{500 \times 501}{2}$

For series (*ii*) we have formula as follows

$$\sum_{K=1}^{n} K^2 = 1^2 + 2^2 + 3^2 + 4^2 + ... + n^2$$

$$= \frac{(n) \times (n+1) \times (2n+1)}{6}$$

ΣK^2 read as **"summation k square"** in which we can put $k = 1, 2, 3,$ ...... upto n, square each number and then add. It is also known as **Sum of Squares of first n natural numbers**.

Now for the sum of series (*ii*) we put

$n = 30$ we get

$\therefore$ if (*ii*) Sum $= \frac{30 \times 31 \times 61}{6}$

For series (*iii*) we have formula as follows

$$\sum_{K=1}^{n} K^3 = 1^3 + 2^3 + 3^3 + 4^3 + ... + n^3$$

$$= \left[\frac{(n)(n+1)}{2}\right]^2$$

where, ΣK^3 read as **"summation k cube"** which means that we put k = 1, 2, 3, ... upto n, and add the cube of each natural number.]

For series (*iii*) we have to put

$n = 30$ we get

$\therefore$ sum of (*iii*) equal to $\left(\frac{30 \times 31}{2}\right)^2$

(iv) if $a, b, c,$ are in **A.P.**

then $b - a = c - b$ **(*i.e.* the common difference is same)**

or $2b = a + c$

(v) if a, b, c, are in **G.P.**

then $\frac{b}{a} = \frac{c}{b}$ **(*i.e.* the common ratio is same)**

or $b^2 = ac$

(vi) if a, b, c, are in **H.P.** then

$$b = \frac{2ac}{a+c}$$

(vii) REMEMBER for any two numbers a and b if **A, G, H** are **A. M.**, **G. M.** and **H. M.** between them respectively then

(a) $A > G > H$

(b) **A, G, H** .... will always be in **G. P**

Therefore $G^2 = AH$

Till now we have gone through basic concepts for this topic and now we will do the types of questions based on these results. The types of questions are as follows.

Types of Questions

Type I To find series or number of terms of (**A. P., G. P., H. P.,**) when first few terms or last term or general term and common difference or common ratio is given.

Type II To find the sum of series of (**A. P., G. P., H. P.,**) or if sum is given then find out the series, common difference, common ratio or total number of terms.

Type III Related to **A. M., G. M., H. M.,**

Type IV Practical Applications

EXAMPLES

Type I

Example 1. Find the 10th term of the series 1, 4, 7, 10 ....

Sol : Here the given series is in A. P.

as $4 - 1 = 7 - 4 = 10 - 7$

i.e., the common difference is same between two successive terms and is equal to 3.

Here $a = 1 \quad d = 3.$

we know $\mathbf{T_n = a + (n-1)\,d}$ **...(i)**

for 10th term, put $n = 10$ in **(i)**

we get $T_{10} = a + 9d$

or $T_{10} = 1 + 9 \times 3 = \mathbf{28}$

Example 2. Find the 8th term of the series 3, 6, 12, ...

Sol : Here the given series in **G.P.** as the ratio between two successive terms i.e., $\frac{6}{3}, \frac{12}{6}$ is equal to 2

Here $a = 3 \quad r = 2$

we know in G. P. $\mathbf{T_n = ar^{n-1}}$ ...(i)

$\therefore$ for 8th term put $\mathbf{n = 8}$ **in (i)**

we get $T_8 = ar^7 = 3 \times (2)^7 = 384$

Example 3. If the 7th term of H.P. is 8 and 8th term is 7, then its 15th term is

Sol : We know that nth term of H.P. is given by

$$T_n = \frac{1}{a+(n-1)d}$$

$$T_7 = \frac{1}{a+6d} = 8 \quad \text{(given)} \qquad ...(1)$$

$$T_8 = \frac{1}{a+7d} = 7 \quad \text{given} \qquad ...(2)$$

$$T_{15} = \frac{1}{a+14d} = ?$$

So we have to find 'a' and 'd' to get the answer and for this we have two equations in 'a' and 'd' as (1) and (2)

From 1, $a + 6d = \frac{1}{8}$ ...(3)

From 2, $a + 7d = \frac{1}{7}$ ...(4)

Solving (3) and (4) we get

$$d = \frac{1}{56}$$

$$\therefore \quad T_{15} = \frac{1}{a+14d} = \frac{1}{a+7d+7d} = \frac{1}{\frac{1}{7}+\frac{1}{8}} = \mathbf{\frac{56}{15}}$$

Example 4. How many terms are there in the series 5, 8, 11, 14 320

Sol : Here the series is in **A. P.** with

1st term $a = 5$

common difference $d = 3$

and last term $T_n = 1 = 320$

we have to find *n*

we know that $T_n = a + (n-1)\,d$ (in **A.P.**)

Putting the values of T_n, a, and d we get

$$320 = 5 + (n-1)\,3$$

$$(n-1)3 = 320 - 5 = 315$$

$$n - 1 = \frac{315}{3} = 105$$

$$n = 105 + 1 = 106$$

So total terms are **106**

Example 5. If the third term of **A. P.** is 10 and seventh term is 18 than find fifteenth term of the series.

Sol : We know

$$T_n = a + (n-1)\,d$$

$$\therefore \quad T_3 = a + (3-1)\,d = a + 2d = 10 \qquad \textbf{...(1)}$$

$$T_7 = a + (7-1)\,d = a + 6d = 18 \qquad \textbf{...(2)}$$

Subtracting (1) from (2) we get

$$4d = 8 \quad \text{or} \quad d = 2$$

put $d = 2$

in (1) or (2)

we get $a = 6$

$$\therefore \quad T_{15} = a + 14d = 6 + 14 \times 2 = \mathbf{34}$$

Example 6. Find the 10th term of the **G. P.** if 3rd term is 20, and eight th term is 640

Sol : We know that in G. P

$$\text{nth term} = T_n = ar^{n-1}$$

$$\therefore \quad T_{10} = ar^9$$

Now to find out T_{10} we must have the value of a and r and for this we have the terms T_3 and T_8

given $\quad T_3 = ar^2 = 20 \qquad ....(i)$

and $\quad T_8 = ar^7 = 640 \qquad ...(ii)$

dividing (2) by (1) we get

$$\frac{ar^7}{ar^2} = \frac{640}{20}$$

$$r^5 = 32 = (2)^5$$

$$\therefore \quad r = 2$$

[when powers are same we can equate bases]

put this $\quad r = 2$ in (i)

we get $\quad a\,(2)^2 = 20$

$$a = 5$$

put the value of $\quad a = 5 \quad \text{and} \quad r = 2$

in $\quad T_{10} = ar^9 = 5 \times (2)^9 = 2560$

Example 7. Which term of the series 3, 6, 12, ... is 1536.

Sol : The given series is **G. P.** as

$$\frac{6}{3} = \frac{12}{6} = 2$$

i.e., common ratio is same and equal to 2.

let $T_n = 1536$

we know in G. P. $T_n = ar^{n-1}$

here $a = 3 \quad r = 2$

$\therefore \quad 1536 = (3)(2)^{n-1}$

$$2^{n-1} = \frac{1536}{3} = 512 = (2)^9$$

$\therefore \quad n - 1 = 9$

$n = 10$

Hence there are total 10 terms in the series.

Type II

Example 8. Find the sum of first 20 terms of the series 4, 7, 10, 13, ...

Sol : Here the series is **A.P.** as common difference

$$7 - 4 = 10 - 7 = 13 - 10 = 3$$

also $a = 4$

$d = 3$

we have $S_n = \frac{n}{2}[2a + (n-1)d]$ [$\therefore$ n = total number of terms]

$$\therefore \quad S_{20} = \frac{20}{2}[2 \times 4 + (20 - 1) \times 3] = 10(8 + 57) = \mathbf{650}$$

Example 9. How many terms of **A. P.** are required to get the sum equal to 235 if its common difference is 5 and first term is 1

Sol : Let n terms are required to make the sum = 235

$\therefore \quad S_n = 235$

Also here $a = 1$

$d = 5$

we know that $S_n = \frac{n}{2}[2a + (n-1)d]$

$$235 = \frac{n}{2}[2.1 + (n-1)5]$$

$$470 = n[2 + 5n - 5] = n[5n - 3]$$

$$5n^2 - 3n - 470 = 0$$

solving by quadratic equation

we get $n = 10$

and $n = \frac{-94}{10}$ (as sum of terms cannot be negative)

But $n \neq \frac{-94}{10}$

$\therefore \quad n = 10$

Hence 10 terms are required to make the sum = 235 for this series.

Example 10. Find the sum of first n odd natural numbers.

Sol : The first odd natural numbers are 1, 3, 5, 7, 9, ...

Here first term is $a = 1$

common difference is $d = 2$

we know for **A. P.** the sum of n terms is

$$S_n = \frac{n}{2}\{2a + (n-1)d\} = \frac{n}{2}[2 \times 1 + (n-1) \times 2] = n^2$$

Example 11. If T_n of **A. P.** is $2n + 1$, then the sum of first 20 terms.

Sol : For **A. P.** $S_n = \frac{n}{2}[2a + (n-1)d]$

Here $n = 20$

but for a = first term

and d = common

difference we will use $T_n = 2n + 1 - (1)$

for first term put $n = 1$ in (1)

$T_1 = 2 \times 1 + 1 = 3$

for T_2 put $n = 2$ in (1)

i.e., $T_2 = 2 \times 2 + 1 = 5$

for T_3 put $n = 3$ in (1)

$T_3 = 2 \times 3 + 1 = 7$

$\therefore$ $\boldsymbol{a = 3}$

$\boldsymbol{d = 2}$

Hence $S_{20} = \frac{20}{2}[2a + (20-1)d]$

$= 10[2 \times 3 + 19 \times 2] = 10[6 + 38] = 440$

Alternate, given $T_n = 2n + 1$

$\therefore$ T_1 = first term = 3

last term = $T_{20} = 2 \times 20 + 1 = 41$

we know in A. P. that $S_n = \frac{n}{2}$ [first term + last term]

$\therefore$ $S_{20} = \frac{20}{2}[3 + 41] = 10[44] = 440$

Example 12. If the sum of n terms is $3n^2 - 4n$ of an **A. P.** then the common difference of the series is

Sol : Here sum of n terms of A. P. is given as

$$S_n = 3n^2 - 4n \qquad \textbf{...(1)}$$

To find common difference we should have first few terms of the series for this

we put $n = 1, 2, 3,$ in (1)

put $n = 1$ in (1), *i.e.* sum of 1st term

$$S_1 = T_1 = 3.1^2 - 4.1 = -1$$

put $n = 2$ in (2), *i.e.* sum of the first 2 term

$$S_2 = T_1 + T_2 = 3.4 - 4.2 = 4$$

put $n = 3$ *i.e.* sum of the first 3 terms

$$S_3 = T_1 + T_2 + T_3 = 3.9 - 4.3 = 15$$

now $\quad S_2 - S_1 = T_2 = 4 - [-1] = 5$

$$S_3 - S_2 = T_3 = 15 - 4 = 11$$

So we have $\quad T_1 = -1, T_2 = 5, T_3 = 11$

$\therefore$ common difference is $T_2 - T_1 = T_3 - T_2 = 6$

Type III

Example 13. If the numbers 3, A_1, A_2, A_3, A_4, 28 are in A. P. then value of A_2 and A_3

Sol : The series is 3, A_1, A_2, A_3, A_4, 28 ..... is in A.P.

Total number of terms = 6

with first term = $T_1 = 3$

and last term = $T_6 = 28$

let '***d***' be the common difference of this series.

Also $\quad A_1$ = second term

A_2 = third term

$$T_n = a + (n-1)\,d$$

$$\therefore \quad 28 = 3 + (5-1)\,d$$

$$d = \frac{28-3}{5}$$

$$\therefore \quad A_1 = T_2 = a + d = 3 + 5 = 8$$

$$A_2 = T_3 = a + 2d = 3 + 2 \times 5 = 13$$

Example 14. The value of x if 3, x, 0.75 are G. P.

Sol : We know that if a, b, c, are in G. P.

then $\quad \dfrac{b}{a} = \dfrac{c}{b}$

$$b^2 = ac$$

$\therefore$ if 3, x, 0.75 are in G. P.

then $\quad \dfrac{x}{3} = \dfrac{0.75}{x}$

or $\quad x^2 = 0.75 \times 3 = 2.25$

$\therefore \quad x = \sqrt{2.25} = 1.5$

Example 15. If 3, x, y, $\frac{1}{9}$ are in G. P. then value of x and y is.

Sol : Given $3, x, y, \frac{1}{9}$ in G. P.

with first term $= a = T_1 = 3$, and fourth term $= T_4 = \frac{1}{9}$

we know in G.P. $T_n = ar^{n-1}$

$\therefore$ $T_4 = ar^{4-1} = ar^3 = \frac{1}{9}$

also $\boldsymbol{a = 3}$

$\therefore$ $3 \times r^3 = \frac{1}{9}$

or $r^3 = \frac{1}{27} = \left(\frac{1}{3}\right)^3$

$\therefore$ $r = \frac{1}{3}$

Now $x =$ second term $= T_2 = ar = 3 \times \frac{1}{3} = 1$

and $y =$ third term $= T_3 = ar^2 = 3 \times \frac{1}{3} \times \frac{1}{3} = \frac{1}{3}$

Example 16. Find the numbers if A. M. and G. M. between them is 4 and 5 respectively.

Sol : Let the number be a and b

$$\text{A.M} = \frac{a+b}{2} = 5 \quad \text{(given)}$$

$$\therefore \quad a + b = 10$$

$$\text{G.M} = \sqrt{ab} = 4 \quad \text{(given)}$$

$$\therefore \quad ab = 16$$

we know that if A. M and G. M are roots of any equation then equation is

$$x^2 - 2\,(\text{A. M.})x + (\text{G. M.})^2 = 0$$
$$x^2 - (a+b)x + ab = 0$$
$$x^2 - 10x + 16 = 0$$
$$x^2 - 8x - 2x + 16 = 0$$
$$(x-2)(x-8) = 0$$
$$x = 2$$
$$x = 8$$

Hence the numbers are 2 and 8

Example 17. Find two Harmonic mean between $\frac{1}{4}$ and $\frac{1}{16}$.

Sol : Let h_1 and h_2 be two H.M mean between two numbers $\frac{1}{4}$ and $\frac{1}{16}$

$\therefore$ $\frac{1}{4}, h_1, h_2, \frac{1}{16}$ are in H.P. **...(1)**

[i.e., if we insert any number of A.M. between two number the resulting series is an A.P.

Similarly if we insert any number of G. M.'s between two number the resulting series is a G.P.

And if we put any number of H.M. between two number the result is H.P. series.]

Also we know that if we reverse each term of H.P. we get corresponding A.P. therefore by reversing the (1) we get

$$4, \frac{1}{h_1}, \frac{1}{h_2}, 16 \text{ in AP} \quad \text{...(2)}$$

if '*a*' be the first term and '*d*' is common difference of this series than

first term $\quad T_1 = a = 4 \quad$ **...(3)**

second term $\quad T_2 = a + d = \frac{1}{h_1} \quad$ **...(4)**

third term $\quad T_3 = a + 2d = \frac{1}{h_2} \quad$ **...(5)**

fourth term $\quad T_4 = a + 3d = 16 \quad$ **...(6)**

put value of from (3) in (6) we get $\quad a = 4$

$$4 + 3d = 16$$

$$3d = 12$$

or $\quad d = 4$

$\therefore \quad T_2 = a + d = \frac{1}{h_1} = 4 + 4 = 8$

or $\quad h_1 = \frac{1}{8}$

and $\quad T_3 = \frac{1}{h_2} = a + 2d = 4 + 2 \times 4 = 12$

or $\quad h_2 = \frac{1}{12}$

Example 18. A boy saves Rs. 2 on 1st day, Rs. 4 on 2nd day, Rs. 8 on 3rd day, and so on. How much money he will save after 30 days.

Sol : The series of his savings is 2, 4, 8 ..... upto 30 days

which is G. P. as $\frac{4}{2} = \frac{8}{4} = 2$. *i.e.* the common ratio is same and is equal to 2

so here we have to find out the sum of 30 terms of G.P.

we know in G. P. the sum of n terms is

$$S_n = \frac{a(r^n - 1)}{r - 1} \text{ if } r > 1 \; \{\text{here } r = 2 \text{ which is} > 1\}$$

$$\therefore \quad S_{30} = \frac{2(2^{30} - 1)}{2 - 1} = 2\,(2^{30} - 1)$$

EXERCISE

Q.1. If third term of an A. P. is 12 and its seventh term is 24 then its tenth term is

(a) 48 (b) 36 (c) 33 (d) 52

Q.2. If third term of G. P. 5 then product of its first five terms.

(a) 5^5 (b) 5^4 (c) 4^5 (d) 6^5

Q.3. If pth, qth, rth term of any G.P. are in G.P. then $p, q, r,$ are in

(a) A.P. (b) G.P. (c) H.P. (d) none

Q.4. Which term of the series 28, 24, 20 is zero.

(a) 16th (b) 12th (c) 8th (d) 10th

Q.5. The tenth term of the series $\sqrt{3}, \sqrt{12}, \sqrt{27}$ ...

(a) 510 (b) $\sqrt{200}$ (c) $\sqrt{300}$ (d) $\sqrt{312}$

Q.6. Find the third term of G. P. if tenth term of the G. P. is 9 and eighth term of G. P. is 4.

(a) 12 (b) 2 (c) 6 (d) none

Q.7. If a, 4, b, are in A. P. and a, 2, b, are in G. P. then a, 1, b are

(a) A. P. (b) G. P. (c) H. P. (d) none

Q.8. If $2 + 2\alpha + 2\alpha^2 + 2\alpha^3 + \ldots\ldots \infty$ is equal to 4 then α is equal to.

(a) $\frac{1}{4}$ (b) $\frac{1}{2}$ (c) $\frac{1}{8}$ (d) $\frac{1}{64}$

Q.9. In an infinite G. P. each term is equal to thrice the sum of the terms that follows it then common ratio is

(a) 36 (b) $\frac{1}{4}$ (c) $\frac{1}{8}$ (d) 44

Q.10. The sum of the series $1^2 + 1 + 2^2 + 2 + 3^2 + 3 + \ldots\ldots + n^2 + n$ is equal to

(a) $\frac{7n^2 + n^3}{14}$ (b) $\frac{8n^2 + 9n + 6}{2}$ (c) $\frac{2n^3 + 3n^2 + n}{6}$ (d) $\frac{5n^2 + 7n}{8}$

Q.11. The sum of the series 2, 5, 8, 11 .... is 60100 then n is

(a) 200 (b) 400 (c) 300 (d) 500

Q.12. The sum of the series $1 + \frac{3}{5} + \frac{5}{5^2} + \frac{7}{5^3} + \frac{9}{5^4} \ldots.$

(a) $\frac{18}{29}$ (b) $\frac{20}{17}$ (c) $\frac{17}{9}$ (d) $\frac{15}{8}$

Q.13. If the sum of first n positive integers is $\frac{1}{5}$ times the sum of the squares of first n positive integers of then n is

(a) 23 (b) 14 (c) 7 (d) 9

Q.14. If each term of G. P. is sum of its two succeding terms then the common ratio is

(a) $\frac{-1 \pm \sqrt{3}}{2}$ (b) $\frac{1 \pm \sqrt{17}}{4}$ (c) $\sqrt{15}$ (d) $\frac{-1 \pm \sqrt{5}}{2}$

Q.15. If the sum of first n terms of A.P. is 40 with its common difference equal to 2 and last term 13 then value of n is

(a) 12 (b) 10 or 4 (c) 6 (d) 8

Q.16. The sum of two digit numbers which are divisible by 5.

(a) 945 (b) 1050 (c) 350 (d) 1200

Q.17. If $x > 0, y > 0, z > 0$, are in G. P. then $\log x$, $\log y$, $\log z$, are in

(a) A.P. (b) G.P. (c) H.P. (d) A. G. P.

Q.18. The A.M., G.M., and H.M. between the two numbers p and q are equal then

(a) $p < q$ (b) $p > q$ (c) $p = q$ (d) $p \neq q$

Q.19. Find three arithmetic means (A. M.) between 6 and 14

(a) 12, 14, 16 (b) 6, 8, 12, 14, (c) 8, 10, 12 (d) –8, –10, –12.

Q.20. If A, G, H be A.M., G.M., and H.M. between any two number ***a*** and ***b*** then

(a) $\frac{2AG}{H+G}$ (b) $A = G + H$ (c) $A^2 = GH$ (d) $G^2 = AH$

Q.21. The sum of the series 2 .... 98 is 2450 then n is equal to

(a) 49 (b) 67 (c) 117 (d) 1129

Q.22. The length of side of square is 'a' meter. A second square is formed by joining the mid points of sides of this square. A third square is formed by joining mid points of sides of 2nd square and so on. The sum of areas of square up to infinity.

(a) 2 (b) $4a$ (c) $2a^2$ (d) $7a$

Q.23. Find the nth term of the series 1, 3, 6, 10 .....

(a) $n + 1$ (b) n (c) $n - 1$ (d) $\frac{(n)(n+1)}{2}$

Q.24. If H is H.M. between ***a*** and ***b*** then $\frac{H}{a} + \frac{H}{b}$ is

(a) 4 (b) $a + b$ (c) 2 (d) none

Q.25. In A.P. 5 times the 5th term is equal to 8 times the 8th term, its 13th term is

(a) 2 (b) 1 (c) 12 (d) 0

Q.26. The next term of sequence 2, 6, 15, 31 ....

(a) 15 (b) 26 (c) 25 (d) 46

Q.27. The 10th term of the series $1 \times 3 + 3 \times 5 + 5 \times 7 + 7 \times 9$ ....

(a) 399 (b) 499 (c) 599 (d) none

Q.28. If a, b, c, are in G. P. then

(a) $a > b$ (b) $b > c$ (c) $a - b = b - c$ (d) $\frac{a-b}{b-c} = \frac{a}{b}$

SOLUTIONS

1. (c) We know in A. P., the nth term is

$$T_n = a + (n-1)d$$

$\therefore \quad T_{10} = a + 9d$ ...(1)

$\therefore \quad T_3 = a + (3-1)\,d = a + 2d = 12$ ...(2)

$\therefore \quad T_7 = a + (7-1)\,d = a + 6d = 24$ ...(3)

In (1) we need the value of a and d to find T_{10} for this we have (2) and (3)

$a + 6d = 24$ ...(4)

$a + 2d = 12$ ...(5)

(4) – (5) gives $\quad 4d = 12 = d = 3$

put $\quad d = 3$

in (4) or (5) we get

$a + 6 = 12$

$\therefore \quad a = 6$

$\therefore \quad \mathbf{T_{10} = a + 9d = 6 + 9 \times 3 = 6 + 27 = 33}$

2. (a) Let $\boldsymbol{a}$ be first term and $\boldsymbol{r}$ be common ratio of the series

$T_3 = ar^2 = 5 \quad [T_n = ar^{n-1}]$ ...(1)

its 1st five terms are $a, ar, ar^2, ar^3, ar^{40}$ and its product

$= a \times ar \times ar^2 \times ar^3 \times ar^4$ ...(2)

from (1) $\quad ar^2 = 5$

or $\quad (ar^2)^5 = (5)^5$

or $\quad \boldsymbol{a^5 r^{10} = 5^5}$

3. (a) Let a be the first term of series and R be the its common ratio then

$T_p = a\,R^{p-1}$

$T_q = a\,R^{q-1}$

$T_q = a\,R^{r-1}$

given T_p, T_q, T_r are in G. P.

$\therefore \quad (T_q)^2 = T_r \times T_p$

$\{a\,R^{q-1}\}^2 = a\,R^{r-1} \times a\,R^{p-1}$

$a^2\,R^{2q-2} = a^2\,R^{r+p-2}$

$2_q - 2 = r + p - 2$

or $\quad 2_q = r + p$

$\therefore$ p, q, r are in A. P.

4. (c) Let nth term of the series 28, 24, 20 ...... is zero

i.e., Let $\quad T_n = 0$

and we have to find n

Here $\quad a = 28$

$d = -4$ [as the sequence is A. P.]

we know $\quad T_n = a + (n-1)\,d$

$0 = 28 + (n-1)(-4)$

$\therefore$ 8th term or T_8 of this series is zero.

5. (c) If we see the series closely we observe that

first term $\quad T_1 = \sqrt{3 \times 1^2}$

second term $T_2 = \sqrt{3 \times 2^2}$

third term $T_3 = \sqrt{3 \times 3^2}$

similarly tenth term $T_{10} = \sqrt{3 \times 10^2} = \sqrt{300}$

6. (d) If a be the first term of the series and 'r' be its common ratio

then $T_{10} = ar^9 = 9$ ...(1)

and $T_8 = ar^7 = 4$ ...(2)

also for $T_3 = ar^2$, we need the value of a and r

now dividing (1) by (2)

we get
$$\frac{ar^9}{ar^7} = \frac{9}{4}$$
$$r^2 = \frac{9}{4} = \left(\frac{3}{2}\right)^2$$
$$r = \frac{3}{2}$$
$$\therefore \quad a = \frac{2^9}{3^7}$$

and hence
$$T_3 = \frac{2^7}{3^5}$$

7. (c) $a, 4, b$, are in A. P.
$$\therefore \quad 8 = a + b$$

also $a, 2, b$ ......G. P.
$$\therefore \quad a \times b = 4$$

consider
$$\frac{ab}{a+b} = \frac{4}{8} = \frac{1}{2}$$

or
$$\frac{2ab}{a+b} = 1$$

therefore 1 is H.M. between ab

$\therefore \quad a, 1, b$ ...... H.P.

8. (b) The given series $2 + 2\alpha + 2\alpha^2 + \ldots\ldots \alpha$ is a infinite G. P. with $T_1 = 2$ and common ratio $= \alpha$

we know sum of infinite
$$\text{G.P.} = S\,\infty = \frac{a}{1-r}$$
$$\therefore \quad \frac{2}{1-\alpha} = 4, \quad 1 = 2 - 2\alpha \quad \text{or} \quad 2\alpha = 1$$

or
$$\alpha = \frac{1}{2}$$

9. (b) Let a be the first term and r be the common ratio of the series

given $T_1 = 3 \quad [T_2 + T_3 + T_4 + \ldots\ldots\ldots \infty]$

or $a = 3 \quad [ar + ar^2 + ar^3 + \ldots\ldots\ldots \infty]$

or $a = 3a\,[r + r^2 + r^3 + \ldots\ldots\ldots \infty]$

or $1 = 3 \quad [r + r^2 + r^3 + \ldots\ldots\ldots \infty]$

or
$$1 = 3\left(\frac{r}{1-r}\right)$$

or $1 - r = 3r$

or $$r = \frac{1}{4}$$

10. (c) The given series is mixture of two seperate series ($1^2, 2^2, 3^2$ .... n^2) and (1, 2, ... n)

$$S_n = (1^2 + 2^2 + 3^2 + ... + n^2) + (1 + 2 + 3 + ... n)$$

$$= \frac{(n)(n+1)(2n+1)}{6} + \frac{(n)(n+1)}{2} = \frac{2n^3 + 3n^2 + n}{6}$$

11. (a) 2, 5, 8, 11, ... A.P. series $a = 2, d = 3$

$$S_n = \frac{n}{2}[2a + (n-1)d] = 60100$$

$$= \frac{n}{2}[4 + (n-1)3] = 60100$$

or $$3n^2 + n = 120200$$

or $$3n^2 + n - 120200 = 0$$

or $$n = 200 \text{ [after factorization]}$$

12. (d) If we closely observe the series we see that the given series

$1 + \frac{3}{5} + \frac{5}{5^2} + \frac{7}{5^3} + \frac{9}{5^4}$ ..... is product of corresponding terms of an A.P. and G.P.

i.e., the A. P. series is 1, 3, 5, 7, 9, **...(1)** the G.P. series is

$$1, \frac{1}{5}, \frac{1}{5^2}, \frac{1}{5^3} \quad ...(2)$$

the given series is product of T_1 of A. P. $\times$ T_1 of GP and so on now let

$$S = 1 + \frac{3}{5} + \frac{5}{5^2} + \frac{7}{5^3} + \frac{9}{5^4} \quad ...(3)$$

multiply both sides of (3) by common ratio of G. P.

which is $r = \frac{1}{5}$ we get

$$\frac{S}{5} = \frac{1}{5} + \frac{3}{5^2} + \frac{5}{5^3} + \frac{7}{5^4} \quad ...(4)$$

(3) – (4) gives

$$S - \frac{1}{5}S = 1 + \frac{2}{5} + \frac{2}{5^2} + \frac{2}{5^3} + \frac{3}{5^4} \quad ...(5)$$

$$= 1 + 2\left(\frac{1}{5} + \frac{1}{5^2} + \frac{1}{5^3} + ...\infty\right)$$

$$= 1 + 2\left(\frac{\frac{1}{5}}{1 - \frac{1}{5}}\right) \left(\because \frac{1}{5} + \frac{1}{5^2} + \frac{1}{5^3} \text{ is a G.P}\right)$$

$$= 1 + 2\left(\frac{\frac{1}{5}}{\frac{4}{5}}\right) = 1 + 2\left(\frac{1}{5} \times \frac{5}{4}\right) = 1 + \frac{1}{2} = \frac{3}{2}$$

$$S\left(1-\frac{1}{5}\right) = \frac{3}{2}$$

$$S\left(\frac{4}{5}\right) = \frac{3}{2}$$

$$\mathbf{S = \frac{15}{8}}$$

13. (c) Given $\quad 1 + 2 + 3 + ... + n = \frac{1}{5}\ (1^2 + 2^2 + 3^2 + ... n^2)$

or $\quad \frac{(n)(n+1)}{2} = \frac{1}{5}\left(\frac{(n)(n+1)(2n+1)}{6}\right)$

or $\quad 15 = 2n + 1$

$$2n = 14$$

$$\boldsymbol{n = 7}$$

14. (d) Given $\quad T_1 = T_2 + T_3 \quad$ ...(1)

$$T_2 = T_3 + T_4 \quad \text{...(2)}$$

and so on let a be the first term and r be its common ratio than from (1) we get

$$a = ar + ar^2$$

$$r^2 + r - 1 = 0 \text{ (it is quadratic eq.)}$$

$$r = \frac{-1 \pm \sqrt{1+4}}{2} = \frac{1 \pm \sqrt{5}}{2}$$

15. (b) Given $\quad S_n = 40$

$$d = 2$$

last term $\quad T_n = 13$

we know that

$$S_n = \frac{n}{2}\ [\text{first term + last term}]$$

$$40 = \frac{n}{2}\ [a + 13] \quad \text{...(1)}$$

now for a we have $\quad T_n = a + (n-1)d$

or $\quad 13 = a + (n-1)2$

$$a = 13 - (n-1)2$$

$$a = 15 - 2n \quad \text{...(2)}$$

put (2) in (1) we get

$$40 = \frac{n}{2}\ [15 - 2n + 13]$$

or $\quad 80 = n\ [28 - 2n] = 28n - 2n^2$

or $\quad 2n^2 - 28n + 80 = 0$

or $\quad n^2 - 14n + 40 = 0$

or $\quad n^2 - 10n - 4n + 40 = 0$

or $\quad n\ (n - 10) - 4[n - 10] = 0$

$$(n - 10)\ (n - 4) = 0$$

$$n = 10$$

or $$n = 4$$

$\therefore$ either the sum of 10 terms or sum of 4 terms is 40.

16. (a) The series is 10, 15 ...... 95 **...(1)**

Which is An A.P. as difference between any two successive terms is same and is equal to 5.

To find the sum we should know the total numbers of terms of this series. The last term is 95.

$\therefore$
$$T_n = a + (n-1)d$$
$$95 = 10 + (n-1)5$$
$$n = 18$$

Hence total terms are 18

$\therefore$
$$S_{19} = \frac{n}{2}[a + l] = \frac{18}{2}[10 + 95] = 945$$

17. (a) $x, y, z,$ are in GP

$\therefore$
$$\frac{y}{x} = \frac{z}{y} \quad \text{...(1)}$$
$$y^2 = xz \quad \text{...(2)}$$

taking log of both the sides

$$2\log y = \log x + \log z$$

$\therefore$ $\log x, \log y, \log z,$ are in A.P. [if $a, b, c,$ **is in A.P. then** $2b = a + c$]

18. (c) The numbers are p and q

$$\text{A.M.} = \frac{p+q}{2}$$
$$\text{G.M.} = \sqrt{pq}$$
$$\text{H.M.} = \frac{2pq}{p+q}$$
$$\text{A.M.} = \text{G.M.} = \text{H.M.} \quad \text{(given)}$$
$$\frac{p+q}{2} = \sqrt{pq} = \frac{2pq}{p+q}$$

take
$$\frac{p+q}{2} = \sqrt{pq} \quad \text{...(1)}$$

or
$$p + q = 2\sqrt{pq}$$

Squaring both the sides

$$p^2 + q^2 + 2pq = 4pq$$
$$p^2 + q^2 + 2pq = 0$$
$$p^2 + q^2 - 2pq = 0$$
$$(p-q)^2 = 0$$

or
$$p = q$$

19. (c) Let A_1, A_2, A_3 be the three numbers A. M. means between 6 and 14

$\therefore$ 6, A_1, A_2, A_3, 14 ....are in A. P. **...(1)**

where $T_1 = 6$

$$T_2 = A_1$$
$$T_3 = A_2$$
$$T_4 = A_3$$
$$T_5 = 14$$

if d is common difference then

$$T_1 = a = 6$$
$$T_2 = A_1 = a + d$$
$$T_3 = A_2 = a + 2d$$
$$T_4 = A_3 = a + 3d$$
$$T_5 = 14$$

or $T_5 = 6 + 4d = 14$

$$4d = 14 - 6 = 8$$
$$d = 2$$

$\therefore$ $A_1 = T_2 = a + d = 6 + 2 = 8$

$$A_2 = T_3 = a + 2d = 6 + 4 = 10$$
$$A_3 = T_4 = a + 3d = 6 + 6 = 12$$

20. (d) Standard result.

21. (a) Let a and l be 1st and last term of the series

$\therefore$ $a = 2$

$$T_n = 98$$
$$S_n = 2450$$
$$S_n = \frac{n}{2}[a + l]$$
$$2450 = \frac{n}{2}[2 + 98]$$
$$2450 = \frac{n}{2}[100]$$
$$n = 49$$

22. (c)

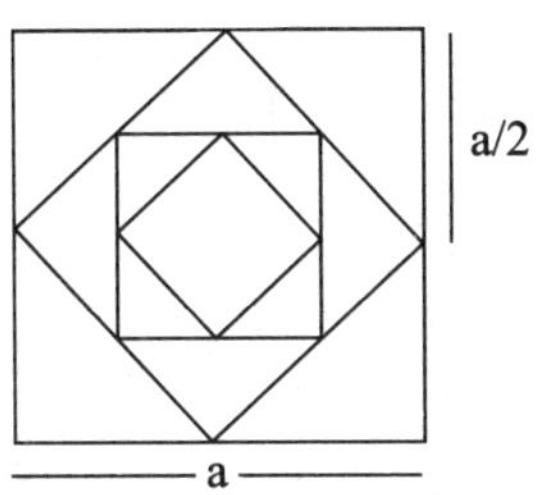

Area of first square $= a^2$

side of second square $= \sqrt{\left(\frac{a}{2}\right)^2 + \left(\frac{a}{2}\right)^2} = \frac{a}{\sqrt{2}}$

(use Pythagoras theorem)

$\therefore$ area of 2nd square $= \frac{a^2}{2}$

$$\text{side of third square} = \sqrt{\left(\frac{a}{2\sqrt{2}}\right)^2 + \left(\frac{a}{2\sqrt{2}}\right)^2} = \frac{a}{2}$$

$\therefore$ $\quad$ area of third square $= \dfrac{a^2}{4}$

So required sum is

$$= a^2 + \frac{a^2}{2} + \frac{a^2}{4} + \ldots\ldots = \frac{a^2}{1-\frac{1}{2}} = \frac{a^2}{\frac{1}{2}} = 2a^2$$

23. (d) Let

$$S = 1 + 3 + 6 + 10 + \ldots\ldots + T_n \quad \ldots(1)$$
$$S = 1 + 3 + 6 + \ldots\ldots + T_n \quad \ldots(2)$$

(1) – (2) gives

$$0 = 1 + 2 + 3 + 4 + \ldots\ldots T_n$$

(here total terms = $n + 1$)

$$\therefore \quad T_n = 1 + 2 + 3 + 4 + \ldots\ldots n \text{ terms}$$
$$= \frac{(n)(n+1)}{2}$$

24. (c) Harmonic mean

$$H = \frac{2ab}{a+b} \quad \ldots(1)$$
$$\frac{H}{a} = \frac{2b}{a+b}$$
$$\frac{H}{b} = \frac{2a}{a+b}$$
$$\frac{H}{a} + \frac{H}{b} = \frac{2a+2b}{a+b} = \frac{2(a+b)}{2} = 2$$

25. (d)

$$5T_5 = 8T_8$$
$$5[a + 4d] = 8\,[a + 7d]$$

[if a and d first term and common difference of the series]

$$5a + 20d = 8a + 56d$$
$$3a + 36d = 0$$
$$3(a + 12d) = 0$$

or $\quad a + 12d = 0 \quad$ or $\quad T_{13} = 0$

26. (d) 2, 6, 15, 31 series is +4, +9, +16

so $\quad$ next term $= +25$

or $\quad$ next term $=$ **46**

27. (a) $1 \times 3 + 3 \times 5 + 5 \times 7 + 7 \times 9$ ... The given series is product of the two A. P. series i.e., 1, 3, 5, 7 ... and 3, 5, 7, 9 ...

$\therefore$ 10th term is the product of T_{10} of (1, 3, 5, 7 ...) and T_{10} of (3, 5, 7, 9 ...) required

$$T_{10} = (T_{10} \text{ of } 1, 3, 5, 7) \times (T_{10} \text{ of } (3, 5, 7, 9.)$$
$$= (1 + 9 \times 2)\,(3 + 9 \times 2) = (1 + 18)\,(3 + 18) = (19)\,(21) = 399$$

28. (d) If a, b, c are in G. P.

then $\quad \dfrac{a-b}{b-c} = \dfrac{a}{b}$

Day 30

Statistics

Today we will do the topic of statistics.

Before going further into the detail of this topic let us consider **one example** which will **help us to understand this topic** in better way. Consider **the marks obtained by 20 students** of a class in mathematics test. **Where maximum marks of the test is 50. The scores of 20 students are as follows**

8, 2, 6, 17, 16, 2, 6, 18, 20, 15, 14, 8, 7, 9, 25, 5, 7, 8, 10, 12.

Now the marks obtained can be **shown** or **collected** in a few different ways for its analysis, which are as follows:

(a) In above we have shown the marks of 20 students written randomly or not in any definete pattern i.e., we have just asked the student and written down his/her marks. **This method of collection and representation of the data is know as ungrouped data.**

(b) The other way of showing the data is to see that **how many times any particular marks is obtained by how many student.**

for example in above data.

MARKS OBTAINED	NO. OF STUDENTS
2	2
8	3
2	2

In this method of collection and representation, we see that **how frequently** or **how many times** the particular observation (In this example observation is marks) is occuring or coming in data. **This is known as discrete frequency distribution of the data. The number of times that any particular item is coming is known as frequency of that observation.**

(c) In yet another way we divide the given or collected data in small groups or intervals then each observation is placed in interval in which it lies.

Let **intervals be 0 – 10, 10 – 20, 20 – 30, 30 – 40, 40 – 50**

i.e., we have devided 50 markes into 5 interval of width 10 each.

Now marks **(2, 6, 7, 8, 9)** will come under interval **0 - 10**. Marks **[10, 12, 14, 17, 18, 15, 16]** will come under **10 - 20** and so on.

In tabular form it is written as follows.

Class Interval	frequency
0-10	5 (as there are marks of 5 student in this interval)
10-20	7 (as the marks fo 7 student lies in the interval)
20-30	and so on.

This method is known as continous **frequency distribution method**.

So we have seen a few ways to represent the collected data.

After the data is **collected** it is analysed for various purposes as per instructions or as per requirement. But we will not go into the detail of these analysis because **from competitive point of view we try to calculate the mean, median, mode deviation etc of the given data.**

We will discuss some of these parameters later on in this chapter. So now we can define statistics and some other basic terms related to this topic.

STATISTICS

It is science of collection, presentation, analysis and interpretation of data.

1. (a) Class Interval:

The small group in which we place each collected observation are known as class interval for example 0-10, 10-20 etc are class interval.

(b) Class Frequency:

The number of observation which comes under any class interval in the frequency of that class interval for example in above marks distribution of the students.

5 is the class frequency of class interval 0–10, 7 is class frequency or class interval 10–20 and so on.

(c) Limit:

Each interval has **two limits** for example in calss interval 0-10, 0 is lower limit and 10 is upper limit.

(d) Class Marks or mid value:

It is the **mid point** of any class interval for example class interval of 10-20 the class marks is $\frac{10+2}{2}=\frac{30}{2}=15$

2. Frequency:

It is the number representing that how many times or how frequently that observation is occuring in data.

3. Range:

It is difference between **highest** value and **lowest** value in a data.

4. Cumulative frequency:

The sum of preceeding frequencies of class intervals.

In this chapter we will discuss two concepts as follows

(1) Measures of central tendency and measure of dispersion

MEASURES

For the sake of understanding consider measure as the limit or ways or methods to find out certain mathematical and statistical terms like mean, median, mode, deviations etc. This is a bit empirical defination (not purely technical) of measure, but this much understanding of this term (measure) is sufficient from competitive point of view.

The different parameters or terms we calculate and analyse for a given data are broadly divided into two major categories also known as measures as follows:

(i) Measures of central tendency

(ii) Measures of dispersion

Now we will discuss these measures, so first the measure of central tendency.

(i) Measure of central tendency:

In this category we have the terms or parameters as follows

(a) Mean **(b) Median** **(c) Mode**

There are a few more measures of central tendency but from competitive point of view they are not as important as above measures.

(ii) Measure of dispersion :

In this category we have terms or parameters as follows

(a) Range **(b) Deviation**

(**NOTE :** When the value of **mean** or **median** or **mode** is calculated it is observed that thses values lies within the data and are considered as true or standard values or they lie very close to standard values. Thats why mean, median or mode are known as measure of central tendency.)

But when we calculate the values of **range, deviation** etc of the data. It is generally observed that these values lie outside the range of data or distribution of data that is why these parameters are known as measure of dispersion. In this measure we try to see that how much our observations are close or how much far from the true values or standard values.

Now we will discuss the each parameter of these two types of measure conceptually.

CENTRAL TENDENCY

(A) (i) Arithmetic Mean : It is a numerical value obtained by dividing the total sum of all observations by total number of all observations so if x_1, x_2, x_3....... **upto** x_n are n **observations** then

arithmetic mean of these n observations which is denoted by $\bar{x}$ is given by

$$= x = \frac{x1 + x_2 + x_3 + ... + x_n}{n} \quad ...(i)$$

$$= \frac{1}{n}\sum_{i=1}^{n} x_1$$

Where $\sum_{i=1}^{n} x_i$ (Read as summation x_i) is a notation to represent the sum of all considered observations.

The mean in (i) is for ungrouped data.

And if the data is grouped i.e., obsevations $x_1, x_2, x_3 \ldots x_n$ are having frequencies $f_1, f_2, \ldots, f_n$ respectively then the mean is given by

$$\bar{x} = \frac{x_1 f_1 + x_2 f_2 + \ldots\ldots + x_n f_n}{f_1 + f_2 f_3 + \ldots\ldots + f_n}$$

$$= \frac{\sum_{i=1}^{n} x_i f_i}{\sum_{i=1}^{n} f_i} = \frac{\sum_{i=1}^{n} x_i f_i}{n}, i = 1, 2, 3\ldots\ldots n$$

where is total number of observation

Another method for mean:

$$\bar{x} = a + \frac{\Sigma f_i d_i}{n}, i = 1, 2, 3\ldots\ldots n$$

where a = assumed mean

$d_i = x_i - a$, is deviation or difference of each observation from the assumed mean.

(ii) Group mean or combined mean: If two groups with mean $\bar{x}_1$ and $\bar{x}_2$ are having n_1 and n_2 number of observations respectively then combined mean of the group is given by

$$\bar{x} = \frac{n_1 \overline{x_1} + n_2 \overline{x_2}}{n_1 + n_2}$$

(iii) Weighted Mean : When the weights of each observation instead of frequency is given then the mean of this set of observation is known as weighted mean. Which is given as follows:

$$\text{weighted mean} = \bar{w} = \frac{w_1 x_1 + w_2 x_2 + w_3 x_3 + \ldots\ldots\ldots + w_n x_n}{w_1 + w_2 + w_3 + \ldots\ldots\ldots + w_n}$$

where $w_1, w_2, w_3, \ldots. w_n$ are weights of observation of $x_1, x_2, x_3, \ldots. x_n$ respectively.

For the sake of understanding consider weights as importance in terms of number given to that observation.

Some facts about arithmetic mean

(a) If every observation is increased or decreased by same number, then the arithmetic mean of new observation also increases or decreases by same number.

(b) If every observation is multiplied or divided by same number or constant then arithmetic means of the new set of observation is obtained by multiplying or dividing the initial arithmetic mean by same number respectively.

(B) Median : It is size or the value of observation which lies in the middle of the observations. This mid point or median is depends whether the total number of observations are odd or even. Now the medians for

(i) Ungrouped Data (frequency not given) : First arrange the data in ascending or descending order and if total observations are n then median.

If **n is odd** then, **median = valueof** $\left(\frac{n+1}{2}\right)^{\text{th}}$ **observation**. and If **n is even** then **median** $= \frac{1}{2}$ **[value of** $\left(\frac{n}{2}\right)$ **th item + value of** $\left(\frac{n}{2}+1\right)$ **th item]**

(ii) For descrete series (frequency is given) : first we calculate the cumulative frequencies of the observation, and then we arrange the cumulative frequencies in ascending or descending order.

then **Median** $= \left(\frac{n+1}{2}\right)^{\text{th}}$ **observation**

where n is cumulative frequency

(iii) For grouped or continuous frequency distribution.

(a) If series is in ascending order :

then **median** $= l + \dfrac{\left(\frac{n}{2} - c\right)}{f} \times i$

where l = lower limit of the midean class

n = sum of all the frequencies

i = width of median class

f = frequency of median class

c = cumulative frequency of class preceeding to median class and the median class is interval where the value of $\frac{n}{2}$ lies.

(b) If the series is in descending order

then **median is** $= u - \dfrac{\left(\frac{n}{2} - c\right)}{f} \times i$

where u = upper limit of the median class.

(c) Mode :

The value of the observation which occurs maximum numbers of time in the data. Generally we locate this value just by inspecting the data

(i) for indiviual series or raw data :

The mode is that observation which is occuring maximum number of time.

(ii) for discrete data (With frequency) :

The mode is observation having highest frequency.

(iii) for grouped data with continuous frequency distribution.

$$\textbf{mode} = l_1 + \left(\frac{f_m - f_1}{2f_m - f_1 - f_2}\right) \times i$$

here first we locate the modal class which is class interval with highest frequency

then l_1 = lower unit of model class

f_m = frequency of modal class

f_1 = Frequency of the class preceding to modal class.

f_2 = frequency of class succeeding to modal class.

i = size of class internal of modal class.

Note: sometimes in data, we have more than one class intervals which are having same frequencies and these may or may not be highest in value. So for these types of situations we have a method, known as grouping method to locate the mode. But in entrance exams this situation is rarely given as question.

(iv) Relation between Mean, Median and Mode

Mode = 3 Median – 2 Mean.

B. Measure of dispersions :

(i) Range : It is difference in values of highest and lowest observation in the given data.

$$\textbf{Range} = \textbf{highest} - \textbf{lowest}$$

(ii) Mean Deviation : It is mean of all the deviations. Here deviation may be calculated by taking mean, median or mode as reference values and therefore it will known as mean deviation about mean or mean deviation about median or mean deviation about mode respectively.

$$\text{Mean} - \text{deviation} = \frac{\Sigma |d_1|}{\Sigma f}$$

Where $d_i = x_i - A$ $[i = 1, 2, 3........n]$ respective deviations of each observation from the reference parameter A. Here A can be mean, median or mode of the given data.

(iii) Standard deviation

(a) for ungrouped data, $\text{S.D} = \sqrt{\frac{\Sigma di^2}{n}}$

where $d_i = xi - A$, i = 1, 2, 3.......n

(b) If the frequency of the variables $x_1, x_2,....x_n$ is $f_1, f_2,....., f_n$ respectively then standard deviation is given by

$$\text{S.D} = \sigma = \sqrt{\frac{\Sigma d_i^2 f_i}{n}}$$

till now we have done a few measures and now we discuss correlation.

Till now we have gone through the basic concept related to this topic. Now we will do the types of questions, generally asked for competitive examination point of view from this topic.

SOLVED EXAMPLES

Example 1. Find the mean of data given below

(i) 15, 20, 25, 10, 5, 30, 12, 8

(ii) If the frequencies of observation of above data are 2, 4, 3, 5, 7, 6, 1, 8 respectively.

Sol : (i) we know mean $= \dfrac{x_1 + x_2 + \ldots\ldots + x_n}{n}$

here total observations are $n = 8$

and $x_1, x_2, x_3 \ldots.$ are 15, 20, 25, 10, 5,.......

$$\therefore \quad \text{mean} = \frac{15+20+25+10+5+30+12+8}{8} = 15.62$$

(ii)

x_i	f_i	$f_i x_i$
15	2	30
20	4	80
25	3	75
10	5	50
5	7	35
30	6	180
12	1	12
8	8	64
	$\Sigma f_i = 36$	$526 = \Sigma f_i x_i$

we know when frequency is given the mean is

$$\text{mean} = \frac{\Sigma f_i x_i}{\Sigma f_i} = \frac{526}{36} = 14.611$$

Example 2. *Find the mode of data given below*

(i) 2, 2, 3, 5, 7, 8, 9, 2, 2, 9

(ii) x : 10, 15, 20, 25, 30, 35

f: 7, 18, 16, 14, 12, 10

(iii) class interval: 0-10, 10-20, 20-30, 30-40, 40-50

frequency : 4 8 11 7 8

Sol : (i) we known mode is that observation in data which comes maximum no. of times in that data.

here 2 is coming four times in data, which maximum in occurance

∴ mode of this data is 2

(ii) In this data frequency of observation 15 is maximum(18)

∴ mode of this data is 15

(iii) here frequency of $C.I$, 20-30 is maximum

∴ 20-30 is modal class

for cumulative frequency distribution, mode is

$$= l_1 + \left(\frac{f_m - f_1}{2f_m - f_1 - f_2}\right)xi$$

where l_1 = lower limit of modal class, here l_1 =20

f_m = frequency of modal calss, here f_m = 11

f_1 = frequency of class preceeding to modal class

here f_1 = 8

f_2 = frequency of class succeding to modal class

here f_2 = 7

i = width of class interval.

mode $$= 20 + \left(\frac{11-8}{22-8-7}\right) \times 10$$

$$= 20 + \left[\frac{3}{7}\right] \times 10 = 24.28$$

Example 3. *Find the median of data given below*

(i) 6, 11, 13, 15, 17, 21, 24

(ii) x : 10, 15, 20, 25, 30, 35

f: 7, 18, 19, 6, 8, 24

(iii) class interval: 0-10, 10-20, 20-30, 30-40, 40-50

frequency : 4 8 21 7 8

Sol : (i) Here total observation are 7, which are odd in number. So to locate the median, the data is arranged either in ascending order or descending order but here data is already given in ascending order. i.e.,

6, 11, 13, 15, 17, 21, 24

here observation are odd

∴ $$\text{median} = \text{value of } \left(\frac{n+1}{2}\right)\text{th item}$$

$$= \text{value of } \left(\frac{7+1}{2}\right)\text{th item}$$

$$= \text{value of 4th item} = 15$$

(ii)

x	f	$c.f$
10	7	7
15	18	25
20	19	44
25	6	50
30	8	58
35	24	82

now value of $\left(\frac{n+1}{2}\right)$th item $= \left(\frac{82+1}{2}\right)$th $= \left(\frac{83}{2}\right)$th $= 41.5$

The value of $\left(\frac{n+1}{2}\right)$th item (which is 41.5) falls in 3rd observation in cumulative frequency table which corresponds to 3rd observation in value of variables ($x_1, x_2, ...$)

∴ median of this data = 20

(iii)

C.I	FREQUENCY	C.F
0-10	4	4
10-20	8	12
20-30	11	23
30-40	7	30
40-50	9	39
	N = 39	

$$\text{median} = l + \frac{\left(\frac{N}{2} - C\right)}{f} x_i$$

Now $\frac{N}{2} = \frac{39}{2} = 19.5$, which coresponds to class interval 20-30 as per cumulative frequency table

∴ 20-30 is median class

$$\therefore \text{ Median} = 20 + \left(\frac{19.5 - 12}{11}\right) \times 10$$

$$= \mathbf{26.818}$$

Example 4. The average marks of 80 students found to be 50. Later it was disclosed that a score of 35 was misread as 65. Find the corrected mean of 80 students.

Sol : we know that $\frac{\sum_{i=1}^{n} x_i}{\Sigma n}$ ...(i) here $\bar{x}$ = 50, Sn = 80

∴ from (i) $50 = \dfrac{\sum_{i=1}^{20} x_i}{80}$ or $\sum_{i=1}^{20} x_i = 80 \times 50 = 4{,}000$

i.e., total marks of whole class is 4,000

now we will deduct the score of 65 form the total marks and add the score of 35

i.e., $(4000 - 65) + 35 = 3970$

∴ corrected mean = $\dfrac{3970}{80} = 49.62$ (Approx)

Example 5. In a class of 40 students, 10 have failed and their average marks is 4. The total marks scored by entire class is 600. find the average of students who have passed.

Sol : As the average marks of 10 students, who have failed are 4

∴ Total marks scored by these 10 students = 10 × 4 = 40

Total marks scored by entire class = 600

marks scored by 40 students who have passed = 600 – 40 = 560

hence their average = $\dfrac{560}{40} = 14$

Example 6. In a certain distribution, the mean, and median are 24 and 26 respectively. Find the value of the mode.

Sol. we know that

mode = 3 median – 2 mean

∴ mode = 3(24) – 2(26) = 72 – 52 = 20

Example 7. If arithmetic mean of 10 observation is 120 and that of 8 observation is 90. The combined mean of 18 observation is

Sol : We know that mesured for combined mean of two groups is given by

$$\bar{x} = \frac{n_1 \bar{x}_1 + n_2 \bar{x}_2}{n_1 + n_2} \quad \text{here } n_1 = 10, \quad \bar{x}_1 = 120 \text{ and } n_2 = 8, \quad \bar{x}_2 = 90$$

$$\therefore \quad \bar{x} = \frac{10 + 120 + 8 \times 90}{10 + 8} = \frac{1200 + 720}{18} = \frac{1920}{18} = 106.66$$

Example 8. Find the mean deviation of 2,4,6,8,10,12,14 about mean.

Sol : Mean deviation about mean is given by

$$\text{M.d} = \frac{\sum_{i=1}^{n} |d_i|}{n}, \text{ where } |d_i| = |x_i - \bar{x}|$$

so we will find the mean of givem set of observations which is

$$\bar{x} = \frac{2 + 4 + 6 + 8 + 10 + 12 + 14}{7} = \frac{56}{7} = 8$$

then to find deviations (d_i, l = 1 to 7) we will subtract each given observation from mean and for mean deviation we will take the mean of deviation

$$\text{mean deviation} = \frac{\sum_{i=1}^{n}|d_i|}{n} = \frac{\sum_{i=1}^{n}|x_i - \bar{x}|}{n}$$

$$= \frac{|x_1-\bar{x}|+|x_2-\bar{x}|+|x_3-\bar{x}|+|x_4-\bar{x}|+|x_5-\bar{x}|+|x_6-\bar{x}|+|x_6-\bar{x}|+|x_7-\bar{x}|}{7}$$

$$= \frac{|2-8|+|4-8|+|6-8|8-8|+|10-8|+|12-8|+|14-8|}{7}$$

$$= \frac{6+4+2+0+2+4+6}{7} = \frac{24}{7}$$

Example 9. Find the standard deviation of the data

C.I :	0 - 5	5 - 10	10 - 20	20 - 30	30 - 40
f. :	12	18	16	19	14

Sol :

C.I	Class Marks	f	$c.f$	$f_i x_i$	$x_i - \bar{x}$
0 - 5	2.5	12	12	30	14.84
5 - 10	7.5	18	30	135	9.84
10 - 20	15	16	46	240	2.34
20 - 30	25	19	55	475	– 7.66
30 - 40	35	14	69	490	17.66

$$\text{Mean} = \bar{x} = \frac{\Sigma f_i x_i}{\Sigma f_i} = \frac{1370}{79} = 17.34$$

$$\text{S.D} = \sqrt{\frac{D(x_1-\bar{x})^2}{n}} = \sqrt{\frac{693.05}{79}} = \sqrt{8.77} = 2.96$$

EXERCISE

Q.1. If the algebric sum of 10 observation is 20, then the mean is

(a) 2 (b) 4 (c) $\frac{1}{2}$ (d) $\frac{1}{4}$

Q.2. The median of the series is 16, two additional observation 6 and 21 are added to the given series. The median of the new series is

(a) 14 (b) 15 (c) 16 (d) none

Q.3. Coefficient of correlation between two variables x and y as

x	1	2	3	4	5
y	5	4	3	2	1

(a) –1 (b) 0 (c) – 2 (d) none

Q.4. The standard deviation of the series 22, 26, 28, 20, 24, 30 is

(a) $\frac{29}{4}$ (b) 14 (c) 3.41 (d) none

Q.5. If the mean of following distribution is 2.6, then the value of y is

x	1	2	3	4	5
f	4	5	y	1	2

(a) 14 (b) 8 (c) 4 (d) 6

Q.6. The mode of the series 2, 6, 72, 9, 2, 8, 7, 17, 10, 2, 11, 21, 12

(a) 2 (b) 6 (c) 9 (d) 11

Q.7. The median of observations 5, 6, 7, 9, 4, 3, 2, 8, 19 is

(a) 4 (b) 9 (c) 6 (d) 8

Q.8. Which one of the following is not median

(a) 5th decile (b) 5th percentile (c) 2 quartile (d) none of these

Q.9. Weigthed mean is given

(a) $\frac{\Sigma w_i n_i}{\Sigma w_i}$ (b) $\frac{\Sigma w_i}{\Sigma w_i n_i}$ (c) $\frac{\Sigma w_i f_i}{\Sigma w_i n_i}$ (d) none

Q.10. A distribution consists of three groups having 50, 60, 70 items with means 20, 24, 28 respectively. The mean of the distribution is

(a) 56 (b) 45 (c) 30 (d) 25

Q.11. Quartile deviation is nearly equal to

(a) $\frac{3}{4}\sigma$ (b) $\frac{2}{3}\sigma$ (c) $\frac{7}{9}\sigma$ (d) $\frac{\sigma}{2}$

Q.12. If mean = [3 median – mode] y, then the value of y is

(a) 2 (b) $\frac{1}{4}$ (c) $\frac{1}{2}$ (d) none

Q.13. The standard deviation of 25 number is 40. If each of the number is increased by 5, then the new standard deviation will be

(a) 40 (b) 45 (c) $\frac{45}{2}$ (d) none

Q.14. The mode of the following data 0, 1, 6, 7, 2, 3, 7, 6, 6, 2, 6, 0, 5, 6, 0

(a) 0 (b) 7 (c) 2 (d) 6

Q.15. The standard deviation of first n-natural members is

(a) $\frac{n^2-6}{36}$ (b) $\frac{n}{12}$ (c) $\sqrt{\frac{n^2-1}{12}}$ (d) none

Q.16. If the mean of n=observations $1^2, 2^2, 3^2, 4^2, \ldots n^2$ is $\frac{46n}{11}$ item n is equal to

(a) 11 (b) 12 (c) 169 (d) 184

Q.17. The range of following set of observations.

2, 3, 5, 9, 8, 7, 6, 5, 7, 4, 3

(a) 7 (b) 8 (c) 5 (d) none

Q.18. A man purchases green pen worth Rs. 100 at the rate of Rs. 5 per pen, red pens worth Rs. 100 at rate of Rs. 10 per pen, and black pens worths Rs. 100 at rate of Rs. 20 per pen. The average cost of one pen is

(a) $\frac{29}{728}$ (b) $\frac{17}{201}$ (c) $\frac{25}{39}$ (d) $\frac{60}{7}$

Q.19. The mean of Ist n – natural numbers is

(a) $\frac{n+1}{2}$ (b) $\frac{n(1+1)}{2}$ (c) $\frac{n}{2}$ (d) $\frac{n^2+1}{2}$

SOLUTIONS

1. (a) we know that mean if variable x is given by

$$\bar{x} = \frac{\sum_{i=1}^{n} x_i}{\Sigma n}$$

here $n = 10$ $\quad \sum_{i=1}^{n} x_i = 20$

$$\therefore \quad \bar{x} = \frac{20}{10} = 2$$

2. (b) The median will again be 16 as 6 is less than median and 21 is more than median, therefore the position of median (which is mid point of data) will remain unchanged.

3. (a) Source $x + y = 6$ for all the pairs of observations

$$\therefore \quad b_{xy} = b_{yx} = -1$$

hence $\quad r = -1$

4. (c) Mean of given observation is mean $= \frac{22+26+28+20+24+30}{6} = 25$

$$\text{also S.D} = \sqrt{\frac{(22-25)^2+(26-25)^2+(28-25)^2+(20-25)^2+(24+25)^2+(30-25)^2}{6}}$$

$$= \sqrt{\frac{70}{6}} = \sqrt{\frac{35}{3}}$$

5. (b) we know that mean $= \frac{\Sigma f_i x_i}{\Sigma f_i}$ or, $2.6 = \frac{4+10+3y+4+10}{12+y}$

or, $\quad 2.6 = \frac{28+3y}{12+y}$ or, $(12+y)(2.6) = 28+3y$ or, $y = 8$

6. (a) Since mode is that value or observation in data, which is coming maximum number of times in data.

It clear from the data that 2 is coming maximum number of times (4) times.

Hence 2 mode of this data.

7. (c) For finding the median of given data first we have to arrange the data either in ascending order or in descending order, then median is located as per number of observations in data is even or odd.

After arranging the data in ascending order. We get as follows 2,3,4,5,6,7,8,9,10

here the number of observations are odd as total observations are 9.

therefore median is $\left[\frac{n+1}{2}\right]$th observation or median $= \left[\frac{9+1}{2}\right]$th observation

$$= 5\text{th observation}$$
$$= 6$$

8. (d) rest all the medians

9. (a) $\dfrac{\Sigma w_i x_i}{\Sigma w_i}$

10. (d) We know that formula for combined mean is given by $\overline{x} = \dfrac{n_1\overline{x_1} + n_2\overline{x_2} + n_3\overline{x_3}}{n_1 + n_2 + n_3}$ where $\overline{x_1}$,

and $\overline{x}_2$ are the means of groups containing n_1, n_2 and n_3 items respectively.

$\overline{x}_3$ here $\quad n_1 = 50 \quad n_2 = 60 \quad n_3 = 70$

$\quad \overline{x_1} = 20 \quad \overline{x}_2 = 24 \quad \overline{x}_3 = 28$

$\therefore$ required combined mean $= \dfrac{50 \times 20 + 60 \times 24 + 70 + 28}{50 + 60 + 70}$

$$= \frac{1000 + 1940 + 1960}{180}$$

$$= \frac{4500}{180} = \mathbf{25}$$

11. (b)

12. (c) We know that

$$\text{mode} = 3\text{ median} - 2\text{ mean} \qquad \text{...(i)}$$

$\therefore$ from (i)

$$2\text{ mean} = 3\text{ median} - \text{mode}$$

or

$$\text{mean} = \frac{3}{2}\text{median} - \frac{1}{2}\text{mod}e$$

$$= [3\text{ median} - \text{mode}]\frac{1}{2} \qquad \text{...(ii)}$$

now comparing (ii) with the given result which is

$$\text{mean} = [3\text{ median} - \text{mode}]y$$

we get

$$y = \frac{1}{2}$$

13. (a) standard deviation and variance of the data is same, when each observation in the data is increased or decreased by the same number.

14. (d) 6 [$\therefore$ its frequency is max.]

15. (c) $\text{S.D} = \sqrt{\dfrac{\Sigma n^2}{n} - \left(\dfrac{\Sigma n}{n}\right)^2}$ [Formula for S.D] now Sn^2 = sum of square of Ist n natural numbers

$$= \frac{n1(n+1)(2n+1)}{6}$$

$$\therefore \qquad \text{S.D} = \sqrt{\frac{(n)(n-1)(2n+1)}{6n} - \frac{n^2(n+1)^2}{4n^2}}$$

$$= \sqrt{\frac{n^2-1}{12}}$$

16. (a) mean of $1^2, 2^2, 3^2, ..., n^2$

$$\text{mean} = \frac{1^2+2^2+3^2+..........+n^2}{n}$$

$$= \frac{n(n+1)(2n+1)}{6n}$$

$$= \frac{(n+1)(2n+1)}{6}$$

but given $$\text{mean} = \frac{46n}{11}$$

$$\frac{46n}{11} = \frac{(n+1)(2n+1)}{6}$$

or, $22n^2 - 243n + 11 = 0$

or, $(n-11)(22n-1) = 0$

or, $n = 11$, as n cannot be fraction

17. (a) range = 9 – 2 = 7

[range = largest value – lowest value]

18. (d) total no. green pens $\frac{100}{5} = 20$ = pens

total no. of red pens $\frac{100}{10} = 10$ = pens

total no. of black pens $\frac{100}{20} = 5$ = pens

so this is question of combined mean of three groups having 20, 10 and 5 items respectively

required mean = $\frac{1000+100+100}{20+10+5} = \frac{60}{7}$

19. (a) the first n natural numbers are 1, 2, 3, ..., n.

$$\therefore \qquad \text{required mean} = \frac{1+2+3+4+5+...+n}{n}$$

$$= \frac{(n)(n+1)}{2.n} = \frac{n+1}{2}$$

Section V

Test Yourself

1. There are two urns. One contains two white balls and four red balls, the other contains three white and nine red balls. All balls are of the same shape and size. From each urn, one ball is drawn. What is the probability of getting both the balls of the same colour? **(CLAT 2017)**

   (a) 7/12 (b) 1/2 (c) 1/24 (d) 1/12

2. Three fair dice are thrown simultaneously. What is the probability that the sum of numbers on their tops, is at least 6? **(CLAT 2018)**

   (a) 5/108 (b) 1/24 (c) 103/108 (d) 17/18

3. John was writing an entrance test. Every question has multiple choice questions with four probable answers and one correct answer. The probability that he knows the answer is 80%. If he correctly answered a question then what is the probability that he was guessing? **(SNAP 2017)**

   (a) 1/5 (b) 1/17
   (c) None of the mentioned options are correct (d) 1/27

4. In how many ways can we arrange the word CAPSTONE so that all the vowels come together? **(SNAP 2017)**

   (a) 1440 (b) 4320 (c) 2160 (d) 3480

5. On her birthday, Vrinda gets a gift box which contains 6 chocolates and 8 cup cakes. She eats 7 of them randomly. What is the probability that she ate 3 chocolates and 4 cupcakes? **(SNAP 2017)**

   (a) 0.16231 approximately (b) 0.0565S approximately
   (c) 0.4079 approximately (d) 0.01165 approximately

6. $\sin^2 5 + \sin^2 10 + \ldots\ldots. \sin^2 80 + \sin 85 = ?$ **(SSC CHSL LDC 2011)**

   (a) $39\frac{1}{2}$ (b) $40\frac{1}{2}$ (c) 40 (d) $39\frac{1}{\sqrt{2}}$

7. The expression $\dfrac{\tan 57° + \cot 37°}{\tan 33° + \cot 53°}$ is equal to: **(SSC Graduate level Tier II 2012)**

   (a) tan 33° cot 57° (b) tan 57° cot 37° (c) tan 33° cot 53° (d) tan 53° cot 37°

8. The value of is $\dfrac{\cot 30° - \cot 75°}{\tan 15° - \tan 60°}$ **(SSC CHSL LDC 2012)**

   (a) 0 (b) 1 (c) $\sqrt{3}-1$ (d) – 1

9. The value of will be: $\operatorname{cosec}^2 60° + \sec^2 60° - \cot^2 60° + \tan^2 30°$ **(SSC CGL Tier-I 2016)**

   (a) 5 (b) $5\frac{1}{2}$
   (c) $5\frac{1}{3}$ (d) $5\frac{2}{3}$

10. If $(\theta + 23°) = \cos 58°$, then what is the value of cos 5? **(SSC CAPF Delhi Police Exam 2017)**

(a) $\frac{1}{2}$ (b) $\frac{1}{\sqrt{2}}$ (c) $\frac{\sqrt{3}}{2}$ (d) 0

11. If $\text{cosec}^2\theta = \frac{625}{576}$, then what is the value of $\left[\frac{(\sin\theta - \cos\theta}{\sin\theta + \cos\theta}\right]$? **(SSC CGL Tier-I 2017)**

(a) 1 (b) $\frac{31}{17}$ (c) $\frac{17}{31}$ (d) $\frac{14}{25}$

12. What is the value of $\left[\tan^2(90-\theta) - \sin^2(90-\theta)\right]\text{cosec}^2(90-\theta)\cot^2(90-\theta)$?

(SSC CGL Tier-II 2018)

(a) 0 (b) 1 (c) – 1 (d) 2

13. What is the value of $\left(\sin 60° + \frac{2}{\sqrt{3}}\right)$? **(SSC CGL Tier-II 2018)**

(a) $\frac{(2\sqrt{3}+1)}{\sqrt{2}}$ (b) $\frac{7}{2\sqrt{3}}$ (c) 2 (d) $2+\sqrt{3}$

14. The top of a broken tree touches the ground at a distance of 15 meter from its base. If the tree is broken at a height of a 8 meter from the ground, then the actual height of the tree is.

(SSC CGL Tier-I 2016)

(a) 17 meter (b) 20 meter (c) 25 meter (d) 30 meter

15. A helicopter, at an altitude of 1500 meter, finds that two ships are sailing towards it, in the same direction. The angle of the depression of the ships as observed on the helicopter are to representatively. distance between the two ships, in meter is: **(SSC CGL Tier-I 2016)**

(a) $1000\sqrt{3}$ (b) $\frac{1000}{\sqrt{3}}$ (c) $500\sqrt{3}$ (d) $\frac{500}{\sqrt{3}}$

16. The distance between the tops of two buildings 38 meters and 58 meters high is 52 meters. What will be the distance (in meters) between two buildings? **(SSC CGL Tier-II 2018)**

(a) 46 (b) 42 (c) 44 (d) 48

17. The angles of the elevation of the top of a tree 220 meter high from two points lie on the same plane are and What is the distance (in meter) between the two points?

(SSC CGL Tier-II 2018)

(a) 461,02 (b) 142.03 (c) 441.09 (d) 161.05

18. If 7 times the Seventh term of an Arithmetic Progression (AP) is equal to 11 times it Eleventh term, then the 18th term of the AP will be **(SSC CGL Tier-I 2016)**

(a) 1 (b) 0 (c) 2 (d) –1

19. what is the sum of the first 13 terms of an Arithmetic Progression if the first term is –10 and last term is 26? **(SSC CGL Tier-I 2017)**

(a) 104 (b) 140 (c) 84 (d) 98

20. The 3rd and 7th term of an Arithmetic progression are –9 and 11 respectively. What is the 15th term? **(SSC CGL Tier-I 2017)**

(a) 28 (b) 87 (c) 51 (d) 17

21. What is the unit digit of the sum of first 111 whole numbers? **(SSC CGL Tier-II 2018)**

(a) 4 (b) 6

(c) 5 (d) 0

22. An aeroplane when 3000 m high passes vertically above another aeroplane at an instant when the angles of elevation of the two aeroplanes from the same point on the ground are 60° and 45° respectively. The vertical distance between the two aeroplanes is **(MAT Feb 2010)**

(a) 1268 m (b) 1500 m (c) 3000 m (d) 1200 m

23. An aeroplane flying horizontally 1 km above the ground is observed at an elevation of 60°. If after 10 seconds, the elevation be 30°, the uniform speed of the aeroplane is

(MAT Feb 2010)

(a) $240\sqrt{3}$ km/hr (b) $240/\sqrt{3}$ km/hr (c) $120/\sqrt{3}$ km/hr (d) $120\sqrt{3}$ km/hr

24. From the top of a cliff 25 m high the angle of elevation of a tower is found to be equal to the angle of depression of the foot of the tower. The height of the tower is **(MAT Feb 2010)**

(a) 25 m (b) 50 m (c) 75 m (d) 100 m

25. Namrata wants to visit four cities A, B, C and D on an official trip. The probability that she visits A just before B is **(MAT Feb 2010)**

(a) 1/2 (b) 1/12 (c) 1/6 (d) 1/4

26. A lot of 12 bulbs contains 4 defective bulbs. Three bulbs are drawn at random from the lot, one after the other. The probability that all three are non-defective is **(MAT Feb 2010)**

(a) 14/55 (b) 8/12 (c) 1/27 (d) None of these

27. The supervisor of a packaging unit of a milk plant is being pressurised to finish the job closer to the distribution time, thus giving the production staff more leeway to cater to last minute demand. He has the option of running the unit at normal speed or at 110% of normal -"fast speed". He estimates that he will be able to run at the higher speed 60% of time. The packet is twice as likely to be damaged at the higher speed which would mean temporarily stopping the process. If a packet on a randomly selected packaging runs has probability of 0.112 of damage, what is the probability that the packet will not be damaged at normal speed? **(Xat 2010)**

(a) 0.81 (b) 0.93 (c) 0.75 (d) 0.60

(e) None of the above

28. The odds that A speaks the truth are 3 : 2 and the odds that 'B' speaks the truth are 5 : 3. In what percent of cases are they likely to agree each other on an identical point? **(MAT Sept 2009)**

(a) 47.5% (b) 37.5% (c) 63.5% (d) None of these

29. Chanda Negi got a 4-digit pass code (which is formed out of the digits 0, 1, 2, 3, 4, 5, 6, 7, 8, 9) of her ATM card from SBI Bank. But after the 50th day she lost the pass code and also forgot the number. How many maximum number of trials she may have to take to get the right number? (0 can be the beginning of the code number.) **(MAT Sept 2009)**

(a) 10^4 (b) 10! (c) 9^4 (d) 9!

30. A number lock on a suitcase has 3 wheels each labelled with 10 digits from 0 to 9. If opening of the lock is a particular sequence of three digits with no repeats, how many such sequences will be possible? **(MAT Sept 2009)**

(a) 720 (b) 760 (c) 680 (d) 780

31. A portion of a 30 m long tree is broken by a tornado and the top strikes the ground making an angle of 30° with the ground level. The height of the point where the tree is broken is equal to

(MAT Sept 2009)

(a) $30/\sqrt{3}$ m (b) 10 m (c) $30\sqrt{3}$ m (d) 60 m

32. Out of eight crew members three particular members can sit only on the left side. Another two-particular members can sit only on the right side. Find the number of ways in which the crew can be arranged so that four men can sit on each side. **(MAT Sept 2009)**

(a) 865 (b) 864 (c) 863 (d) 1728

DIRECTIONS ***for questions (32 to 37):*** Study the following information carefully and answer the questions that follow.

A basket contains 3 red, 4 blue and 5 green marbles.

33. If three marbles are picked at random, what is the probability that all are blue? **(ATMA 2010)**

(a) $\frac{5}{12}$ (b) $\frac{1}{3}$ (c) $\frac{3}{22}$ (d) $\frac{1}{55}$

34. If two marbles are picked at random, what is the probability that both are either blue or both are green? **(ATMA 2010)**

(a) $\frac{3}{4}$ (b) $\frac{8}{33}$ (c) $\frac{5}{11}$ (d) $\frac{6}{11}$

35. If three marbles are picked at random, what is the probability that none is red? **(ATMA 2010)**

(a) $\frac{4}{11}$ (b) $\frac{1}{4}$ (c) $\frac{21}{55}$ (d) $\frac{2}{33}$

36. If two marbles are picked at random, what is the probability that at least one is green? **(ATMA 2010)**

(a) $\frac{15}{22}$ (b) $\frac{5}{22}$ (c) $\frac{1}{6}$ (d) $\frac{7}{22}$

37. If four marbles are picked at random, what is the probability that two are red and two are green? **(ATMA 2010)**

(a) $\frac{2}{33}$ (b) $\frac{7}{12}$ (c) $\frac{6}{11}$ (d) $\frac{1}{3}$

38. In how many different ways can 3 boys and 5 girls be made to stand in a row such that all the girls are standing together and all the boys are standing together? **(ATMA 2010)**

(a) 225 (b) 1440 (c) 15 (d) 720

39. Find the sum of 10 terms of the arithmetic progression series {10,12, 14, ...}?

(a) 190 (b) 380 (c) 140 (d) 260

40. In how many different ways can the letters of the word 'FITTEST' be arranged?

(a) 5040 (b) 840 (c) 1680 (d) 2520

SOLUTIONS

1. (a)

Required probability

$$= \frac{2}{6} \times \frac{3}{12} \times \frac{4}{6} \times \frac{9}{12}$$

$$= \frac{6+36}{72}$$

$$= \frac{42}{72} = \frac{7}{12}$$

2. (c) Required probability

$$= 1 - (\text{getting sum at most 5})$$

so total cases of at most 5

are (1, 1, 1), (1, 1, 1) (1, 2, 1)
(2, 1, 1) (1, 1, 3) (1, 3, 1)
(3, 1, 1) (1, 1, 2) (2, 1, 1)
(2, 1, 1)

So ten Cases are there

$\therefore$ Required probability

$$= 1 - \frac{10}{216} = \frac{206}{216}$$

$$= \frac{103}{108}$$

3. (b)

Required probability

$$= \frac{\frac{1}{4} \times \frac{1}{5}}{1 \times \frac{4}{5} + \frac{1}{4} \times \frac{1}{5}} = \frac{1}{17}$$

4. (b)

Required number of ways = 6! × 3!

$$= 720 \times 6 = 4320$$

5. (c) Required probability

$$= \frac{8_{C_4} \times 6_{C_3}}{14_{C_2}}$$

$$= .4079$$

6. (b) $\sin^2 5 + \sin^2 6 - \sin^2 85$
$(\sin^2 5 + \sin^2 85) + (\sin^2 6 + \sin^2 84)$
$(\sin^2 44 + \sin^2 46) + \sin^2 45$

$$= (\sin^2 5 + \cos^2 5) + (\sin^2 6 + \cos^2 6)$$

$$= (\sin^2 44 + \cos^2 44) + \sin^2 45$$

$$= 1 + 1 + \ldots.\ 40 + \text{times} + \left(\frac{1}{\sqrt{2}}\right)^2$$

$$= 40\frac{1}{2} = 40\frac{1}{2}$$

7. (b) $\dfrac{\tan 57° + \cot 37°}{\tan 33° + \cot 53°}$

$$= \frac{\cot 33° + \tan 53°}{\tan 33° + \cot 53°}$$

$$\frac{\frac{1}{\tan 33°} + \tan 53°}{}$$

$$= \frac{1 + \tan 53° \tan 53°}{\tan 33° \times \tan 53° - 1} \times \frac{\tan 53°}{\tan 33°}$$

$$= \tan 53° \times \cos 23°$$

$$= \tan 37 \ \cot 57$$

8. (4)

$$\cot 30 = \cot (90 - 60)$$

$$= \tan 60$$

$$\cot 75° = \cot (90 - 15)$$

$$= \tan 15°$$

$$\therefore \quad \frac{\cot 30° - \cot 75°}{\tan 15° - \tan 60°}$$

$$= \frac{\tan 60° - \tan 15°}{\tan 15° - \tan 60°}$$

$$= -1$$

9. (3) $\operatorname{cosec}^2 60° + \sec^2 60° - \cot^2 60° + \tan^2 30°$

$$\left(\frac{2}{\sqrt{3}}\right)^2 + 2^2 - \left(\frac{1}{\sqrt{3}}\right)^2 + \left(\frac{1}{\sqrt{3}}\right)^2$$

$$= \frac{4}{3} + 4 = \frac{16}{3} = 5\frac{1}{3}$$

10. (2)

$$\sin (\theta + 23) = \cos 58°$$

$$\sin \theta + 23° = (\cos \theta + 32°)$$

$$= \sin 32°$$

$$\Rightarrow \quad \theta + 23° = 32°$$

$$\theta = 9°$$

$$\cos 5\theta = \cos 45° = \frac{1}{\sqrt{2}}$$

11. (3)

$$\operatorname{cosec}^2\theta = \frac{625}{576}$$

$$\operatorname{cosec} \theta = \frac{25}{24}$$

$$\therefore \quad \sin\theta = \frac{24}{25}$$

$$\cos\theta = \sqrt{1-\sin^2\theta}$$

$$= \sqrt{1-\left(\frac{24}{25}\right)^2} = \sqrt{\frac{49}{625}}$$

$$= \frac{7}{25}$$

$$\therefore \quad \frac{\sin\theta-\cos\theta}{\sin\theta+\cos\theta}$$

$$= \frac{\frac{24}{25}-\frac{7}{25}}{\frac{24}{25}+\frac{7}{25}}$$

$$= \frac{\frac{17}{25}}{\frac{31}{25}} = \frac{17}{31}$$

12. (2) $[\tan^2(90-\theta)-\sin(90-\theta)]$
$\text{cosec}^2(90-\theta)-\cot^2(90-\theta)$

$$= (\cot^2\theta-\cos^2\theta)\sec^2\theta.\tan^2\theta$$

$$= \left(\frac{\cos^2\theta}{\sin^2\theta}-\cos^2\theta\right).\sec^2.\tan^2\theta$$

$$= \cos^2\theta\left(\frac{1-\sin^2\theta}{\sin^2\theta}\right).\sec^2\theta\tan^2\theta$$

$$= \frac{\cos^2\theta\times\cos^2\theta}{\sin^2\theta}\times\sec^2\theta.\tan^2\theta$$

$$= 1$$

13. (2)

$$\sin 60^\circ = \sin 60^\circ + \frac{2}{\sqrt{3}} = \frac{\sqrt{3}}{2}+\frac{2}{\sqrt{3}}$$

$$\frac{3+4}{2\sqrt{3}} = \frac{7}{2\sqrt{3}}$$

14. (3)

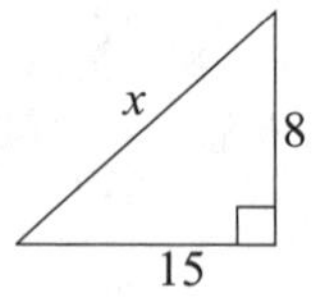

$$8^2 + 15^2 = x$$
$$x^2 = 289$$
$$x = 17$$

Height of tree $= 8 + 17 = 25\text{m}$

15. (1)

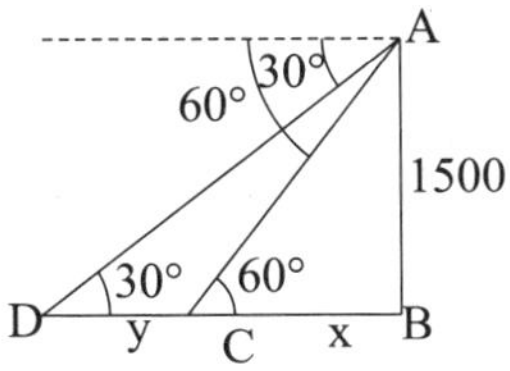

In ΔABC

$$\frac{1500}{x} = \tan 60°$$
$$\frac{1500}{x} = \sqrt{3}$$
$$x = \frac{1500}{\sqrt{3}} \quad \ldots(1)$$

In ΔABD

$$\frac{1500}{x+y} = \tan 30 = \frac{1}{\sqrt{3}}$$
$$x + y = 1500\sqrt{3} \quad \ldots(2)$$

(1) and (2)

$$y = 1500\sqrt{3} - \frac{150}{\sqrt{3}}$$
$$= \frac{3000}{\sqrt{3}} = 1000\sqrt{3}$$

16. (4)

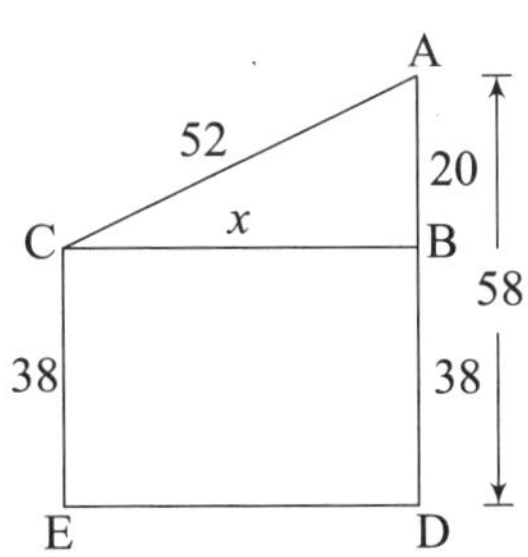

In ΔABC

$$x^2 + 20^2 = 52^2$$
$$x^2 + 400 = 2304$$
$$x^2 = 2304$$
$$\Rightarrow \quad x = 48$$

17. (4)

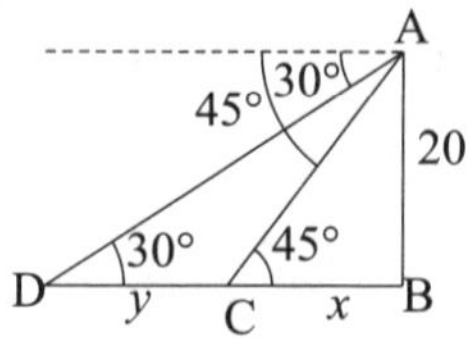

In ΔABC

$$\frac{220}{x} = \tan 45° = 1$$

$$\therefore \quad x = 220 \quad \ldots(1)$$

In ΔABD

$$\frac{200}{x+y} = \tan 30° = \frac{1}{\sqrt{3}}$$

$$\therefore \quad x + y = 220\sqrt{3}$$

$$220 + y = 220\sqrt{3}$$

$$= 220\sqrt{3} - 220$$

$$= 200(\sqrt{3} - 1)$$

$$= 220 \times 732$$

$$= 161.05$$

18. (a)

$$7a_7 = 11a_{11}$$

$$7(a + 6d) = 11(a + 10d)$$

$$7a + 42d = 11a + 110d$$

$$4a = -68d$$

$$a = -17d$$

$$a_{18} = a + 17d$$

$$= -17d + 17d = 0$$

19. (1)

$$S = \frac{n}{2}[a + l)$$

$$= \frac{13}{2}[-10 + 26)$$

$$= \frac{13}{2} = 104$$

20.

$$a + 2d = -9 \quad \ldots(1)$$

$$a + 10d = 11 \quad \ldots(2)$$

Solving (1) and (2)

$$a = -19 \text{ and } d = \text{J}$$

$$\therefore \quad a_{15} = a + 14d$$

$$= -19 + 14 \times 5 = \text{A} + 70$$

$$= 51$$

21. (3) $0 + 1 + 2 \ldots 110$

$$= 1 + 2 \ldots 110$$

$$= \frac{110 \times 111}{2} = 55 \times 111$$

$$= 6105$$

Hence unit digit $= 5$

22. (a)

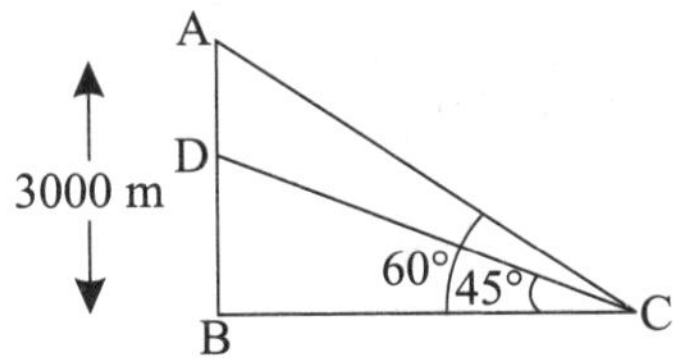

In rt triangle DBC,

tan 45° = DB/BC = 1, so DB = BC.

Also in rt triangle ABC = tan 60° = AB/BC

= AB/DB = $\sqrt{3}$, DB= 3000/ $\sqrt{3}$ =1732 m. Hence AD = 3000 –1732 = 1268 m. Option (a).

23. (a)

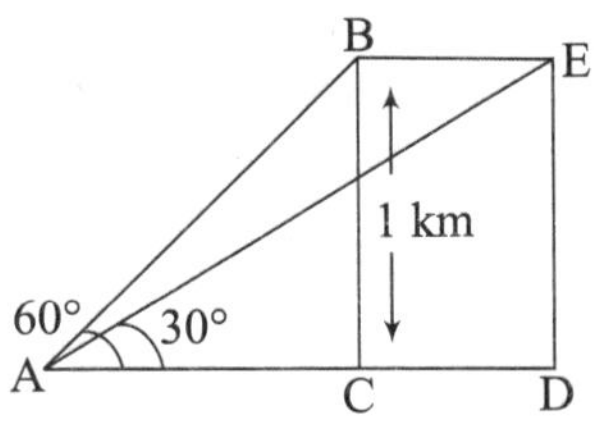

In rt triangle ADE,

tan 30° = ED/AD = 1/ $\sqrt{3}$,

→ AD = $\sqrt{3}$ km,

In rt triangle ABC, tan 60° = BC/AC = $\sqrt{3}$,

AC = $\sqrt{3}$/3, so CD = BE = AD – AC

= $\sqrt{3}$ – $\sqrt{3}$/3 = 2 $\sqrt{3}$/3 km.

$$\text{So speed of the aeroplane} = \frac{\dfrac{2\sqrt{3}}{\sqrt{3}}}{\dfrac{10}{60 \times 60}} \text{ km/hr}$$

= 240 $\sqrt{3}$ km/hr. Option (a)

24. (b)

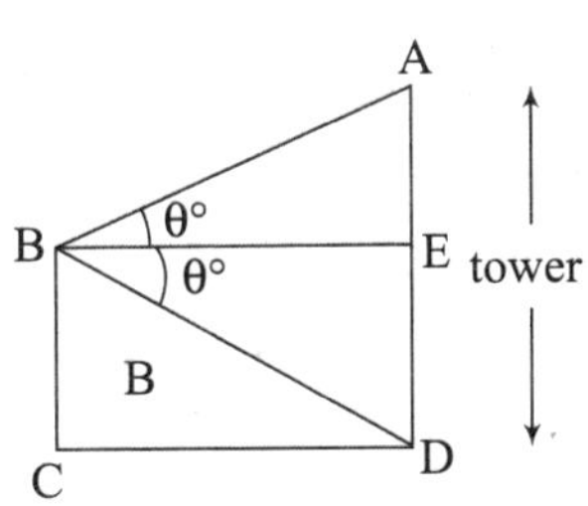

In rt triangle ABE tan θ = AE/BE, also
in rt triangle BED tan θ = ED/BE
so equating the two we get AE = ED. ED = 25 m i.e. height of the cliff.
So height of the tower = 50 m. Option (b).

25. (d) The required probability = $\frac{3!}{4!}$ =1/4. Option (d).

26. (a) The required probability = $\frac{8}{12}\times\frac{7}{11}\times\frac{6}{10}$ = 14/55. Option (a).

27. (b) Let probability of damage at normal speed = x

∴ Probability of damage at higher speed = $2x$

Hence
$$x \times 4 + 2x \times 6 = .112$$
$$4\,x + 1.2\,x = .112$$
$$1.6\,x = .112$$
$$x = \frac{.112}{1.6} = 0.7$$

Hence probability that package will not be damaged at normal speed = 1 – 0.7 = .93

28. Probability that A speaks the truth = $\frac{3}{5}$

Probability that A does not speak the truth = $\frac{2}{5}$

Probability that B speaks the truth = $\frac{5}{8}$

Probability that B does not speak the truth = $\frac{3}{8}$

Cases that are likely to agree each other on identical point = $\frac{3}{5}\times\frac{5}{8}+\frac{2}{5}\times\frac{3}{8}$

$= \frac{21}{40}$ = 52.5%. Hence none of these.

29. (a) 4 digits and every place has 10 options, so clearly the answer is $10 \times 10 \times 10 \times 10 = 10^4$

30. (a) Number of such sequences possible = 10 × 9 × 8 = 720.

31. (b)

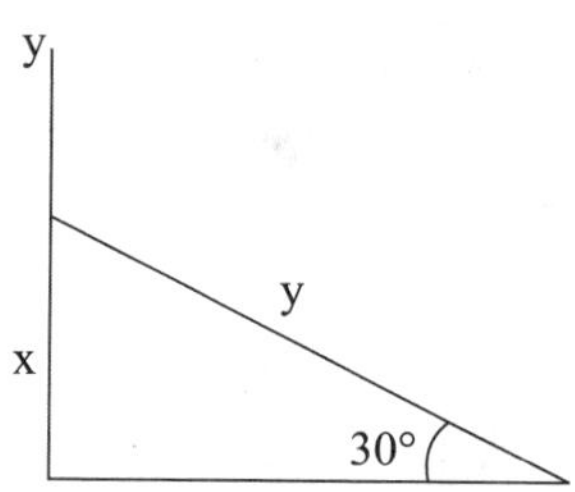

The height of the tree = 30 m. So, $x + y = 30$ m

Also $\sin 30^0 = \frac{x}{y} \Rightarrow \frac{1}{2} = \frac{x}{y} \Rightarrow x = 2y$, so $3x = 30$,

$$x = 10 \text{ m.}$$

32. (d) Three particular members can sit only on the left side in ${}^4P_3$ ways. Also, two particular members can sit only on the right side in ${}^4P_2$ ways. Also, the remaining three people can sit in 3 remaining seats in 3! ways. So required answer is ${}^4P_3 \times {}^4P_2 \times 3! = 1728$. Hence answer is 4th option.

33. (d) Required probability $= \frac{{}^4C_3}{{}^{12}C_3} = \frac{1}{55}$. Option (d).

34. (b) Required probability $= \frac{{}^4C_2 + {}^5C_2}{{}^{12}C_2} = \frac{8}{33}$. Option (b).

35. (c) Required probability $= \frac{{}^9C_3}{{}^{12}C_3} = \frac{21}{55}$. Option (c).

36. (a) Required probability $= 1 - \frac{{}^9C_3}{{}^{12}C_2} = \frac{15}{22}$. Option (a).

37. (a) Required probability $= \frac{{}^4C_3}{{}^{12}C_3} = \frac{1}{55}$. Option (a).

38. (b) Required no. of ways = 2! × 3! × 5! = 1440 ways. Option (b).

39. (a) Sum of an AP = n/2 {2a + n (n–1)d}

Here a = 10, d = 2, n = 10. On solving we get sum = 190. Option (a).

40. (b) Required no. ways $= \frac{7!}{3!} = 840$.

VEDIC MATHS SUPPLEMENT

Vedic maths is a very important topic of quantitative aptitude. It is not only helpful in calculations but also helpful in solving typical number system question **like finding last two digits of 847^{148}, square root of 1048576, square of 493, etc.** With the help of Vedic maths you can solve these types of questions in fraction of seconds.

Students, you should understand that number system is an important topic of **quantitative aptitude** and questions of number system are always there in any competitive exam. So in order to clear any competitive exam you should have a good understanding of number system concepts which is not possible without Vedic maths.

So in short, to clear any competitive exam you should have a good hold on Vedic maths. So in this supplement of vedic math, we will help you to understand Vedic maths in a ***very easy and simple way***. Given below are the concepts of multiplication with the help of Vedic maths.

VEDIC MATHS CONCEPT I (MULTIPLICATION BELOW BASE)

Vedic maths is useful when a number is near to **10, 100, 10000**.....let us suppose we have to ***multiply 97 by 99***. We will do it in following steps:

STEP I: We will see **99 and 97** are nearer to which rounded number say **10, 100, 1000, 100000,** etc.obvously they are near to **100**. So we take **100 as base.**

STEP II: Both **99 and 97** are less than **100**. So now the **first question** is they are less than base *i.e.*., 100 by how much?

Of course **99 is 1 less than 100** and **97 is 3 less than 100**

STEP III: in order to multiply **99 by 97** we will write as shown below:

99 – 01

97 – 03

We will read it as **"99 is less than the base by *i.e.*, 100 by 1 and 97 is less than the base *i.e.*, 100 by 3."** One very important point is to be noted that **negative sign** is there as numbers are **less than the base** and secondly we have written **1 as 01** and **3 as 03** as **we have to write the number of digits equal to the number of zero's in base**. As **100 have 2** zeros so we have written **1 as 01** and **3 as 03**. Now you can easily understand if we had to multiply **999 by 997** we would have taken the **base as 1000** and would have written as shown below

999   – 001

997   – 003

We will now solve in next step

STEP IV: Step IV is shown below:

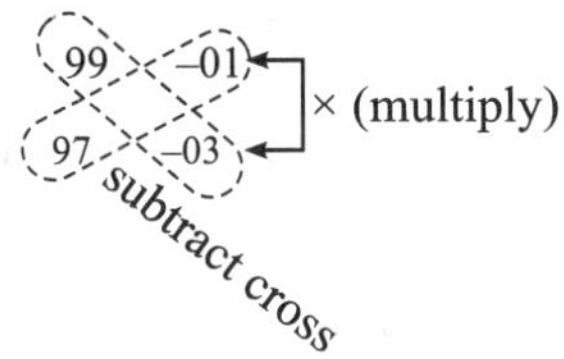

This is the last step. In this as shown above **we will subtract cross (03 from 99 or 01 from 97)** and **multiply (–01 by –03 *i.e.*, –01 × –03 = 03) to get 9603** as final answer. The beauty of the method is that when you subtract cross as shown you will get the same result in both the pairs **means 99 – 3 = 96 and 97 – 1 = 96 both are same. You can do subtraction in any of the pair**. Hope fully you have understood the concept. Now do the following example:

Example 1:

Solve the following with Vedic maths

(i) Multiply **98 by 94**

(ii) Multiply **99 by 89**

(iii) Multiply **96 by 91**

(iv) Multiply **97 by 88**

(v) Multiply **94 by 86**

Sol :

(*i*) Multiply **98 by 94.**

As both the numbers are near to 100. So we take **100 as base**. Further **98 is less than base by *i.e.*, 100 by 2 and 94 is less than base *i.e.*, 100 by 6.** *Now as discussed we will solve as follows*:

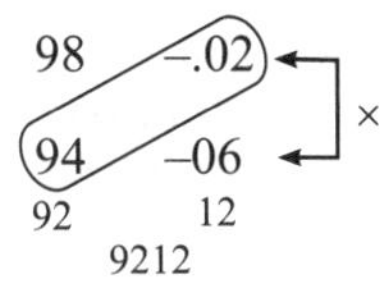

As shown above **9212** is the answer.

(*ii*) Multiply **99 by 89.**

As both the numbers are near to 100. So we take **100 as base**. Further **99 is less than base by *i.e.*, 100 by 1 and 89 is less than base *i.e.*, 100 by 11**. *Now as discussed we will solve as follows:*

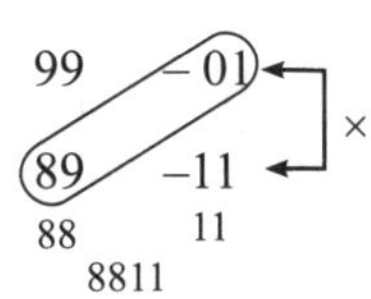

As shown above **8811** is the answer.

(*iii*) Multiply **96 by 91**.

As both the numbers are near to 100. So we take **100 as base**. Further **96 is less than base by *i.e.*, 100 by 4 and 91 is less than base *i.e.*, 100 by 9.** Now as discussed we will solve as follows:

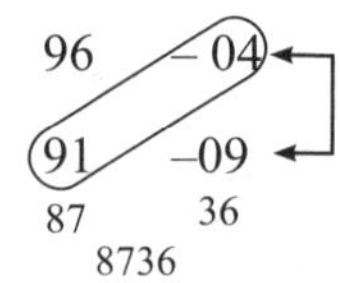

As shown above **8736** is the answer.

(*iv*) Multiply **97 by 88**.

As both the numbers are near to 100. So we take **100 as base**. Further 97 *is less than base by **i.e., 100 by 3 and 88 is less than base i.e., 100 by 12**. Now as discussed we will solve as follows*:

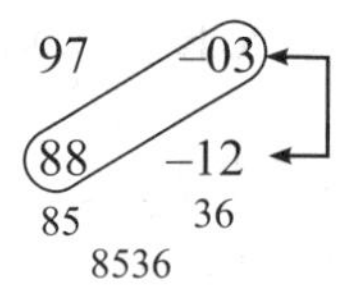

As shown above **8536** is the answer.

(*v*) Multiply **94 by 86**.

As both the numbers are near to 100. So we take **100 as base**. Further **94 *is less than base by i.e., 100 by 6 and 86 is less than base i.e., 100 by 14. Now as discussed we will solve as follows***:

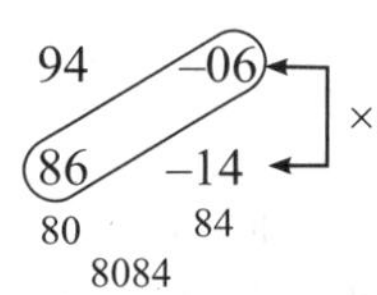

As shown above **8084** is the **answer.**

Hopefully you have understood the above method. Now what if the number of digits after multiplication on right hand side is more than the number of zeros in the base. The method of solving in such situation is given in following example:

Example 2:

Solve the following with Vedic maths

(i) Multiply **91 by 86**

(ii) Multiply **92 by 84**

Sol:

Multiply **91 by 86.**

As both the numbers are near to 100. So we take **100 as base**. Further **91 *is less than base by i.e., 100 by 9 and 86 is less than base i.e., 100 by 14.*** *Now as discussed we will solve as follows*:

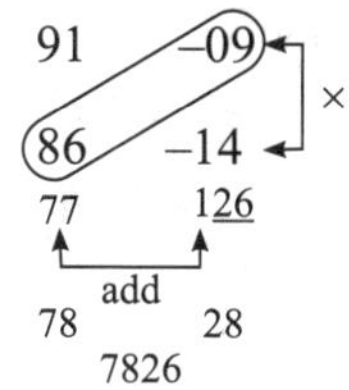

As shown above as the ***base is 100***, so we have to keep ***2 digits on right hand side*** after multiplication. ***The third spare digit i.e., is carried forward*** and added in 77 and we get the answer as **7826.**

(*vii*) Multiply **87 by 89.**

As both the numbers are near to 100. So we take **100 as base**. Further **87** ***is less than base by i.e., 100 by 13 and 89 is less than base i.e., 100 by 11***. *Now as discussed we will solve as follows*:

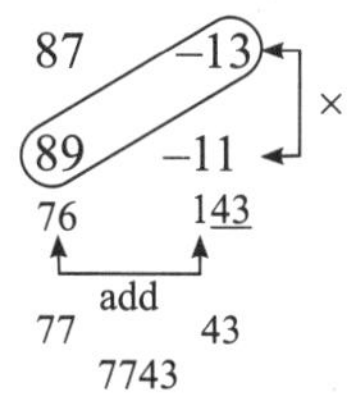

As shown above as the ***base is 100***, so we have to keep ***2 digits on right hand side*** after multiplication. ***The third spare digit i.e., is carried forward and added in*** **76** and we get the answer as **7743**.

VEDIC MATHS CONCEPT II (MULTIPLICATION ABOVE BASE)

Multiplication above base in Vedic maths is almost similar to **multiplication below base. As discussed** Vedic maths is useful when a number is near to **10, 100, 10000**.....let us suppose we have to ***multiply 103 by 108*** we will do it in following steps:

STEP I: we will see 103 and 108 are nearer to which rounded number say 10,100, 1000, 100000, etc. obvously they are near to 100. So we take 100 as base.

STEP II: Both 103 and 108 are more than 100. So now the first question is they are more than base *i.e.*, 100 by how much?

Of course 103 is 3 more than 100 and 108 is 8 more than 100

STEP III: in order to multiply 103 by 108 we will write as shown below:

103 + 03

108 + 08

We will read it as **"*103 is more than base by i.e., 100 by 3 and 108 is more than base i.e., 100 by 8.*"** Here we will take positive sign as numbers are **more than the base** and as discussed in previous concept we have written **3 as 03** and **8 as 08** as ***we have to write the number of digits equal to the number of zero's in base***. As **100 have 2** zeros so we have written **3 as 03** and **8 as 08**. Now you can easily understand if we had to multiply **1001 by 1003** we would have taken the **base as 1000** and would have written as shown below:

1001 + 001

1003 + 003

We will now solve in next step

STEP IV: Step IV is shown below:

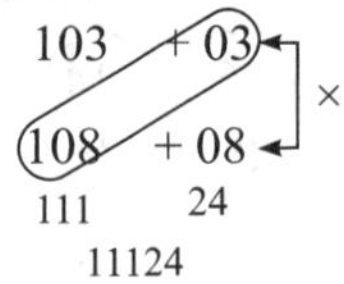

This is the last step. In this as shown above **we will add cross (08 to103 or 03 to 108)** and **multiply (+ 03 by + 08 *i.e.*, + 03*x* + 08 = 24) to get 11124** as final answer. Here again you can see. The beauty of the method that when you add cross as shown you will get the same result in both the pairs **means 103 + 8 = 111 and 108 + 3 = 111 both are same. You can do addition in any of the pair**. Hope fully you have understood the concept. Now do the following example:

Example 1:

Solve the following with Vedic maths

(i) Multiply **102 by 106**

(ii) Multiply **101 by 111**

(iii) Multiply **104 by 109**

(iv) Multiply **103 by 112**

(v) Multiply **106 by 114**

Solutions

(*viii*) Multiply **102 by 106.**

As both the numbers are near to 100. So we take **100 as base**. Further **102 *is more than base by i.e., 100 by 2 and 106 is more than base i.e., 100 by 6.*** *Now as discussed we will solve as follows*:

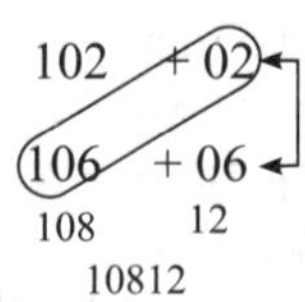

As shown above **10812** is the answer.

(*ix*) Multiply **101 by 111.**

As both the numbers are near to 100. So we take **100 as base**. Further **101 *is more than base by i.e., 100 by 1 and 111 is more than base i.e., 100 by 11***. *Now as discussed we will solve as follows*:

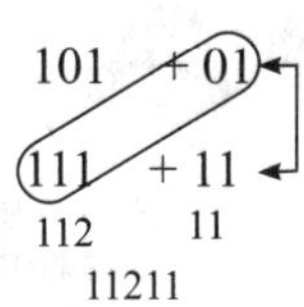

As shown above **11211** is the answer.

(*x*) Multiply **104 by 109**.

As both the numbers are near to 100. So we take **100 as base**. Further **104 *is more than base by i.e., 100 by 4 and 109 is more than base i.e., 100 by 9.*** *Now as discussed we will solve as follows*:

```
104   + 04
109   + 09
113     36
   11336
```

As shown above **11336** is the answer.

(*xi*) Multiply **103 by 112**.

As both the numbers are near to 100. So we take **100 as base**. Further 103 *is more than base by* ***i.e., 100 by 3 and 112 is more than base i.e., 100 by 12****. Now as discussed we will solve as follows*:

```
103   + 03
112   + 12
115     36
   11536
```

As shown above **11536** is the answer.

(*xii*) Multiply **106 by 114**.

As both the numbers are near to 100. So we take **100 as base**. Further **106** ***is more than base by i.e., 100 by 6 and 114 is more than base i.e., 100 by 14. Now as discussed we will solve as follows*****:**

```
106   + 06
114   + 14
120     84
   12084
```

As shown above **12084** is the **answer.**

Hopefully you have understood the above method. Now what if the number of digits after multiplication on right hand side is more than the number of zeros in the base. The method of solving in such situation is given in following example:

Example 2:

Solve the following with Vedic maths

(i) Multiply **109 by 114**

(ii) Multiply **113 by 111**

Solutions

(*xiii*) Multiply **109 by 114.**

As both the numbers are near to 100. So we take **100 as base**. Further **109** ***is more than base by i.e., 100 by 9 and 114 is more than base i.e., 100 by 14.*** *Now as discussed we will solve as follows*:

```
109   +09
114   +14
123    126
   12426
```

As shown above as the ***base is 100***, so we have to keep ***2 digits on right hand side*** after multiplication. ***The third spare digit i.e., 1 is carried forward*** and added in 123 and we get the answer as **12426**.

(*xiv*) Multiply **113 by 111.**

As both the numbers are near to 100. So we take **100 as base**. Further **113 *is more than base by i.e., 100 by 13 and 111 is more than base i.e., 100 by 11***. *Now as discussed we will solve as follows*:

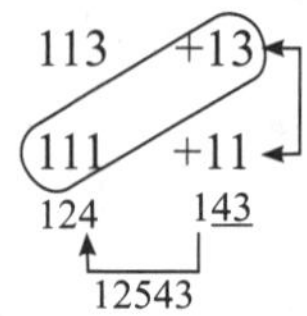

As shown above as the ***base is 100***, so we have to keep ***2 digits on right hand side*** after multiplication. ***The third spare digit i.e., is carried forward and added in* 124** and we get the answer as **12543**.

VEDIC MATHS CONCEPT III (MULTIPLICATION IN HIGHER BASES)

So now it is hoped that you have understood the concept and now you can solve the questions having multiplications of higher bases. As discussed, the two points that you should keep in mind are:

(i) You should take the base (in rounded number say 10,100, 1000, 100000, etc.) which is nearer to the numbers to be multiplied e.g., if you have to multiply 99999 by 99997 you will take the base as 100000 as both 99999 and 99997 are nearest to 100000.

(ii) While *writing more than or less than the decided base* the you should ***write the number of digits equal to the number of zero's in base***. So multiplication of 99999 by 99997 is shown below.

Now we will solve the following examples

Example 1:

Solve the following with Vedic maths

(i) Multiply **1004 by 1006**

(ii) Multiply **9999 by 9987**

(iii) Multiply **10004 by 10009**

(iv) Multiply **1024 by 1002**

(v) Multiply **1014 by 1011**

Solutions

(*xv*) Multiply **1004 by 1006.**

As both the numbers are near to 1000. So we take **1000 as base**. *Now as discussed we will solve as follows*:

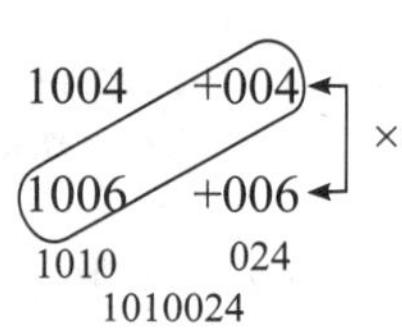

As shown above **1010024** is the answer.

(*xvi*) Multiply **9999 by 9987.**

As both the numbers are near to 10000. So we take **10000 as base**. *Now as discussed we will solve as follows*:

```
9999    -0001
9987    -0013
9986     0013
  99680013
```

As shown above **99680013** is the answer.

(*xvii*) Multiply **10004 by 10009**.

As both the numbers are near to 10000. So we take **10000 as base.** *Now as discussed we will solve as follows*:

```
10004   +0004
10009   +0009
10013    0036
  100130036
```

As shown above **100130036** is the answer.

(*xviii*) Multiply **1024 by 1002**.

As both the numbers are near to 1000. So we take **1000 as base**. *Now as discussed we will solve as follows*:

```
1024    +024
1002    +002
1026     048
  1026048
```

As shown above **1026048** is the answer.

(*xix*) Multiply **1014 by 1011**.

As both the numbers are near to 1000. So we take **100 as base**. ***Now as discussed we will solve as follows*:**

```
1014    +014
1011    +011
1025     154
  1025154
```

As shown above **1025154** is the **answer.**

Now we will do the application of Vedic maths that are very helpful min competitive exam's.

APPLICATIONS OF VEDIC MATHS

There are two main applications of Vedic maths. These are explained below:

1. To find the square of a number
2. To find the square root of a number

We will discuss each of them one by one

(i) To find the square of a number: Vedic maths is useful in finding the square of any number. In fact you have to do a lot of practice of this. This is the basic requisite to clear any competitive exam. suppose if you have find the square of 88 you will solve it as follows:

Simply you have to Multiply **88 by 88.**

As both the numbers are near to 100. So we take **100 as base**. Further **88 *is less than base by i.e., 100 by 12***. *Now as discussed we will solve as follows*:

88 – 12
88 – 12
76 144

7744
7744

As shown above as the ***base is 100***, so we have to keep ***2 digits on right hand side*** after multiplication. ***The third spare digit i.e., 1 is carried forward and added in* 76** and we get the answer as **7744**.

(**Note: $88^2 = 7744$ is very important number as it is the only perfect square of four digits in which first two and last digits are same or you can say of aabb type. This fact has been asked in many competitive exams in different ways**)

As shown in order to find the square of a number you have to just multiply the two same numbers which is very easy with the help of Vedic maths one more reason to do the practice is if you have done sufficient practice of finding squares of different numbers with Vedic maths, then only you can find the square root easily.

Now do the following example:

Example 1:

Find the square of following with Vedic maths

(i) 98 (ii) 81
(iii) 97 (iv) 104
(v) 121 (vi) 124
(vii) 118 (viii) 76
(ix) 106 (x) 91
(xi) 79 (xii) 111

Solution: As now it is hoped that you have understood the concept clearly. So below the answer s of above examples are given directly.

(i) 9604 (ii) 6561
(iii) 9409 (iv) 10816
(v) 14641 (vi) 12976
(vii) 13924 (viii) 5776
(ix) 11236 (x) 8281
(xi) 6241 (xii) 12321

(ii) To find square by changing the base: Now we will learn how to find the square of a number when the number is not near to rounded number say **10,100,1000,100000,** etc. rather it is near to somewhere between two rounded numbers like 50 (between

10 and 100), 500 (between 100 and 1000), etc.

Let us understand this with the help of an example. Suppose we have the square of 53, now if we take base as 100 then we have to multiply 47 by 47 which is as tough as multiplying 53 by 53. So in this case we will change the base or you can say take the temporary base which is multiple of or divisor of main base. In this case we will change the base as 50 and will write ***as base 50 with respect to base 100*** and will solve as follows:

Simply you have to multiply **53 by 53.**

As both the numbers are near to 53. So we take **50 as base and will write *as base 50 with respect to base 100***. Further **53 *Is more than the temporary base by i.e., 50 by 3* and** *we are solving it* ***with respect to base 100*** so we will write 3 as 03. *Now as discussed we will solve as follows*:

53 + 03
53 + 03
56 09
56/2 09
28 09
2809

As shown above 53 + 03 = 56 and 03 × 03 = 09 and since we are solving ***with respect to base 100*** and base ***50 is half of base 100***. So we divide 56 by 2 and get **2809** as the final answer.

Now we will do few more questions.

Example 1:

Find the square of following with Vedic maths by changing the appropriate base

(i) 46 (ii) 54
(iii) 502 (iv) 496
(v) 203 (vi) 304

Solutions:

(i) Simply you have to multiply **46 by 46.**

As both the numbers are near to 50. So we take **50 as base and will write *as base 50 with respect to base 100***. Further **46 *is less than the temporary base by i.e., 50 by 4* and** *we are solving it* ***with respect to base 100*** so we will write 4 as –04. *Now as discussed we will solve as follows*:

46 – 04
46 – 04
42 16
42/2 16
21 16
2116

As shown above 46 – 04 = 42 and 04 × 04 = 16 and since we are solving ***with respect to base 100*** and base ***50 is half of base 100***. So we divide 42 by 2 and get **2116** as the final answer.

(ii) Simply you have to multiply **54 by 54.**

As both the numbers are near to 54. So we take **50 as base and will write *as base 50 with respect to base 100***. Further ***54 is more than the temporary base by i.e., 50 by 4 and*** *we are solving it* ***with respect to base 100,*** so we will write 4 as 04. *Now as discussed we will solve as follows*:

54 +04
54 +04
58 16
58/2 16
29 16
2916

As shown above 54 + 04 = 54 and 04 × 04 = 16 and since we are solving ***with respect to base 100*** and base ***50 is half of base 100***. So we divide 58 by 2 and get **2916** as the final answer.

(iii) Simply you have to multiply **502 by 502.**

As both the numbers are near to 500. So we take **500 as base and will write *as base 500 with respect to base 1000***. Further ***502 is more than the temporary base by i.e., 500 by 2 and*** *we are solving it* ***with respect to base 1000,*** so we will write 2 as 002. *Now as discussed we will solve as follows*:

502 +002
502 +002
504 004
504/2 004
252 004
252004

As shown above 502 + 002 = 504 and 002 × 002 = 004 and since we are solving ***with respect to base 1000*** and base ***500 is half of base 1000***. So we divide 504 by 2 and get **252004** as the final answer.

(*iv*) Simply you have to multiply **496 by 496.**

As both the numbers are near to 500. So we take **500 as base and will write *as base 500 with respect to base 1000***. Further ***496 is less than the temporary base by i.e., 500 by 4 and*** *we are solving it* ***with respect to base 1000,*** so we will write 4 as –004. *Now as discussed we will solve as follows*:

496 –004
496 –004
492 016
492/2 016
246 016
246016

As shown above 496 – 004 = 492 and –004*x* –004 = 016 and since we are solving ***with respect to base 1000*** and base ***500 is half of base 1000***. So we divide 492 by 2 and get **246016** as the final answer.

(*v*) Simply you have to multiply **203 by 203.**

As both the numbers are near to 200. So we take **200 as base and will write *as base 200 with respect to base 100***. Further ***203 is more than the temporary base***

by i.e., 200 by 3 and *we are solving it* ***with respect to base 100.*** So we will write 3 as + 03. *Now as discussed we will solve as follows*:

203	+ 03
203	+ 03
206	09
206 × 2	09
412	09

41209

As shown above 203 + 03 = 206 and + 03 × + 03 = 009 and since we are solving ***with respect to base 100*** and base ***200 is twice of base 100***. So here we will multiply 206 by 2 and get **41209** as the final answer.

(*vi*) Simply you have to multiply **304 by 304.**

As both the numbers are near to 300. So we take **300 as base and will write *as base 300 with respect to base 100***. Further **304 *is more than the temporary base by i.e., 300 by 4 and*** *we are solving it* ***with respect to base 100,*** so we will write 4 as +04. *Now as discussed we will solve as follows*:

304	+ 04
304	+ 04
308	16
308×3	16
924	16

92416

As shown above 304 + 004 = 308 and 04 × 04 = 016 and since we are solving ***with respect to base 100*** and base ***300 is thrice of base 100***. So here we will multiply 308 by 3 and get **924016** as the final answer.

This was about finding the square of numbers by Vedic maths. Now you have also learnt how to find the square of a number by changing the base when the number whose square is to be found is not near to rounded number say **10,100, 1000, 100000** etc, rather it is near to somewhere between two rounded numbers like 50 (between 10 and 100), 500 (between 100 and 1000) etc.

Students one thing you should remember that you should do sufficient practice of finding the squares of number as discussed above. For that you should practice square of 76-99,101-124,51-74 and 49-31 as many time as possible. Believe us; it will help you a lot in real time exam situation and also for the understanding of typical advance number system questions.

Now we will understand the second application of Vedic maths *i.e.*, to find the square root of a number.

2. **To find the square root of any perfect square:** It is an **important application of Vedic maths**. To understand it considers the following: here below given are the squares from 21–29. Now just **observe the symmetry around 25^2** and **particularly of last two digits**. One **above 25^2 is 24^2** and one **below** is **25^2 is 26^2** and both have **last two digits** as **76**. Now if we go further **one above 24^2 is 23^2** and **one below is 26^2 is 27^2** and **both have last two digits as 29.**

21^2 = **441**
22^2 = **484**
23^2 = 529
24^2 = 576
25^2 = **625**
26^2 = 676
27^2 = 729
28^2 = 784
29^2 = 841

This symmetry also goes further from **51-75 in forward order and again 76-100 in reverse order** and so on and repeats till infinity. ***In nutshell last two digits repeats in an interval of 25, first in forward and then in reverse order. It means all the square in this number system will have last two digits from squares of 1-25.*** That is why it is always expected from students to remember squares of 1-25. This is the property that will help us to find square root of any number.

To find the **square root of any perfect square** you have to observe following two factors:

1. **Last two digits**
2. **Range of number**

Now let us understand this concept. Suppose you have to find the **square root of 11664**. As discussed you have to see two factors **one is last two digits and other is range of number**. Now since last **two digits are 64** which is nothing but **square of 8 and** about **second factor is range the number is in five digit** and slightly above **100^2=10000.** So obviously the **11664 is the square of a number which is slightly above 100 and in which square of 8 is used**. Quite obvious it is of **108**. (As we would have done **108+08** and further would have multiplied **08 by 08** on pretext that it is **more by 8 than base 100**).

Now solve the following example

Example 1: Find the square roots of following with Vedic maths

(i) 11449 **(ii)** 13689 **(iii)** 11236
(iv) 9604 **(v)** 2916 **(vi)** 2116

Solutions:

(i) Here you have to find the **square root of 1144**9. As discussed you have to see two factors **one is last two digits and other is range of numbe**r. Now since **last two digits are 49** which is nothing but **square of 7** and about **second factor** is **range the number is in five digits** and slightly above **100^2=10000.** So obviously the **11449 is the square of a number which is slightly above 100 and in which square of 7 is used**. Quite obvious it is of **107**. (As we would have done **107+07** and further would have multiplied **07 by 07** on pretext that it is **more by 7 than base 100**).

(ii) Here you have to find **the square root of 13689**. As discussed you have to see two factors **one is last two digits and other is range of number**. Now since last **two digits are 89** which occur only in the **square of 17** from1-25 and about second factor is **range the number is in five digits** and slightly above **100^2 = 10000.** So obviously the **13689 is the square of a number which is above 100 and in which square of 17 is used**. Quite obvious it is of **117**. (as we would have **done 117+17+2** (of 289) and further would have multiplied **17 by 17** on pretext that it is **more by17 than base 100).**

(iii) Here you have to find the **square root of 11236**. As discussed you have to see two factors **one is last two digits** and **other is range of number**. Now since last **two digits are 36** which is nothing but **square of 6** and about second factor **is range the number is in five digits** and slightly above **100^2=10000.** So obviously the **11236 is the square of a number which is slightly above 100 and in which square of 6 is used**. Quite obvious it is of **106.** (As we would have done **106+06** and further would have multiplied **06 by 06** on pretext that **it is more by 6 than base 100**).

(iv) Here you have to find the **square root of 9604**. As discussed you have to see two factors **one is last two digits** and **other is range of number**. Now since last **two digits are 04** which is nothing but **square of 02** and about second factor is **range the number is in 4 digits** and slightly below **100^2=10000.** So obviously the **9604 is the square of a number which is slightly below 100 and in which square of 02 is used**. Quite obvious it is of **98**. (As we would have done **98-02** and further would have multiplied **-02 by -02** on pretext that it is **less by 2 than base 100**).

(v) Here you have to find the **square root of 2916**. As discussed you have to see two factors one is **last two digits** and other is **range of number**. Now **since last two digits are 16** which is nothing but **square of 04** and about second factor is **range the number is between 50^2=2500 and 60^2 =3600.** So obviously the **2916 is the square of a number which is between50 and 60 and in which square of 04 is used**. Quite obvious it is of **50+4=54**. (As we would have done **54+04/2** and further would have multiplied **04 by 04** on pretext that **it is more by 4 than temporary base 50**).

(vi) Here you have to find the **square root of 2116**. As discussed you have to see two factors one is **last two digits** and other is **range of number**. Now since **last two digits are 16** which is nothing but **square of 4** and about second factor is **range the number is between40^2=1600 and 50^2 =2500**. So obviously the **2116 is the square of a number which is between40 and 50 and in which square of 04 is used**. Quite obvious it is of **50-4=46**. (As we would have done **46-04/2** and further would have multiplied **-04 by -04** on pretext that it is **less by4 than temporary base 50**).

EXERCISE

Q.1. Solve the following with Vedic maths

(i) Multiply **141 by 101** **(ii)** Multiply **1021 by 1001**

(iii) Multiply **109 by 119** **(iv)** Multiply **10006 by 10012**

(v) Multiply **100008 by 100018**

Q.2. Solve the following with Vedic maths

(i) Multiply **99 BY 82** **(ii)** Multiply **98 by 78**

(iii) Multiply **96 by 83** **(iv)** Multiply **999 by 988**

(v) Multiply **99999 by 99996**

Q.3. Find the square of following with Vedic maths

(i) 989 (ii) 206 (iii) 997

(iv) 1004 (v) 1021 (vi) 1024
(vii) 116 (ii) 404 (iii) 521
(iv) 496 (v) 9999 (vi) 10018

Q.4. Find the square roots of following with Vedic maths
(i) 100320256 (ii) 14161 (iii) 3844
(iv) 12544 (v) 3136 (vi) 3364

Q.5. How many four digit perfect square are there in which first two and last two digits are same. Explain with the help of Vedic maths?

SOLUTIONS

Sol. 1. (i) 14241 (ii) 1022021 (iii) 12971
(iv) 100180072 (v) 10002600144

Sol. 2. (i) 8118 (ii) 7644 (iii) 7968
(iv) 987012 (v) 9999500004

Sol. 3. (i) 978121 (ii) 42436 (iii) 994009
(iv) 1008016 (v) 1042441 (vi) 1048576
(vii) 13456 (ii) 46864 (iii) 271441
(iv) 246016 (v) 99980001 (vi) 100360324

Sol 4. (i) 10016 (ii) 119 (iii) 62
(iv) 112 (v) 56 (vi) 58

Sol. 5. The required numbers are of **aabb** type. Here to find the solution we have to see two factors:

1. **Last two digits**
2. **Range of number**

Now first we will see the range. *First four digit perfect square* is of **32 i.e., 32^2 = 1024** and *last four digit perfect square* is of **99 i.e., 99^2 = 9801**, now the desired numbers are in range of squares from **32 to 99.**

As discussed" ***last two digits repeats in an interval of 25**, first in forward and then in reverse order. It means all the square in this number system will have **last two digits of squares from 1-25**.*"So from the numbers of **aabb** type first we will check for **bb** *i.e.*, last two digits from squares of 1-25.

Now last two digits can be 00 and in the range from **32-99** these can be of **40,50,60, 70,80,90** but none of these are of **aa** type *i.e.*, none of these have ***first two digits same*** and hence are not desired numbers.

Now *except 00* in the range *from square of 1-25*, **bb** can only be **44** *i.e.*, **square of 12 which is equal to 144**. So it means that to get a perfect square in this range, 12^2 must be used and it is used thrice in this range *i.e.*, in **squares of 62, 38 and 88 (** in ***62 which is 12 more than base 50***, in ***38 which 12 less than base 50*** and in ***88 which is 12 less than base 100***). Out of three only **88^2 is the required number as in this only first two digits are also same** as **88^2 =7744** while **38^2=1444 and 62^2 is 3844**). Hence only **one** four digit perfect square is there in which first two and last two digits are same.